T0222772

Lecture Notes in Computer Science 10671

Commenced Publication in 1973
Founding and Former Series Editors:
Gerhard Goos, Juris Hartmanis, and Jan van Leeuwen

More information about this series at http://www.springer.com/series/7407

Roberto Moreno-Díaz · Franz Pichler
Alexis Quesada-Arencibia (Eds.)

Computer Aided Systems Theory – EUROCAST 2017

16th International Conference
Las Palmas de Gran Canaria, Spain, February 19–24, 2017
Revised Selected Papers, Part I

 Springer

Editors
Roberto Moreno-Díaz
University of Las Palmas de Gran Canaria
Las Palmas de Gran Canaria
Spain

Alexis Quesada-Arencibia
University of Las Palmas de Gran Canaria
Las Palmas de Gran Canaria
Spain

Franz Pichler
Johannes Kepler University Linz
Linz
Austria

ISSN 0302-9743   ISSN 1611-3349   (electronic)
Lecture Notes in Computer Science
ISBN 978-3-319-74717-0   ISBN 978-3-319-74718-7   (eBook)
https://doi.org/10.1007/978-3-319-74718-7

Library of Congress Control Number: 2018930745

LNCS Sublibrary: SL1 – Theoretical Computer Science and General Issues

Printed on acid-free paper

This Springer imprint is published by Springer Nature
The registered company is Springer International Publishing AG
The registered company address is: Gewerbestrasse 11, 6330 Cham, Switzerland

Preface

The concept of CAST as a computer-aided systems theory was introduced by Franz Pichler in the late 1980s to refer to computer theoretical and practical development as tools for solving problems in system science. It was thought of as the third component (the other two being CAD and CAM) required to complete the path from computer and systems sciences to practical developments in science and engineering.

Franz Pichler, of the University of Linz, organized the first CAST workshop in April 1988, which demonstrated the acceptance of the concepts by the scientific and technical community. Next, Roberto Moreno-Díaz, of the University of Las Palmas de Gran Canaria, joined Franz Pichler, motivated and encouraged by Werner Schimanovich, of the University of Vienna (present Honorary Chair of Eurocast), and they organized the first international meeting on CAST, (Las Palmas February 1989), under the name EUROCAST 1989. The event again proved to be a very successful gathering of systems theorists, computer scientists, and engineers from most European countries, North America, and Japan.

It was agreed that EUROCAST international conferences would be organized every two years, alternating between Las Palmas de Gran Canaria and a continental European location. Since 2001 the conference has been held exclusively in Las Palmas. Thus, successive EUROCAST meetings took place in Krems (1991), Las Palmas (1993), Innsbruck (1995), Las Palmas (1997), and Vienna (1999), before being held exclusively in Las Palmas in 2001, 2003, 2005, 2007, 2009, 2011, 2013, and 2015, in addition to an extra-European CAST conference in Ottawa in 1994. Selected papers from these meetings were published as Springer *Lecture Notes in Computer Science* volumes 410, 585, 763, 1030, 1333, 1798, 2178, 2809, 3643, 4739, 5717, 6927, 6928, 8111, 8112, and 9520, respectively, and in several special issues of *Cybernetics* and *Systems: An International Journal*. EUROCAST and CAST meetings are definitely consolidated, as shown by the number and quality of the contributions over the years.

EUROCAST 2017 took place in the Elder Museum of Science and Technology of Las Palmas, during February 19–24, and it continued with the approach tested at previous conferences as an international computer-related conference with a true interdisciplinary character. The participants profiles are presently extended to include fields that are in the frontier of science and engineering of computers, of information and communication technologies, and the fields of social and human sciences. The best paradigm is the Web, with its associate systems engineering, CAD-CAST tools, and professional application products (Apps) for services in the social, public, and private domains.

There were specialized workshops, which, on this occasion, were devoted to the following topics:

1. Pioneers and Landmarks in the Development of Information and Communication Technologies, chaired by Pichler (Linz), Stankovic (Nis), Kreuzer Felisa and Kreuzer James (USA)
2. Systems Theory and Applications, chaired by Pichler (Linz) and Moreno-Díaz (Las Palmas)

3. Stochastic Models and Applications to Natural, Social, and Technological Systems, chaired by Nobile and Di Crescenzo (Salerno)
4. Theory and Applications of Metaheuristic Algorithms, chaired by Affenzeller and Jacak (Hagenberg) and Raidl (Vienna)
5. Embedded Systems Security, chaired by Mayrhofer (Linz) and Schmitzberger (Linz)
6. Model-Based System Design, Verification, and Simulation, chaired by Nikodem (Wroclaw), Ceska (Brno), and Ito (Utsunomiya)
7. Systems in Industrial Robotics, Automation and IoT, chaired by Stetter (Munich), Markl (Vienna), and Jacob (Kempten)
8. Applications of Signal Processing Technology, chaired by Huemer (Linz), Zagar (Linz), Lunglmayr (Linz), and Haselmayr (Linz)
9. Algebraic and Combinatorial Methods in Signal and Pattern Analysis, chaired by Astola (Tampere), Moraga (Dortmund), and Stankovic (Nis)
10. Computer Vision, Deep Learning, and Applications, chaired by Penedo (A Coruña) and Radeva (Barcelona)
11. Computer- and Systems-Based Methods and Electronic Technologies in Medicine, chaired by Rozenblit (Tucson), Hagelauer (Linz), Maynar (Las Palmas), and Klempous (Wroclaw)
12. Cyber-Medical Systems, chaired by Rudas (Budapest), Kovács (Budapest), and Fujita (Iwate)
13. Socioeconomic and Biological Systems: Formal Models and Computer Tools, chaired by Schwaninger (St. Gallen), Schoenenberger (Basel), Tretter (Munich), Cull (Corvallis USA), and Suárez-Araújo (Las Palmas)
14. Intelligent Transportation Systems and Smart Mobility, chaired by Sanchez-Medina (Las Palmas), Celikoglu (Istanbul), Olaverri-Monreal (Wien), Garcia-Fernandez (Madrid), and Acosta-Sanchez (La Laguna)

In this conference, as in previous ones, most of the credit for the success is due to the workshop chairs. They and the sessions chairs, with the counseling of the international Advisory Committee, selected from 160 presented papers, after oral presentations and subsequent corrections, the 117 revised papers included in this volume.

The event and this volume were possible thanks to the efforts of the chairs of the workshops in the diffusion and promotion of the conference, as well as in the selection and organization of all the material. The editors would like to express their thanks to all the contributors, many of whom are already Eurocast participants for years, and particularly to the considerable interaction of young and senior researchers, as well as to the invited speakers, Prof. Christian Müller-Scholer from Hamburg, Prof. Manuel Maynar, from Las Palmas, and Prof. Jaakko Astola from Tampere, for their readiness to collaborate. We would also like to thank the director of the Elder Museum of Science and Technology, D. José Gilberto Moreno, and the museum staff. Special thanks are due to the staff of Springer in Heidelberg for their valuable support.

September 2017
Roberto Moreno-Díaz
Franz Pichler
Alexis Quesada-Arencibia

Organization

Organized by

Instituto Universitario de Ciencias y Tecnologías Cibernéticas
Universidad de Las Palmas de Gran Canaria, Spain

Johannes Kepler University Linz,
Linz, Austria

Museo Elder de la Ciencia y la Tecnología
Las Palmas de Gran Canaria, Spain

Conference Chair

Roberto Moreno-Díaz, Las Palmas

Program Chair

Franz Pichler, Linz

Honorary Chair

Werner Schimanovich, Austrian Society for Automation and Robotics

Organizing Committee Chair

Alexis Quesada Arencibia
Instituto Universitario de Ciencias y Tecnologías Cibernéticas
Universidad de Las Palmas de Gran Canaria
Campus de Tafira
35017 Las Palmas de Gran Canaria, Spain
Phone: +34-928-457108
Fax: +34-928-457099
e-mail: alexis.quesada@ulpgc.es

Invited Lectures

Trust Communities – Social Mechanisms for Technical Systems

C. Müller-Schloer

Leibniz Universität, Hannover

Abstract. Trust, fairness, reputation, forgiveness but also laws and sanctions are indispensable ingredients of human societies. Despite Dawkins' "The Selfish Gene", we are primarily social and cooperative animals, and this is for a good reason: Cooperative societies are more successful than collections of selfish agents, and social behavior pays off not only for the group as a whole but also for the individuals.

Technical systems resemble more and more our human societies: They are complex, consist of semi-autonomous subsystems (or "agents"), which are largely unknown and self-interested, and interact with each other. These systems are highly dynamic and open, hence impossible to rigidly define at design time and difficult to control at runtime. We will increasingly rely on self-organization mechanisms to make them adaptive and robust against changing requirements and environment. Therefore it is an obvious approach to transfer and utilize social mechanisms into technical systems.

The lecture discusses the prisoners' dilemma and the Tragedy of the Commons to motivate the introduction of trust into technical systems. It shows how different forms of self-organized implicit and explicit Trust Communities (iTCs and eTCs) can improve the robustness of a multi-agent organization. An open Grid Computing system serves as an example.

While iTCs and eTCs greatly improve the system robustness in the presence of egoistic agent attacks there are certain global situations like a trust-breakdown (so-called Negative Emergent Behavior NEB), which require a more massive interference by higher-level authorities. The lecture will explain how norms and sanctions can help to recover from NEB states.

In an outlook, we will compare the social mechanisms introduced so far with Elinor Ostrom's (Nobel laureate in Economic Sciences 2009) rules for Enduring Institutions and discuss their applicability to future technical systems.

More detailed information can be found in the book "Trustworthy Open Self-organising Systems" [1], and there in the chapters [2, 3].

References

1. Reif, W., Anders, G., Seebach, H., Steghöfer, J.P., André, E., Hähner, J., Müller-Schloer, C., Ungerer, T. (eds.): Trustworthy Open Self-Organising Systems, Autonomic Systems series, vol. 7. Springer International Publishing, Cham (2016)
2. Kantert, J., Edenhofer, S., Tomforde, S., Hähner, J., Müller-Schloer, C.: Normative control - controlling open distributed systems with autonomous entities. In: Reif, W., Anders, G., Seebach, H., Steghöfer, J.P., André, E., Hähner, J., Müller-Schloer, C., Ungerer, T. (eds.) Trustworthy Open Self-Organising Systems. Autonomic Systems series, pp. 87–123, vol. 7. Springer International Publishing, Cham (2016)
3. Edenhofer, S., Kantert, J., Tomforde, S., Bernard, Y., Klejnowski, L., Hähner, J., Müller-Schloer, C.: Trust communities: an open, self-organised social infrastructure of autonomous agents. In: Reif, W., Anders, G., Seebach, H., Steghöfer, J.P., André, E., Hähner, J., Müller-Schloer, C., Ungerer, T. (eds.) Trustworthy Open Self-Organising Systems. Autonomic Systems Series, pp. 125–150, vol. 7. Springer International Publishing, Cham (2016)

Surprising Applicability of Abstract Discrete Structures – A Signal Processing View

Jaakko Astola

Signal Processing Laboratory, Tampere University of Technology,
Tampere, 33101, Finland

1   Plenary Lecture Summary

The usefulness (and necessity) of mathematics in engineering becomes evident just by browsing a high-school physics book. Almost every branch of mathematics have some engineering applications, sometimes, via advanced theories in physics and chemistry. Traditional engineering mathematics was mainly "continuous" but the emergence of digital systems discrete mathematics into engineering curricula. However, especially a research engineer must still have solid working knowledge of mathematical analysis and probability theory.

A good example of a discipline needing a strong background in mathematics is signal processing. We deal with (natural) information gathered by sensors and in the modeling is performed using e.g. real and complex functions and random processes. The data is processed digitally and designing such systems requires almost all available tools in discrete mathematics – Finite Groups, Rings, Fields and Linear Spaces; Boolean Algebras; Trees and Graphs; Combinatorics; Automata and Formal Languages and so on.

We look at couple of the topics where discrete structures play an important role. With each topic, I look at an engineering problem, and discuss the mathematical structures and theories we use to study the problem. There is usually a "natural or obvious" mathematical discipline to apply but, interestingly, interpreting the problem within a seemingly unrelated theory may give a better solution. It is good to have many interpretations because typically, the practical solution is a hybrid method to balance cost, complexity, and efficiency.

In image processing the rich theory of Boolean functions has proven to be very useful in designing very nonlinear image filtering techniques. In 1980's it was notices that the visual quality of images corrupted by impulsive "salt and pepper" could be improved by applying the median filter, that is replacing each pixel value by the median of values surrounding it. For a binary image, median filter is in fact a Boolean operation. Treating a nonbinary image as a 3-dimensional body bounded from above by a surface it could be processed as a stack of binary images obtained as horizontal slices of the body. These ideas led to a large variety of image processing methods called morphological and stack filters. Today, they are widely used as components of more sophisticated image processing systems.

The Boolean (or discrete) functions appearing in image processing problems usually have large number of variables and need to be represented so that fast manipulation is possible. It turns out that one of the most efficient structures of representing such functions is so called decision diagram. It can be described as a rooted directed graph where for each binary vector of input variables one moves from node to node starting from the root and ending to a leaf indicating the value of the function for that particular vector of input variables. At each node on the route one chooses the edge dictated by the value of input variable.

In a typical digital communication scenario, fixed length strings bits are sent over a channel that corrupts them by changing bits some probability, and to be able to correct corrupted bits, redundancy is added to the binary strings. If we call the subset of digital messages with added redundancy a code, then its reliability or depends only on the Hamming distances between code-strings. To minimize the probability of uncorrectable errors the code needs to have the minimum Hamming distance between codewords as large as possible. Thus, the problem can be described in very simple terms. On the other hand, to design powerful codes deep algebraic structures and theories are required. For instance, polynomial rings over finite fields and tools from algebraic geometry have been used to produce codes with large minimum distance between the codewords.

The same discrete structures, namely linear spaces and polynomial rings, that are used in coding theory can be used to obtain efficient representations of certain Boolean functions. In many applications of control-systems the actions are described by Boolean functions having large number of variables. However, in practice, only a very small fraction of possible combinations of input variables can ever occur. This kind of sparse partially defined Boolean functions can be efficiently implemented using so called linear decomposition. The idea of the linear decomposition is that the huge but sparse binary space is first transformed by a linear transform to a smaller binary space with such a linear transform that is one-to one for those points of the original space for which the Boolean function is defined. The original function is then described as cascade of the linear function and a nonlinear function having greatly reduced number of variables. The key problem is to find the linear transform that is one-to-one for the set of defined points and with minimal dimension of range space. This can be efficiently computed using polynomial rings over finite fields.

References

1. Wendt, P., Coyle, E., Gallagher, N.: Stack filters. IEEE Trans. Acoust. Speech Signal Process. **34**(4), 898–911 (1986)
2. Maragos, P., Schafer, R.: Morphological filters – Part I: their set-theoretic analysis and relations to linear shift-invariant filters. IEEE Trans. Acoust. Speech Signal Process. **35**(8), 1153–1169 (1987)
3. Maragos, P., Schafer, R.: Morphological filters – Part II: their relations to median, order-statistic, and stack filters. IEEE Trans. Acoust. Speech Signal Process. **35**(8), 1170–1184 (1987)

4. Yli-Harja, O., Astola, J., Neuvo, Y.: Analysis of the properties of median and weighted median filters using threshold logic and stack filter representation. IEEE Trans. Signal Process. **39**(2) 395–410 (1991)
5. Astola, J., Kuosmanen, P.: Fundamentals of Nonlinear Digital Filtering. CRC Press, New York (1997)
6. Bryant, R.E.: Graph-based algorithms for Boolean functions manipulation. IEEE Trans. Comput. **35**(8), 667–691 (1986)
7. Astola, J.T., Stanković, R.S.: Fundamentals of Switching Theory and Logic Design. Springer, The Netherlands (2006)
8. Sasao, T.: Memory-Based Logic Synthesis. Springer, Heidelberg (2011)
9. Sasao, T.: Multiple-valued index generation functions: reduction of variables by linear transformation. J. Multiple-Valued Logic Soft Comput. **21**(5–6), 541–559 (2013)
10. Nechiporuk, E.I.: On the synthesis of networks using linear transformations of variables. Dokl. AN SSSR **123**(4), 610–612 (1958)
11. Astola, J., Astola, P., Stanković, R., Tabus, I.: An algebraic approach to reducing the number of variables of incompletely defined discrete functions. In: International Symposium on Multiple-Valued Logic (ISMVL-2016), Sapporo, Japan, May 2016

Contents – Part I

Systems Theory, Socio-economic Systems and Applications

Contents – Part II

Model-Based System Design, Verification and Simulation

Applications of Signal Processing Technology

Algebraic and Combinatorial Methods in Signal and Pattern Analysis

Computer Vision, Deep learning and Applications

Computer and Systems Based Methods and Electronic Technologies in Medicine

Intelligent Transportation Systems and Smart Mobility

Invited Lectures

Hospital 2090: How Nowadays Technology Could Contribute to the Future of the Hospitals

M. Maynar[1,2(✉)], T. Zander[1,2], J. Ballesteros[3], Y. Cabrera[3], and M. A. Rodriguez-Florido[1,4]

[1] Chair of Medical Technologies and Institute for Researching in Biomedicine and Health (iUIBS), Las Palmas de Gran Canaria University (ULPGC), Paseo Blas Cabrera Felipe "físico" s/n, 35016 Las Palmas de Gran Canaria, Canary Islands, Spain
educa@motivando.me
[2] Grupo Hospiten Tenerife, Rambla de Santa Cruz 115, 38001 Santa Cruz de Tenerife, Canary Islands, Spain
[3] Fundación Canaria Ágora, c/ Victor Hugo 36, 35006 Las Palmas de Gran Canaria, Canary Islands, Spain
[4] Canary Islands Institute of Technology, c/ Cebrian 4, 35006 Las Palmas de Gran Canaria, Canary Islands, Spain

Abstract. Everyone is an user of the Healthcare related services. Evolution of technology has changed the mode of practicing medicine, mainly, in surgical areas. In general, no one is wondering how the technology has contributed to the improvement of the medical procedures. In this paper, we propose some technologies used at present in other areas that could be used in the daily workflow of the hospitals, and how these technologies may contribute to the improvement of the physicians' work and the health system. This idea emerges from our clinical and technological experience.

Keywords: Education · Medical technology · Emerging technologies

1 Background

Health services are currently used by the 100% of the world's population and, in the occidental countries, people average life expectancy reaches 90 years old. Therefore, all the contributions to the progress of the health are welcome. Technology has become a way to move forward in health.

The technology has changed the mode of practicing medicine, mainly, the surgeries. In general, no one is wondering how the technology has contributed to the improvement of the medical procedures, but everybody assumes that this improvement is only due to the surgeon ability.

Nowadays, technology is present in almost every professional area, but in medicine, there is a gap between the most popular technologies and its application to the health.

R. Moreno-Díaz et al. (Eds.): EUROCAST 2017, Part I, LNCS 10671, pp. 3–10, 2018.
https://doi.org/10.1007/978-3-319-74718-7_1

In this paper, we show how technologies at present could contribute to the improvement of the daily work in the hospital; we call this tech hospital as Hospital 2090. The manuscript is organized in different sections following this one. First, we are introducing the Hospital 2090 idea, including subsections that propose potential applications of different technologies in a typical hospital workflow, and finally a conclusions section to wrap up all our proposals.

2   The Hospital 2090 Idea

The main idea under the Hospital 2090 concept is based on a real integration of all the professional areas (engineering, physics, economics, telecommunications, etc.) in the daily workflow of the hospital.

2.1   Virtual and Augmented Reality

The VR (Virtual Reality) and AR (Augmented Reality) devices and applications have significantly grown in the last two years. Many big companies, like Google, Facebook or Microsoft, and some other small startups, emerged by the occurrence of these technologies, have invested on developing approaches in different areas (games, education, design, etc.) (Fig. 1).

Fig. 1. An illustration of how doctors could learn using VR technology.

However, although this technology is becoming popular on areas cited above, is not included yet at the Hospital. We can enumerate many applications for health and its uses in the Hospital [1]: Education, surgical planning, simulation, physician support, etc.

2.2 Design, Store and Visualization Based on the Cloud

Building Information Modeling (BIM) is a digital representation of physical and functional characteristics of a facility. A BIM is a shared knowledge resource for information about a facility forming a reliable basis for decisions during its life-cycle; defined as existing from earliest conception to demolition (Figs. 2 and 3).

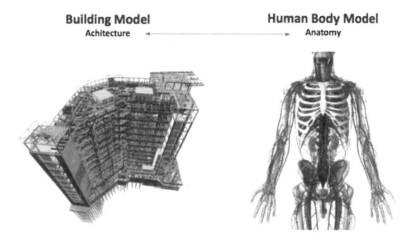

Fig. 2. Comparison between a building and the human anatomy.

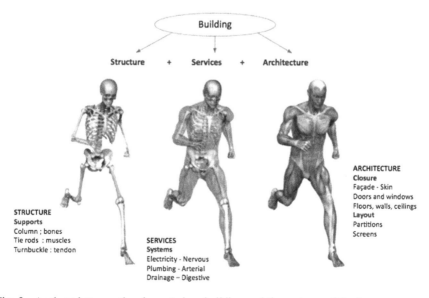

Fig. 3. Analogy between the elements in a building and the systems of the human anatomy.

The amount of information that a health-related professional, from student to physician, can accumulate during his or her professional life is huge. This architecture methodology, BIM, could be used for medical education, medical knowledge management and the store of medical information. The underlying idea of this item is to develop a software package that allows, with the support of the anatomy teachers, to make interactively a human body and to build, with some constraints to control the connectivity between the anatomy pieces, the human anatomy, storing the associated information that the user could be interested in.

Therefore, the user can access to this information at any time, so the user will not study to forget it, but keep his/her knowledge it safe and updated.

2.3   Information and Communications Technology

We are now in the communications era, where ICT applications are driving many of our daily tasks (Fig. 4).

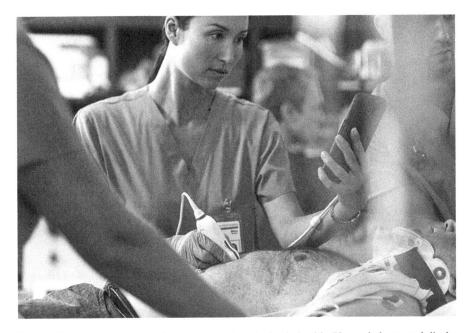

Fig. 4. The mobile devices are becoming in a key device in health. Picture belongs and display the Philips approach for including the "Bring Your Own Device" (BYOD) idea in health.

In a Hospital, there is important information that is not usually available for the health system's user. Think, for example, in the physicians' CV, the queue for the medical consultation, etc.

Then, we could mimic many applications that people uses daily (i.e. apps for knowing the next bus, etc.) focused to different issues at, or about, the Hospital.

2.4   Internet of Things

The Internet of Things (IoT) define the interconnection via the internet of computing devices embedded in everyday objects, enabling them to send and receive data (Fig. 5).

Fig. 5. Illustration of the interconnection in a hospital.

This concept could be used at the Hospital. Think, for example, devices (i.e. wearables) that are connecting the workflow of the Hospital (data of patients, emergencies, etc.) with the physicians when he or she is arriving to his or her clinical position or office.

2.5   Clinical Videoconference

Big hospitals in Los Angeles, New York, Tokyo or Madrid cannot have medical consultations all around the World. Today, technology may grant the patient to establish a connection to the doctor, independently of which part of the world where they are.

Some areas of the Hospital could be prepared, in which nurses could receive the patient and then explain the medical doctor the reason of the patient's consult. After that the medical doctor, with the help of the nurses, could auscultate or ask the patient, and therefore diagnose him (Fig. 6).

Fig. 6. Snapshot of a clinical videoconference in our group.

2.6   Automatic Aided Devices

In medicine these kinds of devices sound as something strange or uncommunicative for the patient and the physician.

However, this could be viewed like a capability that allows surgeons to aid the patient after the surgery while he or she is in other side of the hospital or even in other country or in other place (i.e. his or her home). Think for example in a robot that move around the hospital and its head is a monitor where the medical doctor is displayed in a video chat (Fig. 7).

Fig. 7. Snapshot of the AMIGO suite. Courtesy of Prof. Ron Kikinis (Surgical planning lab at Brigham & Women's Hospital).

Also, the patient may stay in the operating room during the whole procedure and all the tests could be done in the same room without having to take him out from the operating room [2]. All the controls may be in an extra room, with people working there not needing hygiene protocols, but with access to all the patient information through sensors, sound and displays.

Consulting live information of the patient, talking to a colleague that is somewhere else, displaying tomographies and 3D reconstructions, will help us being more precise in each case.

We should not forget the devices that can be installed at the patient's home, to monitor different medical data or to get some tests. The citizens, using this kind of technology, could drive their own clinical tests.

2.7   The Fourth Space

We call the fourth space to the slot of time where people now aged 18 years old people, will reach when they will be 100 years old. We have to teach to this people to manage their lifetime, their health, their economy, etc. They are going to be health services' users for a long time, and the retirement should be re-planned for the sustainability of the health service (Fig. 8).

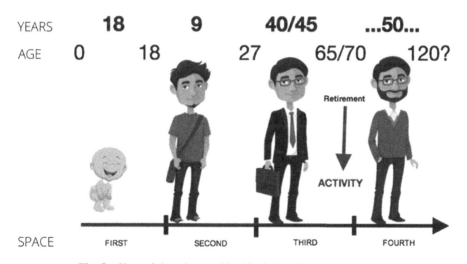

Fig. 8. Slots of time that we identify during lifetime of a person.

This people will need to change their activities during the life, not only the physical activity but also their educational. For example, nowadays a postman workload has been dramatically reduced because most of the written communication is done online. Then, he could do a good job in transporting and supporting all the information taken by the automatic aided devices in the houses, instead in the hospital.

3 Conclusions

Hospital 2090 is not a project in the future, it is an idea that can be implemented, step by step, using nowadays technology. This hypothetical situation in hospitals can be reached by close cooperation among the different professional areas (engineering, physics, etc.) involved in healthcare environments and medicine.

A new education methodology within the health system is needed to understand the utility, mode of use and application of technology in health. As a result of the implementation of this idea, there will be an improvement of the services in health, optimization of the physician's time and a reduction of the health system's costs.

Acknowledgments. The authors of this paper would like to thank to the innovation techno-logical plan MOTIVA (www.motivando.me) for its support and to the Obra Social "laCaixa" and CajaCanarias Foundation for partially funding this work under the project FAST (Formación Avanzada en Sanidad con Tecnología).

References

1. StereoInMotion. www.stereoinmotion.com
2. Brigham and Women's Hospital website about the Advanced Multimodality Image Guided Operating (AMIGO) suite. www.brighamandwomens.org/Research/amigo/

Pioneers and Landmarks in the Development of Information and Communication Technologies

Konrad Zuse's First Computing Devices

Roland Vollmar[✉]

Karlsruhe Institute of Technology, Institute of Theoretical Informatics,
76131 Karlsruhe, Germany
vollmar@ira.uka.de

The structure of this paper is in line with the historical course.

After a short introduction to the development of mechanical computers the principles of Konrad Zuse's developments of the so-called Z1 – Z4 will be discussed.

Very early the desire to support arithmetics resulted in abaci resp. similar devices (jap. Soroban, chin. Suan-pan). But, as F.L. Bauer mentioned, only the place-value system based on the Indians and communicated to Europe by the Arabs allowed the construction of simple mechanical devices, esp. with gear wheels. As far as we know in 1623 Wilhelm Schickard constructed a wooden 4-species machine; all his copies have been destroyed in the Thirty-Years' war and not before the fifties of the 20th century his concept has been found in a letter to Johannes Kepler. A much more elegant and durable machine is a 2-species machine of Blaise Pascal which was built without knowledge of Schickard and at least one copy has been finished in 1642. A minimum of eight copies survived, one in the Musée des Arts et Métiers in Paris – misleadingly labelled as the first of such a machine in the world. At the end of the 17th century Gottfried Wilhelm Leibniz designed a 4-species calculator. It seems that it didn't work correctly; only in the midst of the 20th century Joachim N. Lehmann found a small bug. By the way, Leibniz was aware of the dyadic system and the easiness to compute in it. In the next centuries a multitude of calculators operating with gear wheels, sprocket wheels or step rolls on the decimal base have been built, later also electromechanical driven. In some sense a completely other approach has been pursued by Charles Babbage (1791–1871): He designed two machines which can be considered as the forerunners of modern computers, the so-called Difference Engine and the Analytical Engine. The origin of Babbage's thoughts was his anger about many errors found in mathematical tables, caused as well by wrong calculations as by the type-setting process. Consequently he designed a mechanical calculator based on the method of differences combined with a printing unit. The realization was prevented by the lack of sufficiently exact tools, trouble with his mechanics and not at least by missing a complete financial support by the government. One reason for the stop of payment was the announcement of Babbage that it would be reasonable not to finish the Difference Engine, instead he proposed a new concept (about 1835): The Analytical Engine. To quote Williams [4]: "Although the Difference Engine is an important development in the story of computation, it was Babbage's ideas for another machine – the Analytical Engine - that signal a new concept in computing, namely a computing machine being directed by an external program." I abstain from the details, therefore in

© Springer International Publishing AG 2018
R. Moreno-Díaz et al. (Eds.): EUROCAST 2017, Part I, LNCS 10671, pp. 13–18, 2018.
https://doi.org/10.1007/978-3-319-74718-7_2

short: The machine was equipped with an arithmetical, unit, a storage and some input-output devices. The program and the data should be inputted through punch cards, known from the mechanical looms. It was possible to work with loops in the program. Both engines should work in the decimal system with gear wheels. During the lifetime of Babbage also for the Analytical Engine only some small parts were built.

Konrad Zuse rejected the honor to be called "the father of the computer", instead he proposed to award Babbage. In this connection it is important to emphasize that Zuse was not aware of Babbage's projects during his own early developments. Only applying a patent he received the knowledge of him. Generally speaking at this time Babbage and his work was relatively unknown.

Konrad Zuse (1910–1995) was not predestined to develop automatic calculators. He started with a study of mechanical engineering, changed to the faculty of architecture and at last he graduated as a civil engineer, i.e. he had no special education or some training in electrotechnics. Zuse's eventually successful constructions of an automatic calculator had to follow several steps. As an employee of an aircraft factory Konrad Zuse was bored by similar statical calculations. Such in 1935 he began his considerations for an automatic processing of the forms he has to use.

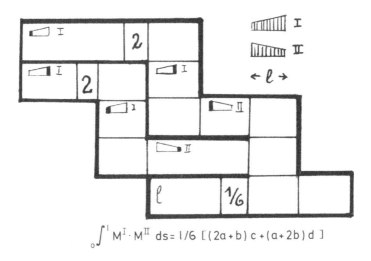

$$\int_0^1 M^I \cdot M^{II} \, ds = 1/6 \, [\, (2a+b) \, c + (a+2b) \, d \,]$$

Fig. 1. Example of a calculating form ([5]) (with the permission of Springer Nature)

In these forms only the corresponding numbers should be inscribed and the process of calculation should follow the structure. In some sense an automatisation would become possible but it would be restricted to the special forms. Zuse observed that a one-dimensional input is superior to a two-dimensional one and that to insist on the decimal notation would be too complex for him. Therefore he left the usual track and he developed another type of calculator based on the dual system with floating point numbers – called Z1. The choice of the dual system allowed him instead of decimal gears simple basics, namely metal plates with cutouts (realized by hand with a jigsaw) which were arranged in several layers and were moved by vertical rods only in one level. To get an impression in Fig. 2 a switching element is sketched.

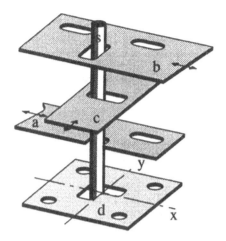

Fig. 2. A simple switching element ([3]) (with the permission of Springer Nature)

To harness the complexity of the basic elements Konrad Zuse developed a so-called "abstrakte Schaltgliedtechnik" (very similar to the propositional logic) which allowed him to use the same design as well as for his mechanical as the electromechanical constructions (by relays). The arithmetical unit and the storage unit (with 64 words) were separated. The data were inputted as decimal numbers through a console and internally translated into binary numbers. The program (called "Rechenplan") was inputted by a punched tape. In principle in 1936 the functioning was given but with a very limited reliability esp. due to the tolerance and sychronization problems. In electrical devices the signal transmission is very simple, but not so with mechanics. Nevertheless the soundness of the concept was proved. To overcome the before-mentioned problems for the arithmetical unit (which has a much more complex design than the storage unit) he used electromechanical relais for the arithmetical unit but furthermore a mechanical storage unit. This machine has been named Z2. Shortly after the finishing of the Z2 he started the work on the Z3 which was completey built by relais, but had the same architecture as the Z1. It is of overwhelming interest, being the "first fully automatic program controlled and freely programmable in a binary floating point working computing device" (F.L. Bauer [in [5]][1]). Konrad Zuse completed it in 1941 and on May 12, 1941 it has been presented to some scientists of the "Deutsche Versuchsanstalt für Luftfahrt" (German Research Center for Aviation). To underline the significance of this machine a quotation

---

[1] Translated by the author.

[[4]]: "...the Z3 was the first machine in the world that could be said to be a fully working calculator with automatic control of its operations." The overall architecture of the Z3 looks rather modern: The storage unit for 64 floating point numbers of 22 bits has been separated from the control unit, the floating point arithmetical unit, and the input-output devices. All the parts are synchronized by signals originating from the control unit (Fig. 3).

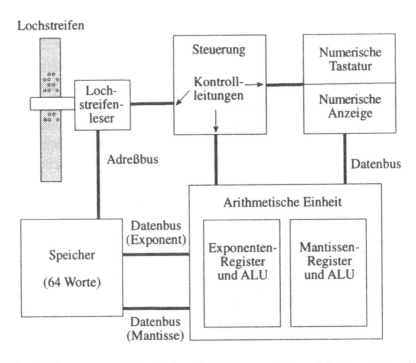

Fig. 3. The structure of the Z3 ([3]) (with the permission of Springer Nature)

Only nine instructions are implemented: Two input/output instructions ("read from the console", "display the result"), two storage instructions and five arithmetical instructions (addition, subtraction, multiplication, division and root extraction). An addition could be completed in about 0.5 s and the multiplication in 3 s.

On the first glance the lack of control instructions may attract the attention. And this is the reason that the Z3 was widely considered as a non-universal computer, if one has Turing's universality notion in mind. Konrad Zuse insisted that the addition of a conditional jump could have added without special efforts, but that he was skeptical about the controllability of such augmented programs. The programming of a loop could be attained by the gluing of the punched tape. From a pure academic interest is a clever construction of Rojas ([2]) who showed that by a tricky programming an indirect addressing would be possible such that in principle a universal Turing machine could be simulated – assumed enough storage is available.

During the World War II the Z3 has been destroyed by the air raids of Berlin.

The structure of the next machine built by Zuse (Z4) was by and large the same as that one of Z3. It had a relay arithmetical unit but a mechanical storage again, only the word length had been increased to 32 bits. It is interesting to hear about the reason to use a mechanical storage: It took less than one cubic meter of space, instead of a very large room if built by relays. He brought essential parts of the Z4 on an adventurous flight through Germany in the last weeks of the war into the southern part of Germany (into the villages Hinterstein und Hopferau [in the Allgäu]) where he worked on its completion, but only in 1950 it has been sent from Neukirchen (near Fulda) to the ETH Zürich, which rented it. Among others Zuse had to built in the possibility for conditional jumps. The Z4 worked there quite reliably for several years and with its help the Mathematical Institute of the ETH gained its international standing. It is now an exhibit in the Deutsche Museum in München. The further developments of Konrad Zuse, now at his own company, the ZUSE KG, will be summarized by the words of Williams [4], esp. because they cannot be considered belonging to the same level of originality as the earlier ones: "One of his first projects [after the establishing of the ZUSE KG] was to produce another relay-based, punched-tape driven, machine for the Leitz Optical Company [in Wetzlar]. This was very similar to the refurbished Z4 in that it had a conditional branch instruction which could cause it to exit from loops. This machine, known as the Z5, was the forerunner of 42 more relay based machines, called Z11s, which were in production until the mid-1950s." But yet another machine should be explicitly mentioned, the Z22: In 1956 Zuse designed in strong cooperation with Fromme a vacuum-tube-based computer. It is not exaggerated to consider this machine as the seed for the establishing of Informatics at German universities. Thanks to the financing of the German Science Foundation (DFG) such machines have been delivered to many universities; due to its simple, but elegant construction principle it was in vogue among the young scientists and for many informaticians the first contact with computers was with the Z22.

Zuse's construction of control computer devices and his early idea of an array computer will only be mentioned.

A talk about Zuse's developments excites nearly unavoidably the question "who constructed the first computer?". But before I shall discuss this topic another work of him should be at least mentioned, also under the aspect that hardware and software are two sides of a medal. With the so-called "Plankalkül" he designed the first high-level programming language in the world. This concept was completed in 1945, but unfortunately there was no publication of the full text before 1972. Unlike the other well-known problem oriented languages as e.g. FORTRAN or ALGOL 60, which were primarily aimed to program numerical problems, with the Plankalkül rather symbol manipulation should be easily noted. Zuse demonstrated this with programs for sorting algorithms and graph algorithms as well as for chess problems. Only one quotation should attest the high esteem of the Plankalkül ([1]): John Backus comments on the Plankalkül: "Like most of the world [...] we were entirely unaware of the work of Konrad Zuse. Zuse's 'Plankalkül', which he completed in 1945, was, in some way, a more elegant and advanced programming language than those that appeared 10 and 15 years later. (IEEE Annals of the History of Computing, Band 20, 1998, Heft 4, Seite 69 [...])" To our question: "Let me emphasize that there is no such thing as 'first' in any activity associated with human invention. If you add enough adjectives to a description you can always claim your own favorite. For example the ENIAC is often claimed to be the 'first electronic, general purpose, large scale, digital computer' and you certainly have to add all these adjectives before you have a correct statement. If you leave any of them off, then machines as the ABC, the Colossus, Zuse's Z3, and many others (some not even constructed as Babbage's Analytical Engine) become candidates for being 'first' ". Corresponding to the title here we abstain from mentioning the scientific work not directly connected to the development of computing devices.

References

1. Bruderer, H.: Konrad Zuse und die Schweiz Oldenbourg, München (2012)
2. Rojas, R..: How to make Zuse's Z3 a universal computer. IEEE Ann. Hist. Comput. **20**(3), 51–54 (1998)
3. Rojas (Hrsg.), R.: Die Rechenmaschinen von Konrad Zuse. Springer, Heidelberg (1998). https://doi.org/10.1007/978-3-642-71944-8
4. Williams, M.R.: A History of Computer Technology, 2nd edn. IEEE Computer Society Press, Los Alamitos (1997)
5. Zuse, K.: Der Computer - Mein Lebenswerk. Springer, Heidelberg (1984). https://doi.org/10.1007/978-3-662-06513-6

Ramon Llull's Ars Magna

Thessa Jensen[(⊠)] [iD]

Aalborg University, Rendsburggade 14, 9000 Aalborg, Denmark
thessa@hum.aau.dk

Abstract. The Ars Magna of Ramon Llull must be seen as one of the first attempts to formalize language, thought processes, and creating a basis for rational discussions. It consists of so-called principles, concepts, which are defined and combined through the use of four main figures. These are explained and an example of their application is given. Llull's contribution and influences are shown and traced through history. Finally, an overview is given of computer science related fields in which Llull's ideas have had or could have had valid contributions.

Keywords: Ramon Llull · Ars Magna · Conceptual systems

1 Vita

Llull was born 1232 in Palma de Mallorca, a melting pot of different cultures and religions at the time. Being educated at the king's court, Llull learned the trade of the troubadour as well as reading and writing in Catalan. He became a devout Christian later in life only after he had married. Christ showed himself to Llull on several occasions, and eventually Llull resumed his rather debauched life and decided to dedicate the rest of his life to three purposes. These were to become a missionary and die for Christ, to develop, write and teach the Ars Magna, and finally, to build monasteries which should teach various languages of the infidels [1, 2].

The rest of his life, Llull spent travelling around the Mediterranean in an effort to convince Muslims, especially, of the truth of the Christian faith. He soon discovered that the main challenge was to explain the divine Trinity to non-Christians. He realized that cultural and language barriers had to be taken into account when anyone tried to explain about the Christian faith. Instead of focusing on the differences between the three main religions, Llull sought out similarities, even going so far as to copying the worshipping style of Muslims.

Legend has it that Llull was stoned to death in the city of Tunis in 1316 by an angry mob of Muslims who were unable to dismantle his arguments for the truth of the Christian faith. His dead body was brought back to Mallorca, and the people of Mallorca have since tried to have Llull canonised as their saint.

© Springer International Publishing AG 2018
R. Moreno-Díaz et al. (Eds.): EUROCAST 2017, Part I, LNCS 10671, pp. 19–24, 2018.
https://doi.org/10.1007/978-3-319-74718-7_3

2   Ars Magna

A few years before his death Llull began writing the most thorough and final version of his Ars Magna, the *Ars Generalis Ultima* [3]. It consists of several books, explaining the different parts of the Ars, which are the principles or concepts, the questions, definitions, and finally, the possible combinations. The four main figures are Figure A, Figure T, the Third Figure and the Fourth Figure.

Figure A, which is called the divine figure, contains nine divine or basic concepts. It is made up by two concentric circles on top of each other. The inner circle holds nine compartments with the letters B to K (J is missing in the Latin alphabet). The outer circle has the nine concepts written on it (see Fig. 1). The various combinations of letters are by substituting each letter with its corresponding concept are turned into quaestios, inquiries. These are expanded by the use of the definitions or the use of concepts from the other figures.

Figure T is made up by three triangles on top of each other. The triangles are coloured, representing the three different religions: the green triangle with the letters B-C-D represents Islam; the red triangle with the letters E-F-G Christianity; and the yellow triangle with the letters H-I-K Judaism. Each triangle points at its designated letters, which are placed in a circle around the triangles. Each letter is assigned its corresponding concept, which for the figure T are all concerned with a triad of elements, such as difference, concord, opposition (green triangle); beginning, middle, end (red triangle); majority, equality, minority (yellow triangle). On the outer circle, each triad is further explained in its possible combinations. E.g. beginning can be a cause, a quantity or a temporal indication.

The Third Figure shows every possible combination of two of the nine letters. The place of a given pair of letters is interchangeable. Thus, the pair of BC and CB are to be seen as the same pairing. Each letter stands for the concepts found in figures A and T, but can also denote concepts from older figures, such as the judiciary figure or the figure of elements, found in earlier editions of the Ars Magna.

The Fourth Figure, also called the syllogism figure, consists of three moving circles, each with nine compartments for one of the nine letters denoting concepts from Figure A and T, or like with the Third Figure from older figures of earlier versions of the Ars. The Fourth Figure is further extended in the Tabula Generalis, a list of all possible combinations of pairings of three from the Figures A and T.

Finally, Llull provides an extensive list of definitions and explanations, both for the particular concepts as well as their various combinations. The Ars Magna is thought of as a tool to enable a rational and logical discussion among peers, that is among learned theologians of the three religions. The chosen concepts were common for the theological and philosophical trained scholars of that time. Through the design and use of the different figures and concepts of the Ars, Llull enabled a discussion which was removed from the holy scriptures and their interpretations. Instead, his Ars revolves around a common understanding of the very nature of belief, life, and God as such.

Fig. 1. The four figures of the Ars Magna. From upper left to lower right: Figure A, Figure T, the Third Figure and the Fourth Figure [3].

3   Providing an Example

Using the Fourth Figure and the Tabula Generalis the following is an example on how the figures are applied. The main concept will be the concept of 'Good' denoted by the letter B in Figure A.

Each concept in the Ars Magna has three states. Thus, 'Good' is defined by producing good (bonificative), an active state in which the concept creates goodness; doing good (bonifies) in which the concept interacts with the world around it, influencing other concepts, things, and people by doing good; and finally, affected by good (bonifiable), a passive state in which the concept affects other concepts, things, and people. Lull perceives the concepts of his Ars as a network, in which each concept can and will be affected by and interact with all the other concepts as well as the world beyond.

A node in this network could be the combination of the letters BCD, retrieved by turning the three concentric circles in Third Figure and finding the combination of the three letters in the Tabula Generalis, column 1, which shows all the possible combinations of the letters with regard to Figure A and T. Thus, the clean pairing of BCD means using the three corresponding letters from Figure A, which would be goodness, greatness and eternity. The other pairings in column 1 are denoting the use of both Figure A and T, by placing the letter T in front of subsequent letters which then denote concepts from the Figure T. Thus, the combination BTCD, column 1, compartment 10, denote the following concepts: goodness (Figure A), concordance (Figure T) and contrariety (Figure T).

The Tabula Generalis is followed by an expanded list of explanations or questions, which gives examples on how to interpret and discuss the given combination of concepts. The scholar would only need to find the matching section, clearly denoted by 'Column BCD, camera BTCD'. In this section, the scholar is presented with an extensive explanation of how the three concepts work in the world using the elements as illustrations. This is a short excerpt: "The tenth question asks whether goodness contains any concordance and contrariety, and the answer is that it does in some subjects habituated with goodness and in which fire agrees with air through heat, and air with water through moisture, and water with earth through cold, and fire with earth through dryness." [4] It shows how Llull systematically works through the consequences of applying different concepts to each other in different contexts. This excerpt explains how the concordance of different elements provide goodness in the form of heat, moisture, cold and dryness.

Llull's Ars is remarkable in its exhaustive, objective, and systematic approach to the question of faith and–through it–life in all its different aspects. The Ars provides not only examples on how nature is understood at Llull's time, it also contains small manuals on agricultural or seafaring issues, just to name a few.

4   The Influence of the Ars

Llull's Ars Magna was to become an inspiration for later philosophers and scientists. His influence can, at times, be hard to track since his Ars and many of his other writings were blacklisted by the Inquisition. Thomas Le Myésier (?–1336) is his pupil and compiled the first collection of Llull's work. Nicolaus de Cusa (1401–1464) is named as one of the first to develop Llull's Ars into a new understanding of faith and science [5]. Agrippa of Nettesheim (1468–1535) writes the first commentary on Llull's Ars, calling it the art of discourse [6]. Giordano Bruno (1548-1600) repeatedly references Llull in his own attempts to create a consistent system of knowledge and faith [6]. Bruno's work on Llull will become one of Leibniz' inspiration to examine Llull's Ars as a basis of his own attempt to create a similar system.

Johann Heinrich Alsted (1588–1638) attempts to unite Llull's work with the logic of Aristoteles in his *Clavis artis lullianae*. Alsted expands the number of concepts and figures used in the Ars in an effort to create a new encyclopedia which would describe the existing as well as invent new knowledge [8]. Finally, Athanasius Kircher (1602-1680) should be mentioned, because his work on trying to establish a method to develop a mechanical method to create German words and sentences to be applied in scientific discussions was heavily influenced by Llull's Fourth Figure with movable concentric circles on top of each other. Kircher's 'fünffacher Denckring der Teutschen Sprache' was designed as five circles with letters and syllables drawn on each ring. Turning the circles would create a lexicon of the German language, according to Kircher [9]. Also, Kircher's work is used by Leibniz and his attempts on creating a universal language.

Llull's Ars is mentioned by Descartes, Newton, and even Peirce. Mostly, in a rather dismissive tone which could be due to a large number of alchemistic and mystical works which have falsely been attributed to Llull. Due to the work of Ivo Salzinger (1669–1728) Llull's original works were collected and in part published. As the head of the Lullian University of Mallorca, Salzinger collected, translated and published Llull's works in eigth volumes. It has since been republished in 1965. Like Le Myésier, Salzinger's books kept the knowledge about Llull's work and ideas alive and made it possible to examine its potential for later generations.

One of the other great lullian scholars is Anthony Bonner, who through several translations and commentaries has made Llull's philosophy accessible to the modern scientist. For further reading, his *Doctor Illuminatus: A Ramon Llull Reader* is highly recommended [1].

5   The Relevance of the Ars Magna in Computer Science

When Gottfried Wilhelm Leibniz (1646–1716) as a twenty-year old writes his dissertation *De Arte Combinatoria* he draws heavily on Llull's Ars Magna and the idea of creating a system of networked concepts, which through mechanical rule of reasoning shows a universal system of knowledge. This idea continues throughout his life, the development of a universal language which can be applied to find and proof knowledge in an objective, rational way. The basic foundation is to be found in an encyclopedic list of concepts, syllogisms, variables and constants, bound together by rules and procedures.

These basic ideas can be traced from Leibniz to the work of Sowa [10] and the development of conceptual graphs in language understanding and artificial intelligence. Sowa cites Llull as an inspiration for his work, as do Sørensen and Sørensen [11] in their explanation of the conceptual pond, a tool for understanding the reception, understanding, and evaluation of knowledge in a teaching environment.

Likewise, Umberto Eco explains Llull's Ars in his book *The Search for the Perfect Language*. It tells the quest for discovering the basic language which would enable people to understand or translate everything easily from one language to another. Artificial intelligence, knowledge representation and language understanding all seek to formalize language and knowledge in ways which could be translated into a computer programme.

Fidora and Sierra's anthology [12] on Ramon Llull's importance shows how profound the relevance and influence of Llull's thoughts still are in a large range of fields. From Peirce's pragmatic thinking over social choice theory, which Llull anticipated several hundred years earlier than previously thought. In his novel *Blanquerna*, written in 1283, Llull explains the process on how nuns should elect their abbess by a majority rule.

From logical analysis to adaptive reasoning, these theories can be traced back to the Catalan lay monk, who tried to create an approach to the missionary quest which included the ethical consideration of the other, a systematic, combinatoric, rational and logical Ars, which should be used as a tool for discourse and discussion.

References

1. Bonner, A.: Doctor Illuminatus: A Ramon Llull Reader. Princeton University Press, Princeton (1993)
2. Platzeck, E.W.: Das Leben des seligen Raimund Lull: die "Vita Coetanea" und ausgewählte Texte sum Leben Lulls aus seinen Werken und Zeitdoumenten. Patmos-Verlag (1964)
3. Lullus, R.: Ars Generalis Ultima. Minerva GmbH (1970)
4. Lullian Arts Homepage. http://lullianarts.narpan.net/Ars-Magna/5.htm#Column_bcd_camera_btcd_Tenth-questions. Accessed 28 May 2017
5. Colomer, E.: Nikolaus von Kues und Raimund Llull - aud den Handschriften der Kueser Bibliothek. Walther de Gruyter & Co., Berlin (1961)
6. Bonner, A.: Der neue Weg Ramon Llull. In: Llull, R. (ed.) Buch vom Heiden und den drei Weisen, pp. 26–31. Verlag Herder, Freiburg im Breisgau (1996)
7. Yates, F.: Lull and Bruno, vol. 1. Collected Essays. Routledge and Kegan Paul, London (1982)
8. Schmidt-Biggeman, W.: Topica Universalis. Felix Meiner Verlag, Hamburg (1983)
9. Couturat, L. (ed.): Opuscules et fragments inèdits de Leibniz. Georg Olms Verlagsbuch-handlung, Hildesheim (1966)
10. Sowa, J.F.: Conceptual graphs as universal knowledge representation. Comput. Math Appl. **23**(2–5), 75–93 (1992)
11. Sørensen, C.G., Sørensen, M.G.: The conceptual pond–A persuasive tool for quantifiable qualitative assessment. In: Emerging Research and Trends in Interactivity and the Human-Computer Interface. IGI Global (2013)
12. Fidora, A., Sierra, C. (eds.): Ramon Llull: From the Ars Magna to Artificial Intelligence. Artificial Intelligence Research Institute, Barcelona (2011)

The 19th-Century Crisis in Engineering

Hartmut Bremer$^{(\boxtimes)}$

Institut für Robotik, Johannes Kepler Universität Linz, Linz, Austria
hartmut.bremer@jku.at

Abstract. At the end of the 19th century, *applied* (or technical) mechanics, as one of the basics of technical development, found itself a desolate state, due largely to the refusal of its practitioners to recognize the influence of kinetics on (spatial) motion. They had failed to keep up with developments in the science underlying their craft and were unable to keep pace with the speeds of such systems as the steam engine. On the other hand, *theoretical* (or rational) mechanics was already well established, mainly in conservative astrodynamics. Into this critical situation, two scientists began to build a bridge, from two different sides: August Föppl (1854–1924) and Felix Klein (1849–1925).

1 Introduction

The nineteenth century – brilliant in culture in all its aspects. Considering *music* for instance we have Ludwig v. Beethoven (1770–1827) at the turn of century, followed by Richard Wagner (1813–1897), Anton Bruckner (1824–1896) and Johannes Brahms (1833–1897), just to mention some of them. Or *visual art*: Gustav Klimt (1862–1918) with his famous drawing "The Kiss" and Alfons Mucha (1860–1939) with a whole museum dedicated to him in Prague. They demonstrate the typical drawing style of that time and also influenced architecture ("Jugendstil"). But also Adolph von Menzel (1815–1905), a historical painter, belongs in that area. His drawing "The Flute Concert" (Fig. 1) shows Friedrich II (1712–1786), king of Prussia, playing the flute, with C.P.E. Bach (1714–1788) at the cembalo. Among the enthusiastic listeners only Pierre-Louis Moreau de Maupertuis (1698–1759) seems bored, inspecting the ceiling. Maupertuis, a mathematician and natural scientist, plays an important role in *Natural Science* also in the 19th century: Hermann v. Helmholtz, whose reputation began 1847 with the generalization of the principle of energy conservation [9] later also used Maupertuis' Principle [2].

Maupertuis was the president of Friedrich's Prussian Academy, with Leonhard Euler (1707–1783) as director of the mathematical class. Euler gave Maupertuis' Principle a mathematical base, but the jealous François-Marie Arouet (1694–1776), vulgo Voltaire, mobbed Maupertuis so heavily that he gave up in 1753. His successor was Jean-Baptiste de Boyer (1703–1771), marquis d'Argens. The philosopher d'Argens does not play a significant role in this story, but the director of the mathematical class became Joseph-Louis Lagrange (1736–1813). 1787, one year after Friedrich's death, Lagrange changed to Paris, directly

© Springer International Publishing AG 2018
R. Moreno-Díaz et al. (Eds.): EUROCAST 2017, Part I, LNCS 10671, pp. 25–32, 2018.
https://doi.org/10.1007/978-3-319-74718-7_4

towards the french revolution. He survived in a good shape and was even ennobled by Napoleon (1769–1821). With the heritage of Euler and Lagrange we are
entering the 19th century and its state in natural sciences, especially *Rational
Mechanics*.

Fig. 1. A. v. Menzel: *The Flute Concert of Frederic the Great in Sanssouci* (1852)

2   Engineering

After Friedrich II there came a lot of Friedrich-Wilhelms till Wilhelm I (1797–
1888) who (under the influence of his chancellor Otto v. Bismarck (1815–1898))
became german emperor after the french-german war 1870/71. His successor
Friedrich III died already after 99 days and Wilhelm II (1859–1941) overtook
government. Wilhelm II was a fan of any kind of engineering, especially mechanical engineering.

The 19th-century engineering started mainly with Newcomen's steam engine
from 1712 with low efficiency – just about one cycle per 15 min. Around 60 years
later it was improved by James Watt who separated heating and condensation.
Once more around 60 years later George and Robert Stephenson construed the
famous railroad engine "Rocket" with already 47 km/h velocity. The railroad
became catalyzer of the first industrial revolution, and here we are at the end of
the 19th century: steam engines everywhere, stationary e.g. in weaving mills with
its sophisticated control via perforated wooden or cardboard strips (Vaucanson

and Jacquard, resp.) leading to the punchcards from Herman Hollerith which eventually lead to the foundation of the International Business Machines Corporation (IBM) in 1926. Mobile e.g. in railroad and ships: Wilhelm II expanded his imperial navy with one new ship nearly every year. Steam engines need a lot of coal, but the improvement was already on the way: Diesel with his patent from 1893. Rudolph Diesel (1858–1913) was not a friend of military applications. He disappeared 1913 on a boat trip to England under mysterious circumstances. His engine, however, had already been realized on a danish merchant ship one year before: the "Selandia". As pictures of that time show: no more chimneys. The 20th century innovation is thus scheduled: Diesel ships (and railroad electrification, and digital control, and ...).

But Wilhelm's hobby was dangerous: some neighbours were not amused ("rule, britannia, britannia rule the waves..."). It ended, as well known, 1914 in a catastrophe.

3   The Crisis

Technical education in Europe started 1794 with the École Polytechnique, 1805 reorganized as a military school by Napoleon. A lot of famous scientists teached there, for instance Lagrange from 1797 on. The foundation of technical schools in the following decades mainly concentrates on middle Europe. However, we should have in mind that especially mechanics was split into two branches, the theoretical (rational) and the applied (technical) one. Rational mechanics was already in good condition. Technical mechanics, however, had its roots mainly in inventions and practical experience (e.g. Watt and Stephenson). With the upcoming fast moving systems it does not take wonder that engineering came into a desolate state. Here, two people began to build a bridge between rational and technical mechanics from two different sides: the mathematician Felix Klein and the engineer August Föppl. Thereby, the *ingenious mathematicians* [Lagrange, Gauss, Riemann etc.] *had been far away from technical applications; the translation of their ideas into a homespun form which is needed here is by no means an easy task,* [10].

4   A Bridge Between Rational and Technical Mechanics

4.1   Felix Klein, The Mathematician

Felix Klein (1849–1925) obtained 1872 his first chair in Erlangen, 23 years old, where he published the main research goals in geometry (kown as "Erlanger Programm"). Two years later, in München (his second chair), he introduced lectures on mathematics for engineers, parallel to mechanics. After five years he eventually went back to Göttingen via Leipzig. His statement: *The exact university sciences have lost contact with the living world in a threatening manner. They look for fame in a self-chosen isolation,* [17].

Due to Klein's influence, Karl Heun obtained 1902 a chair in Karlsruhe. Karl Heun (1859–1929) had studied mathematics and philosophy in Göttingen and taught since 1886 in Munich as a private lecturer. With his income there he was soon unable to feed his family and changed 1890 to Berlin as a senior teacher at the Erste Realschule (middle school). In Berlin he met a group of engineers with their unsurmountable problems in fast moving machinery. His statement: *In view of complex kinetic problems, intensive knowledge of Lagrange* [rational mechanics], *Gauss* [numerical integration] *and Riemann* [projective geometry] *will lead to success rather than tedious searching in integral tables and engineer calendars* [10].

One of Heun's assistents was Georg Hamel. Georg Hamel (1877–1954) obtained his doctoral degree in Göttingen and joined Heun in Karlsruhe 1902 where he finished his the habilitation thesis in 1903. Therein he states: *The possibility to obtain the facts so easy and clear is mainly due to Lagrange...* [8].

4.2   August Föppl, The Engineer

August Föppl (1854–1924) had studied civil engineering in Darmstadt, Stuttgart and Karlsruhe. He taught at the vocational school in Leipzig from 1877 on and overtook 1894 the mechanical chair in Munich. One has to be aware of the situation at that time to understand his wording *I do not deny the right to look at mechanics as part of mathematics and formulate it correspondingly. But it is an offense to treat people this way if they want practical instruction* [7]. It is remarkable that he already introduced vector analysis – apparently similar to Klein's projective geometry approach.

As mostly, leaving traditional paths earns hostility (e.g. "...cannot be recommended", or worse: "if one reverses the distortions which the development had to suffer by the author one could take advantage from already known results", [21]). Föppl's overwhelming success, however, is demonstrated by his textbooks with more than onehundred thousand copies, and with generations of students who attended his lectures. Among these we find Ludwig Prandtl (1875–1953), the famous fluid dynamicist, and Max Schuler (1882–1972) who developed the gyro compass. Prandtl became the first director of the Institute of Applied Mechanics in Göttingen which had been founded by Felix Klein in 1905, and Schuler became Prandtl's successor.

4.3   The Bridge Foundation

It was due to Felix Klein and August Föppl that outstanding scientists like Heun and Hamel (and themselves, of course) could eventually bring engineering and natural science back together. Heun's and Hamels's contributions were strongly based on Lagrangian mechanics.

Lagrange had won the big prize of the french academy for the year 1764 [15]. He used there his δ-operation which he had formulated in 1760 (*essay on a new method finding the maxima and minima...*, [13], already communicated to Euler in a letter from 1755 when he was 19 years old [12], very much appreciated by Euler). In [13] he applies his *new method* to the Brachistochrone problem. In *application de la méthode* [14] he considers Maupertuis' Principle of least action. And then, in 1764, he frees himself from the stanglehold of the least action principle with its restrictive conservative forces with the *Lagrange-Principle* $\int \delta \mathbf{r}^T (\ddot{\mathbf{r}} dm - d\mathbf{f}) = 0$ (which up to the time being has been confused with d'Alembert's Principle [1] which has next to nothing in common). *The only difficulty here consists in finding the analytical expression for the forces [df_i] which one assumes to act...and to insert the minimum number of undetermined variables [q_i] such that their differentials denoted by δ are totally independent from each other.* This is a breakthrough and a milestone in dynamics, also impressively demontrated in his *Analytical Mechanics* [16] where he calls it the "general formula of dynamics" being the basis of all his following considerations (including his famous equations of the second kind which have been generalized by Hamel 1903 for the non-holonomic case [8]). Clearly, Lagrange's Principle contains Euler's momentum theorem from 1736, $\ddot{\mathbf{r}} dm = d\mathbf{f}$ [4]. Fourteen years later, in 1750, Euler published the momentum of momentum theorem [5]. Once more fourteen years later, Lagrange offered his results. Transforming all these contributions to a representation in an arbitrary moving reference frame R which rotates with $\boldsymbol{\omega}_R$ one obtains for a single rigid body, along with $\int \widetilde{\mathbf{r}} \widetilde{\mathbf{r}}^T dm = \mathbf{J}$ and $\mathbf{p} = m\mathbf{v}, \mathbf{L} = \mathbf{J}\boldsymbol{\omega}$ (**v** and $\boldsymbol{\omega}$: absolute translational and angular velocity, resp.):

$$\dot{\mathbf{p}} + \widetilde{\omega}_R \mathbf{p} - \mathbf{f} = 0 \qquad (\sim \text{Euler } 1736) \qquad (1)$$

$$\dot{\mathbf{L}} + \widetilde{\omega}_R \mathbf{L} - \mathbf{M} = 0 \qquad (\sim \text{Euler } 1750) \qquad (2)$$

$$\delta \boldsymbol{\rho}^T (\dot{\mathbf{p}} + \widetilde{\omega}_R \mathbf{p} - \mathbf{f}) + \delta \boldsymbol{\pi}^T (\dot{\mathbf{L}} + \widetilde{\omega}_R \mathbf{L} - \mathbf{M}) = 0 \qquad (\sim \text{Lagrange } 1764) \qquad (3)$$

$((\dot{}) = d()/dt$ (time derivative), $(\widetilde{})$: spin tensor, **f** and **M**: force and moment, resp.; $\delta \boldsymbol{\rho}$ and $\delta \boldsymbol{\pi}$: virtuals assigned to $\dot{\boldsymbol{\rho}} = \mathbf{v}$ and $\dot{\boldsymbol{\pi}} = \boldsymbol{\omega}$, resp.; all terms represented in frame R. Since Eq. (3) is a salar equation, one obtains a N-body representation just by summing up. This is great.) But now comes a problem: eleven years later, in 1775, Euler postulated Eqs. (1) and (2) to be independent [6]. Hence, how can $\int \delta \mathbf{r}^T (\ddot{\mathbf{r}} dm - d\mathbf{f}) = 0$ lead to two axioms in Eq. (3) while it contains only one, namely $(\ddot{\mathbf{r}} dm - d\mathbf{f}) = 0$ – just by premultiplying $\delta \mathbf{r}^T$ and integrating? At first: dm is not a mass point but $dm = \rho dV$ with ρ: density and dV: volume element ($dV = dxdydz$ if considered cartesian) which enables integration. Considering "mass points" as molecules or atoms leads to quite other questions (Bohr, Heisenberg, Schrödinger...). We are operating here, so to say, on a macroscopic scale. In the same sense, molecular Brownian motions for instance enter the Lagrange Principle on that macroscopic scale with also macroscopic assumptions via δW_{therm} (thermodynamic virtual work) etc. Next, what are the forces $d\mathbf{f}$ like? In case of a solid, these include the gluing forces which hold the body together. Considering a rigid body, the constraint that dm holds its position and orientation w.r.t. a body fixed frame during motion yields the constraint forces $(d\mathbf{f}^{\text{constr}}/dV)^T = \nabla^T \boldsymbol{\lambda}_{ij}$ (∇: Nabla-operator, $\boldsymbol{\lambda}_{ij} = \boldsymbol{\lambda}_{ji}$: "Lagrangian multipliers") which in case of a solid convert (by "relaxation of the constraints") to $(d\mathbf{f}^{\text{elast}}/dV)^T = \nabla^T \boldsymbol{\sigma}_{ij}$ with the symmetric stress tensor $\boldsymbol{\sigma}_{ij} = \boldsymbol{\sigma}_{ji}$. However, the problem is the other way round: a rigid body does not really exist. Hence, using the symmetric stress tensor which is obtained this way has to be looked at axiomatic (called the "Boltzmann Axiom" by Hamel, representing the missing second axiom). Axiomatic because it refers to initial presumptions which do for instance not take superimposed relative rotations of dm into account (polar media, e.g. [3]). Do such rotations exist? When does, accordingly, an unsymmetric stress tensor join the game? The corresponding forces should then be observable, or measurable, a problem field where Lagrange's Principle is predestined for.

However, one should be aware of the fact that everything we are dealing with is based on an abstraction process of human thinking, also called modeling [19]. Models describe reality. One of the most fundamental models in mechanics is the rigid body as a limiting case where the stresses convert to constraint forces anyhow. Dealing with "rigid body dynamics", we may therefore proceed with Eqs. (1) to (3) without further restrictions, but having in mind that the constraint forces are essential to form (or to define) the rigid body.

4.4 Results

In 1889, Carl Gustav de Laval (1845–1913) published experimental results from rotor dynamics, but nobody believed him. According to common sense a rotor should become unstable once the centrifugal forces exceed the bearing forces. August Föppl was the first one to correctly explain the strange rotor behaviour in 1895, probably not noticed by his contemporaries. Hence, the Royal Society commissioned Henry Homann Jeffcott (1877–1937) in 1916 to solve the problem. It took him 3 years – we have thus 30 years of basic discussion. Since then, the considered rotor is called Jeffcott Rotor in the English speaking area, elsewhere Laval Rotor (and nowhere Föppl Rotor). By the way: The Alfa-Laval Company, founded 1883 in Stockholm by Laval and his co-worker Oskar Lamm, still exists and celebrated its 130th birthday in 2013 – a centennial success story.

Obviously, a rotor represents the most basic example for fast spatial motion. Föppl's textbook ([7], six volumes) contains a lot of corresponding applications from the engineering point of view with deep insight, based on fundamental mathematical tools. The five years older Felix Klein, from the mathematical viewpoint, published (together with his assistant Arnold Sommerfeld (1868–1951)) a famous book on *gyrodynamics* [11] where he characterizes the gyro "an idea forming moment", always looking for mechanical interpretation and verification. One may say that gyrodynamics is the representative of spatial dynamics: it consists basically of a rotating rigid body, or of a system of rigid bodies which undergo rotational and, consequently, translational motion ("Kreiselsysteme" [18]).

As already mentioned, Klein founded 1905 the Institute of Applied Mechanics in Göttingen with its directors Ludwig Prandtl and Max Schuler, two Föppl students. Under their guidance, Kurt Magnus (1912–2003) approved his PhD thesis in 1937 and his habilitation thesis in 1943. After WWII, Magnus was deported to the USSR, along with 20000 germans, where he was forced to do research on rocket control. After 7 years he was freed. Via Freiburg and Stuttgart he came to Munich and founded 1966 a second chair of mechanics. His book on *gyrodynamics* from 1971 comprises the development up to that time [18]. In his foreword he underlines *I abstained from specialized mathematical tools when the desired equilibrium between effort and success would have been lost.* In common interpretation: If one wants to plant a flower one should not use a bulldozer. Otherwise one would once more risk *to reach fame in a self-chosen isolation.*

5   CAST – Computer Aided Systems Theory

Consider once more Eq. (3). As Lagrange postulates one has to insert the independent q_i $[\Rightarrow \delta\boldsymbol{\rho} = (\partial\boldsymbol{\rho}/\partial\mathbf{q})\delta\mathbf{q} \wedge \delta\boldsymbol{\pi} = (\partial\boldsymbol{\pi}/\partial\mathbf{q})\delta\mathbf{q}]$ and *to separately set all terms which are multiplied by δq_i equal to zero* [15]. The question arises what the Jacobians $(\partial\boldsymbol{\rho}_i/\partial\mathbf{q})$ and $(\partial\boldsymbol{\pi}_i/\partial\mathbf{q})$ look like.

Thirty years of "epoch-making research in the area of numerical dynamics" have recently been celebrated. The claim is asserted there to use an inertial frame for best procedure. Then, $\boldsymbol{\omega}_R$ is zero and $\partial\boldsymbol{\rho}/\partial\mathbf{q} \to \partial\mathbf{r}_c/\partial\mathbf{q}$ ($\mathbf{r}_c$: mass center position) does not cause problems while $\boldsymbol{\pi}$ remains a quasi-coordinate. Its determination shall then be computed via $\partial\tilde{\pi}_{ik}/\partial\mathbf{q} = [\partial\mathbf{A}_{ij}/\partial\mathbf{q}]\mathbf{A}_{kj}$ (where $\mathbf{A}$: transformation matrix from body-fixed to inertial representation). Because this is tedious, a symbolic manipulation computer code is recommended.

One can argue against this foregoing with two characteristic examples:

(1) Consider Eq. (3) for a rotor with fixed point (inertial frame): $(\partial\boldsymbol{\pi}/\partial\mathbf{q})^T[\dot{\mathbf{L}} - \mathbf{M}] = 0$. Computing the Jacobian $(\partial\boldsymbol{\pi}/\partial\mathbf{q})$ via the proposed procedure fills pages. However, because $\boldsymbol{\omega} = \dot{\boldsymbol{\pi}} = (\partial\boldsymbol{\pi}/\partial\mathbf{q})\dot{\mathbf{q}} + (\partial\boldsymbol{\pi}/\partial t)$, partial differentiation w.r.t. $\dot{\mathbf{q}}$ yields the identity $(\partial\boldsymbol{\pi}/\partial\mathbf{q}) = (\partial\boldsymbol{\omega}/\partial\dot{\mathbf{q}})$. This is the coefficient matrix of $\boldsymbol{\omega}$ w.r.t. $\dot{\mathbf{q}}$ which is already known from kinematics. No additional calculation is needed at all.

(2) This holds, of course, for every coordinate representation. What about restriction to inertial systems? Consider an edge mill via $\mathbf{A}_\alpha[\dot{\mathbf{L}} + \mathbf{M}^{\text{react}}] = 0$ to calculate the reaction torque. (Since one is interested in the torque at the millstone a corresponding post-transformation $\mathbf{A}_\alpha$ is needed). Even in case of constant mill stone rotation the calculation fills pages. On the other hand, using a moving frame which rotates with the (constant) angular velocity $\boldsymbol{\omega}_R$ of the whole equipment, one obtains $\mathbf{M}^{\text{react}} = -\tilde{\boldsymbol{\omega}}_R\mathbf{L}$ just in one line [19].

Hence, should one really put everything into a numerical program and don't care any more – the computer will do it – or is Truesdell right: *the computer: ruin of science and threat to mankind* [20]? CAST-specialists will probably not go so far. No doubt, the computer is useful (or even needed for large systems), but a strict warning is advisable: *put brain in motion before starting computer.*

The 19th century crisis is overcome. Lets wait for the next.

References

1. d'Alembert, J.: Traité de dynamique, 2nd edn. David, Paris (1758)
2. du Bois-Reymond, E.: Hermann von Helmholtz: Gedächnisrede (1897). Univ. Bibl. Heidelberg (2014)
3. Cosserat, E., Cosserat, F.: Théorie des corps déformables. Hermann, Paris (1909)
4. Euler, L.: Mechanica sive motus scientia analytice exposita (1736). German Translation by J. Ph. Wolfers, vol. 2. C.A. Koch, Greifswald (1848, 1850)
5. Euler, L.: Découverte d'un nouveau principe de la mécanique (1750). Mém. Acad. Sci. Berlin **6**, 185–217 (1752, printed)
6. Euler, L.: Nova methodus motum corporum rigidorum determinandi (1775). Mém. Acad. Sci. Petropol. **20**, 208–238 (1776, printed)
7. Föppl, A.: Vorlesungen überTechnische Mechanik. B. G. Teubner, Leipzig (1899)
8. Hamel, G.: Die Lagrange-Eulerschen Gleichungen der Mechanik. ZAMP, 1–57 (1904)
9. von Helmholtz, H.: Über die Erhaltung der Kraft. Reimers, Berlin (1847)
10. Heun, K.: Die kinetischen Probleme der wissenschaftlichen Technik. Jahresberichte der DMV, IX, 2 (1900)
11. Klein, F., Sommerfeld, A.: Theorie des Kreisels. B. G. Teubner, Leipzig (1897)
12. de Lagrange, J., Correspondance (1755). Reprint: Oeuvres de Lagrange, vol. 14, pp. 146–151 and pp. 152–154. Gauthiers-Villars, Paris (1892). http://gdz.sub.uni-goettingen.de
13. de Lagrange, J.: Essai d'une nouvelle méthode pour déterminer les maxima et les minima des formules intégrales indéfinies (1760). Reprint: Oeuvres de Lagrange, vol. 1, p. 389. Gauthiers-Villars, Paris (1873). http://gdz.sub.uni-goettingen.de
14. de Lagrange, J.: Application de la méthode exposée dans le mémoire précédent a la solution de différents poblèmes de dynamique (1760). Reprint: Oeuvres de Lagrange, vol. 1, p. 419. Gauthiers-Villars, Paris (1873). http://gdz.sub.uni-goettingen.de
15. de Lagrange, J.: Récherches sur la libration de la lune (1764). Reprint: Oeuvres de Lagrange, vol. 6. Gauthiers-Villars, Paris (1873)
16. de Lagrange, J.L.: Méchanique analytique. Desaint, Paris (1788). German by H. Servus. Springer, Berlin (1897)
17. Magnus, K.: Mechanik bei B. G. Teubner. B. G. Teubner, Leipzig (1986)
18. Magnus, K.: Kreisel, Theorie und Anwendungen. Springer, Berlin (1971). https://doi.org/10.1007/978-3-642-52162-1
19. Pfeiffer, F., Bremer, H. (eds.): The Art of Modeling Mechanical Systems, pp. 100–110. Springer, Heidelberg (2016). https://doi.org/10.1007/978-3-319-40256-7
20. Truesdell, C.: An Idiot's Fugitive Essays on Science. Springer, New York (1984). https://doi.org/10.1007/978-1-4613-8185-3
21. Weingarten, J.: Rezensionen. Archiv d. Math. u. Phys., pp. 342–352 (1901) and pp. 239–243 (1908). B. G. Teubner, Leipzig, Berlin

Nicolas Rashevsky: Mathematical Biophysicist

Paul Cull[(⊠)]

Computer Science, Kelley Engineering Center,
Oregon State University, Corvallis, OR 97331, USA
pc@cs.orst.edu

Abstract. Nicolas Rashevsky was a pioneer in applying mathematics to biology. He is perhaps best known for his creation of neural net models and his applications of these models to a wide variety of problems in physiology and psychology.

His journal, the Bulletin of Mathematical Biophysics (renamed Bulletin of Mathematical Biology), and his two volume tome, *Mathematical Biophysics* [1], inspired others to use mathematical models in biology.

Here we give a brief description of Rashevsky's career and discuss his lasting (or not so lasting) influence.

1 Introduction

Nicolas Rashevsky (1899–1972) was a pioneer in the applications of Mathematics to Biology. He proposed reorganizing how biology was done with the following slogan:

MATHEMATICAL BIOPHYSICS : BIOLOGY
:: MATHEMATICAL PHYSICS : PHYSICS

{He should not be confused with Samuel Reshevsky, the chess grandmaster, nor with Nicolai Lobachevsky, one of the creators of non-Euclidean geometry.} For many years, he was associated with the University of Chicago where he was the founder of the Committee on Mathematical Biology. He was also the founder of the Bulletin of Mathematical Biophysics (now re-named the Bulletin of Mathematical Biology). Here we will give a brief description of his career. We'll note at the outset that he was a somewhat controversial figure, but we will focus on his positive contributions.

R. Moreno-Díaz et al. (Eds.): EUROCAST 2017, Part I, LNCS 10671, pp. 33–40, 2018.
https://doi.org/10.1007/978-3-319-74718-7_5

The following is a concise resume of Rashevsky's career [2]:

Early Career in Europe

- BORN: Chernigov, Russia – 1899
- EDUCATION: University of Kiev
                 Physics (graduate degree)
- TEACHING: University of Kiev
                 Robert College (Constantinople)
- 1921: Professor of Physics
                 Russian University at Prague.

Career in the United States

- 1924–1933: Westinghouse Research Laboratories
- 1934: Rockefeller Fellow
           University of Chicago
- 1935–1940: Depts. of Psychology and Physiology
- 1940–1947: Section on Mathematical Biophysics
- 1947–1964: Professor and Chairman
                 Committee on Mathematical Biology
- 1965–1972: Professor of Mathematical Biology
                 University of Michigan.

2 Rashevsky's Research

Here, we will give a brief description of Rashevsky's Research. He declared that his field was MATHEMATICAL BIOPHYSICS with the slogan:

MATHEMATICAL BIOPHYSICS: BIOLOGY
        :: MATHEMATICAL PHYSICS: PHYSICS

This slogan meant that he proposed to devise theories for Biology in the same sense that Mathematical Physics furnishes theories for Physics. He added two essential caveats that were necessary to make the theories relevant:

- The Theory has to be Quantitative
- The Theorist has to be in contact with Experiments.

2.1 Diffusion: An Example of Rashevsky's Approach

The first example in his book Mathematical Biophysics [1] concerns diffusion of substances in biological systems. To create a model, he had to face some problems and make some assumptions:

- Not enough data to specify ALL parameters
- Assume some quantities are constant
- When possible assume variables are linearly changing
- Assume some variables obey linear differential equations.

His models produce predictions which look *like* experimental data. He DID compare his predictions with experimental data. This approach is now so standard, that no one would mention who started using this approach.

2.2   Mathematical Biophysics (The Book)

Most of Rashevsky's research was first published in journals, but he then made his results more accessible by assembling them into books. His most important book is *Mathematical Biophysics: Physico-Mathematical Foundations of Biology* which appeared in several editions.

– 1938 First Edition (U. of Chicago Press) [3]
– 1948 Second Edition (U. of Chicago Press) [4]
– 1960 Expanded Two Volumes (Dover) [1].

The final edition contains almost 1000 pages. Among its various topics are the following:

– Cell Respiration
– Cell Dynamics
– Discrimination of Intensities
– Reaction Times
– Learning
– Neural Nets
– Relational Biology.

Rashevsky noticed that the techniques he was developing for Biology could also be applied to problems in the Social Sciences. He wrote (or edited) several books devoted to Psychology and to History.

– 1947 and 1949 Mathematical Theory of Human Relations [5]
– 1951 and 1959 Mathematical Theory of Social Behavior [6]
– 1960 Mathematical Study of the Causes and Origins of War [7]
– 1968 Looking at History through Mathematics [8].

3   Neural Nets

Neuron Model

The basic idea was to use a pair of linear differential equations and a nonlinear threshold operator:

$$Input := I(t) \tag{1}$$

$$\frac{d\,e}{d\,t} = A\,I(t) - a\,e \tag{2}$$

$$\frac{d\,j}{d\,t} = B\,I(t) - b\,j \tag{3}$$

$$Output := H(e - j - \theta) \tag{4}$$

$$\theta \text{ is the threshold and } H(x) \text{ is the Heaviside operator.} \tag{5}$$

Here e and j could represent excitation and inhibition or the amount or concentration of two substances within a neuron. After the work of Hodgkin and Huxley in the 1950s, these could be identified as sodium and potassium concentrations. The Heaviside operator, $H(x)$ takes positive values to 1, and non-positive values to 0. This gives an easy way to model the all-or-none firing of a neuron [9,10]. These model neurons could then be assembled into nets to create more complicated models. For example, the following cross-couple net was often used:

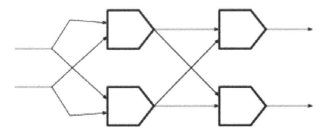

An amazing number of phenomena from physiology and psychology can be described by this model with 4 interconnected neurons [1,6,11].

McCulloch-Pitts – Discrete Neural Nets [12]

Each *neuron* has one of two states: 0 or 1.
Neuron i has state $x_i(t)$ at time t.

$$x_n(t+1) = H\left(\sum c_{in} x_i(t) - \Theta_n\right) \tag{6}$$

- W. S. McCulloch – Professor, University of Illinois
- W. Pitts – undergraduate, University of Chicago.

Rashevsky Neurons vs McCulloch-Pitts Neurons

McCulloch-Pitts

- Mathematical Approach
- Discrete Model
- Prove Theorems e.g. **All** functions can be computed by feed-forward nets
- Theorems may *not* lead to practical construction.

Rashevsky

- Physicist's Approach
- Differential Equations
- Approximation of All functions implied
- Gave examples of small nets which capture experimental results.

4   Rashevsky and Administration

RED Scare [13]

– faculty of Committee on Mathematical Biology refuses to sign "loyalty" oath
– all non-tenured faculty fired
– Committee reduced to Rashevsky and Landahl in one office without a telephone
– after the Scare died out, Committee regrew,
  and even hired some of the fired faculty.

NIH: National Institutes of Health

– Funds almost all Biology and Medicine research in the US
– Rashevsky tried to get a Study Group on Mathematical Biology established (FAILED)
– Convinced NIH that more mathematics training was needed for biologists [14]
– Training Grant (about 15 years)
  supported both PhD students and post-docs from Biology and Medicine.

New Chairman [15]

– Rashevsky had to step down as Chairman at age 65
– Rashevsky wanted Landahl to replace him
– Outside Chairman to be appointed
– Rashevsky resigns
  and moves to the University of Michigan.

Bulletin of Mathematical Biology

– 1939 established as "house journal" (Bulletin of Mathematical Biophysics)
– expanded (published by U. of Chicago Press)
– encouraged papers by outsiders
– revisions rather than rejections
  (the published models were often developed during the review process)
– still publishing
  (now published by Springer).

Society for Mathematical Biology

– Rashevsky created Mathematical Biology, Inc.
– 1973 Landahl established Society for Mathematical Biology
– Society for Mathematical Biology continues,
  the Bulletin of Mathematical Biology is their official journal.

5   Later Research

Principle of Optimal Design [1]:

> For a set of prescribed biological functions an organism has the optimal possible design with respect to economy of material used, and energy expenditure, needed for the performance of the prescribed function.

He was able to show that various properties of the vascular system obeyed optimality laws. For example, the optimal angle of branching for veins could be calculated and actual veins seem to show this optimal branching angle. Rosen [16] took up this work and wrote a whole book devoted to a study of biological optimality. For a biologist's take on this, see Gould [17].

Relational Biology: Generalized Postulate of Relation Forces [18]:

> The development of an organismic set proceeds in such a manner as to maximize the total number of relations and the number of different kinds of relations during the total course of development.

What does this mean? I'm not sure. Rashevsky attempted to use this to show why there should be only two sexes.

Mathematical Biology of Automobile Driving

One of Rashevsky's last research topics was automobile driving [19]. He learned to drive late in life, and he noticed that driving was like a non-linear control system. He was able to adapt the cross connected differential equation models that he had devised for neurons to serve as models of driving behavior. Contrary to the claims that he never did experiments, he arranged for experiments to be carried out on a closed course using traffic cones. He then used the data from these experiments to show good agreement with his differential equation models.

6   Conclusion

Rashevsky was a pioneer in applying Mathematics to Biology

– Largely forgotten
– Misremembered as proposing models and ignoring data
– Lives on through former students
  although most are deceased
  few academic grandchildren because Committee on Mathematical Biology was closed
– Lives on
  His Journal, the Bulletin of Mathematical Biology
  is still being published by the Society for Mathematical Biology (Springer).

Rashevsky in Italy 1962 [20]

For further information see my paper *The mathematical biophysics of Nicolas Rashevsky* [21] which is available on the Web. Also see the longer study by Abraham [22] and the recent book by Shmailov [23].

References

1. Rashevsky, N.: Mathematical Biophysics: Physico-Mathematical Foundations of Biology, vol. 2. Dover, New York (1960)
2. BMB: A biographical sketch of Nicolas Rashevsky. Bull. Math. Biophys. **27**, 3–4 (1965)
3. Rashevsky, N.: Mathematical Biophysics: Physico-Mathematical Foundations of Biology. University of Chicago Press, Chicago (1938)
4. Rashevsky, N.: Mathematical Biophysics: Physico-Mathematical Foundations of Biology, 2nd edn. University of Chicago Press, Chicago (1948)
5. Rashevsky, N.: Mathematical Theory of Human Relations: An Approach to Mathematical Biology of Social Phenomena, 2nd edn. Principia Press, Bloomington (1947/1949)
6. Rashevsky, N.: Mathematical Biology of Social Behavior, rev. edn. University of Chicago Press, Chicago (1951/1959)
7. Richardson, L.F.: Arms and Insecurity: A Mathematical Study of the Causes and Origins of War. The Boxwood Press and Quadrangle Books, Chicago (1960). Rashevsky, N., Trucco, E. (eds.)
8. Rashevsky, N.: Looking at History Through Mathematics. MIT Press, Cambridge (1968)
9. Rashevsky, N.: Outline of a physico-mathematical theory of excitation and inhibition. Protoplasma **20**, 42–56 (1933)
10. Cull, P.: General two factor models. Bull. Math. Biophys. **29**, 405 (1967)
11. Householder, A.S., Landahl, H.D.: Mathematical Biophysics of the Central Nervous System. Principia Press, Bloomington (1945)

12. McCulloch, W.S., Pitts, W.H.: A logical calculus of the ideas immanent in nervous activity. Bull. Math. Biophys. **5**, 115–133 (1943)
13. Rashevsky, N.: The History of the Committee on Mathematical Biology of the University of Chicago. The Committee on Mathematical Biology, Chicago (1963)
14. Lucas, H.L. (ed.): The Cullowhee Conference on Training in Biomathematics. The Institute of Statistics of North Carolina State University, Raleigh, North Carolina (1962)
15. Rashevsky, N.: Address by N. Rashevsky. The Committee on Mathematical Biology, Chicago (1964)
16. Rosen, R.: Optimality Principles in Biology. Butterworth Publishers, Stoneham (1967)
17. Gould, S.J.: The Panda's Thumb. W.W. Norton, New York (1981)
18. Rashevsky, N.: Organismic Sets: Some Reflections on the Nature of Life and Society. Mathematical Biology Inc., Holland (1972)
19. Rashevsky, N.: Mathematical biology of automobile driving: V. Bull. Math. Biophys. **32**, 173–178 (1970)
20. Rashevsky, N. (ed.): Physicomathematical Aspects of Biology: Proceedings of the International School of Physics "Enrico Fermi" XVI Course. Academic Press, New York (1962)
21. Cull, P.: The mathematical biophysics of Nicolas Rashevsky. BioSystems **88**, 178–184 (2007)
22. Abraham, T.: Nicolas Rashevsky's mathematical biophysics. J. Hist. Biol. **37**, 333–385 (2004)
23. Shmailov, M.: Intellectual Pursuits of Nicolas Rashevsky; The Queer Duck of Biology. Birkhaüser, Basel (2016)
24. Rashevsky, N.: Some Medical Aspects of Mathematical Biology. Charles C. Thomas, Springfield (1964)
25. Rashevsky, N.: Mathematical Principles in Biology and Their Application. Charles C. Thomas, Springfield (1961)
26. Rashevsky, N.: Advances and Applications of Mathematical Biology. University of Chicago Press, Chicago (1940)
27. Rosen, R.: Anticipatory Systems. Pergamon Press, New York (1985)
28. Rosen, R.: Life Itself. Columbia University Press, New York (1991)
29. Rapoport, A., Chammah, A.: Prisoner's Dilemma. University of Michigan Press, Ann Arbor (1965)

McCulloch's Relation to Connectionism and Artificial Intelligence

Gabriel de Blasio$^{2(\boxtimes)}$, Arminda Moreno-Díaz^1, and Roberto Moreno-Díaz^2

^{1} Escuela Técnica Superior de Ingenieros Informáticos,
Universidad Politécnica de Madrid, UPM, Madrid, Spain
`amoreno@fi.upm.es`
^{2} Instituto Universitario de Ciencias y Tecnologías Cibernéticas,
ULPGC, Las Palmas, Spain
`gdeblasio@dis.ulpgc.es, rmoreno@iuctc.ulpgc.es`

Abstract. It is normally accepted that the beginnings of modern computing connectionism can be traced to McCulloch and Pitts' paper of 1943 [1]. The important points of their historical contributions are however mislead by the drift that developments on theoretical computer architectures took after the 50's. The so called Artificial Neural Nets and subsequent connectionist philosophy were actually fixed by Rosenblatt's Perceptrons and his detractors, plus the more recent addenda of multi-layer perceptrons and back propagation adjusting techniques. They clearly used the basic idea of threshold logic and computation but evolved away from McCulloch-Pitts proposals, towards and in a computer tool of many times questionable power, just as parametric classifiers.

What is apparent, however, is that the Macy Foundation Meetings from 1943 to 1945, started by Wiener and McCulloch and all chaired by the latter, provided for the roots of many of present day concepts and ideas for the so called Computational Neurosciences.

On another side, Artificial Intelligence appears in the 50's by the hand of McCarthy and Minsky, mostly influenced by Mathematicians and Logicians like Gordon Pask, Von Neumann and Donald McKay. McCulloch, again, stood aside, in spite of his strong friendship and relations to all of them, even helping seriously in the creation of the Project MAC and the Artificial Intelligence Laboratory (AI Lab) of MIT in Tech Square. He always thought, however, that what was later called Good Old-Fashioned Artificial Intelligence (GOFAI) and, as consequence, its successor, Knowledge Engineering, will contribute nothing to brain understanding, but, rather, they were in most cases "toys or even little monsters".

Through a quick reminder of McCulloch's activity, we shall try to show how the basic contributions to the computer-brain paradigm of McCulloch came mostly from the two important early papers: the one in 1943 and "How We Know Universals" in 1947. We shall end by reminding some questions that still remain open since his last meetings in Europe (Lisbon, July 1968): on command and control, consciousness, intention, multi-functionality and reliability in the nervous system.

© Springer International Publishing AG 2018
R. Moreno-Díaz et al. (Eds.): EUROCAST 2017, Part I, LNCS 10671, pp. 41–48, 2018.
https://doi.org/10.1007/978-3-319-74718-7_6

1 Epistemology

In his "*Recollection of the Many Sources of Cybernetics*" [2] a paper which is considered by some as McCulloch's scientific testament, he claims he had become a "cybernetician" very early when he was a freshman at Haverford College. He was already fascinated by why we know and "create" mathematics. To a question of his teacher in philosophy on "what is all you wanted" he answered: "I want to know what is a number, that a man may know it, and a man, that he may know a number". "Friend, thee will be busy as long as thee lives", was the reply of the teacher. And so, he practically was.

In his epistemological pursuit, from Logic Philosophy and Psychology, he turned to Medicine to learn about the brain, sane and insane. From 1930 to 1941 he joined Düsser de Barenne at Yale Medical School, where he worked on the sensory projections and on cortico-cortical connections. He learned and adopted from de Barenne his Kantian-like ideas on the "neurophysiological a priori" (a priori brain structures and functions to handle sensory data). He was already convinced that to understand the brain in its epistemological side, one has to use at least symbolic logic to treat the firing activity in ranks of neural nets, keeping "meaning attached to symbols".

His epistemological Neurophysiology, transmitted to his student in Chicago, Jerry Lettvin, culminated years later, already at MIT. There, with Humberto Maturana and Walter Pitts, they produced the landmark paper "What the Frog's Eye Tells the Frog's Brain" [3]. This and the subsequent "Two Remarks on the Visual System of the Frog" [4] are the clearest and, perhaps, final expression of what he meant by "neurophysiological epistemology", complement of his Gestalt conceptions of perception of the earlier paper "How we Know Universals" [5], expressions both of a new way of approaching the understanding of brain functions.

In 1960, McCulloch retakes his original quest [6] in a somehow rhetorical disquisition, where he mixes his own history with the history of logic, and ends in his acceptance of Bertrand Russell definition of a number as the "class of all classes..." as a proper answer to the first part of his question. Then he claims that the probabilistic logic of Cowan and his and Pitts formulation of neural nets, expressed in Venn Diagrams where "dots are changed to probability (p's)" and a general formula answers the second part. The paper is at all lights unsatisfactory since it ignores brain, biological, cultural and social evolution and adaptation. In fact, we believe that one of McCulloch essays important failures is his systematic neglect of evolution in all aspects. He somewhere even asserts that "he is interested in how the brain is and what it does, not in how is that it came to be what it is".

2 A Calculus for Nervous Activity

From 1941 to 1952, McCulloch was at the university of Chicago setting the Department of Psychiatry. He kept on his work on cortico-cortical connections and sensory mapping, as well as on command and control of posture and motion. On the neurophysiological aspect, his collaborators were Jerry Lettvin and Patrick Wall. He attended Rashevsky's Seminars on Mathematical Biology and had the idea that brain "information" are the signals flowing through ranks of neurons, forming a "hierarchy" of layers of neurons.

There, he met Walter Pitts and tried to "formalize" one of his repeated objectives, that later became leitmotif of many of his papers: circularities or loops in nervous nets and in nets having a hierarchical architecture; mainly how to handle the regenerative activity of such loops. Pitts' strong logic-mathematical background was responsible for the conclusion that, to treat that, they have to use modal logic. From here, it came McCulloch obsession to find the "appropriated logical calculus" for loops.

By translating to logic some basic neurophysiological facts, Pitts worked out a formal expression of what McCulloch had in mind. McCulloch dressed up everything with appropriate biological and philosophical arguments. Thus in 1943 they produced their key and best seller paper "A Logical Calculus of the Ideas Immanent in Nervous Activity", which innovative hybrid nature found place only in Rashevsky's Journal [1]. Inputs and outputs of every neuron in a net have the nature of a proposition. Neurons handle "symbols", their "meaning and intention" being attached in the context of them. Captivated by Pitts' modal logic, McCulloch proposed a type of "neurophysiological projection of logic". He practised one of his favourite assertions (originally due to Mark Twain?): "You have to have the facts before you can pervert them". With all these pretensions included in his formally innocent proposal, it is not surprising that McCulloch was very much disappointed with the misleadings of Von Neumann, that considered McCulloch-Pitts Nets simply as a particular case of a more general Automata Theory. From there, it comes McCulloch's late complains of reductionism: "Von Neumann tried to reduce an automaton like me to his Automata Theory. Nonsense".

That is, we believe, the origin of all wrongly attributed relations of this and posterior works of McCulloch-Pitts to later Perceptrons and theirs subproducts, the so-called artificial neural network theories and practices, adaptive learning and the similar.

A new paper on the right direction is a theoretical proposal on the structure, functions and behaviour of a global property of the nervous system: Pitts and McCulloch's "How we Know Universals" [5]. It did also find a publication chance only in Rashevsky's Journal.

3   Foundations of Computational Neuroscience

In 1942 McCulloch met Norbert Wiener with their mutual friend Arturo Rosemblueth. The crucial paper for the emergence of Cybernetics was presented at the first Macy Foundation meeting in New York City: Behaviour, Purpose and Teleology, published the following year by Norbert Wiener, Arturo Rosemblueth and Julian Bigelow. Rosemblueth and McCulloch had reached with the Josiah Macy Foundation an important agreement to organize a yearly interdisciplinary meeting. Before they started, there was late in 1942 a meeting of engineers, physiologists and mathematicians at Princeton, referred by Wiener in the introduction of his book of 1949, Cybernetics [7]. There, McCulloch says that he met John von Neumann.

The Macy Foundation Conferences started under the name "Conferences on Circular, Causal and Feedback Mechanisms in Biological and Social Systems", which was changed to "Conferences on Cybernetics" in 1949. These series of stimulating and constructive Conferences run until 1953. They established a new conception for treating living and non-living machines, which with more or less successes, failures and redefinitions, has come to our days. It would be fruitful to dig into that remarkable source of ideas and inspiration but only some of the Transactions are available [8]. As Von Foerster said "The Conferences have became an oral tradition, a myth". Today, the consequences of that myth can be found in McCulloch and other attendants' essays. Many of the attendants to the Conferences may be considered the real foundations of the use of computational terms and concepts in the sciences of the nervous system and viceversa. Some of the guest attendees were W. Ross Ashby, Julian Bigelow, Jan Droogleever-Fortuyn, W. Grey Walter, Rafael Lorente de Nó, Donald MacKay, Warren McCulloch (Chairman of the Conferences), J.M. Nielsen, F.S.C. Northrop, Antoine Remond, Arturo Rosemblueth, Claude Shannon, Heinz Von Foerster, John Von Neumann or Norbert Wiener. From that time, McCulloch was advisor and friend of the operational research pioneer Stafford Beer.

From the Macy Conferences on, there were a number of crucial subjects and problems, raised and discussed in the many sessions. Among them, there were the concepts of regulation, homeostasis and goal directed activity, the transmission of signals and communication in machines and nervous systems. In what refers to nervous systems organization, the ideas of reverberating and closing loops to explain brain activity were established there. These ideas generated concepts and theories on circular causality in Economics and in the polling of public opinion.

The analysis of conflict between motives in psychiatry led to the developing of concepts like heterarchy of values in mental processes. Also, the ideas of content addressable memory, active or reverberating memories and the consideration of learning as changes in transition probabilities among states, were inspired from Biology to became terms applicable to machines. In sum, a considerable and rich flow of new ideas and concepts to be applied both to machine and to living systems.

In spite the basic perspective differences between McCulloch and Von Neumann (one in physiological psychiatry, the other in logico-mathematical computation), they were personal friends, that meet regularly and enriched mutually. McCulloch delighted in recounting an anecdote from their all-night meetings in Chicago, that he said started Von Neumann interest in reliability. Once, in one of these meetings, McCulloch, Von Neumann and other colleagues were having their usual whisky dosage. McCulloch stopped the conversation and commented something on its effect: "The thresholds of neurons are now very, very low. Nevertheless, neurons are still computing reasonably reliable. What can there be in the brain, in its modular structure and links, which makes it such a reliable piece of machinery, in spite of failure in threshold levels and components?".

A confusing piece of work called "Agathe Tyche: The Lucky Reckoner" [9] offers a fair overview of much of his philosophy with respect to ways of building reliable machinery from unsafe components. The classic by Winograd and Cowan "Reliable Computation in the Presence of Noise" [10] and almost all of his later work on reliable computing were the result of McCulloch's expansion of Von Neumann's original concepts.

By 1952, McCulloch moved to MIT at the instance of his friend Norbert Wiener, taking his collaborators with him. As already mentioned, the main papers on epistemological neurophysiology were produced there. A sad, but amazing disagreement with Wiener provoked, in the end of the 60's, the final of their friendship and what was worse, the "social and scientific" death of Walter Pitts, who had felt adopted by Wiener but betrayed by McCulloch.

4   Logic and Closed Loops for a Computer Junket

Back in 1952, McCulloch was concerned with abductive reasoning and abductive computing ("machinery", he said), as distinct from deduction (logical nets of formal neurons) and induction (as in learning nets). He assigned this type of function to the reticular formation of the vertebrate's nervous system, that he thought to be the "the abductive organ that commits the entire animal to one rather to any other of a small number of incompatible modes of behaviour". He had the anatomical and physiological basis mostly from the Scheibel's [11], but realized that neither the theory of coupled oscillators nor the theory of iterated nets could help in the problem. He recurs to the subject with William Kilmer in 1964 [12], but they finally turn to computer simulations, that we believe were not appropriate but obfuscated what they were after, a convergence problem in diagnosis.

Oscillation in nets have been for long time a type of leitmotif in all of his essays. He insisted that closed loops will account for, among other, vestibular nystagmus (based in Lorente de Nó's anatomical findings), causalgia, i.e. pain after amputation, early stages of memory and conditioning, anxiety, and the effects of shock therapy. In the 1960's, he used to draw on his blackboard a formal neuron "talking to itself" (a loop), that will oscillate with one mode 0101... under input 1 (see Fig. 1). He insisted that the neuron "was saying that the number of 1's that have been in the input is odd", a perfect illustration of the "propositional" nature that he and Pitts associated to the responses of a formal neuron. On the side of the blackboard (Fig. 1) is a redrawing of the same neuron in a way that permitted to show the realizability of all possible modes of oscillation for a net of N neurons having a fixed anatomy and the introduction of functional matrices and the concept of an Universal Neural Net [13]. The number of possible modes of oscillations [17] grows extremely with the number of neurons of the net. So, Warren was satisfied that it were "plenty of room" for all "temporary memory" even in small fixed anatomy.

Fig. 1. McCulloch, Moreno -Díaz, Sutro, MIT, 1967

By then, 1964–1969, McCulloch, Da Fonseca, Moreno-Díaz and Kilmer, were working with Louis Sutro on what he called "a computer junket to Mars" [14]. This paper summarizes the problems he had at that time: loops, oscillations and triadic relations with Moreno-Díaz; the reticular formation with Kilmer and dynamic models of memory with Da Fonseca.

In the summer of 1968 there was the last encounter with W.S. McCulloch in Europe, hosted by Da Fonseca, to Sesimbra, a summer place south of Lisbon. There, Warren, sitting on the floor all night, talked and argued while eating shrimps ("pure protein", he claimed) and drinking Scotch. He showed no interest in the so called "artificial intelligence" not in the current flourishing of multi-layer Perceptron-like networks. His arguments were still epistemological, on significa-tion, on intention, that attracted completely the attention of Da Fonseca, as good psychiatrist. Moreno-Díaz tried to express Neural Nets by Triadic Rela-tions in "intenso", that is, relations defined by rules, starting from their func-tional matrices. And Pepe Mira, who was by then just a graduated physicist working on his thesis, wanted to decipher the extraordinary computational plas-ticity of the brain, specially in the sensory cortex, following the magnificent and almost unknown work of Justo Gonzalo. All their questions are still in the air. Dominating everything was McCulloch's strong assertion about thoughts and brains: "not brain nor machines ever think. Persons think".

5 Epilogue

After McCulloch death the following year (1969), his widow Rook undertook the task of collecting his papers. With the help of many of the friends and disciples she achieved an almost complete success. Edited by Rook, in four volumes, with a preface by Heinz Von Foerster, they were published 1989 by Intersystems, Salinas Ca., a publisher that disappeared soon after [15]. Very few collections are known to survive, so that the web of the American Association of Cybernetics call them a "gold mine". They are, anyway, available for scholars at the American Philosophical Society in Philadelphia, PA. McCulloch's own selection of his preferred papers around 1964 is however available in "Embodiments of Mind" [16] (Fig. 2).

Fig. 2. Moreno-Díaz, von Foerster, Lettvin, Las Palmas Conference, 1995.

In November 1995, twenty-five years after his death, an international conference was held to commemorate the figure and contributions of McCulloch. The conference was held in Las Palmas de Gran Canaria. It was attended by friends, his daughter Taffy Holland and former collaborators and scientists, among them, Von Foerster, Lettvin, Kilmer, Da Fonseca, Gestelland and Sutro. Thanks to Von Foerster, a copy of the four volumes of the collected works is available in Las Palmas.

The general conclusion of the conference is valid today. McCulloch was a great neurophysiologist and a first-line thinker. Many of his questions are profound, classic and everlasting. In spite that they were often expressed in a rhetoric, obscure way, they are a source of meditation, in fact, even in Artificial Intelligence and Distributed Computation. Most of the serious problems he attacked (an effective theory of triadic relations, the reticular formation, i.e., a true theory of abduction and diagnosis, and all of the other epistemological questions) were not well stated questions, but we still find them provocative, inspiring and stimulating.

It's definitely worthwhile to go back to his essays to enjoy his richness of language, his scholarship and, mostly, his optimistic believe in Science.

Acknowledgments. This research has been supported in part by project MTM2014-56949-C3-2-R.

References

1. McCulloch, W.S., Pitts, W.: A logical calculus of the ideas immanent in nervous activity. Bull. Math. Biophys. **5**(4), 115–133 (1943)
2. McCulloch, W.S.: Recollections of the many sources of cybernetics. ASC Forum **6**(2), 5–16 (1974)
3. Lettvin, J.Y., Maturana, H.R., McCulloch, W.S., Pitts, W.H.: What the frog's eye tells the frog's brain. Proc. IRE **47**(11), 1940–1951 (1959)
4. Lettvin, J.Y., Maturana, H.R., McCulloch, W.S., Pitts, W.H.: Two remarks on the visual system of the frog. In: Rosenblith, W.A. (ed.) Sensory Communications, pp. 757–776. MIT Press/Wiley, Cambridge/NewYork (1961)
5. Pitts, W., McCulloch, W.S.: How we know universals; the perception of auditory and visual forms. Bull. Math. Biophys. **9**(3), 127–147 (1947)
6. McCulloch, W.S.: What is a number that man may know it, and a man that he may know a number. Gen. Semant. Bull. **26**(27), 7–18 (1960)
7. Wiener, N.: Cybernetics. The MIT Press, Cambridge (1949)
8. Pias, C.: Cybernetics - Kybernetic. The Macy-Conferences 1946–1953. Diaphanes Verlag, Zürich (2003)
9. McCulloch, W.S.: Agathe Tyche: of nervous nets - the lucky reckoners. In: Mechanization of Thought Processes, Stationery Office, London, pp. 611–625 (1959)
10. Winograd, S., Cowan, J.D.: Reliable Computation in the Presence of Noise. The MIT Press, Cambridge (1963)
11. Scheibel, M.E., Scheibel, A.B.: Structural substrates for integrative patterns in the brain stem reticular core. In: Jasper, H.H., et al. (eds.) Reticular Formation of the Brain, pp. 31–55. Little Brown, Boston (1958)
12. McCulloch, W.S., Kilmer, W.: Towards a theory of the reticular formation. In: Fifth National Symposium on Human Factors in Electronics, San Diego, California, pp. 7–25 (1964)
13. Moreno-Díaz, R., McCulloch, W.S.: Circularities in nets and the concept of functional matrices. In: Proctor, L. (ed.) Biocybernetics of the Nervous System, pp. 145–150. Little & Brown, Boston (1969)
14. McCulloch, W.S.: Logic and closed loops for computer junket to mars. In: Caianiello, E.R. (ed.) Neural Networks, pp. 65–91. Springer, Heidelberg (1968). https://doi.org/10.1007/978-3-642-87596-0_7
15. McCulloch, W.S.: Collected Works, vol. 4, Edited by Rook McCulloch. Intersystems Publications, Salinas (1989)
16. McCulloch, W.S.: Embodiments of Mind. The MIT Press, Cambridge (1965)
17. Moreno-Díaz, R., Mira-Mira, J. (eds.): Brain Processes, Theories and Models. The MIT Press, Cambridge (1996)

Kurt Gödel: A Godfather of Computer Science

Eckehart Köhler[✉] and Werner Schimanovich

University of Vienna, Vienna, Austria
eckehart.koehler@univie.ac.at, werner.depauli@gmail.com

Abstract. We argue that Kurt Gödel exercised a major influence on computer science. Although not immediately involved in building computers, he was a pioneer in defining central concepts of computer theory. Gödel was the first to show how the precision of the formal language systems of Frege, Peano and Russell could be put to work to prove important facts about those language systems themselves, with important consequences for mathematics. As Hilbert's collaborator, Paul Bernays put it, in his famous Incompleteness Proof Gödel did the "homework" that the people in Göttingen working on Hilbert's Program to prove the consistency of mathematics missed. (Hilbert's Metamathematics assumed that all mathematical proofs could be treated as coding problems, and enciphering is applied arithmetic.) The core of Gödel's Proof also gave exact definitions of the central concept of arithmetic, namely of the recursion involved in mathematical induction (with help from the great French logician Jacques Herbrand). This immediately led to whirlwind developments in Göttingen, Cambridge and Princeton, the working headquarters of major researchers: Paul Bernays and John von Neumann; Alan Turing; and Alonzo Church, respectively. Turing's and von Neumann's ideas on computer architecture can be traced to Gödel's Proof. Especially interesting is the fact that Church and his lambda calculus was the main influence on John McCarthy's LISP, which became the major language of Artificial Intelligence.

1 The Incompleteness Proof

Although Kurt Gödel was in fact, unlike his colleagues John von Neumann or Alan Turing at Princeton, not involved in the design of computers, we argue that he has the right to claim some credit for computers' development. Gödel's most famous work was his proof that Hilbert's wish to secure the safety of mathematics by a formal consistency proof could not be done the way Hilbert wanted. That way was called Hilbert's Program, and it led Gödel direct to computing.

The deeper background of Hilbert's Program lay in taking advantage of "formalizing" mathematics in the new style of Gottlob Frege, Giuseppe Peano, and

---

The title of this article is a play on Gödel's name. In the Austrian and Bavarian dialect of German, "Göd" means godfather, and "Gödel" means literally "little godfather", although it is actually used to refer only to godmothers in the regions of its use. In any case, Kurt Gödel did take a godfatherly interest in children, in particular to Oskar Morgenstern's son in Princeton.

© Springer International Publishing AG 2018
R. Moreno-Díaz et al. (Eds.): EUROCAST 2017, Part I, LNCS 10671, pp. 49–65, 2018.
https://doi.org/10.1007/978-3-319-74718-7_7

Bertrand Russell, who founded mathematical logic[1]. Such a formalization was already envisioned by Ramon Llull (the influential Spanish scholastic) and Gottfried Wilhelm Leibniz. With Llull's Characteristica universalis we obtain a Calculus ratiocinator! The idea was to codify all arguments, using only words with univocal meanings, enabling conclusive proofs to be drawn, as in mathematical argumentation.

Leibniz demanded that the inconclusive debates of Medieval and Renaissance philosophy should stop—instead he demanded "calculemus !": debates should be settled by precise proofs. David Hilbert's goal was to settle the virulent debate on Foundations of Mathematics by proving the consistency of (an acceptable axiomatization of) mathematics, which was to be axiomatized in the style of Hilbert's Foundations of Geometry—(1899)—and without resorting to meanings [Inhalte].

Although not a Formalist, Gödel fully appreciated the precision of formalisms. He loved exactness, and knew that only strict rules allowed sharp mathematical theorizing. He especially favored Gottlob Frege's contributions to syntax over those of Giuseppe Peano and Bertrand Russell. Frege's Begriffsschrift was the first really exact "finitary" formal system in Hilbert's sense. The idea behind Frege's Begriffsschrift was to display all logical reasoning used in mathematics as a calculus—which could itself in turn be mathematically analyzed, as Hilbert saw.

Gödel was not a Formalist in the sense of Hilbert. Hilbert claimed that mathematics could ignore the meanings of the terms used in arguments. In contrast, Gödel held that the essence of mathematics lies in its contents, and that humans were able to perceive truths about those contents by intellectual intuition. This became the core of Gödel's famous Platonism[2].

However, in certain ways, Hilbert was a Platonist as well. For Hilbert notoriously believed in completeness: that all meaningfully formulated mathematical problems are in principle solvable[3]. This involves the assumption that the human mind will ultimately be able to solve all open problems!

- Thus, for example, Hilbert rejected DuBois-Reymond's "Ignorabimus" (a critique of Platonism): that some things we can never know[4].
- Hilbert famously refused to be ejected from "Cantor's Heaven"—the realm of transfinite sets.

---

[1] Frege [1]; English translation by Bauer-Mengelberg in [2]. Whitehead and Russell [3].

[2] Gödel was a "convinced Platonist" in 1925 already. Cf. Köhler [4]; and Wang [5], (esp. pp. 17f.).

[3] NB: Gödel also believed in such completeness! Despite his Incompleteness Proof, Gödel always thought that human reason was (in principle) capable of solving all problems; this couldn't be done in any single system, but intuition goes beyond all formal systems anyway—just as Brouwer claimed.

[4] Hilbert's most famous call for completeness, including an explicit opposition to (DuBois-Reymond's) "Ignoramibus", came at the end of the introduction to his list of "Mathematische Probleme" (published in [6]).

Hilbert's Program[5] was developed after he already took these positions on DuBois-Reymond and Cantor. And in Hilbert's Program, completeness again plays a crucial role, because the consistency proof it wants for mathematics needs to exhaustively comb through all the propositions of a powerful axiom system covering all of mathematics!

Hilbert's strong belief in completeness is Platonist, because it assumes we can know arbitrarily strong proof methods in order to decide all open conjectures. This presupposes that such conjectures have objectively definite solutions. Of course, in Hilbert's Program he thought we only needed finitary proofs for consistency; but Gödel's Proof put an end to that. However, even the weak finitary proof rules which Hilbert originally tried represent what Gödel called "weak Platonism", since belief in their correctness was beyond the scope of (strictly empirical) Natural Science.

1.1   Gödel's Technique in his Incompleteness Proof (1931)

Gödel Enumeration. Hilbert had already envisioned that proofs could be "arithmetized", once one axiomatized any part of mathematics in a formal system in the manner of Frege, Peano, or Russell—Leibniz saw this, too[6].

Letter combinations, in particular propositions and entire proofs, could be encoded. Encoding is traditionally treated as (applied) arithmetic. The study of such arithmetized proofs were called "Metamathematics" by Hilbert and his school; and this became the basis of Hilbert's "Proof Theory".

Gödel's breakthrough was to rediscover Leibniz's encoding method and to apply it to Hilbert's Program: to each well-formed expression assign a unique "Gödel number" uniquely factorizable into prime numbers. Gödel thus had a "machine code" for all mathematical proofs[7]. Gödel thereby won a "programming language" for Metamathematics! He used a substitution operation typically used for encoding, which was easy to include in the arithmetization. He could define "provable", "complete" and "incomplete" within this programming language, all having Gödel numerals, hence computable! (Such assignments will of course include one to a sentence saying of itself that it was unprovable, the famous "Gödel-sentence" G.)

By assigning numerals to proof procedures, actions (processes taking time to execute) are represented by (static) data, allowing the conscious mind to hold

---

[5] Hilbert [7]; reprinted with notes by Bernays in [8]; translated into English by Ewald in [9] and in Mancosu [10]. An authoritative evaluation of Hilbert's Program is by Kreisel [11]; revised and reprinted in Benacerraf and Putnam [12].

[6] Leibniz's encoding method was first published by Louis Couturat (1903), which Gödel may not have known in Vienna; but Gödel almost certainly knew Lewis [13] where Couturat's discovery of Leibniz's encoding method is sketched on pp. 11f.

[7] An accessible general survey of Gödel's Proof is provided by Casti and DePauli [14]. The proof itself was in Gödel [15]; translated into English by van Heijenoort in [2]); reprinted with detailed commentary by Kleene in Feferman et al. [16]. Useful information about Gödel's life and work appeared in Köhler et al. [17] and Buldt et al. [18].

them fast in memory (like a melody), reflect on them and revise them. For this reason, Douglas Hofstadter saw in Gödelization the fundamental mechanism of consciousness; reflection is exactly what the simple sensation of William James's "stream of consciousness" lacks, by comparison.

1.2 "Numeralwise Representable"

Now what Gödel needed to do was to prove that his arithmetization really does the job it is supposed to. This was the "grunt" work, the slogging work, which, surprisingly none of Hilbert's collaborators in Göttingen had got around to. Bernays admitted in a later conversation with the Proof Theoretician Gaisi Takeuti: "*What we had intended to do but had neglected, the diligent Gödel did*"[8].

Gödel proved that with Whitehead and Russell's Principia Mathematica (more elementarily: from Peano Arithmetic[9] plus a little coding theory), we could accurately calculate that every sentence of PM had a unique Gödel numeral (was "numeralwise representable"). Inversely, every Gödel numeral would "correctly" represent expressions in proofs—and thus could be freely substituted into arithmetic formulas.

It was this "grunt work" that immediately convinced Hilbert's collaborators in Göttingen, especially Bernays and von Neumann, but also other collaborators, such as Ackermann, Behmann, and Herbrand. Some years before, Paul Finsler had already published the same idea which Gödel used, but he had not "done his homework" arithmetizing the syntax of the axiom system as Gödel did, and no one paid attention[10].

1.3 Reflection

The key idea now was to formulate a weird sentence "I am unprovable" and to show it unprovable. The sentence is syntactically correctly formed, hence the axiom system is incomplete.

- By clever use of Gödel numerals, the substitution operation, and elementary arithmetic, Gödel succeeded.
- The sentence was a self-reflection (a "fixed-point") saying of itself (actually, of its own Gödel numeral) that it was unprovable.

This weird sentence is a specialized version of the Liar Paradox (the sentence saying "I am untrue"), but with the important difference that "**provable**" is used

---

[8] Takeuti [19] (p. 26).
[9] Introduced by Peano [20]. Frege [34] and then Dedekind [35] had the axioms beforehand; for commentary cf. footnote 20.
[10] Finsler [21].

instead of "**true**". This is important, because provability is definable within the Gödel's system, whereas "true" is not definable ("I lie" is not definable)[11].

This sentence is an example of a "self-reflective" sentence, referring to itself. Self-reflection is "unnatural" for arithmetic, because arithmetic does not normally deal with semantic matters. But when arithmetic involves coding, the "unnatural" becomes possible[12].

Overnight, Gödel's reputation was made. All logic research was re-oriented toward this amazing discovery, which shocked the Hilbert School very much. Many considered Gödel to be the greatest logician since Frege—no, "the greatest since Aristotle!"[13].

Gerhard Gentzen proved the consistency of Arithmetic using "finitary" (constructive) arguments in Hilbert's sense; but Gentzen could not prove consistency within the system itself (a kind of "bootstrapping", one might say), which made Hilbert's Program so attractive to von Neumann and others, but which Gödel's Proof now showed was impossible[14].

1.4   Computability/Recursion

In the course of trying to fulfill Hilbert's Program (even though failing), Gödel was the first to precisely define a main concept of Computer Science using the precision of mathematical logic, namely that of **computable number**. Gödel saw that, in order to work at all, Hilbert's Program needed such a precise definition to truly "arithmetize" the axiom system—he needed a complete "numeral-wise" representation. In turn, he needed to prove that all Gödel-number assignments to sentences were **computable**.

An exact definition of computability within formal Peano Arithmetic was needed to show that all sentences of the system being axiomatized are really captured by the encoding procedure. For this purpose, Gödel defined what was later called "primitive-recursive" functions (Kleene's terminology, not Gödel's) and Gödel proved the necessary lemmas about them. Then Gödel showed that all his Gödel-number assignments are indeed "primitive-recursive" numbers. This completed the "grunt-work" of Gödel's Proof.

---

[11] This is easy to explain: within Hilbert's Program, provability was a strictly finitary notion, hence arithmetizable within Peano Arithmetic. In contrast, truth (e.g. the truth of Gödel's sentence G) always occupies a standpoint from outside the system and is thus not representable within it. Of course there are stronger notions of provability which approach truth at an upper limit. Cf. the references below to Gentzen (footnote 14). Turing's Ordinal Logics clarifies the situation more pointedly; cf. footnote 32.

[12] Much later Jeff Paris and Leo Harrington found a "natural" sentence of Peano Arithmetic (in Ramsey Theory) also unprovable: "A Mathematical Incompleteness in Peano Arithmetic", in Barwise [22].

[13] Thus John von Neumann, cited by Goldstine [23], p. 174; cf. footnote 35.

[14] Gentzen [24, 25].

The term "recursion" is familiar to all number theorists, referring to the repeated applications of algorithms in mathematical induction. In computer jargon, this is what a subroutine does when it repeatedly "calls itself" in computations, i.e. applies itself to successive iterations of variable-assignments. Recursion Theory is a cornerstone of Computer Science, and one can claim with some justice that Gödel's work initiated this branch of the science, later developed by Kleene et al.[15]. (See Sect. 2 below.)

Davis[16] argues that, in the course of his proof, Gödel immersed himself deeply into a virtual machine (to use current computer jargon), in whose "machine language" Gödel programmed. Gödel defined computability using the formalism of Principia Mathematica (PM), and then he wanted to "numeralwise represent" formulas of PM within PM itself. Gödel's "grunt work" showing how to do this involved close attention to syntactical operations within PM—so Gödel was forced to develop a makeshift programming language, analyzing all the computations on the syntax of PM. The kind of programming language Gödel was using was, Davis says, not a "command language" like C or Fortran, but a "functional language" like LISP. Whereas command languages specify actions to be executed (leaving open what actions have preceded), functional languages define operations the computer is expected to execute (without assuming what else has gone on). That's what Gödel (1931) did. (It's ironic that von Neumann always used command languages, whereas Gödel anticipated the more sophisticated functional languages like Church's lambda conversion and McCarthy's LISP).

1.5 "Computable", Not "Recursive"

Gödel never used "recursive function theory" when referring to computability! For example, Davis [38] reported on a conversation he had with Gödel at some time in the period 1952–54. Gödel corrected Martin Davis in the common room[17] of the IAS against a misleading usage of the term "Recursive Function Theory".

Gödel said that "recursive function theory" should refer to the kind of work done by Péter[18]. Péter used "recursive" to mean "inductively defined", in the tradition of Dedekind's work on arithmetic functions.

Robert I. Soare complained about "intolerable confusion" going back to Kleene [26], who should have used "computable" instead of "recursive"[19].

---

[15] Kleene [26], Rogers [27], Davis [28]. The last has been called "one of the few real classics in computer science" in an advertising blurb.

[16] Davis [29] (pp. 120f.).

[17] Oswald Veblen had early established an Afternoon Tea at exactly three o'clock every weekday, "we explain to each other what we do not understand" (J. Robert Oppenheimer); cf. Dyson [30] p. 90.

[18] See [31,32]; translated as *Recursive Functions*, Academic Press, New York 1967. (NB: "Zs" is pronounced in Hungarian like the French "j".) Péter refers "recursive function" to Dedekind [35]; important here Sect. 9, "126. Satz der Definition durch Induktion".

[19] Soare [33].

Kleene's usage for "primitive recursion" and "general recursion" was historically inaccurate. Correct usage for "calculable" (i.e. algorithmic) should not be "recursive", but "computable". Turing had always used "computable", not "recursive". Since the '90s, usage has shifted back the way Soare recommends. Davis [28] (p. 64) still furthered confusion: "Having proved the equivalence of computability and recursiveness, we shall henceforth use the two terms more or less interchangeably." And nevertheless, not all algorithms (computations/calculations) are recursions[20]—only those using mathematical induction are genuinely "recursive".

2   Gödel's Lectures in Princeton

World-famous at the age of 25, Gödel was soon in great demand as a lecturer. Word had soon spread to Princeton, where the new Institute for Advanced Study (IAS) opened in 1933—with Albert Einstein as first to be hired. The mathematicians included Oswald Veblen and John von Neumann[21]. Close by in the Mathematics Department of Princeton University was a philosophically-minded ex-student of Veblen's, the mathematical logician Alonzo Church. (There had also been Emil Post, an ex-post-doctoral student of Veblen's, but he had just left to teach mathematics in a New York high school. And Haskell Curry had just left Princeton for Göttingen).

---

[20] The recursion concept of Dedekind [35] is the same as *Vererbung* [inheritance, a property of certain relations, in particular that of inheritance or ancestry] of Frege [34]. Dedekind acknowledged Frege's priority in capturing all the axioms of Arithmetic and in precisely characterizing mathematical induction. The axioms are now named after Peano [20]—who had admittedly taken them from Dedekind! (The difference was "merely" that Peano used a symbolic language, much of which was taken over by Whitehead and Russell [3]). So Frege was *twice* cheated out of the credit he deserves for being the first to formally specify the axioms of arithmetic.

[21] Von Neumann (János/"Jancsi" in Hungary, Johann in Germany, John/"Johnny" in the US) had believed very strongly in Hilbert's Program and was otherwise strongly involved with Hilbert's multifarious interests (e.g. axiomatizing parts of Physics: von Neumann wrote the classic *Mathematische Grundlagen der Quantenmechanik* [36]). In [37], Gödel used von Neumann's axioms in a version prepared by Paul Bernays, called NBG in the literature (for Neumann-Bernays-Gödel). NBG extends Set Theory to include "classes" (which were predicate-extensions, what Cantor called, in effect, "sometimes inconsistent sets" because they can not automatically be elements of others sets (or better of any class at all) by the rules of their construction. Today we call these classes proper-classes.). Gödel's Incompleteness Proof [15] had devastated von Neumann rather much, explaining von Neumann's chivalrous, boundless and everlasting awe of Gödel. In 1953, von Neumann told those IAS professors opposing Gödel's promotion to Permanent Member: "If Gödel doesn't deserve a professorship at the Institute for Advanced Study, none of us do".

In 1934, Gödel was invited by Veblen to lecture on his Incompleteness Proofs at the IAS. Attending the lectures at Fine Hall[22] on the Princeton campus were Alonzo Church and two of his students, Stephen Cole Kleene and John Barkley Rosser. The latter two took careful notes which were widely distributed among logicians in the '30s and '40s, for whom they became de-rigeur reading; they were finally published by Davis (1965)[23].

In these lectures, Gödel (*1934) modified his earlier definition of computability, adapting a proposal of Jacques Herbrand's, and introduced "general-recursive" functions (Kleene's term), which became standard in the subsequent literature in Recursion Theory. Herbrand had corresponded with Gödel in 1931 after Paul Bernays in Göttingen showed him galley proofs of Gödel's Incompleteness Proof (1931)[24].

John von Neumann, professor at the IAS, apparently did not attend Gödel's lectures, despite his earlier intimate involvement with Hilbert's Program in Göttingen. But he knew Gödel's Proof all too well, and he strongly supported Gödel's work. It was said, however, that, after Gödel's Proof, von Neumann never read logic again—that is, at least, until Turing (1936)!

At Princeton, Gödel strongly influenced the young American school of mathematical logic, especially Church, Kleene and Rosser, but also Post and others. Supported by Oswald Veblen, von Neumann and Church, Princeton became a center, later attracting Alan Turing 1936–38 from Cambridge, England, to write a dissertation under Church, knowing Gödel was a continual Visitor. Von Neumann even offered Turing an assistantship at the IAS ! But Turing returned to Cambridge. (Turing was a member of the Apostles discussion society there, which had included Russell, G.E. Moore, Wittgenstein, Ramsey, Maynard Keynes and other great thinkers.)

2.1 Equivalence of Computabilities

By 1936, several formally precise definitions of computability had been devised: Gödel's primitive and general recursion (1931, 1934), Turing's Machine (1936),

---

[22] At the time of Gödel's lectures, the IAS was located in Fine Hall on Princeton's main campus together with the mathematicians; only later did the IAS move to Fuld Hall on Princeton's west side, where Gödel later had an office with a picture window and a stunning view of well-tended trees. In Fine Hall, famous mathematicians such as Oswald Veblen, Hermann Weyl, John von Neumann, Marston Morse, Deane Montgomery, later John Nash, had offices.

[23] Kurt Gödel: *"On Undecidable Propositions of Formal Mathematical Systems"*, notes from lectures given in English at the Institute for Advanced Study, Princeton, hectographed and distributed by Kleene and Rosser, first published in Davis [38]. Gödel discusses Herbrand's "private communication" in his footnote 34 (written before the 1965 publication by Davis), still indicating doubt about the "adequacy" of general recursiveness as the correct explication; but in Gödel's ultimate Postscriptum in Davis (1965), Gödel claims Turing's definition is "unquestionably adequate'".

[24] Herbrand [39]; English translation by van Heijenoort [2]. Herbrand had written a letter to Gödel on 7 April 1931 concerning recursion, translated by Wilfried Sieg, with detailed notes, in Feferman *et al.* [40].

but also Church's "Lambda-Conversion" (1936)[25]. With so much talent in Princeton, it was quickly determined that Gödel's concept of computability (called "general recursion" by Kleene [26]) was *equivalent* to these other contenders. Kleene and Rosser quickly proved that Gödel-Herbrand computability was equivalent to Church's λ-calculability. And Turing proved this of Turing-computability![26]

Church's λ-calculus was a logic of functions, i.e. operators which abstracted calculation procedures when applied to expressions, and which could unrestrictedly be applied to themselves, hence were preordained to study recursion[27]. It was close in style and content to Haskel Curry's Combinatory Logic, developed for his dissertation under Hilbert's supervision in Göttingen[28]. Curry's work has been influential in developing programming languages, but much more influential still is John McCarthy's famous development of LISP out of Church's λ-calculus[29].

Gödel later considered that Turing-computability serves as the basis of all concepts of machines [5], that it was an "absolute concept", and somehow characteristic of all mechanisms. There are problems with this view, however.

2.2   Are All "Machines" Turing Machines?

Hao Wang[30] describes several discussions of Gödel on machines, particularly relating machines with Materialism. Gödel views "machines" as too weak to encompass Mind, and he tends to play Mind against Matter in dubious ways. (Gödel later speculates somewhat inconsistently that biology may use processes not to be subsumable under Turing Machines.).

The problem is that Gödel's idea makes sense only for "finitary" (Hilbert) or "digital" devices (to use modern jargon). "Analog" devices, on the other hand, most famously the solar system as modelled by Newton's Mechanics, are not general-recursive. Hence, in Classical Physics, machines were by no means

---

[25] Church [41,42].

[26] Turing [43,44]. The equivalence proofs by Kleene, Rosser, and Turing are all included in the collection by Davis [38].

[27] This sounds dangerous, and sure enough, Kleene found contradictions which he and Church worked hard to avoid in later versions of the λ-calculus. Kleene himself resorted to different formalisms in his own work on recursion.

[28] Curry and Feys [45], Curry et al. [46]. For a survey, cf. Bimbó [47].

[29] McCarthy [48,49]. In McCarthy's history of LISP at http://www-formal.stanford.edu/jmc/history/lisp/node3.html, drafted in 1996, he writes "Another way to show that LISP was neater than Turing machines was to write a universal LISP function and show it is briefer and more comprehensible than the description of a universal Turing machine. This was the LISP function $EVAL[e, a]$, which computes the value of a LISP expression e—the second argument a being a list of assignments of values to variables. (a is needed to make recursion work.)...". "Russell noticed that EVAL could serve as an interpreter for LISP, promptly hand coded it, and we now had a programming language with an interpreter...".

[30] Wang [5], Sect. 6.1.

restricted to digital devices. This was emphasized to Gödel by Georg Kreisel—but without effect![31].

Turing himself already recognized that computability did not end with "Turing Machines". In his dissertation on Ordinal Logics (transfinite hierarchies), which exercised much too little influence, Turing extended the notion of computation to include anything that analog devices could "measure"[32]. This would make all mechanical devices governed by Newtonian Mechanics computable in a wider sense.

Georg Kreisel's critique of Gödel's assimilating all machines to Turing Machines was recapitulated in his 1978 memoire on Gödel [51]. Kreisel (born in Graz, who also attended Wittgenstein's "seances" at Cambridge University) had studied Physics before Logic and Mathematics, and was aware of the work of Poincaré and Kolmogorov, pointed out to Gödel that Newtonian Mechanics covers "analog" (continuum) processes, such as are found in planetary dynamics and hydrodynamics. In his posthumous memoire on Gödel, Kreisel [51] relates of his discussions with Gödel. Gödel had apparently been stunned by Kreisel's position, finding no reply. But in discussions with Wang, Gödel never modified this position to account for the problems raised by Kreisel.

Kolmogorov [55] had already pointed out that Newtonian Mechanics was "deterministically chaotic" in the sense that abitrarily small "perturbations" in the solar system can, with non-vanishing probability, cause extreme instability—called the "butterfly effect" by Edward Lorenz in his "Chaos Theory". This was contrary to the erroneous belief, since Laplace[33], that the solar system is stable and predictable. Zermelo and Poincaré had studied these possibilities already before World War I in analyses of Boltzmann's Statistical Mechanics. The upshot was the famous KAM-Theorem of A.N. Kolmogorov, Vladimir Igorevich Arnol'd and Jürgen Moser. For a survey of this work, see Broer[34].

3   "Von Neumann Machines"

John von Neumann is widely and with considerable justice held to be the "father" of modern Computer Architecture[35].

Von Neumann's "fathership" began when Herman Heine Goldstine fortuitously ran across him on a visit to the Ballistic Research Laboratory at Aberdeen

---

[31] Kreisel [50,51].

[32] Turing [52]; reprinted in Davis [38], and in Copeland [53]. In his introduction to Turing, Copeland compares Turing's Ordinal Logics very nicely with Gödel's and Church's work. For example, he quotes Turing's claim that, with Ordinal Logics, "one can approximate 'truth' by 'provability' as well as you please.".

[33] Laplace [54].

[34] Kolmogorov [55], Arnol'd [56], Broer [57].

[35] This claim is well supported by Goldstine [23] (Goldstine was the principal engineer for von Neumann's 1951 IAS machine in Princeton; then he worked for IBM). A later authority is Aspray [58].

Proving Ground in Aberdeen MD in summer 1944[36]. Goldstine was a young engineer working for the Moore School of Electrical Engineering at the University of Pennsylvania in Philadelphia on the first US electronic computer, the ENIAC, then under construction by John Mauchly and John Presper Eckert. Soon the great mathematician was invited to the Moore School, and Mauchly and Eckert added him to the committee for developing their improved new machine, the EDVAC. And von Neumann indeed contributed considerably to its design. Von Neumann's advantage over the hardware-oriented electrical engineers Mauchly and Eckert was his familiarity with the logical analysis of computation developed by Gödel, and particularly Turing[37]. The ENIAC's programming was "hard-wired", memory was used only to store intermediate ("volatile") results of computations[38]. The main advance of the EDVAC was to store programs into main memory, as a kind of "higher-order data". Von Neumann was well aware that Gödel and Turing had successfully represented all computational "routines" as data (coded and) stored in main memory. (Dynamic) proof procedures (actions or processes taking time to execute) were now encoded as (static) data (e.g. Gödel numerals semi-permanently stored in memory); this enabled recursive updating during run-time.

Actually, the idea of storing programs as data was already known to Joseph Marie Jacquard and Charles Babbage, who had conditional branching and loops (Davis [29]).

Von Neumann's computer architectures for the EDVAC and for the later IAS machine (the MANIAC), designed in 1946, finished in 1951, became widely known through two reports, the first by von Neumann alone, the second co-authored by him with Arthur Burks and Herman Goldstine[39]. The second report, the "Preliminary Discussion", was pointedly sent (for the IAS) by von Neumann to the Patent Office in Washington, D.C., to ensure the design was put in the public domain. This, the Institute for Advanced Study Computer, the MANIAC, was the real "von Neumann machine"!

---

[36] Originally established by Oswald Veblen during World War I to allow US mathematicians to contribute to the American war effort, giving work to the young Norbert Wiener, among others.

[37] Copeland ([53], pp. 21ff.) provides several citations from von Neumann himself describing how Turing's work influenced his thinking on computer design, in particular storing programs in memory.

[38] In 1947 it was converted to stored-program control after the example of EDVAC; and it was upgraded to 100 words of (IBM) magnetic-core memory in 1953 before the plug was pulled in 1955.

[39] "1. von Neumann [59]. Entirely reprinted in Stern [60]; also excerpted in Brian Randell [61]. 2. Burks et al. [62]"; reprinted in von Neumann [63]; also excerpted in Randell [61]. (This was typed up in Fuld Hall 219, the office next to Gödel's.). Davis [29] (pp. 191ff.) justly makes Turing out to be the "real father" of stored-program architecture; cf. von Neumann's own credits to Turing mentioned in footnote 37. But between Leibniz and Turing there came Gödel [15]: Gödel numerals encoded programs (i.e. proof procedures) as data already in 1931.

No doubt the most important advance of the IAS machine was von Neumann's integration of newly available random-access memory (RAM) devices. The Turing Machine had famously proposed a sequentially accessible tape (obviously from telegraphy practice) as memory; and IBM punched cards were also sequential. The ENIAC used mercury delay-line memory units, again sequential. Memory search-times could vary considerably, depending on total memory size and on where in memory the search begins; thus searching faced serious management problems. With the Williams tube[40], access times for different memory addresses were almost equalized, enabling "random access" memory management and making computations vastly more efficient. Von Neumann requested RCA[41] in nearby Camden, New Jersey, to construct something similar, and RCA responded with its Selectron tube, which went into the original planning of the MANIAC. RCA failed to develop its Selectron in time, so the MANIAC engineers bought off-the-shelf CRTs instead and engineered them to simulate the Williams tube. Some of the subsequent clones of the MANIAC, in particular the IBM 701, used genuine Williams tubes. Von Neumann's memory management worked in every case.

The IAS computer was subsequently "cloned" by several governmental agencies (the Ballistic Research Laboratory, the Los Alamos Laboratory, the Argonne National Laboratory, etc.) and by several universities (University of Illinois at Urbana, University of Chicago's Argonne National Laboratory, the Technische Hochschule in Munich etc.) in the US and abroad (including Sweden and the Soviet Union). After 1952, von Neumann became a consultant to governmental agencies and various companies, especially IBM[42], whose first electronic mainframe, model 701, the IAS computer strongly influenced: the 701's designers, Nathaniel Rochester and Jerrier Haddad, stated that their goal was a machine "like the IAS but with good input-output"[43]. Thus, von Neumann played a key role in assuring IBM's success—and thus of the entire computer industry.

3.1   The Great Patent Fight over Computer Design

Von Neumann's "First Draft" on the EDVAC was written at the request of its builders, Mauchly and Eckert, and widely distributed. The report thus entered the "public domain" as "state of the art", making the design unpatentable. The two engineers soon regreted having given permission to von Neumann to

---

[40] A cathode-ray tube (CRT) patented 1946 in the UK by Freddie Williams and Tom Kilburn. (Thanks to my nephew, Gordon Cichon of the Computer Science Department at the University of Munich, for pointing out this considerable difference between Turing Machines and von Neumann Machines.).

[41] RCA employed Vladimir Zworykin, the leader of RCA's television-development team for David Sarnoff, hence RCA was a leader in CRT technology.

[42] Distressing Eckert, IBM negotiated an advisory contract with von Neumann 1 May 1945 already.

[43] Aspray [58] (p. 93). Since the 701 was contracted for by the Defense Department, it was called the "Defense Calculator". Its business sibling were the IBM 702 and the famous IBM 650, the first mass-produced computer, all influenced by the MANIAC.

distribute the report (and as sole author!), since they became very interested in commercially developing their work and starting a business.

Once it became aware of its professors' commercial intentions, the University of Pennsylvania intervened in 1946, claiming all patent rights for itself— something institutions routinely do to this day, exceptions being unusual. Eckert and Mauchly[44] refused, claiming the rights for themselves, so they were asked to resign. (At the time, the legal status of intellectal property rights with respect to institutions and the developers who worked at those institutions was not at all legally settled. This quickly changed.)

Trying to settle the dispute, patent lawyers from the Washington, D.C., Patent Office met with Eckert, Mauchly, von Neumann, and University staff at the Moore School of Electrical Engineering in 1947 to decide whether the EDVAC could be patented. Von Neumann maintained quite justly that he was a co-inventor, and he wanted the invention in the public domain[45]. The lawyers for the Patent Office ruled in 1947 the EDVAC had been "public" too long already, hence could not be patented.

Meanwhile, the Eckert-Mauchly Computer Corp. (EMCC) had been founded in 1946 to develop and market the UNIVAC, the first commercially successful mainframe computer. Sidestepping the negative ruling on the EDVAC, EMCC resorted to legerdemain, applying for a patent on the earlier ENIAC, a much weaker machine. Von Neumann died in 1957 and was no longer an interested party. The Patent Office took several years to react, seemingly forgetting its ruling on the EDVAC. But in 1964 patent was actually issued for the ENIAC design; and this patent was regarded as being very valuable by Eckert and Mauchly, covering many design features of all electronic computers, which had meanwhile represented a multi-million dollar industry, led particularly by the UNIVAC.

By 1950, EMCC had been sold to Remington Rand, and this company in turn merged with Sperry to become Sperry Rand in 1955. Eckert and Mauchly by then worked for a large corporation, in charge of its industry-leading UNIVAC Division. To avoid lawsuits with the even larger IBM and its model 701, based on von Neumann's design, Sperry Rand entered into a patent-sharing arrangement with IBM in 1956. But Sperry Rand insisted on defending the licensing rights it claimed were covered by the ENIAC patent against the many other competitors in the burgeoning industry. They included several mighty corporations: RCA,

---

[44] Mauchly was senior, but Eckert was more entrepreneurial.

[45] It's interesting that the "violently anti-communist" von Neumann (self-description) put such faith in the public sector, which libertarians (e.g. Friedrich von Hayek) nowadays hardly consider, wishing everything to be privatized. An explanation may lie in the fact that von Neumann was born in the Austrian Monarchy, where the public sector since Empress Maria Theresia and her son, the Emperor Josef II, traditionally played a strong role, as it does in France; whereas Great Britain and its colonies in North America early on became laissez-faire in the Whig tradition. (However, the libertarian Hayek also was born in that same monarchy, and his ancestor was also ennobled by Emperor Franz Josef I.). Despite the negative ruling by the patent lawyers on the EDVAC, Eckert and Mauchly could have tried suing, but that would have been expensive, and they would be facing their ex-employer.

National Cash Register, General Electric, Honeywell, Control Data, Burroughs, and Philco-Ford, all fully capable to pay licensing fees. (Only AT&T with its Bell Labs and Western Electric units, was out of the picture, because it was forbidden by its regulator, the FCC, from entering the computer industry.)

So Sperry Rand started defending its licensing rights within the industry by suing Honeywell in 1967. This lawsuit, "Honeywell, Inc. v. Sperry Rand Corp.", became the biggest, longest, most expensive trial in US Federal Court history, taking six years of litigation. The trial itself ran from June 1971 until March 1972. Sperry Rand, Eckert, and Mauchly waited for judge Larsen's ruling until 1973, and he ruled against them. So the fundamental design of the comparatively primitive ENIAC entered the public domain—just as von Neumann would have wished[46].

The court ruled that John Atanasoff (Iowa State College) had actually developed the first electronic computer, he had established the "state of the art" in 1941 already (and did not care about patent rights). Furthermore, Mauchly had visited Atanasoff and inspected his computer in detail at Iowa State back in 1941. Some of its features were even more advanced that those of the ENIAC, e.g. it used binary rather than decimal arithmetic, it stored programs in memory, etc.

Presiding was U.S. District Court Judge Earl Larson at the Federal District Court in Minneapolis MN, headquarter city of Honeywell—much to the disappointment of Sperry Rand, headquartered in Washington, D.C., where Sperry originally filed. Judge Larson ruled that Honeywell had indeed "infringed" on Sperry Rand's patent; but that the patent itself was invalid, having been wrongly issued—partly because of Atanasoff's prior computer development, partly due to late patent applications.

References

1. Frege, G.: Begriffsschrift, eine der arithmetischen nachgebildete Formelsprache des reinen Denkens. Nebert, Halle (1879)
2. Van Heijenoort, J.: From Frege to Gödel: A Source Book in Mathematical Logic 1879-1931. Harvard University Press, Cambridge (1967)
3. Whitehead, A.N., Russell, B.: Principia Mathematica I, 2nd edn. Cambridge University Press, Cambridge (1915)
4. Köhler, E.: Gödel und der Wiener Kreis. In: Kruntorad, P. (ed.) Jour Fixe der Vernunft: Der Wiener Kreis und die Folgen. Hölder-Pichler-Tempsky, Vienna (1991)
5. Wang, H.: Reflections on Kurt Gödel. MIT Press, Cambridge (1987)
6. Hilbert, D.: Mathematische Probleme, Paris (1900). (Published in Archiv der Mathematik und Physik, 3rd series, vol. I, 1901)
7. Hilbert, D. Neubegründung der Mathematik (Erste Mitteilung), Abhandlungen aus dem mathematischen Seminar der Hamburgischen Universität, vol. I, pp. 157–177 (1922)
8. Bernays, P., Hilbert, D.: Gesammelte Abhandlungen III. Springer, Berlin (1935)

---

[46] Various parts of the story of the lawsuit are told by 1. Burks [64]. 2. Stern [60]. 3. Mollenhoff [65]. 4. Aspray [58].

9. Ewald, W.B.: From Kant to Hilbert: Readings in the Foundations of Mathematics. Oxford University Press, Oxford (1996)
10. Mancosu, P.: From Brouwer to Hilbert: The Debate on the Foundations of Mathematics in the 1920s. Oxford University Press, Oxford (1998)
11. Kreisel, G.: Hilbert's programme. Dialectica **12**, 346–372 (1958)
12. Benacerraf, P., Putnam, H.: Philosophy of Mathematics: Selected Readings. Prentice-Hall, Englewood Cliffs (1964)
13. Lewis, C.I.: A Survey of Symbolic Logic. University of California Press, Berkeley (1918)
14. Casti, J.L., DePauli, W.: Gödel: A Life of Logic. Perseus Publishing, Cambridge (2000)
15. Gödel, K.: Über formal unentscheidbare Sätze der Principia Mathematica und verwandter Systeme I. Monatshefte für Mathematik und Physik **38**, 173–198 (1931)
16. Feferman, S., Dawson Jr., J.W., Kleene, S.C., Moore, G., Solovay, R., van Heijenoort, J.: Kurt Gödel: Collected Works, vol. I, Publications 1929–1936. Oxford University Press, Oxford (1986). (ISIS: J. Hist. Sci. **77**, 691–692)
17. Köhler, E., Weibel, P., Stöltzner, M., Buldt, B., Klein, C., DePauli-Schimanovich-Göttig, W.: Kurt Gödel: Wahrheit und Beweisbarkeit, Band 1: Dokumente und historische Analysen. öbv und hpt Verlagsgesellschaft, Wien (2002)
18. Buldt, B., Köhler, E., Stöltzner, M., Weibel, P., Klein, C., DePauli-Schimanovich-Göttig, W.: Kurt Gödel: Wahrheit und Beweisbarkeit, Band 2: Kompendium zum Werk. öbv und hpt Verlagsgesellschaft, Wien (2002)
19. Takeuti, G.: Memoirs of a Proof Theorist: Gödel and Other Logicians. World Scientific Publishing Co., Singapore (2003)
20. Peano, G.: Arithmetices principia, nova methodo exposita. Bocca, Turin (1889)
21. Finsler, P.: Formale Beweise und die Entscheidbarkeit. Math. Z. **25**, 676–682 (1926)
22. Barwise, J.: Handbook of Mathematical Logic, pp. 1133–1142. North-Holland, Amsterdam (1977)
23. Goldstine, H.: The Computer from Pascal to von Neumann. Princeton University Press, Princeton (1972)
24. Gentzen, G.: Die Widerspruchsfreiheit der reinen Zahlentheorie. Math. Ann. **112**, 493–565 (1936)
25. Gentzen, G.: Neue Fassung des Widerspruchsfreiheitsbeweises für die reine Zahlentheorie. Forschungen zur Logik und zur Grundlegung der exakten Wissenschaften **4**, 19–44 (1938)
26. Kleene, S.C.: Introduction to Metamathematics. Van Nostrand Co., Princeton (1952)
27. Rogers Jr., H.: Theory of Recursive Functions and Effective Computability. McGraw-Hill Book Co., New York (1967)
28. Davis, M.: Computability & Unsolvability. McGraw-Hill, New York (1958)
29. Davis, M.: The Universal Computer: The Road from Leibniz to Turing. W. W. Norton & Co., Inc., New York (2000)
30. Dyson, G.: Turing's Cathedral: The Origins of the Digital Universe. Pantheon Books, New York (2012)
31. Rózsa, P.: Über den Zusammenhang der verschiedenen Begriffe der rekursiven Funktionen. Math. Ann. **110**, 612–632 (1934)
32. Rózsa, P.: Rekursive Funktionen, Akadémiai Kiadó, Budapest (1951); Translated as Recursive Functions. Academic Press, New York (1967)
33. Soare, R.I.: Turing and the discovery of computability. In: Downey, R. (ed.) Turing's Legacy: Developments from Turing's Ideas in Logic. Cambridge University Press, Cambridge (2014)

34. Frege, G.: Grundlagen der Arithmetik, Breslau (1884)
35. Dedekind, R.: Was sind und was sollen die Zahlen?. Vieweg, Braunschweig (1888)
36. von Neumann, J.: Mathematische Grundlagen der Quantenmechanik. Springer, Heidelberg (1932). https://doi.org/10.1007/978-3-642-61409-5
37. Gödel, K.: The Consistency of the Axiom of Choice and of the Generalized Continuum-Hypothesis with the Axioms of Set Theory, Annals of Mathematics Studies, 3. Princeton University Press, Princeton (1940)
38. Davis, M. (ed.): The Undecidable: Basic Papers on Undecidable Propositions, Unsolvable Problems and Computable Functions. Raven Press, Hewlett (1965)
39. Herbrand, J.: Sur la non-contradiction de l'arithmtique. Journal für die reine und angewandte Mathematik **166**, 1–8 (1931)
40. Feferman, S., Dawson Jr., J. W., Goldfarb, W., Parsons, C., Sieg, W.: Kurt Gödel: Collected Works, vol. V: Correspondence H-Z, Oxford University Press, Oxford (2003)
41. Church, A.: An unsolvable problem of elementary number theory. Am. J. Math. **58**, 345–363 (1936)
42. Church, A.: The Calculi of Lambda Conversion. Princeton University Press, Princeton (1941)
43. Turing, A.: On computable numbers with an application to the Entscheidungsproblem. Proc. Lond. Math. Soc. Ser. **2**(42), 230–265 (1937)
44. Turing, A.: On computable numbers with an application to the Entscheidungsproblem. Correct. Proc. Lond. Math. Soc. Ser. **2**(43), 544–546 (1937)
45. Curry, H., Feys, R.: Combinatory Logic, vol. I. North-Holland, Amsterdam (1958)
46. Curry, H., Hindley, J.R., Seldin, J.P.: Combinatory Logic, vol. II. North-Holland, Amsterdam (1972)
47. Bimbó, K.: Combinatory Logic (2016). In Edward Zalta's Stanford Encyclopedia of Philosophy at https://plato.stanford.edu/entries/logic-combinatory/
48. McCarthy, J.: Recursive functions of symbolic expressions and their computation by machine, part I. Commun. Assoc. Comput. Mach. **3**, 184–195 (1960)
49. McCarthy, J.: A basis for a mathematical theory of computation. In: Braffort, P., Hirschberg, D. (eds.) Computer Programming and Formal Systems. North-Holland Publishing Co., Amsterdam (1963)
50. Kreisel, G.: A notion of mechanistic theory. Synthese **29**, 11–26 (1974)
51. Kreisel, G.: Kurt Gödel, 28 April 1906–14 January 1978. Biograph. Mem. Fellows Roy. Soc. **26**, 148–224 (1978)
52. Turing, A.: Systems of logic based on ordinals. Proc. Lond. Math. Soc. Ser. **2**(45), 161–228 (1939)
53. Copeland, B.J. (ed.): The Essential Turing: Seminal Writings in Computing, Logic, Philosophy, Artificial Intelligence, and Artificial Life Plus the Secrets of Enigma. Oxford University Press, Oxford (2004)
54. Laplace, P.-S.: Traité de mécanique céleste, five volumes, Paris (1798–1825)
55. Kolmogorov, A.N.: On the conservation of conditionally periodic motions under small perturbation of the hamiltonian. Dokl. Akad. Nauk SSSR **98**, 525–530 (1954). [in Russian]
56. Arnold, V.I.: Proof of a theorem of A. N. Kolmogorov on the preservation of conditionally periodic motions under a small perturbation of the Hamiltonian. Uspehi Mat. Nauk **18** (1963). [in Russian]
57. Broer, H.: KAM theory: the legacy of Kolmogorov's 1954 paper. Bull. Am. Math. Soc. **41**(4), 507–521 (2003)
58. Aspray, W.: John von Neumann and the Origins of Modern Computing. MIT Press, Cambridge (1990)

59. von Neumann, J.: First Draft of a Report on the EDVAC, Moore School of Electrical Engineering, University of Pennsylvania, Philadelphia (1945)
60. Stern, N.: From ENIAC to UNIVAC: An Appraisal of the Eckert-Mauchly Machines. Digital Press, Bedford (1981)
61. Randell, B.: The Origins of Digital Computers: Selected Papers, 3rd edn. Springer, New York (1982). https://doi.org/10.1007/978-3-642-96242-4
62. Burks, A., Goldstine, H., von Neumann, J.: Preliminary Discussion of the Logical Design of an Electronic Computing Instrument, Part I (1946)
63. von Neumann, J.: Collected Works, vol. 5. Pergamon Press, Oxford (1965). (Ed. by, A.H. Taub)
64. Burks, A.R.: Who Invented the Computer? The Legal Battle That Changed Computing History. Prometheus Books, Amherst (2003)
65. Mollenhoff, C.R.: Atanasoff: Forgotten Father of the Computer. Iowa State University Press, Ames (1988)

Nikola Tesla - A Tribute to His Inventions

Radomir S. Stanković[1(✉)], Milena Stanković[1], and Franz Pichler[2]

[1] Department of Computer Science, Faculty of Electronic Engineering,
University of Niš, Niš, Serbia
Radomir.Stankovic@gmail.com
[2] Johannes Kepler University Linz, Linz, Austria

Abstract. Nikola Tesla (1856–1943) can be considered as one of the most important inventors of the 19-th and early 20-th century. His contributions to the invention and development of electromagnetism and electrodynamics are outstanding. Tesla contributed by various patents to other areas of engineering including telecommunications, remote control, and mechanic engineering. This paper is a brief tribute to his inventions that marked the twentieth century.

1 Introduction

Thomas Alva Edison is well renowned as the creator of the incandescent light bulb and beginning of illumination of the world with electricity. Due to that, he was described as *The Man Who Lit Up the World*, by his biographer Martin Woodside [13]. It should be, however, pointed out that for the DC current strongly advocated by Edison, the maximum transmitting distance is at about 2.5 km, and it can be easily imagined how will now look like big cities and metropolises with forests of electrical wires and many power plants on such small distances from each other and from the consumers of electrical energy. Therefore, Nikola Tesla, who invented necessary facilities and then strongly promoted the AC current, was the man who really highlighted the World, deserving in that way the title *Master of Lighting*, and *Man Out of Time* as pointed by Margaret Cheney and Robert Uth [3,4].

The war of currents run by Edison, in which Tesla did not actually participate from his side, finally ended by building the first hydro-electric power plant in a cooperation of Nikola Tesla and George Westinghouse at Niagara Falls in 1895, where out of 13 patents, 9 were inventions by Tesla. In this way, the introduction of the alternating current into the public use has been performed.

It is interesting to observe that Nikola Tesla and Thomas Edison were the single persons that ever appeared at the cover page of the journal *Electrical Experimenter* published by Hugo Gernsback over seven years, which can be considered as an especial distinction of their work.

2 Who was Tesla?

Nikola Tesla was born in Smiljan, a small village in the Austro-Hungarian Empire. The father of Tesla, Milutin, was an orthodox priest, the same as it was

R. Moreno-Díaz et al. (Eds.): EUROCAST 2017, Part I, LNCS 10671, pp. 66–73, 2018.
https://doi.org/10.1007/978-3-319-74718-7_8

his grand father from the side of his mother Georgina Djuka Mandić. Therefore, he was an Austro-Hungarian citizen, but of Serbian origin, and his baptismal record, issued on June 1856 was written in Cyrillic letters.

The most important recognition that Tesla received was in 1960 at the General Conference on Weights and Measures, where it was decided after rather turbulent discussions that the derived unit for measurement of the strength of a magnetic field would be given the name of Tesla with the associated symbol T. Therefore, $T = \frac{Wb}{m^2}$, where Wb denotes the Weber - the unit for the magnetic flux, as depicted on a banknote of the Serbian currency dinar (Fig. 1). It can be observed that Tesla is the single scholar of Slavic origins who obtained such a recognition and entered this group of very strongly selected scholars and inventors whose names are given to physical quantities in the international System of Units.

Fig. 1. Recto and verso of the Serbian banknote of 100 dinars.

3   Professional Biography of Nikola Tesla

After finishing primary education in Smiljan and obtaining the high school education in Gospić, Nikola Tesla entered the Polytechnic school in Graz, Austria. In December 1878, he move to Marburg (Maribor), Slovenia, and soon after worked for a short time as a Professor in the Gymnasium in Gospić. In January 1880, Tesla joined the Universitas Carolina, (Univerzita Karlova v Praze) in Prague, presently Czech Republic, but at that time within the Austro-Hungarian Empire.

In 1881, Tesla worked as the chief telephony engineer in American Telephone Company in Budapest, and in 1882 as the engineer in the Edison Company in Budapest, and then moved to the chapter of this company in Strazbourg, from October 14, 1883, to February 24, 1884. In June 6, 1884, Tesla arrived in New York, and joined the Edison Company. After unsuccessful attempts to convince Edison into advantages of the alternating current (AC) over the direct current (DC), mainly since Edison was eager to protect his well established investments in DC, Nikola Tesla started in 1886, the Tesla Electric Light & Manufacturing Company. In July 30, 1891, Tesla obtained American citizenship and worked further as an individual entrepreneur and inventor. Among the very many important events in his incredibly reach career, we point out just the following

1. 1895, building the first hydro-electric plant at Niagara Falls,
2. 1897, tear up the contract with Westinghouse,
3. Establishing in June, 1899 the laboratory in Colorado Springs until January 7, 1900,
4. 1901- building the Wardenclyffe Tower, in the laboratory at Long Island.

More details about the life and work of Nikola Tesla can be found in numerous books about Tesla, but we refer in particular to [9].

Table 1 shows the list of institutions where Tesla was awarded the title of a Doctor Honoris Causa and a reduced list of his decorations. A complete list of various awards, decorations, and diplomas can be found in [5].

4   Patents of Nikola Tesla

Nikola Tesla was very careful in protecting his inventions by patents in many countries, with some patents in different countries related to the same inventions. Therefore, the exact number of his patents is difficult to determine. As stated at the web page of the Tesla Museum in Belgrade, Serbia, Tesla held a total of at least 308 patents from 27 different countries on five continents www.nikolateslamuseum.org.

We here point out just 6 patents, four related to the brushless AC electrical motors and two concerning the electrical transmission of power by the high-voltage AC power grid, all patented on May 1, 1888

1. 381968 Electro-magnetic motor,
2. 381969 Electro-magnetic motor,
3. 381970 System of Electrical Distribution,
4. 382279 Electro-magnetic motor,
5. 382280 Electrical transmission of power,
6. 382281 Electrical transmission of power.

With the invention of the polyphase induction motor together with the related dynamos in 1888, Nikola Tesla gave an important contribution to the field of AC-technology. The Westinghouse Company bought from Tesla in 1889 the whole package of AC-patents and manufactured appropriate AC machines

Table 1. List of Doctor Honoris Causa titles and Decorations of Nikola Tesla.

Doctor Honoris Causa	
Technical school, Wiena, Austria	1908
University of Belgrade, Yugoslavia	1926
University of Zagreb, Yugoslavia	1926
Technical school, Prague, Czech Republic	1936
Technical school, Graz, Austria	1937
University of Poitiers, France	1937
Technical school, Brno, Czeck Republic	1937
University of Paris, France	1937
Politechnical school, Bucharest, Romania	1937
University of Grenoble, France	1938
University of Sofa, Bulgaria	1939
Decorations	
Decoration of Saint Sava, 2 class, Serbia	1892
Decoration of Independence of Montenegro	1895
Decoration of Sait Sava, 1 class, Kingdom of Yugoslavia	1936
Decoration of White Lion, 1 class, Czeck Republic	1937
Medal of University of Paris, France	1937
Medal of University of St. Kliment, Sofia, Bulgaria	1939

based on it. The installation of Tesla's polyphase induction dynamos at the Niagara hydro power plant in 1896, can be considered as a highlight of success of AC-technology. The plates fixed on the dynamos showing thirteen patents altogether, contain nine which refer to patents of Nikola Tesla.

Tesla investigated different methods for the generation of high frequency electrical currents. Already in 1890, he developed a high frequency alternator with a special kind of armature which allowed AC generation up to 16 kHz. Later such dynamos came in heavy use for transmitters of oversee-long wave transmitting stations.

Another method for the generation of high frequency currents, considered by Tesla, was the use of a Hertzian spark-gap together with a resonance circuit and a ironless step-up transformer, today considered as a "Tesla coil". Due to this invention, Tesla can be viewed as one of the inventors of wireless transmitters and radio technology, together with Marconi, Lodge, and Popov. However, Tesla's interest was not the wireless transmission of information but that of electrical energy. However, his experiments in New York and Colorado Springs toward that goal, brought no practical success. By lack of financial means the Wardencliffe project at Long Island could not brought to an end and Tesla had in 1905 to close his laboratory there. This was the end of Tesla's plan to establish a "world system" for the wireless distribution of electric energy.

A patent that we selected because of its special importance regarding the actual contemporary applications is *Method of and apparatus for controlling mechanism of moving vessels or vehicles*, No. 613,809, dated November 8, 1898, application filed July 1, 1898, Serial No. 884,934.

In 1898, Tesla demonstrated a boat with a coherer based radio control "teleautomation" to the public (large pool in the great hall) during the First Annual Electrical Exhibition at Madison Square Garden, Testimony in behalf of Tesla, Interference No. 21,701, USA Patent Office, New York, 1902. Tesla demonstrated a wireless secure communication between transmitter and receiver.

This patent was based on a thorough previous work on Tesla coil and oscillation transformer (1889–1892), researches and experiments with currents of high frequency (1889–1898), Tesla Wireless System (1891–1893). Tesla performed related experiments in his Lab in 35 South Fifth Avenue, until March 1895 when the lab burned down, and with continuation at the end of 1895 in the new lab 46, East Houston Street.

It is important to emphasize that the principle Tesla developed in this patent is applicable to *any kind of machine that moves on land or in the water or in the air*. Related to this invention Tesla said in his 1890s letter to Professor B. F. Meissner of Purdue University [3]

"Striking demonstrations, in many instances actually transmitting the whole motive energy to the devices instead of simply controlling the same from distance. In '97 I began the construction of a complete Automaton in the form of a boat, which is described in my original specification #613,809... This application was written during that year but the filing was delayed until July of the following year, long before which date the machine had been often exhibited to visitors who never seized to wonder at the performances... In that year I also constructed a larger boat, which I exhibited, among other things, in Chicago during a lecture before the Commercial Club. In this lecture I treated the whole field broadly, not limiting myself to mechanisms controlled from distance but to machine possessed of their own intelligence. Since that time I have advanced greatly in the evolution of the invention and think that the time is not distant when I shall show an automaton which, left to itself, will act as though possessed of reason and without any willful control from the outside. Whatever be the practical possibilities of such an achievement, it will mark the beginning of a new epoch in mechanics."

Therefore, it is right to conclude that this patent can be viewed as beginning of robotics as it is understood in the modern sense.

In the patent *System of Signaling*, No. 725,605, patented on April 14, 1903, application filed July 16, 1900, while working in his Colorado Springs Lab, Tesla invented multicarrier transmitter with a special receiver tuned to all carriers. A note on this invention Tesla recorded on June 27, 1899 in *Colorado Springs Notes 1899–1900*. This patent actually contains essential features of the spread spectrum wireless communications techniques known as frequency hopping and frequency division multiplexing.

These investigations were continued and the related patent has been issued *Art of Transmitting Electrical Energy Through the Natural Mediums*, Pat. No. 787,412, April 18, 1905, application filed May 16, 1900.

Tesla wrote *this invention consists of generating two or more kinds or classes of disturbances or impulses of distinctive character with respect to their effect upon a receiving circuit and operating thereby a distant receiver which comprises two or more circuits, each of which is tuned to respond exclusively to the disturbances or impulses of one kind or class and so arranged that the operation of the receiver is dependent upon these conjoint or resultant action.*

Recall that spread spectrum systems are intended to protect message from intruders and at the same time decrease disturbing effect of noise in transmission. In Tesla terminology this system *improves individualization and isolation of messages.*

Due to this, Tesla is widely renowned as a forerunner of wireless technology, as confirmed in several publications, see for example [1, 7, 11, 12].

An article entitled *The True Wireless* has been published by Nikola Tesla in *Electrical Experimenter* in the section *Radio Department*. The story of Tesla contribution to the radio development and his dispute with Marconi is well known. In this respect, we can conclude that many omissions in connection with Tesla's research into the radio fundamentals are corrected by the decision of the United States Supreme Court who gave priority to Tesla's patents applied in 1897 as a four-circuit system consisting of "an open antenna circuit coupled through a transformer, to a closed charging circuit at the transmitter, and an open antenna circuit at the receiver similarly coupled to a closed detector circuit". This statement was given in the case *Marconi Wireless Telegraph Company of America vs. United States*, adjudged in the Supreme Court of the United States at October term, 1942. An excellent condensed discussion of the matter can be found in [8].

Both patents *Method of and Apparatus for Controlling Mechanism of Moving Vessels*, Patent No. 613,809 and *System of Signaling*, Patent No. 725,605, contain logic AND circuit (Fig. 2) [9]. In his implementation, the function of the logic AND gate is achieved by the requirement that two wireless receivers have to be both tuned in so in order that the third receiver will respond.

This was the reason that the application for patenting the logic AND circuit by some computer producers, including IBM, was denied by the USA Patent Office, with the explanation referring to these patent of Nikola Tesla.

5 Instead of Conclusions

The work of Tesla has not been unanimously well accepted and recognized, and was sometimes even drastically abused, i.e., used without referring to Tesla, recall the issues with Edison and Marconi, as illustrations of opposing and unauthorized usage, respectively.

Regarding the point of view that Tesla had to his inventions and recognition of his work, the following can be quoted

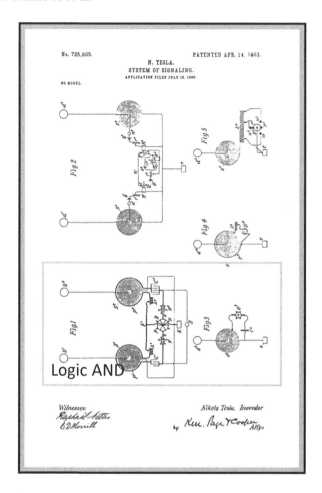

Fig. 2. A page of the Tesla patent No. 725,605 with the logic AND gate.

Let the future tell the truth, and evaluate each one according to his work and accomplishments. The present is theirs, the future, for which I have really worked, is mine.

The renowned Tesla expert, Professor Aleksandar Marinčić of the Faculty of Electrical Engineering, University of Belgrade, Serbia, wrote the following [9]

"Nikola Tesla did not belong to the group of great theoreticians who develop new frontiers of science, or to the class of great practitioners who invent many useful things for our lives. He belonged to a group of pioneers who opened up new field of technology".

Regarding various recognitions of work by Tesla and his contributions to the science and practical applications, we point out the following.

The name of Tesla is given to a minor planet, 2244 Tesla, of a provisional designation 1952 UW1, which is a carbonaceous asteroid with low albedo (reflection coefficient of the light).

Further, Tesla – is the name of a 26 km wide crater on the dark side of the Moon.

Then, *NVIDIA Tesla* is a unified graphics and computing architecture introduced in November 2006 in the *GeForce 8800 GPU* of the company NVIDIA [6].

Notice that Tesla Motors is the name of the electric car producer whose models carry the same name.

References

1. Anderson, L.I., (ed.): Nikola Tesla - Guided Weapons & Computer Technology. Twenty First Century Books (CO), 1 February 1998. ISBN 978-0963601292
2. Breuer, M.: Tesla and AND gates. IEEE Des. Test Comput. **24**(6), 624 (2007). https://doi.org/10.1109/MDT.2007.206
3. Cheney, M.: Tesla - Man Out of Time. Touchstone, USA (2001)
4. Cheney, M., Uth, R.: Tesla - Master of Lighting. Metro Books, New York (2001)
5. Diplomas of Nikola Tesla, Museum of Nikola Tesla. Belgrade, Serbia (2006)
6. Lindholm, E., Nickolls, J., Oberman, S., Montrym, J.: NVIDIA Tesla - a unified graphics and computing architecture. IEEE Micro **28**, 39–51 (2008)
7. Marinčić, A.S.: Nikola Tesla and the wireless transmission of energy. IEEE Trans. Power Appar. Syst. **PAS-101**(10), 4064–4068 (1982)
8. Marinčić, A.: Nikola Tesla contribution to the development of radio. IEEE Microw. Theory Tech. Soc. Newsl. 133 (1993)
9. Marinčić, A.: Nikola Tesla - The Work of a Genius. Gallery of Serbian Academy of Sciences and Art, Belgrade (2006)
10. Marinčić, A.: Nikola Tesla and his contribution to radio development. In: Sakar, T.K., et al. (eds.) History of Wireless, chap. 8, pp. 267–289. Wiley, Hoboken (2006)
11. Marinčić, A., Budimir, D.: Tesla's multi-frequency wireless radio controlled vessel. In: IEEE Conference on History of Telecommunications, HISTELCON 2008, 11–12 September 2008
12. Sarkar, T.K., Mailoux, R.J., Oliner, A.A., Salazar-Palma, M., Sengupta, D.L.: History of Wireless. Wiley, Hoboken (2006)
13. Woodside, M.: Thomas Edison: The Man Who Lit Up the World. Sterling, New York (2007)

Charles Proteus Steinmetz - Pioneering Contributions in Electrical Engineering

Franz Pichler[(✉)]

Linz, Austria
telegraph.pichler@aon.at

1 Introduction

The development of electrical engineering, a century ago, can be considered as one of the most interesting chapters of the history of technology in Europe and North America. The inventors of this time, such as Edison, Tesla, Siemens, Marconi, Morse, Bell, Heaviside, Gramme, and others, are still of strong biographical interest to us.

The ideal profile of an inventor would be the union of a rational thinking scientist with an artist, who brings the subject to the right human dimensions and relations. Leonardo da Vinci and Albrecht Duerer might be mentioned here to represent the ideal combination of a scientist with an artist. This essay should bring the scientific work and the artistic life of a man to our rememberings, a man which is known in Europe only in special scientific circles: Charles Proteus Steinmetz, as he called himself with his Americanized first name.

This lecture should bring the scientific work and the life of a man to remember, a man which is known in Europe only in special scientific circles: Charles Proteus Steinmetz, as he called himself with his American name. Just as the famous Nikola Tesla, Steinmetz also received his fundamental scientific education in old Europe. However, the American continent gave him the chance to apply his talents in full breadth, not forcing him to a certain personal life style. In contrary, he could fully keep his most interesting individuality.

2 Alternating Current Against Direct Current

Similar to the second half of the 20th century being dominated of the computer as the "realization engine of Information Technology", electrical machines (in a very general sense) caused great excitement in the second half of the 19th century. An important step for the practical application of electricity was the invention of the "dynamoelectric principle" by Werner Siemens (1866) - and independently by the British scientist Charles Wheatstone - which made possible the generation of electrical power by dynamos without the use of permanent magnets. Dynamos of this kind replaced voltaic batteries to power galvanisation processes and carbon lights. They were built as direct current machines (D.C. machines). As soon as the invention of the multiphase generator by Tesla (1888) and the invention of the high power transformer (Zipernowsky

© Springer International Publishing AG 2018
R. Moreno-Díaz et al. (Eds.): EUROCAST 2017, Part I, LNCS 10671, pp. 74–81, 2018.
https://doi.org/10.1007/978-3-319-74718-7_9

1885; Stanley 1886) was made, the generation of electrical power and its transportation across long distances became simpler and more efficient. The installation of a high voltage power line in Germany from Lauffen to Frankfurt (1891) by Oscar von Miller and the construction of the Niagara power plant (1895) by the Westinghouse Company - with Tesla as the designer of the alternating current generators established milestones in the practical use of alternating current. In North America, however, the replacement of D.C. technology by A.C. technology was not a simple task. The competing firms, the Edison General Electric Company (in support of D.C. technology) and the Westinghouse Company (which favoured A.C. technology) got into a long fight. It has been reported, that the Edison Company even supported the installation of the electric chair (New York 1889) to demonstrate the danger of the alternating current. The electrical exhibition in Frankfurt (1891) and the electrical exhibition at the world fair in Chicago (1893) brought a decision in favour of A.C. technology. In North America, this result was mainly achieved by the pioneering works of Nikola Tesla and Charles Proteus Steinmetz. In his important lecture on the "Application of complex numbers in Electrical Engineering" (which was published in German in the "Elektrotechnische Zeitschrift" at the same time (ETZ 1893)) Steinmetz was able to show that for alternating current phenomena the laws of Ohm and Kirchhoff were valid just in the same form as for direct current phemomena. By the "symbolic method" of Steinmetz it was possible to represent alternating currents by simple algebraic expressions. The computation of alternating current phenomena became just as easy as for direct current phenomena.

3   Steinmetz the Scientist

Of what kind were the important scientific contributions of Charles P. Steinmetz? How did it happen? Charles Proteus Steinmetz was born on April 9, 1865 in Breslau as Karl Rudolf Steinmetz. From childhood on handicapped by a hook and seemingly having a too big head for his short body and legs, he attended with great success the grammar school and the University of Breslau. There he studied mathematics and astronomy, showing also great interest in physics, philosophy, and the newly up-coming subject of electrical engineering. His doctoral thesis in pure mathematics titled "*Über unwillkürliche selbstreziproke Korrespondenzen im Raum, die bestimmt werden können durch ein dreidimensionales Linearsystem von Flächen der n'ten Ordnung*" had already been approved by the professors when things changed dramatically! Steinmetz was a member of a group of students which were in favour of socialism, which was not approved by the Prussian government and he also was a co-editor of the newspaper of the socialistic party, the "Voksstimme". From an anonymous friend he learned that his imprisonment was planned by the police. To escape, he fled via Vienna to Zurich in Switzerland. There his goal was to live there as a emigrant and to finish his studies at the "Eidgenössische Technische Hochschule". However things changed again. His Danish friend Oscar Asmussen persuaded him to join him in emigrating to the United States of America. Regardless of his poor knowledge of the English language, with no essential financial means but with a letter of recommendation by Mr. F. Uppenborn, the publisher of the internationally recognized scientific journal, the "Elektrotechnische Zeitschrift", he got on board of the steamship "La Champagne" in Le Havre to leave for New York (Fig. 1).

Fig. 1. Charles Proteus Steinmetz: arrival in USA

With the help of the recommendation letter, he immediately found employment at the company Eickemeyer and Osterheld in Yonkers, New York, which manufactured electric machinery. As he worked in the drawing office, however, his boss very soon discovered his mathematical gift, and more and more he was consulted when difficult problems came up. One of the problems in electrical machinery of that time was the heating up of the electromagnets. Steinmetz took a strict scientific approach to that problem and he developed the theory of electromagnetic hysteresis, a theory which has kept its validity until today. In his lecture on January 19, 1892 at the meeting of the American Institute of Electrical Engineers (AIEE) in New York City he reported on his important findings. A rather voluminous paper on the subject of hysteresis in the "Elektrotechnische Zeitschrift" has preserved this result until today (ETZ 1892).

The General Electric Company, which was established in 1892 by the union of several companies of the Edison group, also bought the company of Eickemeyer and Osterheld and Steinmetz was now employee of general Electric. He moved to

Schenectady, headquarter of General Electric, a lovely town on the river Hudson in upstate New York. The General Electric Company supported Steinmetz in his research in the best way, giving him a great deal of independence at the same time. In his own laboratory near his living home at Wendell Avenue he could perform any kind of electrical research he wanted. Of special interest to him was artificial lightning to explore the properties of material when stroke by lightning and to analyze lightning arrestors. In 1902 he got an appointment as a professor for Electrical Engineering at the Union College and he served there as a chairman of the department until the year 1913. In 1903 Union College awarded him the doctor of philosophy (Ph.D.). The numerous textbooks which he published as result of his lecturing helped a generation of students in electrical engineering in their studies of fundamental scientific models of electrical systems. Besides of writing books, he was ambitious in lecturing on different meetings and in publishing special papers on his research (Fig. 2).

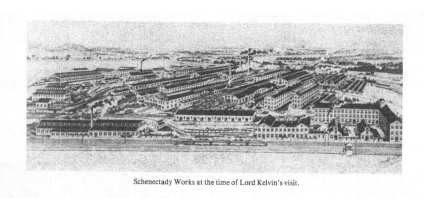

Schenectady Works at the time of Lord Kelvin's visit.

Fig. 2. General electric plant Schenectady 1902

4   Steinmetz in His Private Life

To have a full picture on Charles Proteus Steinmetz we have to look also into his life from a less scientific point of view. To start with that, let us quote Anton Zischka. Zischka writes *"He was a close friend of Marconi and Edison. To communicate with Edison in his later years he would use the Morse-code knocking it to Edisons legs. The American press called Steinmetz because of his artificial lightning experiments the "modern Jupiter". Although he was one of the great scientists he lived all the time the life of a boy: he ordered an exotic green house for his home and since he himself was a cripple he had the strangest animals such as lizards, exotic fishes, and birds in his house. The mirrors of his house were lighted by mercury-light lamps so that the visitors could see themselves with swollen violet lips and looking like having drowned in water. The doors of his house were usually electrified and occasionally he organized "Lightning days" destroying in his laboratory little houses made of cardboards. He loved to go by his canoe, he went regularly to crime-movies (especially he liked to see the actor Douglas Fairbanks) and he read numerous adventure stories. During night he would develop new mathematical formulas and perform computations which brought general Electric millions of dollars"* (Figs. 3 and 4).

Fig. 3. "Camp Mohawk" on Viele's creek

Fig. 4. Steinmetz in his garden

Although we must admit that Zischka reports with some phantasy here, he sketches the situation correctly. Steinmetz lived in fact a boy's life. He loved his "camp Mohawk" on Viele's creek and he loved to work on difficult mathematical problems by drifting in his canoe. He hated formalities in dressing and he would welcome also eminent visitors at "camp Mohawk" in his red bathing suit and wearing a T-shirt. Since he stayed a bachelor for all his life, to have a family, he adopted Mr. and Mrs. Hayden and finally he had also three grandchildren. He never gave up to be fond of socialism and he engaged himself in different social projects of the city of Schenectady (Figs. 5 and 6).

Fig. 5. Research work in the canoe

Charles Proteus Steinmetz passing away on October 16, 1923 after a heart attack was unexpected for everyone. The major American newspapers would bring reports on him. Herbert Hoover, later president of the United States of America on this occasion gave the following tribute:

"His mathematical reasoning broke the path for many of the advances in electrical engineering in recent years and solved problems which were vital to the progress of industry. In his writing he has left engineers a heritage of mathematics that will endue, and as a man he has set us all an example of physical courage and of devotion to our life work."

Fig. 6. Steinmetz and Albert Einstein

5 Summary

Charles Proteus Steinmetz was certainly one of the important scientists and engineers who, by their research and inventive contributions, helped to design electrical systems and to analyze electrical phenomena by sound scientific methods. Similar to Nikola Tesla, Steinmetz received his education at a European University. Contrary to Tesla, who had to fight all his life long to get support for his research and inventions, Steinmetz was lucky to get permanent and generous support for his research by the General Electric Company.

His work and inventions concerned the problems which had to be solved by the advent of the practical use of alternating currents. In electrical machinery, in the construction of dynamos and motors it became necessary to fight against eddy currents which caused losses of power. Steinmetz with his early work on the hysteresis of the iron cores of such machines contributed to the solution. Another important problem which occurred in the use of alternating currents was the design of proper networks for transmission. Steinmetz showed that, by the use of complex numbers a new charac-terization of resistors, inductances and capacitors was possible, such that the Ohm's law and the laws of Kirchhoff were also valid to allow an effective computation. In the history of the General Electric Company the time of 1902, where General Electric was founded and Steinmetz joined the company, until the year 1923, the year in which Steinmetz died, has been called the "Steinmetz era". His inventions, his many books, written for his students at the Union College in Schenectady give enough reasons to consider Charles Proteus Steinmetz as one of the most important electrical engineers of the past (compare with the list "Monographies and Text Books by Steinmetz" of the References).

Besides of his professional interest he had many private and rather artistic interests. They fit to his middle name "Proteus" given to him as nick-name by his friends at the grammar school in Breslau, the multi-faced gnome of Homer's Odyssee. We find Steinmetz here as a ever-lasting young boy, always ready for jokes, reading adventure stories, going to crime movies, but serious engaged in public social affairs - in one word - living a full life.

References

Important Papers Published by Steinmetz (in German)

Das Gesetz der magnetischen Hysteresis und verwandte Phänomene des magnetischen Kreislaufes. ETZ, XIII. Jahrgang, Berlin (1892)
Die Anwendung komplexer Groessen in der Elektrotechnik: ETZ XIV. Jahrgang, Berlin (1893)

Monographies and Text Books by Steinmetz

Theory and Calculation of Alternating Current Phenomena, New York (1897)
Theory and Calculation of Transient Electric Phenomena and Oscillations, New York (1909)
Radiation, Light and Illumination, New York (1909)
Electric Discharges, Waves and Impulses, New York (1914)
Theory and Calculation of Alternating Current Phenomena, New York (1916)
Theory and Calculation of Electrical Circuits, New York (1917)
Engineering Mathematics, New York (1917)
Four Lectures on Relativity and Space, New York (1923)

How Marconi and Gernsback Sparked a Wireless Revolution

James Kreuzer[✉] and Felicia Kreuzer[✉]

c/o Antique Wireless Association Museum, Bloomfield, NY, USA
wireless@pce.net

1 Introduction

The dawn of the wireless era attracted visionaries who recognized the potential of communications far beyond that of cable and telegraph. Two individuals were to play a major part in igniting public interest and gaining support in the growth of a new revolution in communications. Guglielmo Marconi and Hugo Gernsback were to create a revolutionary synergy through the interaction of their individual efforts. The resulting flow of information communications formed a network which would ultimately lead to creating the current information superhighway.

In the early days of wireless, the world's environment was primed and ready for realization of the potential of wireless communications and timing was perfect. Marconi assembled a team of technology specialists, developing equipment which quickly paved the way for strengthening the signal and propelling his vision of forming a system of worldwide wireless to fruition. Gernsback provided equipment and the "how to" expertise to all, uniting experimenters and amateur radio operators through his magazines, E.I. Company offerings, and establishment of the Wireless Association of America and the Radio League of America.

Together their efforts, although their separate visions may have been driven by self-serving ambitions, provided an altruistic system that was to create and unite a network of wireless communication and operators ready for deployment when needed during wartime and beyond. Together, if viewed as a system, Marconi and Gernsback provided advances to communications in the early days of wireless that would not have achieved any comparable degree of success if their efforts had been separated by space or time. This paper will demonstrate how their efforts created a huge impact and accelerated communications growth in those early times.

2 Marconi's Vision

Guglielmo Marconi was born April 25th, 1874 in Bologna, Italy. His Irish mother encouraged him at a young age to experiment and a privileged education led him to being mentored by Professor Righi in his teens (Baker 25). He was the first wireless experimenter to realize and exploit the value of applying wireless messages to a commercial service. He was primarily a businessman with a vision. After realizing that the potential for development in Italy was futile, he used his family connections, and in

© Springer International Publishing AG 2018
R. Moreno-Díaz et al. (Eds.): EUROCAST 2017, Part I, LNCS 10671, pp. 82–89, 2018.
https://doi.org/10.1007/978-3-319-74718-7_10

1896 went to England, secured financial support and applied for and was issued patents (Baker 28). By 1897 he established the Wireless Telegraph and Signal Company in England (Baker 35). The company's first paid wireless message was sent in 1898. His laboratories, gifted engineers and resulting products provided a benchmark that was difficult for others to achieve.

After sending the first Transatlantic communication in 1901, Marconi negotiated an exclusive contract with Lloyds of London and began to establish a network of wireless by outfitting ships and establishing shore stations throughout the world (Maclaurin 37). The United States Navy began to take interest in the potential of wireless and requested a demonstration. Marconi responded by asking the Navy to pay for his costs. Eventually the Navy witnessed the capability of his apparatus during the American Cup races, but Guglielmo Marconi refused to sell his equipment outright, preferring to lease and basically maintain a monopoly (Howeth 35). By 1906, Marconi eventually offered equipment for sale. Expansion efforts and R&D required the backing of financiers who patiently waited for the potential profitability of wireless. Marconi's company continued to operate at a deficit until 1910 (Maclaurin 40). Marconi's vision eventually was to bear fruit and his engineer's patents provided great value for the future (Fig. 1).

Fig. 1. Marconi the businessman and his company expansion (Maclaurin 43)

3 Gernsback's Vision

Hugo Gernsback was born August 16th 1884 in Bonnevoie, Luxembourg. The son of a vintner, he was (similar to Marconi) provided a privileged technical education (Ashley, Lowndes 16). In Gernsback's Radio Craft March 1938 issue he recounted that at age 6 his caregiver gave him an electric bell, wire and a battery – and the green spark generated at the bell's platinum contacts sparked his imagination. He began importing materials and installing bells and house telephones at a young age and at 13 years old received special Papal dispensation to install a bell and telephone system in the nearby convent. He invested his profits in electrical parts and assembled his own laboratory, and was able to duplicate some of Marconi's experiments.

He initially focused on designing batteries which led him to America to market his product in 1904, but his costs were too high and sales were not sufficient to stay in business. He continued to improve the design and found a partner to manufacture and sell ignition batteries for automobiles. This exploit was not profitable enough to succeed, so he shifted his focus back to his former interests in wireless and formed the Electro Importing (E.I.) Company, partnering with Louis Coggeshall, a friend and telegraph operator in 1905. Their plan was to buy materials from Europe, manufacture wireless parts and sell them via mail order catalog. Eventually a store was opened and the business began to thrive. The company had become the first major supplier of low cost wireless goods and apparatus in America. One invention, the Telimco was considered the first complete wireless set to be sold to the public. Gernsback's price was so reasonable that he was investigated for fraud and ultimately cleared when he was able to demonstrate that the set performed as advertised (Radio-Craft 575) (Fig. 2).

Fig. 2. Gernsback in 1909 and the future of radio, frontispiece of Radio for All

At the time Gernsback arrived in New York, there were no schools established for wireless training. It was stressed in his tutorial that "wireless telegraphy must be studied." His E.I. Co. catalogs provided how to instructions, doubling as a primer designed to teach the theory and application of wireless communication and as an outlet for affordably priced parts.

Gernsback was not only an entrepreneur, he was a visionary. At the time, resources were minimal and Gernsback had to create the tools needed to begin communications. He also was to create a supply chain, forming a marketing venue with his magazines. His publications served to promote not only sales, but station and club building and created a youth movement strengthened by their desire to communicate using wireless. His magazines and catalogs presented the new technology in a simple, easy to understand format that appealed to the beginning experimenter. Experimenters and engineers visited his store in Manhattan and many others, through his catalogs and magazines were able to obtain all types of electrical items via mail order. His E.I. Company and catalogs provided an incredible supply chain for building wireless apparatus. Gernsback's many magazines were instrumental in creating, teaching, and uniting wireless operators, advancing the art exponentially. Gernsback organized the Wireless Association of America in 1909, possibly the first national communications club in the world. Later in 1915, when war appeared imminent, Gernsback organized the Radio League of America. This association was to provide a network of trained wireless telegraph operators and engineers ready to take action in the service of their country.

New York City was a hotbed for wireless in the early days. The close proximity to E.I. Company and to other young experimenters and the availability of low cost experimental apparatus provided the tools to communicate. Clubs began to form when the youth of many of the more well to do middle class families became aware of their close proximity. The timing was perfect for Gernsback's arrival. The formation of a company that catered specifically to the wireless experimenter filled a niche created by a sudden demand when the news of major developments in wireless by men such as Marconi began to spread. Gernsback's mail order business's success depended upon the demand for his catalog and the sale of many low cost parts. As his subscriptions increased, so did his ability to perfect a formula that would prepare him for the establishment of a magazine empire which was to debut with the release of the first issue of Modern Electrics in 1908.

Gernsback's magazines became the "Facebook of the times" and authors were drawn to contribute as they witnessed the increasing popularity of his publications. Gernsback's editorials encouraged feedback from his readers and he revised his articles based on their preferences, a revolutionary concept at the time. Although a monthly magazine could not provide instantaneous resolutions, he did create a sort of monthly social media following. His catalogs provided the "how to" instructions that combined theory with practical application in an experimental format. The experiments outlined in his catalogs and magazines provided a clever marketing technique, and served as a stimulus to buy more E.I. Company products. The educational format created a desire to learn, which inevitably led to an increase in sales of parts needed to improve a beginner's simple station. EI's catalog covers were more attractive than the standard catalog of the day, and Gernsback quickly learned that the cover did help sell the contents, a technique he was to apply quite successfully to his future magazines (Fig. 3).

Fig. 3. The first E.I. Co. catalog, 1906

Gernsback managed to use his publishing expertise to launch his first magazine in 1908, Modern Electrics. His fertile imagination fueled a seemingly never-ending stream of ideas that he was able to channel to the public, especially targeting the young boys of the day through his magazines, beginning with Modern Electrics. He had access to and contact with many engineers of the day who not only submitted material to his magazines but were available as consultants to his publishing business to confirm technical accuracy. Gernsback also relied on the creative energies of illustrators such as Frank R. Paul and Howard V. Brown to complement and stimulate the imaginations of his readers. Little by little, month by month, his magazines served to educate and encourage young readers to experiment, build, and perfect their own communication systems using wireless lessons and parts obtained either through E.I. Company mail order catalogs or through their NYC store.

By offering magazine and catalog based education and tools via EI company, Gernsback was to create an army of tinkers who began a revolution that would serve to bring his ideas to life and many times eventually contribute to actual inventions based on the seemingly farfetched or impossible notions of the day. Many were encouraged to tinker, invent and build stations and continue to experiment. One of the perks of joining the Wireless Association of America was to be given information regarding other operators in close proximity. Like minds came together and taught each other, sharing insights based on their own personal experimenter experiences. Clubs were formed without prejudice; it was enough to hold a common interest in wireless.

Gernsback foretold the potential of Government influence in his Modern Electrics editorials, warning that if discretion was not performed in the communications arena, foolish pranks could affect the freedom of future communications. Just as he predicted, bills were introduced in 1910 which aimed to restrict amateur wireless telegraphy. Gernsback pushed amateurs to write to Washington and influenced the press to fight for the free use of wave lengths below 200 m. In 1913, the Alexander Wireless Bill was amended with wording that was uncannily similar to Gernsback's Feb 1912 Modern Electrics editorial. E.I. Company's simple practical tools provided building blocks for young minds eager to experiment at a time when there were fewer distractions and more freedom to tinker. Gernsback's fantastic images provided by creative illustrators who recreated his ideas, helped to spark the imagination of his faithful growing readership (Fig. 4).

In 1901, Marconi's leap across the Atlantic proved the possibility of the growth and potential of wireless communication. He believed that wireless communications was meant for the elite, and the technology must be protected and saved for inventors and the most intelligent engineers of the day. Marconi was a businessman and guarded his research and development carefully. Within a few more years, he was able to improve selectivity and range with the development of the multiple tuner, the set which was used for years for ship communications. Marconi was ever conscious of the need to maintain profitability and similar to techniques used by Bill Gates and Microsoft tried to monopolize the field in the UK by leasing his equipment and providing his own company trained operators. Marconi refused to sell his equipment to the US Navy outright and was basically prevented from entering a market that was controlled primarily by Telefunken and Slaby Arco until World War I was imminent. Marconi's takeover of United Wireless and formation of Marconi Co. of America provided him with a foothold in the American market and when drawn into the war, he was willing and able to offer his stations and services for the war effort.

Gernsback was willing to share his ideas with men coming from all walks of life. His magazines spread the word of the latest developments, primarily in the area of communications. Publishers of Scientific American, Marconigraph which became Wireless Age and many others attempted to mirror his techniques but failed to spark the imagination of their readers the way that Gernsback did. He challenged young people with contests and brought them together with the establishment of the Wireless Association of America. He foresaw the possibility of WW1 and America's eventual involvement. He encouraged operators to send in their station information; collecting it in his 1909 Wireless Blue Book many years before Amateur Radio printed their version in October 1914.

The sinking of the Titanic in 1912 was to emphasize the importance of wireless communications at sea and spurred new regulations that forced ships to have 24 h operators and the ability to maintain equipment with backups if needed. In the early 1920's, Gernsback's book, Radio for All, helped the general public to understand the basics of radio, increasing interest in the field. His monthly magazines morphed into a variety of new releases, and his publishing empire continued to grow as he found outlets for his ever growing interests. Some of his publications were short-lived, while others lasted for many years.

Marconi formed a great team of technology specialists and as his companies spread across the world they attracted the best of the best. By training young wireless operators to maintain and operate his apparatus, he also maintained control of his technology. He encouraged youths to learn wireless while fulfilling their dreams to go to sea and travel the world. Marconi and Gernsback were both self-promoters. Marconi used his contacts and relationships with major newspapers and investors to spread the word of his major achievements and with the Marconigraph and Wireless Age maintained control of the printed word. Gernsback also used his magazines to maintain control over the printed word and spread his ideas, oftentimes wild and farfetched, but with a foundation in scientific tact and practicality.

When Government regulations became imminent, Gernsback organized a movement through his readers and the members of his Wireless Association of America to bring the fight to Washington and keep the frequency band of 200 m and down for the amateurs. Gernsback did not have the backing of investors and major newspapers, his investors were his magazine subscribers who influenced his articles through their feedback. Gernsback took their feedback seriously and implemented their suggestions monthly. He received little or no recognition as an inventor and was not recognized as an influential contributor to the wireless movement in the major accounts of the time. But his effect on the growth and organization of the early amateur movement was tremendous.

4 Gernsback – Father of Science Fiction

Gernsback began writing fantasy stories in April 1911 with the introduction of "Ralph 124C41+". The serial installment provided not only an introduction to science fiction, but produced an outlet for his first forecasts. His far-out ideas had a basis in fact and his technical background grounded them with a practical discipline. He wanted to inspire others to take his ideas and invent new products with practical purposes to improve daily living. By 1926 Gernsback began publishing the world's first science fiction magazine, Amazing Stories. Gernsback's publications served to stimulate minds worldwide and inspired many to reach for possibilities never before achieved.

Fig. 4. Gernsback's magazines

5 Summary

In the early 1900's, communications over distances was primitive and though Marconi led a movement to advance the strength and distance of the wireless signal through his experiments, Gernsback brought the technology to the everyday man, youth, and amateur radio operators. Marconi and Gernsback each made a significant mark on the development of early communications. Technically they did not just improve communications equipment and disseminate information; they inspired the masses and revolutionized the field of wireless communications by creating a social system that collaborated to spread the technology across America and the world without the aid of traditional media methods. For the day, that was an incredible task, nurtured by inspired young tinkerers with a vision provided by a down to earth inventor and a visionary who could foresee future, though oftentimes farfetched possibilities.

In today's world, youth has grown complacent and the ease of communication has produced a generation that takes it for granted. There is often not enough to excite and stimulate young minds. Schools that teach to the test do not inspire innovation and creativity. But there is growing hope throughout the world. Tinkering is encouraged by more and more schools as an important and necessary way to develop young minds in areas of critical thinking, creativity, and innovation. Museums are becoming more involved with promoting STEM based activities, encouraging hands on creative exploration for the next generations. The future sustainability of our world is at stake and must rely on the sparks of a new revolution, incorporating both old and new technological advances. Will these efforts be enough to re-ignite the sparks initiated by Marconi and Gernsback? Can today's youth keep the flame burning?

References

Ashley, M., Lowndes, R.A.W.: The Gernsback Days. Wildside Press, PA, USA (2004)

Baker, W.J.: A History of the Marconi Company. Methuen & Co. Ltd., London (1970)

Gernsback, H.: Modern Electrics (1909–1911)

Gernsback, H.: The Electrical Experimenter (1915)

Gernsback, H.: Radio Craft (1938)

Gernsback, H.: Radio for All. J.B. Lippincott Co., London (1922)

Howeth, L.S.: History of Communications-Electronics in the U.S. Navy, Washington (1963)

Maclaurin, W.R.: Invention and Innovation in the Radio Industry. MacMillan Co, NY (1949)

Marconi, G.: Marconigraph (1911–1913)

Systems Theory, Socio-economic
Systems and Applications

Utilization of a Web Browser for Complex Heterogeneous Parallel Computing Using Multi-core CPU/GPU Systems

Marek Woda[✉] and Adam Hajduga

Department of Computer Engineering, Wrocław University of Technology,
11/17 Janiszewskiego, 50-372 Wrocław, Poland
marek.woda@pwr.edu.pl, adam.hajduga@hotmail.com

Keywords: Heterogeneous parallel computing · HTML5
Computing in a web browser · OpenCL/WebCL

1   Introduction

Since the invention of the first microprocessor has passed many years. Technological developments in CPU construction is primarily based on increasing the performance of devices, their miniaturisation and the reduction of manufacturing costs. Well known Moore's Law, speaking of doubling the number of transistors on a chip at regular intervals (going in hand with reduction of manufacturing costs), proved work well over years (initially assumed rate of eighteen months has been slightly extended to two years). Such a trend, due to the technological constraints cannot be everlasting; right now it can be already observed as it slows down. Limitations in minimum size of the individual components (transistors) and a total power draw of a system, forced to change the direction of the technological development. Instead of boost the clock of a processor, it was decided to multiply its number in a chip. Thanks to clustering of processor cores in a single chip that utilise fast shared cache memory, we still can observe considerable performance boost.

Another approach, recently propagated, involves the use of special, dedicated computation units (such as Graphics Processing Units) to parallelise the operation. In 2006 new graphics chips with unified stream processors were introduced to the market and started to replace outdated one based on specified pixel shader/vertex units. As a result, one can utilise them for general purpose computation while these units are not engaged in rendering graphics. The approach gained its momentum and was adopted as General Purpose on GPU which finally resulted in the project - an open framework to create kernels and communication interface with devices supporting it. Launched by Apple, and then taken over by the Khronos Group, OpenCL standard met with universal acceptance and their own implementations introduced such big players like Intel and AMD.

Advent of Internet era, and common use of plethora, yet more powerful, mobile devices, made Khronos Group to come up with industry standards that

© Springer International Publishing AG 2018
R. Moreno-Díaz et al. (Eds.): EUROCAST 2017, Part I, LNCS 10671, pp. 93–100, 2018.
https://doi.org/10.1007/978-3-319-74718-7_11

remediate the problem with diversity of devices and their operating systems, dealing with displaying accelerated graphics in a browser and with utilisation of potent CPUs/GPUs for general purpose computation. That was the trigger to introduce WebGL and WebCL as a counterparts to well known OpenGL and OpenCL [4,8]. As graphics technology evolves, the OpenGL has also evolved to support advances in hardware and software. More recently, the JavaScript implementation of OpenGL ES 2.0, WebGL, has become an increasingly attractive API for both educational and professional practice [1].

The contribution of this paper is to demonstrate the legitimacy of the use of WebCL/OpenCL frameworks and GPGPU capabilities in the process of creating complex applications that are run in a web browser. The aim of this work is to use WebCL framework to test the possibility of using GPGPU techniques in the creation of complex applications running directly in a browser (and its potential performance boost/degradation in comparison to a desktop application deployment with use of OpenCL). The choice was dictated by the growing importance of interactive, rich Internet applications services, and the problem they are facing - the loading speed and performance penalty introduced by that operation. No direct comparisons of duo OpenCL/OpenGL vs. WebCL/WebGl were made so far. In [7] usability of JScript and WebCL for numerical computations was investigated. Execution times of five algorithms implemented in JS/WebCL and GNU C [3] were scrutinised. Authors [2,5,6] proved that these standards are useful, not only in terms of rendering complex graphics but also capable to deal with complex computations, but none had tried to make complete comparison in terms of performance, features and interoperability of aforementioned standards.

2   Scope of Research

A growing demand for web based (light weight) applications that are computationally intensive and real-time browser based 3D graphic visualisation directed authors to the study involves the design and implementation of environmental research, on the basis of which to be drawn appropriate conclusions. Efforts of the research were focused on investigation of WebCL/WebGL usefulness to implement GPGPU application(s) running in a browser. Authors wanted to:

- deliver a measurable results of comparison: web based application (WebCL based) vs. analogous desktop version (based on OpenCL)
- proove that GPUPU in a web browser is the right direction for computationally intensive applications
- WebCL/WebGL tandem is a worthy competitor of GPGPU processing for browser based applications.

The project consists of a desktop application, written in C++ using frameworks OpenCL and OpenGL, and its mobile counterpart (that can be reached via a web browser), using the equivalents of the above mentioned libraries - WebCL and WebGL. Both systems allow:

– Solving an advanced computational problem without the use of external libraries (natively in C++ and JScript)
– Dealing with the problem by means of use of GPGPU (WebCL/OpenCL)
– Visualisation (WebGL/OpenGL).

Authors used known, though computationally complex problem of generating fractals as a test ground for the research. Studies consequently compared the different aspects such as:

– The execution time of the algorithm(s) along with render-time (rendered frames per unit time)
– Dependency between screen resolution and time to generate a fractal
– Measurement time and system resources utilisation during the initialisation and release of OpenCL/WebCL context.

3   Implementation and Test Environment

During the research a system consisted of two modules was designed and implemented:

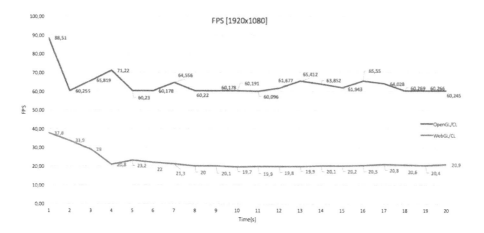

Fig. 1. Real time fractal's animation for FHD resolution

– Desktop application - a module written in C++ using OpenCL
– Web application - that is run directly in a web browser, written in WebCL - a web version of OpenCL.

Both modules had to be 'symmetrical', i.e. be able to perform the same operations and run the same algorithms, but for two different platforms, allowing for comparative testing. These algorithms are as follows:

– Generation an array to store the representation of a fractal (Julia' set) as RGB values (each successive set of four values represents one pixel of a fractal). The algorithm is run for different resolutions (fractal dimensions), each in the appropriate number of iterations, and the results (namely algorithm's execution time) are being averaged out. Calculations for comparative purposes are to be done both with and without OpenCL - using C++ native language capabilities.
– Use of OpenCL/WebCL to generate input objects (textures) for OpenGL/WebGL library. The purpose of the application is to create an animation showing the gradual zooming of a particular fractal. The measured parameter in that case is number of frames per second (fps) of animation during the zooming of a fractal.

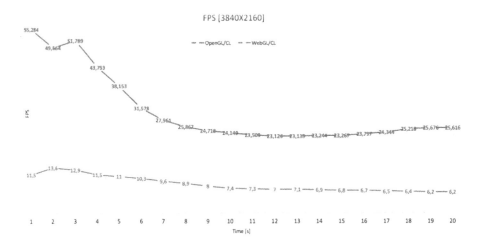

Fig. 2. Real time fractal's animation for 4K resolution

Both modules of the system had to perform similar actions. For this reason, a part of the implementation is similar for both approaches, despite the use of different programming languages. We can distinguish following fragments:

– OpenCL kernel - in both cases, it is a text (*string* stype). In a browser version, the content is stored in a *html*file and is loaded into the module as a JScript
– Initialization of OpenCL/WebCL contexts
– Generator of resolution distribution
– Profiling - measurement of kernel function execution time.

Slightly larger differences constitute an implementation of fractal's graphical representation. The desktop app uses partially Intel's implementation, in which a communication with OpenCL modules is internally implemented. The web app makes use of a separately created WebGL *shader* object, into which OpenGL

Table 1. Test environment

OS	Windows 10
Memory RAM	DDR3 16 GB
Graphics card	NVIDIA GeForce GT 750M
Processor	Intel Core i7 4500U 3 GHz

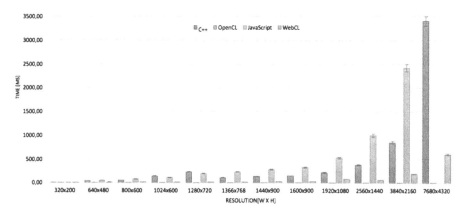

Fig. 3. Comparison of Julias's set computation time for different resolutions

texture objects are delivered. In order to make time measurements, the following methods were used (Table 1):

– $std::chrono::high\_resolution\_clock::now()$ for desktop version - generated result in nanoseconds format, converted to milliseconds
– $Date.now()$ for the browser version - results in milliseconds.

The following tools/technologies were used during the implementation and tests: Microsoft Visual Studio 2015 Community, JetBrains WebStorm 2016.2.1, Mozilla Firefox 32.0.3 (the only browser that natively, fully supported WebCL), OpenCL 2.0, Intel SDK for OpenCL Applications 2016 R2 for Windows 6.1.0.1600.

The results are presented in the form of graphs generated by Microsoft Excel 2016, taking into account the standard deviation (applied directly to the graph) calculated using the *STDEV* function.

The general scheme for using OpenCL in both app versions is as follows:

– Creation of context - *clCreateContext* - OpenCL compatible devices must be passed to the method, that will exchange data with
– Creation of a kernel object - *clCreateKernel* - accepts the program content as a *string* object,
– Buffer initiation through which data is sent (a table of a certain size)
– Passing parameters such as a fractal resolution to the program - *SetArg/clSetKernelArg*

- Creation a queue that synchronises data flow with an OpenCL device - *clEn-queueNDRangeKernel*
- Waiting for the kernel to execute - *clEnqueueReadBuffer*
- After the last operation is completed, results are read from the array of buffer object.

4   Results and Evaluation

Figure 3 depicts computational performance of parallel computing algorithms implemented with use of traditional programming languages C++ for desktop applications and JScripct for web and GPU accelerated ones: OpenCL (desktop) and WebCL (web). Julia's sets were generated for different resolutions using every approach.

Table 2. Computation time of Julia Set (WebCL vs. OpenCL)

Computation	CPU usage [%]				Memory usage [%]			
	WebCL		OpenCL		WebCL		OpenCL	
Resolution	Min	Max	Min	Max	Min	Max	Min	Max
320x200	29,0	31,9	32,1	32,3	1,9	2,2	0,6	1,0
1920x1200	29,0	31,2	31,8	32,5	5,7	6,0	0,6	1,0
3840x2160	30,0	31,6	31,8	32,8	5,9	6,8	0,6	1,0
7680x4320	30,0	32,6	31,0	33,3	11,0	14,1	0,6	1,0

Table 2 presents aggregated results of Julia's set computation time for different resolutions for both OpenCL and WebCL. For lower resolutions (320×200 – 1024×768 px), due to a low computational complexity, there is no remarkable difference between OpenCL and WebCL. However, as resolution increases OpenCL becomes undisputed leader, WebCL performs almost respectively well, but for ultra-high resolutions 4K and 8K, it loses its momentum. C++ performs reasonably well until resolution reaches 2560×1440 px - then its computational performance dramatically drops. The slowest implementation is represented by JScript, which scales quite well until resolution of 1600×900 px. Fractal for 8K resolution in JScript cannot be generated due to depletion of memory resources (JScript limitation) and for C++ such a process takes very long (almost an hour).

Interestingly, CPU load for Julia's set computation, regardless of resolution, for both OpenCL and WebCL is similar and stabilises at around only 30%, however memory utilisation for OpenCL is far better (max 1% of total memory available vs. 14,1% for WebCL).

Table 3 presents aggregated results of frames per second of animated fractal (that could be zoomed by a user) for both OpenCL and WebCL. Surprisingly, at least for authors, OpenGL consumed more CPU cycles (up to 47% vs. max 17,2% with WebGL), during animation of zooming fractal. None the less frame

Table 3. Real time fractal animation (WebGL vs. OpenGL)

Animation	CPU usage [%]				Memory usage [%]			
	WebGL		OpenGL		WebGL		OpenGL	
Resolution	Min	Max	Min	Max	Min	Max	Min	Max
320x200	10,0	15,5	27,0	38,0	2,8	3,0	0,8	0,9
1920x1200	14,1	15,0	41,0	45,0	3,0	3,7	0,8	0,9
3840x2160	15,0	17,2	41,0	47,0	5,0	5,8	0,8	0,9

rate for every resolution (except 320×200) was remarkably higher in favour of OpenGL (see Figs. 1 and 2). On the other hand WebGL, consumed 3%–5% of memory comparing to only 0,9% for OpenGL. Such a phenomenon can be easily demystified as an overhead of a browser (which was quite dated one.)

5   Conclusions

Research results have proved that the use of WebCL affects notably (Fig. 3) the performance of web applications and can utilise multiple computing units simultaneously. The research bares WebCL deficiencies, caused mainly by inadequate development support in recent years. Use of current WebCL leaves much to be desired. One of the factors is the need to use an older version of Firefox browser. None of modern browsers natively support WebCL - a plugin needs to be installed to support it. An additional difficulty is the need to use WebCL resources through https://archive.org/web/. All accessible so far documentation, for unknown reasons, has been recently archived and no longer available. The implementation of WebCL itself does not differ much from implementation of OpenCL. Most of the code written is very alike in both versions. Most operations are performed using a static WebCL object (however Nokia's implementation of WebCL slightly differs from the standard of OpenCL that was used during the research). Another issue is that the debugging process is often tedious and complicated because JScript exceptions thrown by WebCL do not provide desired information what went precisely wrong, and generally other methods needs to be used to identify errors in the code (e.g. display values of variables). It also turned out that the implementation is not flawless. Due to unidentified reasons (numerous exceptions), use of profiling was impossible in WebCL nonetheless in OpenCL it was working fine. As expected, WebCL has diametrically improved application performance - e.g. fractal-generation increased tenfold in comparison to plain JScript. However, a performance boost with OpenCL is incomparable, the difference is colossal - up 800 fold faster for the largest resolutions (4K and 8K) than its web counterpart.

Nonetheless, the predictions are promising - the differences could probably have diminished significantly after upgrading to newer standards of code and newer supported versions of browsers (if any). It is also worth noting that for resolutions UHD and beyond, WebCL performed better than JavaScript implementation (which failed to work due memory constraints) and even C++ (almost

4,5 fold faster for 8K). As far as visualisation of fractals is concerned, the biggest difference in rendering is for FullHD resolution where WebCL renders the images with approx. 22 fps compared to 60 fps with OpenCL. For 4K resolution rendering speed is capped to 10 fps for WebCL whereas to 25 fps for OpenCL.

Interesting results provided measurements of resources consumption (CPU load and memory usage). In terms of fractal visualisation, OpenGL consumes (15–30% depending on the resolution) more CPU cycles than WebGL, whereas memory consumption is at the same level (~1–3% of total memory in favour of OpenGL). Notwithstanding, for the fractal generation CPU load is constant at ~30% regardless of resolution for both OpenCL and WebCL, but memory consumption fluctuates between 1,2% and 14,1% for WebCL, whereas for OpenCL is stable at around 0,7%. This is caused by a browser overhead, unfortunately this gap cannot be closed by a developer.

In conclusion, the use of WebCL makes sense, however software developers shall be aware about its nuisances, more effort shall be put in its popularisation, namely in standardisation along with making available, detailed documentation.

References

1. Angel, E.: The case for teaching computer graphics with WebGL: a 25-year perspective. IEEE Comput. Graph. App. **37**(2), 106–112 (2017)
2. Aho, E., Kuusilinna, K., Aarnio, T., Pietiäinen, J., Nikara, J.: Towards real-time applications in mobile web browsers. In: 2012 IEEE 10th Symposium on Embedded Systems for Real-time Multimedia (ESTIMedia), pp. 57–66. IEEE, October 2012
3. Cushing, R., Putra, G.H.H., Koulouzis, S., Belloum, A., Bubak, M., De Laat, C.: Distributed computing on an ensemble of browsers. IEEE Internet Comput. **17**(5), 54–61 (2013)
4. Herhut, S., Hudson, R.L., Shpeisman, T., Sreeram, J.: Parallel Programming for the web. In: HotPar, June 2012
5. Hoetzlein, R.C.: Graphics performance in rich internet applications. IEEE Comput. Graph. Appl. **32**(5), 98–104 (2012)
6. Jeon, W., Brutch, T., Gibbs, S.: WebCL for hardware-accelerated web applications. In: TIZEN Developer Conference, pp. 7–9, April–May 2012
7. Khan, F., Foley-Bourgon, V., Kathrotia, S., Lavoie, E., Hendren, L.: Using javascript and WebCL for numerical computations: a comparative study of native and web technologies. In: ACM SIGPLAN Notices, vol. 50, no. 2, pp. 91–102. ACM, October 2014
8. Kaeli, D.R., Mistry, P., Schaa, D., Zhang, D.P.: Heterogeneous Computing with OpenCL 2.0. Morgan Kaufmann, Los Altos (2015)

Reversed Amdahl's Law for Hybrid Parallel Computing

Jarosław Rudy$^{(\boxtimes)}$ and Wojciech Bożejko

Wrocław University of Science and Technology,
Wybrzeże Wyspiańskiego 27, 50-370 Wrocław, Poland
{jaroslaw.rudy,wojciech.bozejko}@pwr.edu.pl

Abstract. The paper presents a certain version of Amdahl's law intended for use with heterogeneous systems of parallel computation, which can be used to compare different technologies and configurations of parallel systems in a better way than a simple comparison of achieved speedups. A mathematical formulation of the law, derived from the known Amdahl's law is presented. Next, the proposed law is used in simple comparison of performance of several computation devices. Finally, we use the law to estimate the performance limit of a certain heterogeneous parallel system equipped with GPU device and employing vector processing techniques.

Keywords: Parallel computing · Heterogeneous computing
Amdahl's law · Vector processing

1 Introduction

Speeding up computation is the main reason for employing parallel computing systems. Given p identical processors, speedup $S(p)$ is defined by dividing the time the computation takes using a single processor $T(1)$ by the time it takes using p processors $T(p)$. In 1967 a formula was found to describe the theoretical speedup limit of such a system. This formula is known as the Amdahl's law [1]:

$$S(p) = \frac{1}{(1 - r) + \frac{r}{p}}, \tag{1}$$

where r specifies the part of the computation that can be sped up by use of multiple processors. The remaining $1 - r$ part is called the serial part of the computation and cannot be sped up. The law leads to the following main conclusions (we assume that $T(p + 1) \leq T(p)$ for every $p \in \mathbb{N}^+$):

$$\lim_{p \to \infty} T(p) = (1 - r)T(1), \tag{2}$$

$$\lim_{p \to \infty} S(p) = \frac{1}{1 - r}, \tag{3}$$

$$\forall_{p \in \mathbb{N}^+} : S(p) \leq p. \tag{4}$$

R. Moreno-Díaz et al. (Eds.): EUROCAST 2017, Part I, LNCS 10671, pp. 101–108, 2018.
https://doi.org/10.1007/978-3-319-74718-7_12

Let us note that for $r = 0.95$ the theoretical speedup limit of 20 can only be achieved with infinity processors *i.e.* $S(\infty) = 20$. For 20 processors we get only $S(20) = 10.26$. In practice, this speedup would be even lower, as Amdahl's law ignores the overhead and other technical issues encountered in real-life processors and multi-core systems. This problem was somewhat alleviated by adapting the law to better model modern multi-core processor architecture [4].

Let us also notice that Amdahl's law was intended for systems with homogeneous (identical) processors. This prevents us from directly applying the law for heterogeneous systems, where computation can be done using a mix of various devices (like `Nvidia Tesla` GPU or `Intel Xeon Phi` vector processor). This is important, as use of heterogeneous parallel computing systems has become common in recent years and has various applications ranging from discrete optimization [2,3,8] to cosmology [5].

Thus, considerable effort in recent years was directed at researching properties and techniques for heterogeneous computing. For example, Piao *et al.* proposed JAWS framework for adaptive work sharing between CPU and GPU devices [7]. Similarly, Vialle and Contassot-Vivier employed CPU, Xeon-Phi unit and GPU devices to research the possibility of using various types of parallel computations on a single machine [9]. Furthermore, a number of papers appeared with the purpose of applying or extending the Amdahl's law for the use with heterogeneous computing. For example, Moncrieff *et al.* viewed a heterogeneous system as being made up of several homogeneous subsystems and showed that under certain conditions such system severely outperforms each of its subsystems [6]. Finally, Pei *et al.* extended Amdahl's law by considered the overhead created from the need of data preparation and applied it to several homogeneous and heterogeneous systems.

In this paper we aim to adapt the Amdahl's law to allow easy comparison of capabilities of different heterogeneous parallel systems. This is especially important as each such system might include several different number and types of processors with different processing speeds (clock rates) and different prices, making comparison by speedup only inconvenient and inaccurate.

2   Amdahl's Law Enhancement

Let us notice that in practice the coefficient r from Eq. (1) is dependent on the specific algorithm and problem. Moreover, different parallel mechanisms (like CPU threads, GPU devices or SSE instructions) entail additional sequential overhead. The time needed to create and synchronize the threads and operations on the GPU device is dependent on both specific implementation (*e.g.* the number of threads used) and the specific instance of the problem being solved.

Let I be the instance of a problem we want to solve and C be the configuration of our parallel system (*e.g.* type and number of CPU cores, GPU *etc.*). Then $A(I, C)$ denotes the algorithm which solves I on configuration C (slightly different algorithms for different configurations). Thus, we can define r as:

$$r \stackrel{\text{def}}{=} r(A(I, C)). \tag{5}$$

This means r also depends on the number of processors p. This implies problems with the application of Amdahl's law for heterogeneous systems in its original form. In particular, for some configurations C it is possible to obtain slowdown instead of speedup due to the data preparation overhead mentioned earlier. This issue could be solved if we were to assume a fixed value of r.

Let $C_M = \{C_1, C_2, \ldots, C_k\}$ be a set of possible configurations of some machine M. Also let $\bar{A}_M(I)$ denote a "base" algorithm for solving the problem I on machine M. By "base" we mean that all algorithms $A(I, C)$ for machine M are derived by adjusting $\bar{A}_M(I)$ to the chosen configuration C. With those assumptions (specific machine M and problem I) the coefficient r depends only on the current configuration: $r = r(C)$.

Let r' be the parallelization coefficient of the problem itself. Let us note that due the imperfections and overhead imposed by real-life systems and algorithms, $r(C)$ is never lower than r':

$$\forall_{C \in C_M} : r' = r(C) + \varepsilon(C), \quad \varepsilon(C) \geq 0. \tag{6}$$

Thus, $r(C)$ is a lower bound on the value of r'. Furthermore, $r(C)$ is a real-valued function defined over a finite discrete set of configurations, meaning it has a global maximum we denote C^*:

$$\max_{C \in C_M} r(C) = r(C^*) \Leftrightarrow \forall_{C \in C_M} : r(C^*) \geq r(C). \tag{7}$$

This means that $r(C^*)$ would be the best estimate of r' among all values $r(C)$. The only problem is to determine $r(C^*)$.

First, let us define the speedup for heterogeneous parallel systems using the notion of configuration:

$$S(C) \overset{\text{def}}{=} \frac{T(C_{\text{worst}})}{T(C)}, \quad C, C_{\text{worst}} \in C_M \tag{8}$$

where C_{worst} is the configuration yielding the largest computation time (usually configuration employing single CPU core).

Now, let us go back to Eq. (3), which derives from the original Amdahl's law and describes the speedup limit for the best case (i.e. infinite number of processors). For a heterogeneous machine M this is equivalent to using the best (fastest) available configuration C_{best}:

$$S(C_{\text{best}}) = \frac{1}{1 - r(C_{\text{best}})}. \tag{9}$$

In the above equation $r(C_{\text{best}}) \in [0, 1]$ and thus $r(C_{\text{best}})$ is maximum when $S(C_{\text{best}})$ is maximum. We also know $S(C_{\text{best}})$ is maximum. Thus, $r(C_{\text{best}})$ is also maximum. This means configuration C_{best} is the same as the configuration C^* we were looking for. Thus, after substitituting C^* for C_{best} and transforming the equation we get:

$$r(C^*) = 1 - \frac{1}{S(C^*)}. \tag{10}$$

For simplicity we denote:

$$r(C^*) = r^*, \tag{11}$$

$$S(C^*) = S^*, \tag{12}$$

Let us stop briefly to consider the difference between coefficient r from the original Amdahl's law and the introduced value r^*. Assume we want to solve the problem of summing two arrays A and B, each one holding p elements, into array C. Since each element $C[i] = A[i] + B[i]$ can be computed independently form the others, we can compute all of them at once using p processors and such computation will be p-times faster than using single processor. This implies $r = 1$ (so called *embarrassingly parallel problem*) and here r is problem-dependent, but not machine-dependent. On the other hand, r^* is machine-dependent, so for machine with p processors $S(C^*) = p$ (we assume ideal processors and no super-linear speedup) and thus $r^* = 1 - \frac{1}{p} \neq r$.

Using the definition of $S(C)$ and r^* we can now formulate modified version of the Amdahl's law for any configuration of machine M:

$$S(C) = \frac{1}{(1 - r^*) + \frac{r^*}{p(C)}}, \quad C \in C_M. \tag{13}$$

Since the speedup $S(C)$ can be measured, we "reverse" the equation to obtain the only unknown value $p(C)$:

$$p(C) = \frac{r^*}{\frac{1}{S(C)} + r^* - 1}, \quad C \in C_M. \tag{14}$$

The value $p(C)$ is the number of some *reference cores* needed to match the given configuration C. Thus, we can use $p(C)$ to compare different configurations (even configurations consisting of completely different parallel mechanisms).

However, let us notice that in original Amdahl's law p can approach infinity, and the result is the highest possible speedup. Thus, in our "reversed" law $p(C)$ will equal infinity for $C = C^*$. This feature makes it difficult to use configurations that are far from C^* and C^* itself.

The problem can be alleviated by introducing additional artificial configuration C_α^* and assuming $S(C_\alpha^*) > S(C^*)$. We can actually define $S(C_\alpha^*)$ using α as multiplier:

$$S(C_\alpha^*) \stackrel{\text{def}}{=} \alpha S(C^*), \quad \alpha > 1. \tag{15}$$

Thus, we now define r_α^*, similar to how we defined r^*:

$$r_\alpha^* = r(C_\alpha^*) = 1 - \frac{1}{S(C_\alpha^*)} = 1 - \frac{1}{\alpha S(C^*)} \tag{16}$$

and use it in place of $r(C^*)$ resulting in the final form of the reversed Amdahl's law for heterogeneous parallel systems:

$$p(C) = \frac{r_\alpha^*}{\frac{1}{S(C)} + r_\alpha^* - 1}, \quad C \in C_M. \tag{17}$$

Table 1. Results for machine M_A depending on the value of parameter α

C	$S(C)$	$p(C)_{\alpha=1.1}$	$p(C)_{\alpha=1.4}$	$p(C)_{\alpha=2.6}$
C_1	1.00	1.00	1.00	1.00
C_2	1.96	2.42	2.28	2.11
C_4	3.83	8.82	6.67	4.87
C_8	5.45	49.92	16.56	8.23

Table 2. Results for machine M_B depending on the value of parameter α

C	$S(C)$	$p(C)_{\alpha=1.1}$	$p(C)_{\alpha=1.4}$	$p(C)_{\alpha=2.6}$
C_1	1.00	1.00	1.00	1.00
$C_{1,V}$	7.32	9.46	8.89	8.08
C_G	25.21	188.10	77.60	39.30
$C_{1,G,V}$	26.32	279.60	89.60	42.10

Parameter α can be chosen based on experiments.

In order to illustrate the reversed Amdahl's law we applied it to two machines M_A and M_B. M_A has 8 CPU cores (base configurations designated C_1 to C_8). M_B has 1 CPU (base configuration C_1), a GPU device (C_G) and ability to use vector processing (C_V). Base configurations can be combined to build actual configuration (*e.g.* $C_{1,G}$). The results for selected configurations (compared to sequential configuration C_1) are presented in Tables 1 and 2.

For M_A we know how many cores were actually used, so we can employ this knowledge to choose a value of α for which $p(C_n) = n$. Thus, appropriate value for α seems to be slightly over 2.6. More interesting are results for M_B. For example, configuration $C_{1,G,V}$ yields speedup of 26, but $p(C)$ suggests it is actually equivalent to using 42 reference cores. Thus, the reversed Amdahl's law provides additional information about the performance of heterogeneous parallel systems compared to simple speedup comparison.

Table 3. Results of comparison of machines M_D and M_E

	C	$S(C)$	$p(C)_{\alpha=1.1}$	$p(C)_{\alpha=1.4}$	$p(C)_{\alpha=2.6}$
M_D	C_1	1.00	1.00	1.00	1.00
	C_2	1.89	2.35	2.22	2.04
	C_3	2.65	4.19	3.66	3.08
	C_4	3.46	7.58	5.84	4.34
M_E	C_1	1.62	1.89	1.81	1.71
	C_2	2.91	5.05	4.27	3.46
	C_3	3.34	6.93	5.46	4.14
	C_4	5.01	45.16	15.05	7.52

Next, let us compare two machines D and E, both with 4-core processors, but of different type. The results of such comparison are shown in Table 3. We see that it is possible to express performance of both machines using the same reference cores. We also see that single core of machine M_E is equivalent to 1.71 cores of M_D. Similarly, four cores of M_E are equivalent to 7.52 cores of M_D.

Aside from speedup we can compare parallel systems through efficiency. Based on the original Amdahl's law, efficieny is defined as $E(p) = S(p)/p$, but we can derive equivalent version for the reversed law:

$$E(C) \stackrel{\text{def}}{=} \frac{S(C)}{p(C)}. \tag{18}$$

This allows to compare efficiency of heterogeneous parallel systems. Furthermore, cost of each configuration is generally known, allowing to compare cost-effectiveness of each configuration.

3   Machine Performance Limit

The law formulated in the previous section can be used to estimate computation performance limit of a given heterogeneous parallel system while it employs all of its parallel resources (*i.e.* uses its fastest configuration). Thus, we can calculate how many reference cores the system is worth when using its most powerful configuration *i.e.* the one that yields highest speedup. In this section we would like to show an example of such estimation.

The performance limit we want to obtain should be a single number for a given machine. This performance will depend on the problem, so for the test we should choose some embarrassingly parallel problem (for which $r = 1$ if ideal processors with no overhead were to be used). Three such problems were considered: (1) calculating multiplication of two matrices, (2) calculation of matrix transpose and (3) simulation of Conway's Game of Life cellular automaton. For the sake of brevity we present the results for the problem with highest measured practical value of r (highest achieved speedup). This turned out to be the matrix multiplication problem which had estimated value of r equal to 0.997 (sequential part $1 - r = 0.003$). Compared to this, matrix transpose problem had $r = 0.977$ (sequential part is 0.023, so nearly 8 times larger).

3.1   Matrix Multiplication

For two square matrices A and B we want to compute their product matrix C, defined as follows:

$$c_{i,j} = \sum_{r=1}^{m} a_{i,r} b_{r,j}. \tag{19}$$

The value of each block can be computed independently from others (simultaneous reading of values from the original matrix is allowed). For computing the problem using CPU and GPU we used a naive implementation with computational complexity of $O(n^3)$, where n is the size of the matrix. For GPU computing we employed the CUDA platform.

In order to make use of the vector processing we prepared an implementation of the matrix multiplication algorithm for the SSE2 technology. We assumed data element size of 16 bits (**short int**). This means that data vectors are composed of 8 elements (SSE2 vector processing operates on registers of 128 bits). However, SSE2 is not suited for non-continuous areas of memory – access to 8 subsequent row cells can be done in one SSE2 instruction, but access to 8 subsequent column cells normally requires 8 read instructions.

Let us assume that we want to calculate the value of 8 cells from $c_{i,j}$ to $c_{i+7,j}$ inclusive. In accordance with the formula (19) each of the values from $b_{1,j}$ to $b_{1,j+7}$ must be multiplied by the value $a_{i,k}$ for all k. Cells $b_{1,j}$ to $b_{1,j+7}$ are arranged sequentially in memory and belong to the same row, so their reading can be performed using a single SSE2 instruction. In addition, there is a way to create 8-element vector, where each element has the same value (in this case $a_{i,k}$) – the **_mm_set1_epi16** instruction. In result, the multiplication can be executed for 8 cells at a time. The necessary condition is that the matrix size is divisible by 8. However, the solution can be generalized to matrix of any size: we just need to increase the size of the matrices A, B and C so it is divisible by 8. The rows/columns added in that way can be simply filled with zeroes. Unnecessary rows/columns in C are discarded after the calculation is done.

3.2   Computer Experiment

The above problem was run on a machine with 6-core 12-thread **Intel Core i7-980X** (3.33 GHz) processor and **Tesla K20c** GPU device (with 2496 cores). Moreover, the machine had access to SSE2 technology. The division of data (*e.g.* number of matrix rows) between CPU and GPU was done in such a way as to obtain the highest speedup. The data presented here is an average of 10 runs. Configuration C_1 serves as a reference core. The results are shown in Table 4.

The results show that the top achieved speedup was 294.30, but the estimate of reference cores equivalent to that was 477 (for $\alpha = 2.6$). Using the GPU alone allowed to obtain around 86% of that speedup and was equivalent to about 380 reference cores. However, we also notice that the value of the α parameter has

Table 4. The results for the matrix multiplication problem.

C	$S(C)$	$p(C)_{\alpha=1.1}$	$p(C)_{\alpha=1.4}$	$p(C)_{\alpha=2.6}$
C_1	1.00	1.00	1.00	1.00
C_{12}	5.37	5.45	5.43	5.40
C_{SSE}	8.86	9.08	9.04	8.96
C_{GPU}	255.20	1204.10	669.50	382.60
$C_{12,SSE,GPU}$	**294.30**	**3226.80**	**1027.40**	**477.50**

a major effect on the observed number of reference cores. This means that either the α parameter has to be chosen carefully or that alternative methods that does not rely on this parameter have to be researched.

4   Conclusions

This paper proposed a modified version of the Amdahl's law for heterogeneous systems of parallel computation. The law introduced a metric enabling to compare capabilities of different methods of parallel computation, such as CPU threads, GPU, vector processing and distributed computing. The metric is dependent on parameter α, which needs to be chosen carefully and still requires better methods of estimation. The resulting law allows us to express any configuration by certain number of "universal" reference cores and compare devices with different architectures.

References

1. Amdahl, G.M.: Validity of the single processor approach to achieving large scale computing capabilities. In: Proceedings of the Spring Joint Computer Conference, AFIPS 1967 (Spring), pp. 483–485. ACM, New York (1967)
2. Bożejko, W.: On single-walk parallelization of the job shop problem solving algorithms. Comput. Oper. Res. **39**(9), 2258–2264 (2012). https://doi.org/10.1016/j.cor.2011.11.009
3. Bożejko, W.: Solving the flow shop problem by parallel programming. J. Parallel Distrib. Comput. **69**(5), 470–481 (2009). http://www.sciencedirect.com/science/article/pii/S0743731509000215
4. Hill, M.D., Marty, M.R.: Amdahl's law in the multicore era. Computer **41**(7), 33–38 (2008). https://doi.org/10.1109/MC.2008.209
5. Meng, C., Wang, L., Cao, Z., Feng, L.L., Zhu, W.: Large-scale parallelization based on CPU and GPU cluster for cosmological fluid simulations. Comput. Fluids **110**, 152–158 (2015). parCFD 2013
6. Moncrieff, D., Overill, R.E., Wilson, S.: Heterogeneous computing machines and Amdahl's law. Parallel Comput. **22**(3), 407–413 (1996)
7. Piao, X., Kim, C., Oh, Y., Li, H., Kim, J., Kim, H., Lee, J.W.: Jaws: a javascript framework for adaptive CPU-GPU work sharing. In: Proceedings of the 20th ACM SIGPLAN Symposium on Principles and Practice of Parallel Programming, pp. 251–252. ACM, New York (2015). http://doi.acm.org/10.1145/2688500.2688525
8. Rudy, J., Żelazny, D.: GACO: a parallel evolutionary approach to multi-objective scheduling. In: Gaspar-Cunha, A., Henggeler Antunes, C., Coello, C.C. (eds.) EMO 2015. LNCS, vol. 9018, pp. 307–320. Springer, Cham (2015). https://doi.org/10.1007/978-3-319-15934-8_21
9. Vialle, S., Contassot-Vivier, S.: Algorithmic scheme for hybrid computing with CPU, Xeon-Phi/MIC and GPU devices on a single machine. In: International Conference on Parallel Computing (ParCo 2015), Edinburgh, pp. 25–34 (2015)

A Comparative Study for Real-Time Streaming Protocols Implementations

Iván Santos-González[1]([✉]), Jezabel Molina-Gil[1], Alexandra Rivero-García[1], Antonio Zamora[2], and Rafael Álvarez[2]

[1] Departamento de Ingeniería Informática y de Sistemas, Universidad de La Laguna, 38271 La Laguna, Tenerife, Spain
{jsantosg,jmmolina,ariverog}@ull.edu.es
[2] Departamento de Ciencia de la Computación e Inteligencia Artificial, Universidad de Alicante, Alicante, Spain
{zamora,ralvarez}@dccia.ua.es

Abstract. The airports security and protection model involves using different resources to ensure its infrastructure. The most used are equipment for the inspection of persons and luggage, closed circuit television cameras throughout the airport, anti-intrusion systems and, detection and recording systems. However, all these resources imply a high cost that not all airports can afford.

Currently, there are new proposals for security which propose the use of mobile technology as video surveillance system. Mobile technology provide performance, portability, transmission capacity and the ability to capture videos and images that do it possible, and even more important, it reduces costs.

In particular, this paper present a comparative study of three different systems designed to evaluate times at live streaming, using Android smartphones.

1 Introduction

In this proposal, sharing audio and video media is the main objective. For this reason, the most important point is to guarantee a low latency, a low jitter and efficiency transmissions, considering that occasional losses could be tolerated. The media streaming protocol is defined taking into account the structure of the packets and the algorithms used to send real-time media on a network.

The media streaming protocol is defined taking into account the structure of the packets and the algorithms used to send real-time media on a network. There are several protocols, which allow the implementation of video surveillance system in mobile phones. In particular, WebRTC [1] and RTSP [9], are the most used. This work focuses on carrying out a study of three possible implementations in order to analyse their weaknesses, strengths and conclude which one would be the most suitable to be use on mobile devices. This proposal evaluates times at live streaming with the use of smartphones. It takes into account the establishment time and stream reception time from a single source to a large quantity

R. Moreno-Díaz et al. (Eds.): EUROCAST 2017, Part I, LNCS 10671, pp. 109–116, 2018.
https://doi.org/10.1007/978-3-319-74718-7_13

of receivers. In order to perform the analysis, three Android applications, a web service and two different multimedia services have been implemented. The study includes the implementation of three different proposals that claims to be the innovative solutions for video-streaming.

In particular, this paper present a comparative study to analyse the time requirements for establishing and sending packets in a video-streaming connection for each proposed solutions. One of them based on RTSP protocol and two proposals based on WebRTCP protocol.

The present work is structured as follows. Section 2 introduces two of the most commonly used video streaming protocols, RTSP and WebRTC. The developed implementations are defined in Sect. 3. Section 4 presents the comparative studies between the implemented systems. Finally, conclusions and open issues close the work in Sect. 5.

2   Video Streaming Protocols

In this Section, the analysed streaming protocols are explained. On the one hand, the RTSP scheme is described. On the other hand, WebRTC system based on API calls is presented.

2.1   RTSP Protocol

The Real Time Streaming Protocol (RTSP) is a non-connection oriented application layer protocol that uses a session associated to an identifier. RTSP usually uses the UDP protocol [8] to share the video and audio data and TCP [4] for the control. The protocol is almost twenty years old, and therefore, it is not prepared to be used with the present video qualities and resolutions but it continues offering a good response nowadays. The syntax of the RTSP protocol is similar to the HTTP protocol syntax, and it supports the next operations:

- **Retrieval of media from a media server:** The client can request a presentation description via HTTP or some other method. If the presentation is being multicast, the presentation description contains the multicast addresses and ports to be used for the continuous media. If the presentation is to be sent only to the client via unicast, the client provides the destination for security reasons.
- **Invitation of a media server to a conference:** A media server can be "invited" to join an existing conference, either to play back media into the presentation or to record all or a subset of the media. This mode is useful for distributing teaching applications. Several parties in the conference may take turns "pushing the remote control buttons".
- **Addition of media to an existing presentation:** Particularly for live presentations, it is useful if the server can tell the client about additional media becoming available.

The exact performance and request methods for RSTP protocol are shown in Fig. 1.

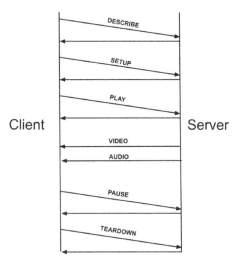

Fig. 1. RTSP request order

2.2   WebRTC Protocol

WebRTC is an API created by the World Wide Web Consortium (W3C) that allows the browser applications to make calls, video chats and the use of P2P files without any plugin. The first implementation of WebRTC was created by Google and released as Open Source. Different organisms as the Internet Engineering Task Force (IETF) to standardize the used protocols and the W3C with the browser APIs have been working on this implementation.

The main components of WebRTC are the next:

– **GetUserMedia:** It allows obtaining video or audio streams from the hardware (microphone or camera). This API call allows users to get a screenshot or share the screen with other users too.
– **RTCPeerConnection:** It allows users to set up the audio/video stream. This consists of a lot of different tasks like the signal processing, the codec execution, the bandwidth administration, the security of the streaming, etc. This API call allows implementing this different task without the intervention of the programmer.
– **RTCDataChannel:** It allows sharing video or audio data between connected users. RTCDataChannel uses a bidirectional communication between peers and allows the exchange of any data type. To do this, RTCDataChannel uses Websockets, which allows users to get a bidirectional communication between the client and the server and allows using a slower and reliable communication over TCP or a faster and non-reliable communication over UDP.
– **GeoStats:** API call that allows getting different statistics about a WebRTC session.

The performance of the WebRTC protocol, used to make a call with the API calls, commented before is shown in Fig. 2.

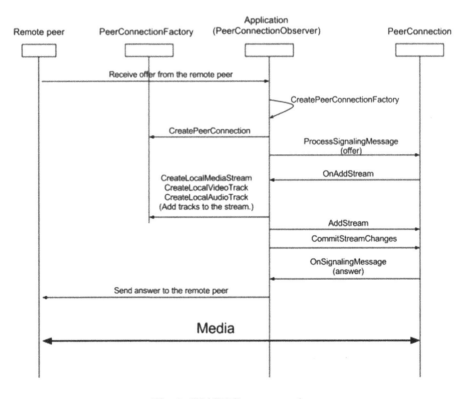

Fig. 2. WebRTC request order

3   Video Streaming Platforms

In this Section, the developed video streaming platforms are explained in detail. On the one hand, the RTSP video streaming platform is presented. On the other hand, WebRTC video streaming platform in two differente version v1 and v2 are described.

The developed Android applications act as server and/or client depending on the situation and it allows the users to connect with multiple users and send their real time stream. In the implementations of these systems the use of the last available version of Android, which was Android 6.0 that corresponds to the API 23 of the Android SDK, was selected. In this last version of Android, and consequently the previous ones, the RTSP and WebRTC protocols are not implemented hence, third party libraries have been used to implement these protocols. At this point, the use of libstreaming open source library [5] that implements the RTSP protocol both the server and the client parts, and the Libjingle library [3] that is an open source library written in C++ language by Google and that allows to establish P2P connections and to develop P2P applications, were chosen. These libraries permit to stream the camera and/or the microphone of an Android device and support the use of the H.264, H.263, AAC and AMR media codecs.

3.1   RTSP Video Streaming Platform

The architecture of the RTSP Video Streaming Platform consists of the use of P2P communications through the RTSP protocol in places where the two users are in the same network, and through the use of a Live555 media server in cases where there is not a dedicated network and it is necessary to use the Internet connection. Live555 media server [7] is a complete open source RTSP server that supports the most common kinds of media files. Moreover, this server can stream multiple streams from the same or different files concurrently, using by default the UDP protocol to transmit the streams, but it can be transmited over the TCP protocol if it is decided. The general flow of this streaming platform can be observed in Fig. 3.

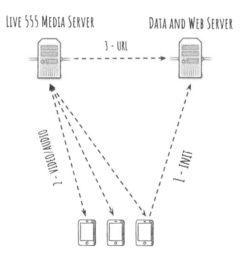

Fig. 3. RTSP video streaming platform general flow

3.2   WebRTC v1 Video Streaming Platform

The architecture of the WebRTC v1 Video Streaming Platform consists of the use of P2P communications through the WebRTSP protocol in places where the two users are in the same network, and through the use of a Kurento media server [6] in cases where there is not a dedicated network and it is required to use the Internet connection. Kurento is a WebRTC media server and it has a set of client APIs that simplifies the development of advanced video applications for web and smartphone platforms. Kurento media server features include a group

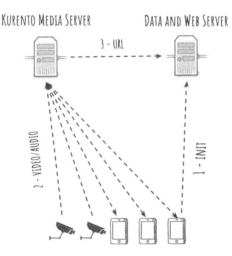

Fig. 4. WebRTC v1 video streaming platform general flow

of communications, transcoding, recording, mixing, broadcasting and routing of audio-visual flows and it also provides advanced media processing capabilities involving computer vision, video indexing, augmented reality and speech analysis. The general flow of this streaming platform can be observed in Fig. 4.

3.3   WebRTC v2 Video Streaming Platform

The architecture of the WebRTC v2 Video Streaming Platform consists of the use of P2P communications through the WebRTSP protocol in places where the two users are in the same network, and without use any media server in cases where there is not a dedicated network and it is unavoidable to use the Internet connection. In this case, a signalling server to communicate the smartphones between them is used, moreover a STUN server is used to obtain the public IP and the port available to make possible the communication and push notifications are used to send a stream invitation to the other. The general flow of this streaming platform can be observed in Fig. 5.

4   Comparative Study

In this section, the results of the different experiments can be observed. These experiments consists in the comparision of the connection establishment time. This measure shows the time since a user launches the streaming and the connection is completely established, and the stream reception. That is to say, the time between the connection establishment and the time when a user receive the first video packet. The espresso [2] testing software has been used to perform these experiments, it measures the time using a time java function automatically. These experiments have been performed during 100 iterations each one in

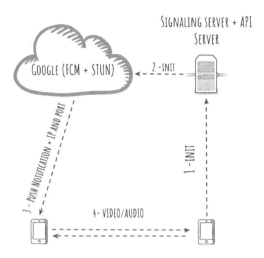

Fig. 5. WebRTC v2 video streaming platform general flow

all streaming platforms and the results of the arithmetic average can be observed in Table 1. These results show how the WebRTC implementations are faster and efficient than the RTSP one. Besides, how it was foreseeable, the WebRTC v2 streaming platform is better than the v1 due to the deleting of the media server in this last version.

Table 1. Comparative study of the implemented streaming platforms.

	RTSP	WebRTC v1	WebRTC v2
Connection establishment time	2304 ms	2025 ms	1835 ms
Stream reception time	2161 ms	1916 ms	1709 ms

The aforementioned results show how the WebRTC v2 streaming platform is almost a 20.4 % faster than the RTSP streaming platform and a 12.1 % faster than the WebRTC v1 platform in the connection establishment time. Moreover, the WebRTC v2 platform is almost a 9.4 % faster than the WebRTC v1 platform in the connection establishment time. On the other hand, in the stream reception time, the WebRTC v2 streaming platform shows an improvement of a 20.9 % respect to the RTSP streaming platform and a 11.3 % respect to the WebRTC v1 streaming platform. Moreover, the WebRTC v2 streaming platform improves the WebRTC v1 streaming platform in a 10.8 % for the stream reception time.

5   Conclusions

This work presents a comparative study between three differente implementations of two of the most used video streaming protocols, WebRTC and RTSP.

The three implementations have been developed keeping the same standards and using the same technologies when it was possible. Two variables were selected to do the comparative study. On the one hand, the connection establishment time, that is the time between a user launches the streaming and the connection is completely established. On the other hand, the stream reception, that is the time between the connection establishment and the time when a user receive the first video packet. The measurements obtained using the variables show how the RTSP protocol is slower than the WebRTC for the two measurements. Moreover, the first version of the WebRTC streaming platform shows a worst behaviour using these measurements than the WebRTC v2 streaming platform, fact that can be explained by the absence of a media server in the second version. As future work, new protocols and new measurements to evaluate the different protocols could be added to this study.

Acknowledgments. Research partially supported by TESIS2015010102, TESIS2015 010106, RTC-2014-1648-8, TEC2014-54110-R, MTM-2015-69138-REDT and DIG02-INSITU.

References

1. Bergkvist, A., Burnett, D., Jennings, C., Narayanan, A.: WebRTC 1.0: real-time communication between browsers. World Wide Web Consortium WD WD-webrtc-20120821 (2012)
2. Android Developers: Testing with expresso. https://developer.android.com/training/testing/ui-testing/espresso-testing.html?hl=es
3. Google Talk of Developers: Libjingle library. https://developers.google.com/talk/libjingle/developer_guide
4. Forouzan, B.A.: TCP/IP Protocol Suite. McGraw-Hill Inc., New York (2002)
5. fyhertz: Libstreaming library. https://github.com/fyhertz/libstreaming
6. Kurento. https://www.kurento.org/
7. Live555. http://www.live555.com/
8. Postel, J.: User datagram protocol. Technical report (1980)
9. Schulzrinne, H.: Real time streaming protocol (RTSP) (1998)

Multi-split Decision Tree and Conditional Dispersion

Monica J. Ruiz-Miró and Margaret Miró-Julià[(✉)]

Departament de Ciències Matemàtiques i Informàtica, Universitat de les Illes Balears,
07122 Palma de Mallorca, Spain
{monica.ruiz,margaret.miro}@uib.es

Abstract. Data Mining emerges in response to technological advances
and considers the treatment of large amounts of information locked up
in databases. The objective of Data Mining is the extraction of new,
valid, comprehensible and useful knowledge by the construction of models that seek structural patterns in the data, to ultimately make predictions on future data. The challenge of extracting knowledge from data
is an interdisciplinary discipline and draws upon research in statistics,
pattern recognition and machine learning among others. Clustering and
classification are two methods used in data mining. The key difference
between clustering and classification is that clustering is an unsupervised
learning technique used to group similar registers whereas classification
is a supervised learning technique used to assign predefined class values
to registers.

1 Cluster Analysis and Classification: General Notions

Clustering is a common technique for identifying natural groups hidden in data.
Clustering is a process that automatically discovers structure in data and does
not require any supervision, it is an unsupervised learning method. The grouping is done in such a way that objects in the same group or cluster are more
similar to each other than to those in other clusters. It is important to use the
appropriate similarity metric to measure the proximity between two objects, but
the separability of clusters must also be taken into account. The goal is to find
clusterings that satisfy homogeneity within each cluster as well as heterogeneity
between clusters [1].

The problem of comparing two or more sets of overlapping data allows for
the identification of different partitions of quantitative data. An approach using
statistical concepts to measure the distance between partitions is presented in
the following manner. The data's descriptive knowledge is expressed by means
of a boxplot that allows for the construction of clusters taking into account conditional probabilities. A method for cluster analysis and statistical based split of
the data was presented. The clusters are formed using similarity measures based
on boxplots. Boxplots are non-parametric, they display variation in the data of a
statistical population without making any assumptions of the underlying statistical distribution. The spacings between the different parts of the box indicate

R. Moreno-Díaz et al. (Eds.): EUROCAST 2017, Part I, LNCS 10671, pp. 117–124, 2018.
https://doi.org/10.1007/978-3-319-74718-7_14

the degree of dispersion and skewness in the data, and also show outliers. A boxplot is a convenient way to graphically display the data and identify outliers. Boxplots are useful when comparing data sets. Therefore, boxplots are a convenient tool to study overlapping data. Once clusters are identified the multiway split selection is forthcoming. The number of splits depends on the number of clusters found.

Classification is a process of categorization, where objects are recognized, differentiated and understood using the training set of data. Classification is a supervised learning technique where a training set and correctly classified observations are available [2]. In order to predict the class labels of a new object, hidden relationships between attributes are discovered. There are many classification techniques available. One of them is the decision tree model. A decision tree is a structure that includes a root node, branches, and leaf nodes. Each internal node denotes a test on an attribute, each branch denotes the outcome of the test, and each leaf node holds a class label. The topmost node in the tree is the root node. Decision tree learning creates a decision tree from the training set, that can be used as a predictive model which maps observations about a new object, represented in the branches, to conclusions about the object's class label, represented in the leaves. In data mining, a decision tree describes data and can be used to classify new objects.

2   The Multi-split Decision Tree Method

The majority of methods and techniques used to construct decision trees are binary based. These algorithms decide the split for binary class values that allows for the optimal decision tree. This is convenient when the class is dichotomous. This paper addresses the problem of constructing decision trees where the multi-split values assigned to the branches are found using statistical concepts that measure the distance between partitions and allow for the formation of clusters. In particular, quartiles and interquartile range are used in the calculation of a dissimilarity measure between clusters. The attribute selection is carried out using discriminant analysis procedures. Given a database with p numerical attributes X_i and a class or variable of interest Y which takes class values $y_n = 1, 2, \ldots, c$, a simple, precise and interpretable tree (predictive model) is built using the MDT (Multi-split Decision Tree) method.

The MDT method builds a decision tree considering multi-splits or multiway splits. A multiway split partitions the database in Z subgroups, $z \geq 2$. In addition to considering multi-splits, the MDT method studies where to carry out the multi-split in such a way as to get the most homogeneous possible subgroups. The behavior of the data will be studied through the dispersion of the p attributes. Actually we are not interested in the dispersion of the attribute X_i but the dispersion for each class value of the attribute X_i. That is, the dispersion conditioned by the class value for each attribute X_i. The purpose of the MDT method is finding possible clusters of classes with similar dispersion in each attribute and calculating the interval or cutting points to separate these clusters of the most discriminatory way possible.

The MDT method is an iterative model, at each iteration model's accuracy is calculated. The objective at each step of the process is to select an attribute that allows for the best split of the data. The phases are:

1. Finding the number of clusters for each attribute.
2. Calculation of the splitting points that better discriminate the clusters.
3. Attribute selection.
4. Identification of tree's leaf nodes.
5. Model's accuracy.

A flow chart of the MDT method is given in Fig. 1.

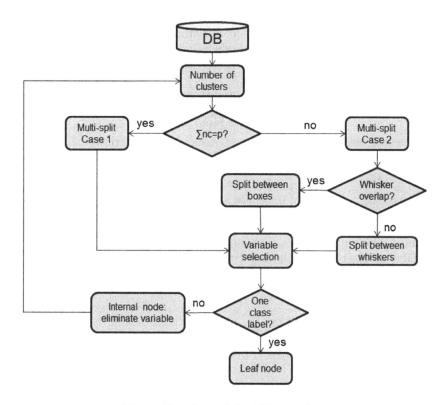

Fig. 1. Flowchart of the MDT method.

Phases 1 and 2, considers the problem of comparing two or more sets of overlapping data as the foundation for identifying different partitions of quantitative data. This problem was addressed in [3], where an approach using statistical concepts to measure the distance between partitions was presented. Boxplots display data variation without making any assumptions of the underlying statistical distribution and are used to graphically display the data and identify outliers. Boxplots are useful when comparing data sets. The most representative part of a boxplot is its box, which contains the central 50% of the data. The amplitude of each box represents the dispersion of the class in the attribute through its interquartile range (IQR). If the boxes, conditioned by the class, overlap or intersect then the dispersion of the variable is similar and the variables belong to the same cluster. The dispersion and Multi-split method studies possible clusters and calculates split values that separates the attribute into clusters in the most discriminatory manner.

A review of phases 1 and 2 is summarized below, followed by the analysis of phases 3, 4 and 5.

2.1 Number of Clusters

The number of clusters is the number of distinct overlapping boxplots. Two boxplots l and m are said to overlap if and only if the interquartile range of the union boxplot is smaller than the sum of the interquartile ranges of boxes l and m. That is,

$$IQR^{\cup} \leq IQR^{l} + IQR^{m}$$

$$|Q_3^{\cup} - Q_1^{\cup}| \leq |Q_3^{l} - Q_1^{l}| + |Q_3^{m} - Q_1^{m}|$$

The overlapping matrix $S = \{s_{lm}\}$ associated to attribute X_i indicates whether boxes l and m overlap can be constructed as follows:

$$s_{lm} = \begin{cases} 1 & \text{if boxes } l \text{ and } m \text{ overlap} \\ 0 & \text{otherwise} \end{cases}$$

This matrix defines a graph, where the nodes represent boxes and the edges represent overlapping boxplots. The number of connected subgraphs represents the number of clusters nc_i associated to attribute X_i. Let NC be a row vector that indicates the number of clusters for every variable $NC = \{nc_i\}, i = 1, 2, \ldots, p$.

2.2 The Best Splits

When a variable appears in a split, it is hard to know if the variable is indeed the most important, or if the selection is due to bias. Unbiased multiway splits are studied in [4]. Given an attribute X_i, the interval o splitting points are those that better discriminate the clusters found though the overlapping matrix S. Two cases are considered: (1) All g attributes have overlapping boxes; and (2) Some of the g attributes have more than one distinct overlapping boxes.

Case 1: All g attributes have only one connected graph, $\sum_h nc_h = g$. Each of the attributes form one cluster. The elements of the splitting vector SP are the splitting points $SP = \{sp_k\}$, $k = 1, 2, \ldots, g - 1$, where the middle point of the intersection of overlapping boxes form the splitting points:

$$sp_k = \frac{Q_3^k + Q_1^{k+1}}{2}.$$

Case 2: Some of the g attributes have more than one connected subgraph, $\sum_h nc_h > g$. There is at least one attribute with more than one cluster. The elements of the splitting vector SP are the splitting points $SP = \{sp_k\}$, $k = 1, 2, \ldots, nc - 1$, where the middle point of the distance that separates each pair of union boxes form the splitting points:

$$sp_k = \frac{Q_3^k + Q_1^{k+1}}{2}.$$

To improve the method, whisker overlap can also be taken into consideration. In this case, the boxplot's union upper limit $UL^{\cup}$ and lower limit $LL^{\cup}$ are defines as follows: $LL^{\cup} = min(LL_1, \ldots, LL_s)$ where $LL_i = Q_1^i - 1.5 \cdot IQR^i$. And $UP^{\cup} = max(UL_1, \ldots, UL_s)$ where $UL_i = Q_3^i + 1.5 \cdot IQR^i$. If whiskers do not overlap, $UP_k < LL_{k+1}$, the splitting points are:

$$sp_k = \frac{UL_k + LL_{k+1}}{2}.$$

Figure 2 illustrates splitting points for Case 2 with overlapping and non overlapping whiskers.

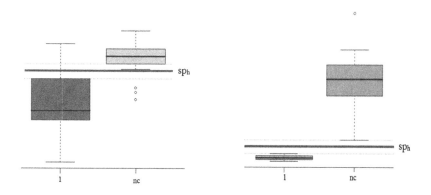

Fig. 2. Splitting points for case 2

2.3   Variable Selection

In order to select the most discriminant variable, classification accuracy is calculated. For each attribute, the dataset is partitioned according to the splitting vector SP and z subgroups are found. For each subgroup, an absolute frequency distribution table that takes into account the class is constructed. Let $f_j max_{z_j}$ be the absolute frequency's maximum value for subgroup z_j. The maximum frequency of the node X_i is $F_i max$.

The variable selected as node of the tree is that with largest maximum frequency $Fmax$:

$$X_i = arg\ max F_i max = arg\ max \sum_j f_j max_{z_j}.$$

2.4   Identification of Leaf Nodes

In a leaf node, all data have the same class label. That is, all data form a cluster. So, the number of clusters at a leaf node is 1, $nc_i(leaf\ node) = 1$.

When performing a multi-way split, each subgroup of data corresponds to a child node and a new problem. If the new problem consists of data of a single class label, the child node will be a leaf with class value c. If there is more than one class labels, the child node becomes a decision node. The selected attribute is deleted and the process is repeated to obtain a new attribute that further discriminates the data of the new problem. The child nodes of the same level of the tree are independent, in this way dynamic multi-way splits are created. A multi-split is dynamic if the intervals of said multi-way split can vary for different nodes. This process is repeated until leaf nodes are obtained for all class labels.

2.5   Model's Accuracy

The validity of the constructed tree is calculated from the percentage of well classified data. It will be said that a data is correctly classified when the real value of the class coincides with value obtained from the constructed tree.

$$accuracy = \frac{\text{correctly classified instances}}{\text{total number of instances}} = \frac{\sum_q f_q max}{n}$$

where q indicates leaf nodes.

3   Implementation of the MDT Method

The MDT method has been implemented using R. R is a language and environment for statistical computing and graphics. It is a open source computer programming project with a large, coherent, integrated collection of tools for data analysis. Different distinct functions were implemented.

- The **numClusters** function finds the number of clusters nc for each variable of the database, it calls the **adj** function that calculates the overlapping matrix for each variable using boxplots.
- The **split** function provides all the splitting points sp_k and returns the splitting vectors SP in the following manner:
  - All variables have one cluster, then $SP = sp_z$ with $k - 1$ components is found halfway between the overlap.
  - Some variable has more than one cluster and whiskers don't overlap, then $SP = sp_z$ with $nc - 1$ components is found halfway in the gap between whiskers.
  - Some variable has more than one cluster and whiskers overlap, then $SP = sp_z$ with $nc - 1$ components is found halfway in the gap between boxes.
- The **categorize** function returns the position of the variable with largest $F_i max$.
- The **classSubgroupList** function returns class labels for all instances of the subgroups. If only one label, it returns the leaf node and the class label.

4   Conclusion and Future Work

Classification analysis examines the interrelationship among several predictor variables and one dependent response variable. Generally, classification analysis handles predictors with nominal measurements and interrelationships of any form. The work presented above, allows for the construction of Multi-split Decision Trees for numerical predictor variables.

The method is based on the analysis of clusters formed using boxplots as similarity measure. Boxplots are non-parametric, they display variation in the data of a statistical population without making any assumptions of the underlying statistical distribution. The spacings between the different parts of the box indicate the degree of dispersion and skewness in the data, and show outliers. Boxplots are a convenient statistical tool to study overlapping data. Once the number of clusters is identified, the multiway split vector SP is easily calculated. The multi-split vector considers more than two child nodes (branches) at each decision node producing smaller and more compact trees.

Variable selection at each node of the tree is determined by the numerical attribute with greater discriminative power for the multi-split vector of that attribute. This choice is made considering statistical methods that allow to determine plausibly and optimally both the variable and the split. The work is based on the idea of analysing the dispersion of attributes conditioned by the values of the response variable to form clusters and thus determine multi-way splits that separate these clusters.

This paper offers a limited vision of one of the many clustering methods available. The following aspects can be considered as future work.

- In the calculation of the number of clusters nc_i, the possibility of using a dynamic overlapping criteria as a function of percentiles should be considered for the analysis of overlapping boxes and the computation of the overlapping matrix $S = \{s_{lm}\}$.
- In the calculation of the splitting vectors, data's skewness should be taken into account. In particular, the analysis of information contained in the whiskers of the boxplot should be further studied. Special consideration should be given to the case when box l is completely contained in box m.
- In the "best" variable selection, other statistical based functions, such as expectation maximization or maximum likelihood, should be looked upon.
- Finally, in the identification of leaf nodes, the problem where labels are partitioned in more than one leaf node should be further addressed and studied.

References

1. Everitt, B.S., Landau, S., Leese, M., Stahl, D.: Cluster Analysis. Wiley, Hoboken (2011)
2. Aggarwal, C.C.: Data Classification: Algorithms and Applications. Chapman and Hall/CRC, New York/Boca Raton (2014)

3. Ruiz-Miró, M.J., Miró-Julià, M.: Dynamic similarity and distance measures based on quantiles. In: Moreno-Díaz, R., Pichler, F., Quesada-Arencibia, A. (eds.) EURO-CAST 2015. LNCS, vol. 9520, pp. 80–87. Springer, Cham (2015). https://doi.org/10.1007/978-3-319-27340-2_11
4. Kim, H., Loh, W.Y.: Classification trees with unbiased multiway splits. J. Am. Stat. Assoc. **96**, 589–604 (2001)

Towards Extension of Open Data Model: Relational Data Model Context

Čestmír Halbich[1(✉)], Václav Vostrovský[2], and Jan Tyrychtr[1]

[1] Department of Information Technologies,
Faculty of Economics and Management,
Czech University of Life Sciences in Prague, Prague, Czech Republic
halbich@pef.czu.cz
[2] Department of Information Engineering,
Faculty of Economics and Management,
Czech University of Life Sciences in Prague, Prague, Czech Republic

Abstract. The main purpose for the use of open data (further OD) is to provide content. From the definition of OD there are two important attributes of OD: – machine readable format and reuse without any legal restriction. The issue of the so called "Open data" has recently been intensely investigated and discussed in detail. However, OD benefit can only be realized if the data are true, so called correct. There are a plenty of models of OD and their metrics, the proposed model extends one of the quality attributes called correctness. In the proposed model it seems preferable to use this aggregated attribute rather than individual attributes of quality.

Keywords: Open data · Quality model · ISO SQuaRE · Correctness
Database · Conceptual model · Logical model

1 Introduction

The issue of the so called "Open data" has recently been intensely investigated and discussed in detail. However, OD benefit can only be realized if the data are true, so called correct. There are a plenty of models of OD and their metrics, the proposed model extends one of the quality attributes called correctness. In the proposed model it seems preferable to use this aggregated attribute rather than individual attributes of quality. The main aim of this paper is to define by the proposed model the possibilities of monitoring the spread of varroa bee through OD. A partial aim is be to propose an acceptable procedure of using of such data as relevant information support in solving problems associated with the spread of varroa bee. Our proposal of innovative varroa mites spread monitoring is based on databases. The proposed methodology is characterised by mental model over the conceptual and relational model. Implementation is based on a physical model followed by testing and validation. We use data resulting from laboratory analyses available at all district representations of the Regional Veterinary Administration and local Czech Beekeepers Association.

OD can become a relevant source of information readily available on the Internet, not only in the public sector but also the commercial sector [1]. However, if data from

© Springer International Publishing AG 2018
R. Moreno-Díaz et al. (Eds.): EUROCAST 2017, Part I, LNCS 10671, pp. 125–133, 2018.
https://doi.org/10.1007/978-3-319-74718-7_15

laboratory analysis are to meet such potential they need to be of certain quality. In general, the quality is defined by international standards (SQuaRE) in the established quality model.

2 Methods

To meet objectives of this paper techniques and procedures of the relational database technology are applied. The proposed solution will be demonstrated and subsequently verified on the issue of OD relating to varroa bee. Other used methods include methods of the relational database technology, in particular methods of data integrity. Pivotal in this proposed procedure are the following issues: authentication, anonymization, integration and aggregation. A very important part of the proposed procedure is anonymity of these open data. This step involves practical use of the resultant data in strategic and tactical decisions by beekeepers.

2.1 The SQuaRE Project

The open data quality characteristics listed above should be implemented by soft-ware, which is used as open data source. In agriculture, a relational data-base (RDB) technology serves as such software in the majority of cases, both on supplier and user side (survey conducted by authors of this article within IGA, FEM CULS in Prague research grant no.: 20131038). The main logical principle of the RDB technology related to the realisation of correctness is data integrity [2–4]. It is crucial to include data integrity rules and constraints when de-signing and consequently realizing a database system which can be used for enforcing data correctness and increasing credibility of stored data [5–7]. From the data integrity point of view, correctness means that data provably relate to objects they belong to, have values that correspond with reality, and contain correct relations to other data (objects from different context).

2.2 Relational Data Model

OD are determined by the fact that they are mostly conceived based on relational database technology. A corresponding relational data model was devised by Codd [8]. Codd uses the term relation in its mathematical meaning. Thus, if there are sets $S_1, S_2, \ldots, S_n$ (which need necessarily not vary, R stands for a relation to n variables where each n variable contain the first element from S_1, second element from S_2, and so on. Mathematicians call a set of all n variables $S_1, S_2, \ldots, S_n$, where $s_j \in S_j$, Cartesian product $S_1, S_2, \ldots, S_n$ and designation $S \times S_2 \times \ldots \times S_n$. Thus R is a subset of Cartesian product. If $s_j \in S_j$ is valid than we say that R is a relation with domains S_j. For each element of an n variable it is possible to define its domain and unique name which is called attribute. A matrix representation of relations which represent a n variable relation to R has the following characteristics [8]:

- Each row represents an n variable from R.
- Order of rows is irrelevant.
- All rows are different.

- Column arrangement is relevant and responds to domain arrangement $S_1, S_2, \ldots, S_n$, through which R is defined (valid for a given relation).
- Importance of individual columns is partially defined by name of the respective domain.

The following crucial definitions related to relational database technology then determine the respective extension of open data quality model.

Definition: Domain relation $S_1, S_2, \ldots, S_n$ is called any subset of Cartesian product $S_1 \times S_2 \times \ldots \times S_n$; n-variable relation to S is what we call any subset of Cartesian exponent S^n, defined by induction: $S^1 = S; S^{(n+1)} = S^n \times S$ for n = 1, 2, ...

Codd in his relational model defines attribute and domains.

Relational scheme is defined as $S(A_1 : D_1, A_2 : D_2 \ldots A_n : D_n)$ and attribute as $A = \{A_1 : D_1, A_2 : D_2 \ldots A_n : D_n\}$.

Relational scheme consists of relation name, attribute names and domain names. Attribute relation is a couple of attribute and domain names Relation model describes data as designated relations of specified values [9]. For example, customer ID identifies with customers's name and address. Relation data model is possible to present on the following relation *Customer: {<(CustomerID, 100), (Name, Jan), (Surname, Novák), (Address, Kamýcká 123)>, <(CustomerID, 101), (Name, Alena), (Surname, Nováková), (Address, Kamýcká 123)>}*. This example specifies a relation: *Customer*; determined couple values: *(CustomerID, 100)*; and *n*-variables: *<(CustomerID, 100), (Name, Jan), (Surname, Novák), (Address, Kamýcká 123)>*.

A relation in a relation data model is represented by a table. Table representation of a relation has the following characteristics:

- Values in tables are atomic.
- Values in tables exist as elements of respective domains.
- Table work use language based on either relational algebra, or relational calculation.
- In each table values from one or more domains clearly identify individual rows (so-called primary keys).
- In some tables values from one or more domains relate to values in other tables (so-called foreign key).
- Within tables it is possible to define subsets of rows or subsets of columns.

Set operations with database based open data are defined by the following relational algebra:

Union, intersection, difference

Binary operation of union is in set theory marked by the symbol $\cup$. The aim of *Union* is to unite all facts from the arguments. The relational union operator is intentionally not as vague as in union operator in mathematics [8]. Union of binary and tertiary relations is prohibited as such a union does not result in a relation. Thus for $R_1 \cup R_2$ the following conditions must be fulfilled (these conditions are essential for difference and intersection, too), [10]:

- Relation R_1 and R_2 are of the same parity (the same number of attributes).
- Domains of the i-attribute R_1 and i-attribute R_2 are identical.

The result of an *intersection* is a relation which includes all n-variables that are in both R_1 and R_2. An intersection symbol is $R_1 \cap R_2$.

The operation of *difference* is marked with the symbol $R_1 - R_2$. The result is a relation which includes all *n*-variables in R_1 but not in R_2.

Union and intersection are commutative and associative. That means that the following are true:

$$A \cup B = B \cup A \text{ a } A \cap B = B \cap A$$
$$A \cup (B \cup C) = (A \cup B) \cup C \text{ a } A \cap (B \cap C) = (A \cap B) \cap C.$$

Cartesian product
Cartesian product of two relations, R_1 and R_2, is recorded in an infix notation as $R_1 \times R_2$. To define final relational schemes it is necessary to use fully quantified attribute names. That means to put relation name in front of an attribute. In this way it is possible to differ between relations $R_1.A$ and $R_2.A$.

Definition: If valid that $R_1(A_1, \ldots, A_n)$ and $R_2(A_1, \ldots, A_n)$, then the Cartesian product $R_1 \times R_2$ is a relation with a scheme containing all fully quantified attribute names from R_1 and R_2: $(R_1.A_1, \ldots, R_1.A_n, R_2.A_1, \ldots, R_2.A_n)$ [10].

2.3   Relational Databases Operations

Projection
Projection is a unary operation designated with the Greek letter PI (Π). Intuitivelly it serves to suppresscolumns. A projection creates a relation above a subset of attributes. A projection of relation R with attributes A onto a set of attributes B, where B is a subset of A. The operation created a relation with the scheme B and elements which originate from the original relation by removing attribute values from $A - B$. Even potential duplicate elements get removed. Notation syntax is as follows:

$$\Pi_{listofattributesdetachedwithacomma}(relation,).$$

Selection
Selection is a unary operation that chooses n-variables that correspond with a given predicate. Similarly to projection, which chooses a subset of attributes, a selection chooses a subset on n-variables. The lower-case Greek letter sigma (σ) is used as a symbol of selection:

$$\sigma_{selectioncondition}(relation).$$

Selection conditions may include [10]:

- constants,
- attribute names,
- arithmetic comparison ($, \neq, <, \leq, >, \geq$),
- logic operators (AND, OR, NOT).

Renaming

Renaming is a unary operation marked with the Greek letter ró (ρ). The result of using this operator on a relation is a relation identical with the original but the relation and its attributes get new names. New relation and attribute names are recorded as index following ρ. New relation names are listed first, then a list of new attribute names separated with a comma and in brackets. Generally it is possible to use renaming in the following forms [10]:

$$\rho_{newrelation(newattributenames)}(\text{relation}),$$

$$\rho_{newrelation}(\text{relation}),$$

$$\rho_{(newattributenames)}(\text{relation}).$$

Join

Join is a binary operation (symbol $\bowtie$) which is used to combine two related n-variables from two relations (tables) into one n-variable. Join is a very important operation for all relational databases with more than just one table as it enables us to work with more than one table at a time. Join corresponds with data from two or more tables based on values in one or more columns from individual tables. Join includes the following types – natural, outer (left and right) and theta join:

Theta join
Theta join enables us to combine selection and Cartesian product into one operation.

Definition: We have two relations $r_1(R_1)$ and $r_2(R_2)$. The formal record is:

$$r_1 \bowtie_\Theta r_2 = \sigma_\Theta(r_1 \times r_2).$$

Natural join

Definition: We have two relations $r_1(R_1)$ and $r_2(R_2)$. Natural join r_1 and r_2 is a relation of scheme $R_1 \cup R_2$, where each attribute A is from $R_1 \cap R_2$. The formal record is:

$$r_1 \bowtie r_2 = \Pi_{R_1 \cup R_2}(r_1 \bowtie r_1.A_1 = r_2.A_1 \wedge \ldots \wedge r_1.A_n = r_2.A_n r_2).$$

Outer join

Definition: We have two relations $r_1(R_1) a\ r_2(R_2)$. The formal record of the right join is:

$$r_1 \sqsupset \bowtie r_2 = (r_1 \bowtie r_2) \cup \big((r_1 - \Pi_{r_1.A_1,\ldots,r_1.A_n}(r_1 \bowtie r_2)) \times \{(null, \ldots, null)\}\big),$$

where each attribute A is from $R_1 \cap R_2$.

Definition: We have two relations $r_1(R_1)$ and $r_2(R_2)$. The formal record of the left join is:

$$r_1 \bowtie \sqsubset r_2 = (r_1 \bowtie r_2) \cup \big((r_2 - \Pi_{r_2.A_1,\ldots,r_2.A_n}(r_1 \bowtie r_2)) \times \{(null, \ldots, null)\}\big),$$

where each attribute A is from $R_1 \cap R_2$.

Relational database technology *offers data integrity* to meet the needs of the above mentioned data quality characteristics. Data integrity can generally be subdivided into three categories:

– entity, domain and referential.

Entity integrity provides unique identification of saved entities by means of the primary key. In relation to open data it means that individual data attributes are definitely matched with an object to which they belong. The qualitative characteristics of consistency and addressability can be realised within this integrity.

Domain integrity ensures data admissibility from the data domain point of view. If data domain is set for every attribute, the risk of both deliberate and accidental defective data entry reduced greatly and the risk of data entry omission is eliminated completely. Domain integrity is possible to use to select current data. It is a convenient means to ensure in-time quality.

Referential integrity ensures correct relations between corresponding tables within described context. This integrity corresponds with the broader open data context and together with the whole entity is useful creating aggregated data.

The purpose of data aggregation is mainly to suppress random effects impact in data with high definition which would impede analyses and interpretation, trend and relation identification (dependence, associations in particular) [11].

Aggregation is also an important means of data anonymisation. Another such method of geographical data anonymisation is random defocusing. It can be defined as adding a randomly generated value (requiring equal distribution) from the interval <−100 m, +100 m> to the X coordinate and adding a randomly generated value (requiring equal distribution) from the interval <−100 m, +100 m> to the coordinate Y. In the given application these are bee hives coordinates. It is well known that for a number of map readers point representation is difficult to digest as they tend to link e.g. Hive location with the projection despite strong warning and thus come to false interpretations.

Atomic data aggregation is possible by means of SQL tools using the clause **group by...**

Accidentally defective data, e.g. Typos when entering data, is possible to dispose of by aggregation, deliberate, e.g. Intentional entry of false data can also be removed by aggregation, but can have various causes. Varroa possitive report is not terminal to beekeepers nor is deliberate false data entry so serious. Possitive bee plague report is lethal for all infected colonies and strict quarantine measures apply within the 5 km radius which severely hinders economy of beekeeping. Deliberate false data entry is much more risky in the second case, and also ensuing damage using false open data can be greater.

When only one member of a community enters false data, this defect is minimised by aggregation. The question is what happens when all members present false data. We can document the effect of aggregation on false data on the following example: The center of the set of points. The mean center is the average X and Y coordinate for a series of points on a map. If only one point of a set has false coordinates, the error in locating the centre is much lesser. In statistics, an outlier is an observation point that is distant from other observations. An outlier may be due to variability in the measurement or it may indicate experimental error; the latter are sometimes excluded from the data set. There are software tools which eliminate outliers.

3   Results

The imperative condition of OD use is their quality which is formally specified by a data quality model included in a set of norms SQua-RE. Due to the specific character of OD this general data model is necessary to modify as follows (Fig. 1):

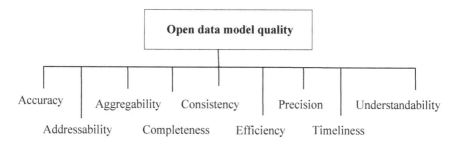

Fig. 1. The extension of open data model

The newly proposed characteristics have following roles in the quality model:

Addressability makes it possible to match partial data attributes with entities they refer to.

Aggregability enables us to aggregate data into higher level categories to increase their informative value.

4   Conclusion

The solutions presented above are based on analytical and preliminary works [12–14], in field of open data in agriculture. The result is an expert system using open data on varroa mites spread. The user interface has as a presentation layer google maps API and is freely accessible to users (beekeepers) on the website of the Czech University of Life Sciences in Prague at www.expert-varroa.czu.cz.

An example can be the solution to the following dilemma: The beekeeper decides whether it would be appropriate to start beekeeping on a larger scale in the village Opočno. Solutions to this problem can be realized by using the following SQL statement:

```
SELECT   Vill,NumbvarDistr,AvervarDistr,NumbPlague,   NumbNosema   FROM
risk_of_beekeeping WHERE Vill='Opočno';
```

Example of SQL query result:

Vill	NumbvarDistr	AvervarDistr	NumbPlague	NumbNosema
Opočno	24	45	0	strong incidence

The obtained data can be interpreted as the following recommendations: In the concerned area it is not advisable to keep bees on a larger scale. This area is determined by a higher level of infection in bees which can cause lower efficiency of this activity. The current OD can be used more effectively during the decision making processes. The key indicators can be analysed from different points of view of aggregation (the numbers or averages for districts, for regions, years etc.).

Acknowledgements. The results and knowledge included herein have been obtained owing to support from the Internal grant agency of the FEM, Czech University of Life Sciences in Prague, IGA č. 20161008 – "The potential of open data in the agricultural sector with regard to their correctness and aggregation".

References

1. Vostrovský, V., Tyrychtr, J., Ulman, M.: Potential of open data in the agricultural eGovernment. AGRIS On-line Pap. Econ. Inf. **7**(2), 103–113 (2015). ISSN 1804-1930
2. Castro, L.M.: Advanced management of data integrity: property-based testing for business rules. J. Intell. Inf. Syst. 1–26 (2014). ISSN 0925-9902
3. Wang, X., Lin, Y., Yao, G.: Data integrity verification scheme with designated verifiers for dynamic outsourced databases. Secur. Commun. Netw. **7**(12), 2293–2301 (2014). ISSN 1939-0114
4. Salman, M., Rehman, N.U., Shahid, M.: Database integrity mechanism between OLTP and offline data. In: Pan, J.-S., Chen, S.-M., Nguyen, N.T. (eds.) ACIIDS 2012. LNCS (LNAI), vol. 7197, pp. 371–380. Springer, Heidelberg (2012). https://doi.org/10.1007/978-3-642-28490-8_39
5. Motro, A.: Integrity = validity + completeness. ACM Trans. Database Syst. (TODS) **14**(4), 480–502 (1989). ISSN 0362-5915
6. Awad, E.M., Gotterer, M.H.: Database Management. Course Technology Press, Boston (1992). ISBN 0-8783-5713-0
7. Zviran, M., Glezer, C.: Towards generating a data integrity standard. Data Knowl. Eng. **32**(3), 291–313 (2000). https://doi.org/10.1016/S0169-023X(99)00042-7
8. Codd, E.F.: The Relational Model for Database Management: Version 2. Addison-Wesley, Boston (1990)
9. Embley, D.W.: Relational model. In: Liu, L., Özsu, M.T. (eds.) Encyclopedia of Database Systems, pp. 2372–2376. Springer, Heidelberg (2009). https://doi.org/10.1007/978-1-4899-7993-3_306-2
10. Group, I.: Introduction to Database Management Systems. Tata McGraw-Hill Education, New York City (2005)
11. Chainey, S., Ratcliffe, J.: GIS and Crime Mapping. Mastering GIS: Technol, Applications & Mgmnt. Wiley, Hoboken, p. 448 (2013). ISBN 9781118685198

12. Rysová, H., Kubata, K., Tyrychtr, J., Ulman, M., Vostrovský, V.: Evaluation of electronic public services in agriculture in the Czech Republic. Acta Universitatis Agriculturae et Silviculturae Mendelianae Brunensis **LXI**(2), 473–479 (2013). ISSN 1211-8516
13. Tyrychtr, J., Ulman, M., Vostrovský, V.: Evaluation of the state of the business intelligence among small Czech farms. Agric. Econ. **61**(2), 63–71 (2015). ISSN 0139-570X
14. Ulman, M., Vostrovský, V., Tyrychtr, J.: Agricultural e-government: design of quality evaluation method. Agris on-line Pap. Econ. Inf. **V**(4), 211–222 (2013). ISSN 1804-1930

Dynamical Feedforward Control of Three-Tank System

Pavol Bisták$^{(\boxtimes)}$

Slovak University of Technology in Bratislava,
Ilkovičova 3, 812 19 Bratislava, Slovakia
Pavol.Bistak@stuba.sk
http://uamt.fei.stuba.sk

Abstract. This paper deals with a new approach to the control design for a third-order nonlinear plant with time-delays. Starting from a pole-assignment proportional-derivative controller for the triple-integrator plant generalized for a constrained control and equiped by a gain scheduling to respect the nonlinear plant dynamics, its structure is used for a dynamical feedforward control that is extended by a nonlinear disturbance observer taking into account filtration of noisy signals and the non-modelled time-delays always present in control circuits. By using an example of a three-tank liquid level control in a hydraulic plant, performance of the resulting two-degree-of-freedom model reference control of nonlinear time-delayed systems with an integral action will be evaluated.

Keywords: Dynamical feedforward · Model reference control
Input constraints · Nonlinear system · Time-delay
Nonlinear disturbance observer

1 Introduction

Although the today's control theory offers numerous sophisticated control approaches that are suitable for complex control tasks, especially in applications to simple control loops one may still find new solutions improving significantly the control performance [4]. Since such simple plants represent a high percentage of all practical applications [1], there still exists a potential for a new research. Especially when considering nonlinear systems we are used to deal with different linearization techniques to deploy linear control system design (the exact linearization method based on a differential geometry [5], e.g.). However, the effect of the nonlinear dynamics typically occurs only at sufficiently rapid transitions appearing under sufficiently strong control actions leading to another typically nonlinear issue - to a control signal saturation. Furthermore, the rapid changes exhibit effect of a nonmodelled dynamics. Its identification and consideration requires usually approaches different from the identification and control design approaches dealing with the dominant dynamics. Thus, for a successful

© Springer International Publishing AG 2018
R. Moreno-Díaz et al. (Eds.): EUROCAST 2017, Part I, LNCS 10671, pp. 134–141, 2018.
https://doi.org/10.1007/978-3-319-74718-7_16

design of nonlinear control loop it is usually not enough to focus on a single traditional method built on a rigorous mathematical framework, but one needs a modular approach integrating several such approaches, each of them specialized in a closer set of problems to be solved.

The proposed work is built on integration of several basic approaches to cover all main issues of a control design: consideration of the control constraints that exhibit the nonlinear plant dynamics and use of the disturbance observer based PI and PID control [4] offering an increased robustness that is finally extended by the two-degree-of-freedom model reference control. The paper will firstly review a constrained PD controller design for a nonlinear system transformed under a modified exact linearization to a triple integrator [2]. This primary control loop will be later considered as a "master" producing an ideal control input, as well as an ideal plant output for the model reference control structure. It fulfills the task of a dynamical feedforward control that is later governed by the second control loop ensuring the plant stabilization and compensation of the model imperfections. The established two-degree-of-freedom model reference control structure is finally completed by an integral action produced by a nonlinear disturbance observer with an inverse plant dynamics. At the end the influence of the allways present non-modelled time-delays is evaluated.

2 Laboratory Hydraulic System Description

Three-tank system (Fig. 1) represent one of the most common examples of a non-linear system. The system consists of three tanks, five valves, three pumps and a reservoir. Each tank has its own outflow controlled by the respective valve. Moreover, two additional valves couple each two neighboring tanks. The three-tanks system variables and parameters are denoted according to Fig. 1. The liquid level in the first tank is denoted as h_1, with the liquid inflow q_1, tank cross-section area A_1 and the valve outflow coefficient c_1. Similarly, the variables and constants in the second and third tank are denoted as h_2, q_2, A_2, c_2 and h_3, q_3, A_3, c_3 respectively. Both two neighboring tanks are coupled through the

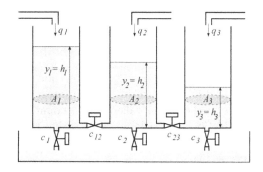

Fig. 1. Three-tank laboratory system and its schematic drawing

valves characterized by the coefficients c_{12} or c_{23}. The system can be modeled by differential equations derived via the application of the mass conservation law. That is

$$\dot{h}_1 = \frac{1}{A_1}q_1 - c_1\sqrt{h_1} - c_{12}\sqrt[\pm]{h_1 - h_2}$$

$$\dot{h}_2 = \frac{1}{A_2}q_2 - c_2\sqrt{h_2} + c_{12}\sqrt[\pm]{h_1 - h_2} - c_{23}\sqrt[\pm]{h_2 - h_3}$$

$$\dot{h}_3 = \frac{1}{A_3}q_3 - c_3\sqrt{h_3} + c_{23}\sqrt[\pm]{h_2 - h_3}$$

$$y_1 = h_1$$

$$y_2 = h_2$$

$$y_3 = h_3 \tag{1}$$

where the symbol $\sqrt[\pm]{\cdot}$ stands for

$$\sqrt[\pm]{z} = \begin{cases} \sqrt{z} & \text{if } z \geq 0 \\ -\sqrt{-z} & \text{if } z < 0 \end{cases} \tag{2}$$

Each valve can be opened or closed and this can be controlled from a computer as well as the power of each pump. In this paper the aim is to control the liquid level in the third tank (i.e. $y = y_3 = h_3$) when only the first pump is acting ($q_2 = q_3 = 0$) and the first and second valves are closed ($c_1 = c_2 = 0$).

3   Design of Nonlinear Feedback Controller with Input Saturation

This section briefly describes the designed control algorithm. The design process is separated into two steps. First the nonlinear hydraulic system is linearized using exact linearization method. Then the control algorithm for the linearized system with input constraints - the constrained PD (CPD) controller is briefly reviewed [2]. The last part of this section is focused on the design of a nonlinear disturbance observer in order to gain a possibility to suppress an input disturbance. A possible control structure is depicted in the Fig. 2.

3.1   Exact Linearization Method

Using the exact linearization method [5] the nonlinear system (1) can be expressed in the form of a triple integrator and then the later derived control can be applied after taken into account the change of control limits caused by linearization feedback. After denoting the Lie derivative of a scalar function y along a vector field $\mathbf{f}$ by $L_f y$ the desired exact linearization feedback can be expressed as

$$q_1 = \frac{u - L_f^3 y}{L_g L_f^2 y} , \tag{3}$$

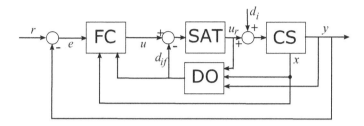

Fig. 2. Simplified constrained feedback control block diagram - feedback controller (FC), saturation (SAT), disturbance observer (DO), controlled system (CS)

where

$$\mathbf{f} = \begin{pmatrix} -c_{12}\sqrt{h_1-h_2} \\ c_{12}\sqrt{h_1-h_2} - c_{23}\sqrt{h_2-h_3} \\ c_{23}\sqrt{h_2-h_3} - c_3\sqrt{h_3} \end{pmatrix} , \quad \mathbf{g} = \begin{pmatrix} \frac{1}{A_1} \\ 0 \\ 0 \end{pmatrix} \tag{4}$$

and new control constraints for u are

$$U_1 = L_g L_f^2 y \, Q_{11} + L_f^3 y$$
$$U_2 = L_g L_f^2 y \, Q_{12} + L_f^3 y \tag{5}$$

where Q_{11} is the minimal value of q_1 and Q_{12} denotes its maximal value.

3.2 Constrained Controller Design for Triple Integrator

The design of the controller that respects input saturations is based on a nonlinear dynamics decomposition [2]. This is done in the phase space and it enables to express the state of the system $\mathbf{x}$ as a sum of subsystem states $\mathbf{x}_i, i = 1, ..., n$

$$\mathbf{x} = \sum_{i=1}^{n} \mathbf{x}_i(p_i, t_i) \quad , \quad \mathbf{x}_i(p_i, t_i) = e^{-\mathbf{A}t_i}\mathbf{v}_i p_i + \int_0^{-t_i} e^{\mathbf{A}\tau}\mathbf{b} d\tau p_i \tag{6}$$

where $\mathbf{A}$ is the system matrix corresponding to the triple integrator, $\mathbf{b}$ is the input vector, $\mathbf{v}_i = (1/\alpha_i^n, ..., 1/\alpha_i)^t$ represent eigenvectors (with different eigenvalues α_i), $U_j, j = 1, 2$ are control limits (5), p_i and t_i are parameters of the decomposition that have to be solved from this system of nonlinear algebraic equations under following conditions

$$\begin{aligned} &\text{if} \quad p_i \in \left[U_1 - \sum_{k=1}^{i-1} p_k \ , \ U_2 - \sum_{k=1}^{i-1} p_k \right] \quad \text{then} \quad t_i = 0 \\ &\text{else} \quad p_i = U_j - \sum_{k=1}^{i-1} p_k \quad \text{and} \quad 0 < t_i \le t_{i-1} \end{aligned} \tag{7}$$

for $i = 1, ..., n$, $j = 1, 2$, $t_0 = \infty$. Then the control law of the CPD controller can be computed as

$$u = \sum_{i=1}^{n} p_i \tag{8}$$

In the case of the triple integrator it is valid $n = 3$. If there are three different eigenvalues with ordering $\alpha_3 < \alpha_2 < \alpha_1 < 0$ one gets the system of three algebraic equations (9) that is necessary to be solved under the condition (7) in order to evaluate the parameters p_1, p_2 and p_3 that according to (8) determine the resulting control law u.

$$\mathbf{x} = \begin{pmatrix} \sum_{i=1}^{3}\left(p_i\left(\frac{1}{\alpha_i^3} - \frac{t_i}{\alpha_i^2} + \frac{1}{2}\frac{t_i^2}{\alpha_i}\right) - \frac{1}{6}p_i t_i^3\right) \\ \sum_{i=1}^{3}\left(p_i\left(\frac{1}{\alpha_i^2} - \frac{t_i}{\alpha_i}\right) + \frac{1}{2}p_i t_i^2\right) \\ \sum_{i=1}^{3}\left(\frac{p_i}{\alpha_i} - p_i t_i\right) \end{pmatrix} \tag{9}$$

To solve this system analytically a simple modification in the nonlinear dynamics decomposition is necessary. For further details readers are referred to [2].

3.3   Nonlinear Disturbance Observer Design

Within a modular approach to constrained PID control one gets a disturbance observer based constrained PID controller by augmenting the CPD controller described in the previous subsection (FC block in Fig. 2) by a disturbance observer (DO block in Fig. 2). One possible solution based on inversion of the dynamics of the first tank is for the three-tank system depicted in Fig. 3 [3]. When solved for an input disturbance d_i from the first equation of (1) it yields

$$\hat{d}_i = A_1\left[\frac{dh_1}{dt} + c_{12}\sqrt{h_1 - h_2}\right] - q_1(t - T_d) \tag{10}$$

Then, by considering a first order filtration it follows

$$\frac{d}{dt}d_f = \frac{1}{T_f}\left[\hat{d}_i - d_f\right] = \frac{1}{T_f}\left[A_1\frac{dh_1}{dt} + A_1 c_{12}\sqrt{h_1 - h_2} - q_1(t - T_d) - d_f\right] \tag{11}$$

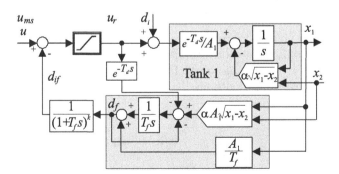

Fig. 3. Reconstruction of an input disturbance d_i by an inverse filtered dynamics of the first tank; $\alpha = c_{12}$, $x_1 = h_1$, $x_2 = h_2$, $u_{ms} = u$

An integration of this equation gives finally

$$d_f = \frac{A_1}{T_f}h_1 + \frac{1}{T_f}\int (A_1c_{12}\sqrt{h_1 - h_2} - q_1(t - T_d) - d_f)dt \qquad (12)$$

and imposes the structure of a nonlinear disturbance observer in Fig. 3. The reconstructed disturbance signal may yet be filtered by additional $k = n - 1$ filters with the time constant T_f.

Compensation of an input disturbance d_i has to be respected also by a shift of the transformed control constraints U_j (5) used by the constrained controller (8) following the requirement

$$u_r = sat(u - d_{if}) \qquad (13)$$

In the Fig. 4 one can see oscillating time responses of the designed feedback controller due to the time delays that were not taken into account during the design.

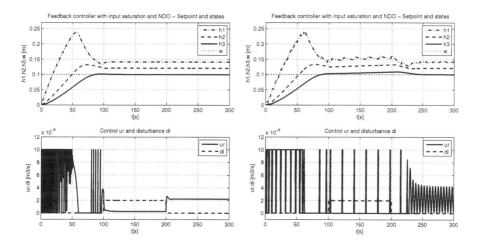

Fig. 4. Feedback CPD controller - output h_3, states h_1 and h_2 and control u_r time responses. Time delays $T_d = 0.1\,\mathrm{s}$ (left) and $T_d = 1\,\mathrm{s}$ (right).

4   Dynamical Feedforward Control

In order to improve the behaviour under time delays the dynamical feedforward control (DFC) has been proposed (Fig. 5). The model with controller (MC block) produces two signals. First one is the output of the model controller $u_m = q_1$ (3) and the second one is the output of the model itself y_m. The signal y_m is compared with the output of the real system y and the established difference between the model and the real system creates the stabilizer control error denoted as y_s

$$y_s = y_m - y \qquad (14)$$

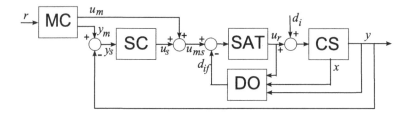

Fig. 5. Dynamical feedforward control block diagram - model control (MC), stabilizing controller (SC), saturation (SAT), disturbance observer (DO), controlled system (CS)

This is multiplied by the gain K_s as a parameter of the second P-controller and that produces the stabilizing control action

$$u_s = K_s \, y_s \tag{15}$$

In the block diagram this is represented by SC block. Then the control action u_{ms} given as a sum of u_m and u_s

$$u_{ms} = u_m + u_s \tag{16}$$

is substracted by the filtered output of the disturbance observer d_{if}. The real control action u_r resulting from MC and DO is

$$u_r = sat(u_{ms} - d_{if}) = sat(u_m + u_s - d_{if}) \tag{17}$$

This influences also the constraints (5) that must be finally modified

$$U_j = L_g L_f^2 y (Q_{1j} - u_s + d_{if}) + L_f^3 y \; ; \; j = 1, \, 2 \tag{18}$$

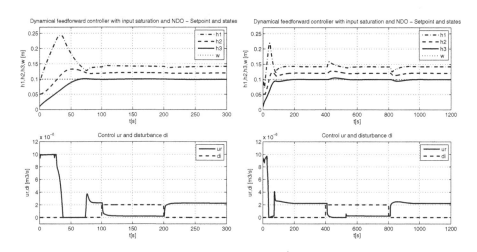

Fig. 6. Dynamical feedforward controller - output h_3, states h_1 and h_2 and control u_r time responses. Time delays $T_d = 1\,\text{s}$ (left) and $T_d = 10\,\text{s}$ (right).

The complete control signal evaluation consists of expressing control u of the constrained PD controller (8) with respect to the constraints (18), then of a calculation of the model control value $u_m = q_1$ (3), later of evaluating the stabilizing control u_s (15) and disturbance d_{if} (12) and finally expressing the real control value u_r (17). Improved time responses can be seen in the Fig. 6 where in comparision with the previous case (Fig. 4) there are no oscillations also for higher time delays.

5   Conclusions

A new design of a 2DOF model reference control has been carried out on the three-tank hydraulic plant including the constrained PD controller, stabilizing controller, disturbance observer respecting time delays. The simulation results have proved the correctness of the developed control applied for the different values of time delays. It is necessary to mention that the model of the controlled system must be well known because the designed controller is sensitive to any plant-model mismatch. The advantage of the 2DOF structure is that it enables to tune the controller's parameters independently as concerning setpoint and disturbance reactions.

Acknowledgments. This work has been partially supported by the grants VEGA 1/0937/14 and APVV-0343-12.

References

1. Åström, K.J., Hägglund, T.: Advanced PID Control. ISA, Research Triangle Park (2006)
2. Bisták, P.: Time sub-optimal control of triple integrator applied to real three-tank hydraulic system. In: Moreno-Díaz, R., Pichler, F., Quesada-Arencibia, A. (eds.) EUROCAST 2015. LNCS, vol. 9520, pp. 25–32. Springer, Cham (2015). https://doi.org/10.1007/978-3-319-27340-2_4
3. Bistak, P., Huba, M.: Three-tank virtual laboratory for input saturation control based on matlab. IFAC-PapersOnLine **49**(6), 207–212 (2016)
4. Huba, M.: Comparing 2DOF PI and predictive disturbance observer based filtered PI control. J. Process Control **23**(10), 1379–1400 (2013)
5. Isidori, A.: Nonlinear Control Systems, 3rd edn. Springer, New York (1995). https://doi.org/10.1007/978-1-84628-615-5

Automatic Inventory of Multi-part Kits Using Computer Vision

A. J. Rodríguez-Garrido, A. Quesada-Arencibia$^{(\boxtimes)}$,
J. C. Rodríguez-Rodríguez, C. R. García, and R. Moreno-Díaz jr

Institute for Cybernetic Science and Technology,
University of Las Palmas de Gran Canaria, Las Palmas de Gran Canaria, Spain
`jgadir2000@gmail.com`, {`alexis.quesada`, `ruben.garcia`,
`roberto.morenodiaz`}`@ulpgc.es`, `jcarlos@iuctc.ulpgc.es`

Abstract. A prototype tool for the detection, segmentation, classification and counting of Lego pieces based on the OpenCV artificial vision library is presented. This prototype arises before the need to automate the complex and tedious task of the inventoried one of Lego kits of the MindStorm serie.

In the process of detection and segmentation there have been used skills of threshold and the algorithm of Watershed segmentation. For the process of classification and count have been used two different approaches in the securing of the vector of characteristics of the image: BOW and Naive; as well as vector machines support (SVM) for the classification.

Keywords: Automation · OpenCV · Lego · Prosecution of images
Segmentation · Extraction of characteristics · Classification

1 Introduction

The intention of our work is to automate the classification and re-count of Lego pieces. The MindStorms serie of Lego is a powerful tool for the educational robotic [1], the conservation of the material in good conditions involves a tough logistics, which completes time and energy, and which nevertheless is necessary. The action to count pieces allows to determine when there are or not missed pieces or if we have all the pieces necessary for the class.

The solution that is proposed belongs to the field of the artificial vision. One works with images of scenes that contain interest objects, in our case Lego pieces of the kit. We must be capable of detecting them, of classifying them and of counting them. This process of visual perception must acquire images, detect and segment the interest objects, and finally classify them and count them (to see Fig. 1).

The solution that is proposed is characterized by a modular structure, represented by a module of "Acquisition of Images" of physical nature and two modules of nature clearly software that there shape a library or bookstore (modules of "Detection and segmentation" and "Classification and count").

© Springer International Publishing AG 2018
R. Moreno-Díaz et al. (Eds.): EUROCAST 2017, Part I, LNCS 10671, pp. 142–149, 2018.
https://doi.org/10.1007/978-3-319-74718-7_17

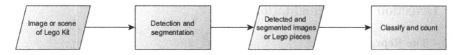

Fig. 1. Process of visual perception.

2 Images Acquisition: The Module of Acquisition of Images

For our intention, it is necessary to be provided with some device that should allow to realize image[1] apprehensions with the Lego pieces placed in a qualified space.

In this respect, the optical device will have to assemble a few certain characteristics in order to guarantee the quality of the above mentioned apprehensions. Therefore, at the time of praising us for an option or other one we will have to value aspects as the space of color that we are going to need (gray or of color), resolution, type of sensor (CCD or CMOS), focal distance (a lens adapted to its place and Surface of work), ... For this intention, we have integrated the following elements:

- Kaiser Platform in which the Lego pieces are placed of the kit to audit. Ideal work condition: Surface of 35 cm^2 and 50 cm focal distance.
- Webcam[2] of Logitech connected to a terminal type PC where the apprehensions are stored and with facility for the remote access and distribution of the same ones.

Fig. 2. Area of work delimited by a rectangle.

In order to assure the maximum level of detail of the image, we establish the following working conditions (to see Fig. 2):

- Images in scale of gray (major contrast).
- 43.9 cm focal distance.
- Surface of work of 8.5 cm^2

---

[1] Examples of optical devices are a camera, video camera, Ethernet camera, web-cam...

[2] Initially it was decided in favor by the Mako high resolution Ethernet camera. However, on not having counted with optical adapted to be able to work with the mentioned camera, it was necessary to use an optical device of worse resolution, the Logitech webcam.

3 Detection and Segmentation

3.1 The Reprocess

Once we have obtained for some of the devices described in the previous paragraph the set of images to be processed, the following step is to be able to detect and segment to (extract) each of the objects from interest (Lego pieces of the kit) presents in the above mentioned images.

For it, it is necessary to realize previously a preprocess of the image, dependent on the context in which the apprehensions have been realized and that allows us to adapt it to the different algorithms or skills of segmentation.

This preprocess is usually intimately tied to the environment or context in which the acquisition of the images is realized. In our proposal, it is a specific procedure, of manual and key adjustment, since it intervenes both in the securing of the scoreboards that uses our segmentation process is realized in a controlled environment, there are used the following skills of preprocess of image (to see Fig. 3).

- Conversion to scale of gray (space of color needed by the elected algorithm of segmentation).
- Equalization of the histogram of the image (it improves the contrast of the image).
- Morphologic Operation of dilation (it eliminates hollows or spaces inside the objects).
- Inverse binary threshold, using like value of threshold obtained by the method of Otsu [2].

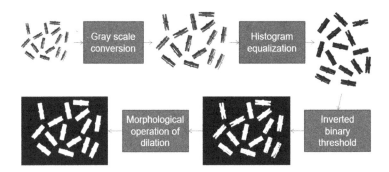

Fig. 3. Steps of the preprocess.

3.2 The Segmentation

The segmentation allows us the extraction and therefore detection of one or more regions or objects of interest of the image (Lego pieces) base on a criterion of discontinuity or similarity. This is; to go on from an image or scene containing the Lego pieces of the kit to be audited to a set of Lego objects or pieces detected and extracted (segmented) in that image. This process in divided into:

1. Securing of the scoreboards of our algorithm of segmentation.
2. Application of our algorithm of segmentation to the image or scene.

As segmentation algorithm, we have used the algorithm of segmentation based on regions Watershed [3, 4] for its confirmed hardiness and the results obtained in our preliminary tests. This algorithm considers an image in scale of gray like a topographic relief, where the level of gray of a pixel is interpreted as its altitude in the relief.

Although the segmentation process concludes with the securing of the set of detected and abstracted images, there has been included the option to normalize the above mentioned set of images, that is to say, all resulting images are of the same size. This decision justifies to the being an aspect that influences in the behaviour and yield of the process carried out by the Classification module and count.

The flow of the process is described in the Fig. 4 and can be appreciated in the Fig. 5.

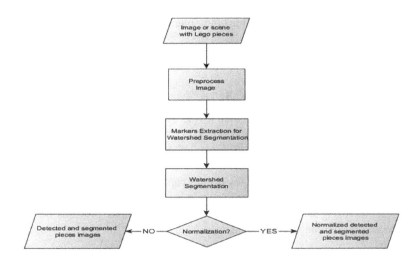

Fig. 4. Flow of the process carried out by the detection and segmentation module.

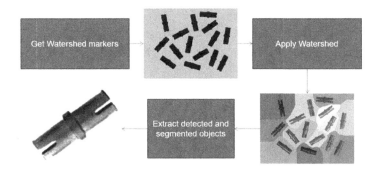

Fig. 5. Different stages of the segmentation.

4   Classification and Count

We come to the module entrusted to realise the process of classification and count of the Lego pieces of the kit to be audited. We will spend of the set of normalized images obtained in the stage before to a list containing the number of pieces for category of the kit recognized, strangers and those that were not possible of classifying to more than one category to be assigned.

This process consists of the phase of securing of vectors of characteristics from the descriptors of image and the phase of classification and count in the strict sense.

4.1   Phase of Securing of the Vector of Characteristics

Image descriptors

The images descriptors are mathematical gadgetry that allow us to identify of univocal form an image, in spite of presenting transformations due to changes of lighting, noise, scale, rotation… Bearing in mind the need for visual invariability opposite to rotations (pieces interspersed in arbitrary position) our solution he supports the use of the following descriptors of images:

– Dense detector and like extractor well SIFT or SURFING
– KAZE and AKAZE
– Moments of HU [5]

This support flexibility allows to be studied by the descriptor who better adapts himself to our problem.

SIFT [6] and SURFING [7] are examples of descriptors based on histogram of gradients of image (direction and magnitude of every pixel). There is an invariant opposite to changes of lighting, but they do not capture geometric information.

KAZE [8] and AKAZE [9] are descriptors who have appeared recently. They are characterized for using not linear spaces of scale which, unlike previous descriptors, respect the natural outline of the objects.

Approaches

For this phase two approaches were used (to see Fig. 6):

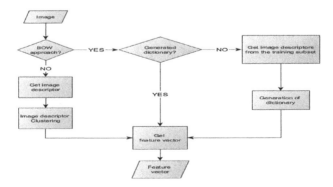

Fig. 6. Flow of the process carried out by the phase of securing of characteristics.

Confirmed standard approach

There is used the generative model BOW [10] to generate the characteristics vector. On having used a words dictionary, we obtain a major standardization of the results and reduce the redundancy in the training information. The construction of the dictionary is realized from the obtained descriptors of images of a subcommittee of images of training.

"Naive" experimental approach

The characteristics vector is obtained from a clustered of the descriptor of image. The main motives of using this approach are both its simplicity and theoretically a less training time with regard to the previous approach. This approach is one of the original contributions of this work.

4.2   Phase of Classification and Count

To the being the mastery of our problem well-known, finite and limited size, we use like classifier the algorithm of supervised learning SVM [11]. Therefore, a phase previous to training is needed. The training information will be obtained from the set of vectors of characteristics associated with the set of images of normalized[3] training of every SVM or classifier.

As soon as the set of sorters was trained (as many as categories of the Lego kit), we proceed to the classification and count of the Lego pieces of the kit to audit. For it, from the vector of characteristics associated with every piece, we obtain the result of the-classifiers. Based on the obtained result we will proceed to increase in a unit (to see Figs. 7 and 8) the value of:

- the category that represents it if only a classifier identifies it.
- the special category "doubtful" if more than one classifier identifies it.
- the special category "known" in case of not being identified.

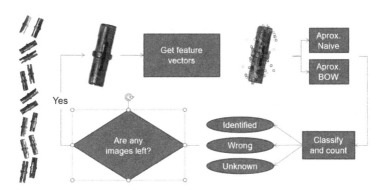

Fig. 7. Diagram of the classification and count.

---

[3] It is important that both sets of images (training and classification) are normalized, that is to say, the images that integrate it are of the same size. In the opposite case, we might obtain vectors of characteristics that they do not represent from trustworthy form to the above mentioned sets.

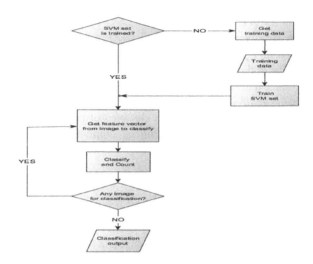

Fig. 8. Flow of the process carried out by the classification phase and count.

5 Contributions

A vision system has appeared of artificial highly configurable, with two approach ways for the classification and count, one of which has been proposed by the authors.

Between others they are used of binary images descriptors novel like KAZE and AKAZE not supported completely at the moment of the design and implementation of the described solution.

As a result of the previous thing we have:

- The possibility of using descriptors of images not supported by the bookstore OpenCV [12] whenever its interface is respected.
- The implementation of a skill of binary clusterization k-medoids, as well as its later use on the part of two approaches of classification and count.

This approach proposed by the authors or "Naive" approach, it is provided with a high valuation of reliability (number of pieces correctly identified) according to the results of a wide battery of tests of combination of methods (procedure used for the securing of the vector of characteristics, image descriptor) and configuration (size of the cluster, type of nucleus SVM). The results, as well as its conclusions, the authors hope to see it published in a future article.

References

1. Rodríguez, J.C.R., Martín-Pulido, E., Padrón, V.J., Alemán, J.A., García, C.R., Quesada-Arencibia, A.: Ciberlandia: an educational robotics program to promote STEM careers in primary and secondary schools. In: Auer, M.E., Guralnick, D., Uhomoibhi, J. (eds.) ICL 2016. AISC, vol. 544, pp. 440–454. Springer, Cham (2017). https://doi.org/10.1007/978-3-319-50337-0_42

2. Otsu, N.: A threshold selection method from gray-scale histogram. IEEE Trans. Syst. Man Cybern. **9**, 62–66 (1979)
3. Beucher, S., Lantuejoul, C.: Use of watersheds in contour detection. In: International Workshop on Image Processing: Real-time Edge and Motion Detection/Estimation (1979)
4. Meyer, F.: Color image segmentation. In: International Conference on Image Processing and its Applications (1992)
5. Hu, M.K.: Visual pattern recognition by moment invariants. IRE Trans. Inf. Theory **8**, 179–187 (1962)
6. Lowe, D.G.: Distinctive image features from scale-invariant keypoints. Int. J. Comput. Vis. **60**, 91–110 (2005)
7. Bay, H., Ess, A., Tuytelaar, T., Van Gool, L.: Speeded-up robust features (SURF). Comput. Vis. Image Underst. **110**(3), 346–359 (2004)
8. Alcantarilla, P.F., Bartoli, A., Davison, A.J.: KAZE Features. In: Fitzgibbon, A., Lazebnik, S., Perona, P., Sato, Y., Schmid, C. (eds.) ECCV 2012. LNCS, vol. 7577, pp. 214–227. Springer, Heidelberg (2012). https://doi.org/10.1007/978-3-642-33783-3_16
9. Alcantarilla, P.F., Nuevo, J., Bartoli, A.: Fast explicit diffusion for accelerated features in nonlinear scale spaces. Trans. Pattern Anal. Mach. Intell. **34**(7), 1281–1298 (2011)
10. Csurka, G., Dance, C.R., Fan, L., Willamowski J., Bray, C.: Visual categorization with bags of keypoints. In: European Conference on Computer Vision (ECCV). Prague (2004)
11. Chih-Chung, C., Chih-Jen, L.: LIBSVM: a library for support vector machines. ACM Trans. Intell. Syst. Technol. (TIST) **2**(3), 27 (2011)
12. OpenCV Homepage. http://opencv.org/. Accessed 17 July 2017

Steganographic Data Heritage Preservation Using Sharing Images App

Zenon Chaczko[1(✉)], Raniyah Wazirali[1], Lucia Carrion Gordon[1], and Wojciech Bożejko[2]

[1] FEIT Faculty of Engineering and IT, UTS University of Technology, Sydney, NSW, Australia
{Zenon.Chaczko,Raniyah.Wazirali,}@uts.edu.au,
Lucia.CarrionGordon@student.uts.edu.au
[2] Department of Automatics, Mechatronics and Control Systems, Faculty of Electronics, Wrocław University of Science and Technology, Wrocław, Poland
wojciech.bozejko@pwr.edu.pl

Abstract. With the advent of smartphones, we have the ability to take a photo and upload it to the internet whenever we desire. Hence, it may be of key importance to include metadata of the image for heritage preservation. This project focuses on heritage concepts and their importance in every evolving and changing digital domain where system solutions have to be sustainable, sharable, efficient and suitable to the basic user needs. Steganography provides a feasible and viable solution to ensure secure heritage preservation of the multimedia content. By embedding information directly into the image, the information about the image will never be lost, as it is not separated from its original source. The aim of the paper is to demonstrate this aspect via an image sharing app that allows users to exchange messages and personalized information that is embedded in the image such that it is inaccessible without knowing their keys.

1 Introduction

Creating images and uploading them has become easy with the introduction of smart phones. Basic metadata of an image could be considered unsuitable for the purpose of personal information due to its easy access through websites and other applications that can extract metadata from an image. The heritage term is defining as the crucial and central part of the research; we can refer it to 'heritage is those items and places that are valued by the community and is conserved and preserved for future generations [1]. The data is often in isolation. However, the data needs to be with the connections and relationships. It gives the meaning of the information. If that heritage is not preserved, the information can be lost forever. The aim of this research paper demonstrates how steganography can be used for embedding critical data into the image that is readable via a dedicated visualization tools [2,3].

© Springer International Publishing AG 2018
R. Moreno-Díaz et al. (Eds.): EUROCAST 2017, Part I, LNCS 10671, pp. 150–156, 2018.
https://doi.org/10.1007/978-3-319-74718-7_18

2   An Overview of Steganographic Sharing

The development of the preservation framework is related with the value of information and based in the improvement of workflow model. Steganography is the practice of concealing information within another piece of information. This information can later be retrieved by someone with the right key. Figure 1 shows the basic architecture of steganography. The most common way of hiding information within images is by selecting an area using a key provided by the user, and then changing the least significant bits of the color of the pixels in that area to that of the bits of information the user wants to hide. This method is undetectable to the naked eye, and can be difficult to discover even programmatically.

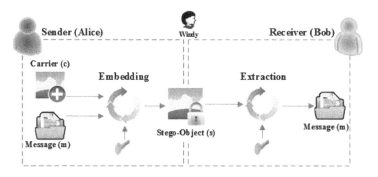

Fig. 1. SDHP workflow model

Serendipity is also a process which leads to a serendipitous finding, and here the insight is the key element. Serendipity as a process starts with something unexpected or odd happening – an event, result, encounter or situation/context – that triggers insight. And when this insight will eventually leads to value creation for the individual, community or company, then we are wittnessing serendipity. In the global business world great insights are rare and therefore so valuable. The competitive edge can often be achieved by only one single insight well executed. Therefore understanding serendipity in all forms becomes a vital part of expertise in SDHP.

The images produced by the Photo App will have messages embedded in them, so it's somewhat irrelevant in the context of an image sharing app. Human cultural heritage, documents and artifacts increase regularly and place Data Management as a crucial issue. Figure 2 shows the Steganography Data Heritage Preservation (SDHP) Workflow Model. The first stage involves exploration and approaches based on review of recent advances. The second stage involves adaptation of architectural framework and development of software system architecture in order to build the system prototype. Increasing regulatory compliance mandates are forcing enterprises to seek new approaches to managing reference

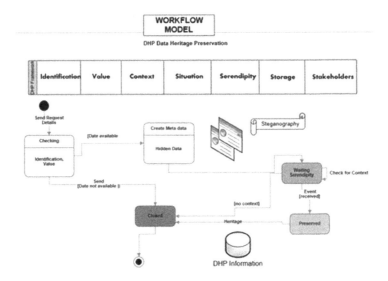

Fig. 2. SDHP workflow model

data. The approach of tracking reference data in spreadsheets and doing manual reconciliation can be time consuming and error prone. As various organizations merge and businesses evolve, reference data must be continually mapped and merged as applications are linked and integrated, accuracy and consistency, realize improved data quality, strategy lets organizations adapt reference data as the business evolves. It is important to highlight the type and structure of data. Through the time preserving digital information has a process for designing a practical system for managing massive amounts of critical data. An example of such a process can be found in [1,5]. The way to improve the understanding of the methodology, the information has to consider two dimensions: access dimension and cognitive dimension. Both of them have the level of importance in terms of the results. As a methodology of treatment digital preservation, it could be risky even when the strategy could develop a clear idea of digital resources and digital artifacts. Steganography is a powerful and effective multimedia tool for the Digital Heritage Preservation. This can be clearly shown in the presented case study of a successful implementation of steganography technique in the Photo App that allows editing photos, adding messages and uploading the gallery for viewing, exchanging messages and searching for images by other users.

2.1   Patterns for DHP

What is heritage conservation? A brief overview Heritage conservation doesn't mean freezing a building in time, creating a museum or tying the hands of property owners so they can't do anything with their properties. Instead, it seeks to maintain and thereby increase the value of buildings by keeping their original built form and architectural elements, favouring their restoration rather than replacement and, when restoration is impossible, recreating scale, period and character.

Heritage preservation and designation increases property values, both of the restored building and surrounding properties.

Heritage preservation can be a draw to tourism and helps businesses attract customers. Retaining the historic integrity of a neighbourhood or downtown attracts people just for that ambiance alone and that attracts business. A small town without a heritage main street attracts no one.

Restoration reduces construction and demolition waste and uses less than half the energy of new construction. Heritage preservation is an investment in our community that rewards us today and leaves an invaluable resource for future generations.

Deep learning (deep machine learning, or deep structured learning, or hierarchical learning, or sometimes DL) is a branch of machine learning based on a set of algorithms that attempt to model high-level abstractions in data by using multiple processing layers with complex structures, or otherwise composed of multiple non-linear transformations.

Deep learning is part of a broader family of machine learning methods based on learning representations of data. An observation (e.g., an image) can be represented in many ways such as a vector of intensity values per pixel, or in a more abstract way as a set of edges, regions of particular shape, etc. Some representations make it easier to learn tasks (e.g., face recognition or facial expression recognition) from examples. One of the promises of deep learning is replacing handcrafted features with efficient algorithms for unsupervised or semi-supervised feature learning and hierarchical feature extraction.

Research in this area attempts to make better representations and create models to learn these representations from large-scale unlabeled data. Some of the representations are inspired by advances in neuroscience and are loosely based on interpretation of information processing and communication patterns in a nervous system, such as neural coding which attempts to define a relationship between various stimuli and associated neuronal responses in the brain [3].

Various deep learning architectures such as deep neural networks, convolutional deep neural networks, deep belief networks and recurrent neural networks have been applied to fields like computer vision, automatic speech recognition, natural language processing, audio recognition and bioinformatics where they have been shown to produce state-of-the-art results on various tasks.

2.2   Steganographic Data Heritage App

Steganography provides a feasible and viable solution to these problems. By embedding information directly into the image, the information about the image will not be lost, if it is separated from its original source. The aim of the capstone is to demonstrate this aspect via an image sharing app that allows users to exchange messages and personalized information that is embedded in the image such that it is inaccessible without knowing their keys, and metadata that can be used by search engines to categorize the image and its contents. The developed application will provide a platform for users to create an account, share images, embed information within the image, and search for images other users have uploaded. This will allow great opportunity for data heritage through embedding the metadata to the multimedia. Therefore, the metadata can travel from software to another without any lost.

The aim of this case study is to create an image sharing application that shows the power of using steganography in multimedia. There are a number of features aimed at both the user taking photos, and the user viewing them. Users will be able to create accounts, post pictures, put comments with the pictures, aimed at general or specific users, search posted photos, and share posted photos.

This case study seeks to develop and implement a method of embedding multiple steganographically hidden messages within a single image. While there are methods to do this out there currently, they can only store a limited number of messages. This project aims to create a method scalable to potentially use all of the applicable space in the image to store different messages.

This project also encompasses an application using the steganographical method we will create. This will provide a working example of how this method can be used in the real world, with some additional features such as a search engine.

2.3   Values in Context

The search for values and meaning has become a pressing concern [2].

The issue is what to conserve and how to conserve. Values are an important determining factor in the current practices and future prospects of the conservation field. The capacity of the conservation field to enrich cultural life and the visual arts in societies worldwide. Try to understand the processes—specific and general—by which material heritage conservation functions in the context of modern society, to look at the kinds of social and cultural dynamics making the greatest impact on conservation's role in society, presently and in the future; and to consider ideas, concepts, and research themes study.

3   Future Projects

Data preservation: Digitalization of the Heritage, the result of proposal is to have like a result of the experimental work, a reliable Framework for measure the digital age of the information and patterns that qualified usability and accessibility of the data. The best pathway for commercialization could be some of them (Fig. 3).

- Commercial Business Structure like a Partnership assuming the cost of the investment and the taxes that generate the buying of the equipment for implementation of the scanning in the digitalization.
- Initial Public Offering IPO, because the application of the data preservation could be focus on Entities from Government and Historical materials and artifacts that sometimes have to be preserved with a public responsibility.

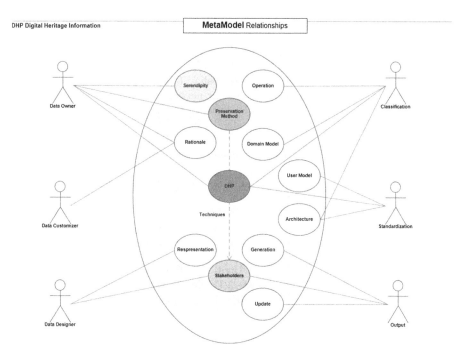

Fig. 3. DHP relationships

– This research could have through the market with POCs proof of concepts, showing the advantages and challenges of the new solution. In this case the relationship between the process and the final patterns there is a model.

4   Conclusion

The context, relation and situation of the Serendipitous Heritage are impressive relevant in the research because it gives the sense of the future of the Knowledge in the World. Through the Socio - Technical, Cultural fields, the process of Preservation will do a contribution for more research. The use of tools and techniques like Steganography, the concepts of Software Architecture will have a real approach and meaningful characteristics for the relevance of the investigation. Steganographic photo sharing app was fairly successful in implementing its goals. It clearly shows the power of steganography as a useful multimedia tools. The Photo App fulfilled all of the Software Requirement Specifications, and a demonstrated ways in which steganography can be utilised outside simple secret messaging.

References

1. Bożejko, W., Hejducki, Z., Wodecki, M.: Applying metaheuristic strategies in construction projects management. J. Civ. Eng. Manag. **18**(5), 621–630 (2012)
2. Challa, S., Gulrez, T., Chaczko, Z., Paranesha, T.N.: Opportunistic information fusion: a new paradigm for next generation networked sensing systems. In: 8th International Conference on Information Fusion, vol. 1, pp. 720–727 (2005)
3. Gulden, J.: Methodical support for model-driven software engineering with enterprise models. Universität Duisburg-Essen 1-332 (2013)
4. Johnson, N.F., Jajodia, S.: Exploring steganography: seeing the unseen. Computer **31**(2), 26–34 (1998)
5. Smutnicki, C., Bożejko, W.: Parallel and distributed metaheuristics. In: Moreno-Díaz, R., Pichler, F., Quesada-Arencibia, A. (eds.) EUROCAST 2015. LNCS, vol. 9520, pp. 72–79. Springer, Cham (2015). https://doi.org/10.1007/978-3-319-27340-2_10
6. UNESCO: Information Document Glossary of World Heritage Terms (1996). http://whc.unesco.org/archive/gloss96.htm
7. Wazirali, R.R., Chaczko, Z., Kale, A.: Digital multimedia archiving based on optimization steganography system. In: Asia-Pacific Conference on Computer Aided System Engineering (APCASE 2014). IEEE Press, pp. 82–86 (2014)

Stereoscopic Rectification Brought to Practical: A Method for Real Film Production Environments

Roman Dudek[3], Carmelo Cuenca-Hernández[1(✉)], and Francisca Quintana-Domínguez[2]

[1] Departamento de Informática y Sistemas,
Universidad de Las Palmas de Gran Canaria, Campus de Tafira,
35017 Las Palmas de Gran Canaria, Canary Islands, Spain
`carmelo.cuenca@ulpgc.es`
[2] Instituto Universitario de Ciencias y Tecnologías Cibernéticas,
Universidad de Las Palmas de Gran Canaria, Campus de Tafira,
35017 Las Palmas de Gran Canaria, Canary Islands, Spain
`francisca.quintana@ulpgc.es`
[3] Universidad de Las Palmas de Gran Canaria, Campus de Tafira,
35017 Las Palmas de Gran Canaria, Canary Islands, Spain
`romandudek@gmail.com`

Abstract. This work proposes a method for geometric and color rectification of stereoscopic images. The method is based on the Optical Flow of every stereoscopic image pair and it follows global and local approximations for both rectifications. Although the method is automatic, it also allows the user to interact in order to obtain the best result. The method has been fully tested and it is suitable to be included in an industrial film postproduction environment.

Keywords: Stereoscopy · Rectification · Alignment · Interactive Application · Optical Flow · GUI

1 Stereoscopic Filming: Background and Main Issues

The need for the stereoscopic tools described in this article, and intended to be used in professional 3D filming, appeared after the James Cameron's Avatar movie released in 2009. In stereoscopic viewing, two separate image streams are delivered, one for each eye. Producing stereoscopic movies can be done synthetically, manually converting monoscopic image streams, or natively filming stereoscopic sequences. In this last case the pair of cameras can be mounted either in parallel or mirror configurations [1], being the last one the most popular configuration in spite of the issues that arise because of the use of the mirror.

This work has been partially supported by the Departamento de Informática y Sistemas of the Universidad de Las Palmas de Gran Canaria.

R. Moreno-Díaz et al. (Eds.): EUROCAST 2017, Part I, LNCS 10671, pp. 157–165, 2018.
https://doi.org/10.1007/978-3-319-74718-7_19

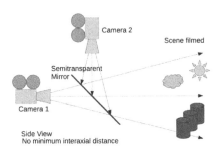

Fig. 1. Stereoscopic pair of cameras in mirror configuration.

Fig. 2. Mirror camera configuration: camera 2 captures the scene as a reflection in a semitransparent mirror.

One of the problems in stereoscopic filming is that the images directly shot from the cameras can be substantially unbalanced, mainly in geometry and color. The stereoscopic image streams need to be aligned, so they differ only in the horizontal parallax. As there is a limit to the precision of the mechanical alignment, further geometric alignment must be performed in postproduction.

Another problem is color imbalance. An important component of this imbalance is static, independent of the observed scene, and it is produced by the differences between both cameras and the imperfect balance of the beam splitting mirror. Another part of the color imbalance is caused by the polarized light behavior and it depends on the observed scene. Although the mirror configuration shown in Figs. 1 and 2 is the most popular, it experiments the problem of the reflection of polarized light, as described by the Fresnel equations [5]. This means that polarized light will be unevenly split between cameras depending on the light polarization angle.

The rectification method described in this works deals with both the geometric and color imbalance of stereoscopic image sequences.

1.1   Geometric Misalignments

A homography is the standard method for finding the relationship between two different views of the same scene [3,4]. As a homography H is not a human readable representation, we describe it in human readable terms, which is equivalent to decomposing the homography into a series of transformations (called sub-homographies) defined by the intuitive parameters: Keystone, Shear, Rotation, Scale and Offset.

$$H = H_{keystone}\ H_{shear}\ H_{rotate}\ H_{scale}\ H_{offset} \tag{1}$$

where

$$H_{keystone} = \begin{pmatrix} 1 & 0 & K_x \\ 0 & 1 & K_y \\ 0 & 0 & 1 \end{pmatrix} \quad H_{shear} = \begin{pmatrix} 1 & 0 & 0 \\ E & 1 & 0 \\ 0 & 0 & 1 \end{pmatrix} \quad H_{offset} = \begin{pmatrix} 1 & 0 & 0 \\ 0 & 1 & 0 \\ O_x & O_y & 1 \end{pmatrix}$$

$$H_{rotate} = \begin{pmatrix} \cos\omega & \sin\omega & 0 \\ -\sin\omega & \cos\omega & 0 \\ 0 & 0 & 1 \end{pmatrix} \quad H_{scale} = \begin{pmatrix} SS_x & 0 & 0 \\ 0 & S & 0 \\ 0 & 0 & 1 \end{pmatrix}$$

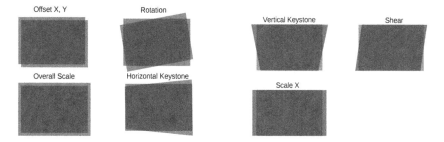

Fig. 3. Homography relations corresponding to possible camera misalignments.

Fig. 4. Homography relations that do not have an interpretation as a camera misalignment.

These five subhomographies contain eight independent parameters, allowing all possible homographies. We can see in Figs. 3 and 4 a graphical representation of the eight possible geometric misalignments. Among these, we only consider the five cases in Fig. 3 (characterized by parameters O_x, O_y, S, ω and K_y). They correspond to possible mechanical misalignments of the cameras and are fixed with our proposed method.

2 Stereoscopic Rectification Method Based on Optical Flow

Our stereoscopic rectification method corrects both color and geometry of the image sequences. On the one hand, it corrects both global and local color imbalance. On the other hand, it also deals with geometric misalisgnments. It consists of the following steps:

Step 1: Global Color Balancing

Color balancing is performed using a classic histogram based method. For any given shot, a sample image pair is taken, measuring the pixels distribution for each view. The 10 percentile and 90 percentile points are found for each image

Fig. 5. Checkerboard mosaic of the original sample images.

Fig. 6. Step 1: Global color balance applied.

of the pair and each color component. Then, one of the views is corrected by applying the linear equation $C' = aC + b$, with a and b chosen so that the 10 and 90 percentile points of the views match.

Figure 5 shows an image built as a checkerboard combination of an original stereoscopic image pair (left and right views). We can see an important color imbalance between the left and right images, as the squares in the checkerboard are very visible. After applying the global color balancing we can see in Fig. 6 that with the color imbalance considerably reduced, the jagged edges of the checkerboard square boundaries become much easier to discern. These are caused by the geometric misalignment of the images. There is still some color imbalance that it will be corrected in the final step to take advantage of the visible checkerboard boundaries during the geometric rectification steps.

Step 2: Global Geometric Rectification

The global geometric rectification step is based on the fact that in a rectified scene only horizontal parallax should be present. The horizontal parallax in the scene depends on the distances of the objects and can be arbitrary, so we can not derive any conclusions about alignment from these values. However, any vertical parallax is a symptom of misalignment that must be rectified for comfortable viewing. Our method, both automatic and human assisted, measures and compensates the vertical parallax in the scene and it is based on the Optical Flow (OF). The automatic part follows this sequence:

1. Initiate the vector $\mathbf{P} = (O_x, O_y, S, \omega, K_y)^T$, with the five geometric rectification parameters corresponding to possible camera misalignments set to their neutral values.

2. Create the homographies H^L and H^R as shown in Eq. 1 using the parameters in $\mathbf{P}$. The rest of the parameters are set to their neutral values ($K_x = 0$, $E = 0$ and $S_x = 1$) for the whole process.
3. Apply the homographies to the stereoscopic pair of images I^L and I^R, creating tentatively rectified images $I'^L = I^L H^L$, and $I'^R = I^R H^R$. Notice that in the first iteration the parameters are neutral, so no actual rectification is applied.
4. Calculate the OF $\mathbf{h}$ from image I'^L to image I'^R.
5. Calculate the disparity measures vector $\mathbf{M} = (O_x, O_y, S, \omega, K_y)^T$. These measures are weighted average based statistics calculated on all the OF vectors $\mathbf{h} = (u, v)^T$, designed in a way that they effectively estimate the different parameters of misalignment.

The overall horizontal and vertical offset disparity measures are calculated as:

$$O_x = \frac{\sum u}{n} \quad O_y = \frac{\sum v}{n} \tag{2}$$

where $n = n_x n_y$ is the number of pixel of the images, and n_x and n_y are the width and height of the images.

The average scale factor detects if the OF vertical components (v) tend to point outwards or inwards, relatively to the central horizontal axis of the field:

$$S = 16 \frac{\sum yv}{nn_y n_y} \tag{3}$$

The rotation disparity measure estimates the relative rotation between the views. Intuitively, this equation detects if the OF vertical components (v) show downwards tendency on the left side of the image while showing upwards tendency on the right side:

$$\omega = \arcsin\left(16\frac{\sum xv}{nn_x n_y}\right) \tag{4}$$

Finally, the keystone disparity measure compares the apparent vertical scale of the objects on the left and right sides of the image.

$$K_y = \frac{\sum xyv}{n_x \sum |xy|} \tag{5}$$

6. Accumulate the disparity measures $\mathbf{M}$ to the parameters $\mathbf{P}$.
7. Repeat from step 2, until all the disparity measures drop below a predefined precision threshold.

As an example of our results, observe in Fig. 6 that before the alignment almost all objects are both horizontally and vertically unaligned, seen as jagged edges on the checkerboard square boundaries. After the geometric alignment, in Fig. 7, the only jagged edges are on seagulls wings, due to their interocular parallax.

Once the automated geometry rectification is finished there is an optional human assisted global geometric rectification. In this stage the user can select reference points with certain properties in order to fix individual parameters:

– One point for horizontal and vertical alignment
– Two points on approximate vertical line for scale alignment
– Two points on approximate horizontal axis for rotation alignment

The OF will be used in this stage to calculate the actual feature match at these points. If the OF fails, even at these well chosen points, the method allows to manually match the image features and derive the values of individual parameters of the homography.

Step 3: Local Geometric Rectification

The horizontal parallax of the objects in the scene is mainly determined by the physical interocular distance of the stereoscopic camera pair. This distance is chosen at filming [2], and it can be different to the usual human interocular distance of approximately 6 cm. Values in the range from 2 cm to 30 cm are commonly used, depending on the distance of the object nearest to the camera in the scene. When this parameter is not properly chosen there is no trivial way to adjust it in postproduction. In our proposal we perform an OF-based warping of one or both images using the horizontal component (u) of the calculated OF $\mathbf{h} = (u, v)^T$. In this way we modify the horizontal parallax of the scene, with a result equal to that of the interocular distance scaled by a factor s:

$$I'(x, y) = I(x + (s - 1)u, y) \tag{6}$$

Notice that the apparent interocular distance can be both reduced or widened, depending on the value of the scale parameter s. There are limits to this extrapolations given by the quality of the OF used, with artifacts increasing with the amplitude of the adjustment.

One special case of the interocular distance modification method happens when the interocular distance is eliminated (by setting $s = 0$). This results in a stereoscopic image pair with both views filmed from exactly the same camera position. We will use this transformation as a useful diagnosis tool to visualize the differences between views in the GUI. On an aligned stereoscopic image pair, the objects parallax is the most obvious difference and it masks other kinds of differences, making them difficult to observe. If we eliminate the interocular distance, the color differences become very obvious, as well as the undesirable vertical parallax or the focus mismatch.

Comparing Figs. 7 and 8 we can see the results of the interocular distance elimination. There are no more jagged edges on the checkerboard square boundaries and the checkerboard remains visible mostly due to the local color imbalance. The image is now prepared in order to apply the local color balancing stage.

Step 4: Local Color Balancing

Some color differences, specially these caused by polarized light (blue sky, shiny surfaces like water, furniture, plant leaves, and even human skin), vary for each

Fig. 7. Step 2: Global geometric rectificated images.

Fig. 8. Step 3: Local geometric aligned images (interocular distance forced to zero).

point of the scene. Balancing them perfectly would require a perfect OF, which is impossible. However, color imbalance tends to form clusters, that is, zones of locally consistent imbalance, and it usually happens on large flat areas like blue sky or water surface. Contrarily, in cluttered zones with fine detail any local color imbalance is hardly noticeable. That gives us an opportunity to use a method similar to the global color balancing, however applied on a per-zone basis, in a continuous smooth manner, with zone mapping assisted by the calculated OF. Our tests proved that the size of the balanced zone can be relatively large and its borders fuzzy and highly tolerant to the OF imperfections.

Let us suppose we want to correct the local color balance of the R view taking as reference the L view. First, we create the L^{warped} view as L warped to the R view. Then, we create the difference between the now per pixel geometrically matched R and L^{warped}, and we apply a Gaussian filter on the result, creating the correction image C:

$$C = G_\sigma \circ (R - L^{warped}) \tag{7}$$

Image C is a continuous soft image representing the overall local difference between left and right views, in the space of the right view. The Gaussian filter radius (σ) controls the degree of continuity of the correction, balancing between the detail of the color correction and the robustness to the OF imperfections. Now, we finally color balance the R view by subtracting the correction image (C):

$$R' = R - C \tag{8}$$

This equation cancels the color difference on large image areas (the definition of "large" depends on the radius of the Gaussian filter), while preserving the

Fig. 9. Step 4: Local color balanced images (interocular distance still forced to zero).

Fig. 10. Step 4: Anaglyph result after recovering the original interocular distance.

image details of both views. Also, notice that any local OF errors will be largely filtered out by the Gaussian applied on top of the OF-dependent L^{warped}.

Comparing Figs. 8 and 9 we can see the effects of the local color balance stage. Remember that we still have the interocular distance set to zero for the purpose of diagnosis. The checkerboard boundaries are virtually indistinguishable.

The final result, shown in Fig. 10, is obtained by dropping the stage of interocular distance modification to zero in order to recover the stereoscopy. That figure is a black and white anaglyph, so it can be visualized in 3D, while reading this document, using basic red/cyan 3D glasses.

3 Conclusions

In our work we proposed, implemented, tested and successfully deployed for industrial use an Optical Flow based method for stereoscopic film postproduction. The method rectifies both geometry and color of stereoscopic image sequences following a series of stages. In both cases global and local approximations are explored in the rectification process. Our results show that after the rectification stages the images are color balanced and geometrically aligned. The final result, presented as an anaglyph of the stereoscopic image pair, shows an acceptable result from the observer's point of view.

In our future work, we will explore an extension of the method for virtual reality media, being the main difference the need to work on spherical image projections instead of planar.

References

1. 3ALITY (2017). www.3alitytechnica.com/3D-rigs/
2. Mendiburu, B.: 3D Movie Making: Stereoscopic Digital Cinema from Script to Screen. Focal Press/Elsevier, Amsterdam (2009)
3. Fusiello, A., Irsara, L.: Quasi-Euclidean uncalibrated epipolar rectification. In: ICPR, p. 14 (2008)
4. Hartley, R., Zisserman, A.: Multiple View Geometry in Computer Vision, 2nd edn. Cambridge University Press, New York (2003)
5. Hecht, E.: Optics. Pearson Education Series. Addison-Wesley, Boston (2002)

Secure UAV-Based System to Detect and Filter Sea Objects Using Image Processing

Moisés Lodeiro-Santiago[1]([⊠]), Cándido Caballero-Gil[1],
Ricardo Aguasca-Colomo[2], Jorge Munilla-Fajardo[3], and Jorge Ramió-Aguirre[4]

[1] Departamento de Ingeniería Informática y de Sistemas, Universidad de La Laguna,
San Cristóbal de La Laguna, Tenerife, Spain
{mlodeirs,ccabgil}@ull.edu.es
[2] Instituto Universitario SIANI, Universidad de Las Palmas de Gran Canaria,
Gran Canaria, Spain
ricardo.aguasca@ulpgc.es
[3] Departamento de Ingeniería de Comunicaciones,
Universidad de Málaga, Málaga, Spain
munilla@ic.uma.es
[4] Departamento de Sistemas Informáticos, Universidad Politécnica de Madrid,
Madrid, Spain
jramio@eui.upm.es

Abstract. In the last years irregular immigration has aroused again due to the fact of war in some regions like middle east and in continents like Africa. Because of that, in this paper we presented a novel solution based in new emerging technology to face off the immigration problem focused on the case of sea-type immigration. The presented system is composed by unmanned aerial vehicles (UAV), also known as RPAs or drones, used to fly autonomously over an area in order to detect unallocated potentials immigration boats. The method used is the image processing using combined algorithms and software like OpenCV2 to detect certain patterns in images to determinate whether it is a possible immigration boat or not. Besides, it has been included some prof of concepts to show the effectiveness of the presented system. As main results we was able to determine and detect using aerial pictures if a boat it is a boat with an 82.3% of accuracy (according to the F1 score accuracy test).

Keywords: UAV · Image processing · Patera boat · Ship detection
Security

1 Introduction

More than four million of Syrians have been forced to flee their country since the Bashar al-Assad regime began in 2011 [1]. Between 2015 and 2016 more than 850,000 people sought refuge in Europe [2], and most of them (220,000 approximately) crossed the frontiers illegally across the Mediterranean, according to the European External Borders Agency, Frontex [3]. Apart from Syria's conflict, a

© Springer International Publishing AG 2018
R. Moreno-Díaz et al. (Eds.): EUROCAST 2017, Part I, LNCS 10671, pp. 166–174, 2018.
https://doi.org/10.1007/978-3-319-74718-7_20

lot of people from Africa each day tries to cross the frontier to look for a better life in the Canary Islands or in the mainland [4,5]. The most used method to try to reach the Canary Islands coast is with large fishing boats (known as pateras). The number of migrations using boats increases each year in comparison with the number of people that cross the frontiers walking [6].

Throughout the years, many ship detection proposals have been presented based on the use of image processing combined with different strategies such as multi-scale techniques [7], photogrammetric analysis of images acquired by UAV [8], fuzzy logic [9], optical satellite images [10], etc. These works show different situations where image processing and drones or Unmanned Aerial Vehicles (UAVs) can be helpful to detect small ships by combining different technologies.

In order to propose a solution to the problem, here we define a new system that combines real-time image processing to detect patterns in the sea, and UAVs. On the one hand, image processing is a methodology based on some open Artificial Intelligence and Mathematics libraries to apply various filters to images, OpenCV and Numpy. On the other hand, UAVs are aircraft vehicles controlled either remotely by pilots or autonomously by onboard computers. The image processing libraries are used for the management of images and the application of filters to acquire a final matrix that represents some possible pateras in the image. Besides, the described system is combined with a real-time marine traffic database that let us know the actual GPS positions of registered ships. The developed system has been designed to filter unregistered objects, based on the object size estimated using trigonometric operations on the altitude and drone's camera meta-data like focal and camera angle. After applying various OpenCV filters and discarding small-sized objects to each frame, we can get a detection rate of match of real pateras of 84% in the worst case and 100% in the best case. In comparison with other systems, we use the image processing process in a computer to take advantage of the computing capacity instead of using the smartphone's CPU. Also, as aforementioned, in order to discard false-positive cases we use a ship GPS position database to compare all the GPS positions of registered ships in the area, with the GPS position given by the drone after a detection of a suspicious ship. The drone takes pictures each 3 s and sends it using a three-way security channel. The first security level applied in our system is provided by the LTE connection that uses the SNOW 3G algorithm for the integrity protection and stream cypher of the UMTS technology [11].

The present paper is structured as follows: Sect. 2 shows a study of the current system and how it works. In addition, the algorithm used for detection is briefly included. In Sect. 3 we present some proof of concepts (PoC) performed with the algorithm seen in the previous section. Finally in Sect. 4, the conclusions and future work are used to close the current work.

2   System and Image Recognition

The system used for the purpose presented consists of two parts; On the one hand we have a web coordination application that is responsible for giving a visual option to a user operator to see in real time the position of the boats in a certain zone (previously registered). Apart from the positions it is also possible to visualize data of course, speed, etc. This data is collected by means of an API of the page MarineTraffic that facilitates this information by means of a request to that API. The communication schema of the system is displayed below (see Fig. 1):

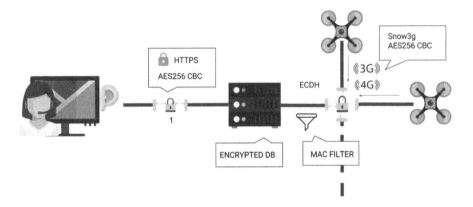

Fig. 1. Global communication schema

- Step 1: We have a control panel, as discussed previously where an operator logs in. The connection to this control panel will be made through a strict protocol HTTPs to ensure the confidentiality and security of the information (and to avoid theft of this one by possible boats with malicious intentions). Subsequently, the operator will be able to visualize the boats in the panel (see Fig. 2a).
- Step 2: In turn and once you start the system, the panel is attentive to any new connection and sending information coming from the drones. The information sent by each drone is the current GPS position, flight altitude and an aerial photo of the zone. This information is transmitted to the main server through a secure route via mobile connection (mobile antennas) of 3g or 4g that make use of a Zuc [12] or Snow3g [13] encryption (respectively) and a different pre-shared key encryption for each drone with AES256CBC. In addition, the server has an additional layer of MAC filtering protection to avoid sending erroneous packets to the server.
- Step 3: This step is done after step 2, almost in parallel. It consists of the image processing in order to detect possible boat/patera shapes. In case the system recognises a strange or unusual shape, a warning will appear on the control panel (see Fig. 2b).

The main purpose of the image recognition for this paper is the creation of an image classification system that allows the identification of marine objects and

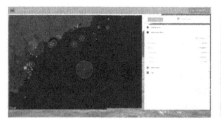
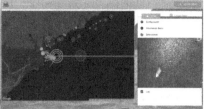

(a) Operator web panel displaying the current known boats positions

(b) Boat detection alert

Fig. 2. Web operator panel

boat shapes that identify boats, discarding possible false positives by applying several filters. Prior to the detection process it is necessary to have an image of which we want to start from the scratch. This is performed by taking an aerial picture that comes from a drone as explained in the previous section. Each drone can have a different camera (different angle, focal length, etc.) so internally in the application, there must be a register of which camera has each one to be able to later perform object filtering based on sizes (calculated trigonometrically using the height, knowing the angle of the camera and size of the photo to be able to get a ratio of pixels per meter).

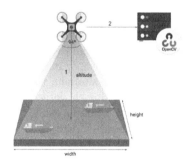

Fig. 3. Drone detection and communication with the main server

The detection process has three phases mainly. The first phase is responsible for making a treatment of the image to remove "noise" from the picture and keep the figures that really interest us. The noise can be represented as small particles, malformations or colours that in similarity could create regular shapes. Eliminating noise (for this particular case) is achieved by applying a series of steps/functions in a logical order. These steps are what we will see below:

– Read base64 input: The method to send an image compressed by the network is not sending a binary file. Instead of that, the image is compressed in a string representation of it in a base64 format.

- Calculate relative size and ratio: This is useful for, depending on the flight height of which the drone, be able to establish a relation of the size of the objects. We always assume that the drone's camera is in a plane parallel to the sea (perpendicular to the sea plane) which it is acceptable considering the limited pitch and roll angle using a multi-copter UAV type.
- Image conversion to gray-scale. This process is very common in image recognition because it is important to detect patterns regardless of the colours of the image and/or spectrum (see Fig. 4a). The gray-scale is applied using the RGB to Gray function: $RGB[A]toGray : Y \leftarrow 0.299 \cdot R + 0.587 \cdot G + 0.114 \cdot B$
- Apply an OpenCV 2D filter with a kernel that allows us to sharp certain shapes or patterns of the image. The kernel is represented with the following matrix (rows separated by semicolons) $[-1, -2, -1; -1, 9, -1; -1, -2, -1]$. This step returns a bit mask that will be used in the next step.
- With this mask, we create the inverse mask by modifying the bits of the array of the original mask with the bitwise_not function.
- Then a "morphologyEx" function is applied to perform a morphological transformation on the image with 3 iterations making use of a MORPH_OPEN operation. After that, a dilate function is applied.
- Finally, before moving to the detection and classification process, we apply a threshold (see Fig. 4b) in binary mode by applying the OTSU method [14] that is shown in the following equation: $\sigma^2 w(t) = q_1(t)\sigma_1^2(t) + q2(t)\sigma_2^2(t)$.

Once we have the "clean" image we can go on to detect possible objects within the image. To perform a detection based on contours we apply the "findContours" function by applying the threshold previously seen and call the function that allows us to detect basic shapes (square, circle, triangle, etc.). For all the detected contours we apply the following algorithm:

1. In the case of a circle or a square (or similar), the shape is discarded.
2. A squared area is established around each detected object. This outline will have a different colour to the object's inner colour.
3. From the resulting area, we extract the first 3 pairs of coordinates $p_1(x_1, y_1), p_2(x_2, y_2)$ and $p_3(x_3, y_3)$ (equivalent 3 of the 4 edges of the rectangle).
4. With these 3 points, we calculate the orientation of the shape calculation the distances d_1 y d_2 between the points p_1, p_2 y p_3 making use of the hypotenuse between the pairs. Then, these distances are divided by the pixels that occupy each meter taking into account the height, angle of the camera and size of the photo (see Fig. 3).
5. If the case of $A = (d_1 > d_2$ and $d_1 \div d_2 > 0.812)$ or $B = (d_1 \leq d_2$ and $d_1 \div d_2 > 0.812)$ means that d_1 is equivalent to the larger side of the object and it has a rectangular shape. In case of A or B distance are lower in compare with the threshold of 0.812, the contours are discarded.
6. In case the distance A or distance B are lower than the threshold of 0.812, the contours are discarded due to the shape ratio.

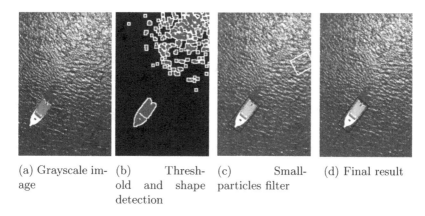

(a) Grayscale image   (b) Threshold and shape detection   (c) Small-particles filter   (d) Final result

Fig. 4. Main steps of recognition (Color figure online)

3   Proof of Concepts

During the development of the project, several tests were carried out to determine the reliability of detection of the system. Among the tests carried out, are the proof of concepts (PoC) based on detections at different heights and different hours of the day. This has been done to check the effects that can have the objects at different relation size and solar rays over the sea. In addition, tests were carried out with images extracted from the internet with different sizes and different qualities (from very bad as an image can be extruded from Google Maps to images of better resolution) as we will see bellow;

Case 1: Aerial image at 65 m height. In order to determine the reliability of the system, we have opted to analyse the image represented in the Fig. 5 where are shown 13 elements of which (by their size mainly) 8 are considered pateras and 5 left, not. These data are represented using a confusion matrix [15] that is a statistical classification problem method used commonly in machine learning. The confusion matrix is shown in the following Table 1 and the results in the Table 2.

Table 1. Confusion Matrix

		Predicted	
	Total: 13	Patera Boat	No Patera Boat
Actual Class	Patera Boat	7	1
	No Patera Boat	2	3

Case 2: Low quality images. In this second case have used different images from many sites such as Google Maps or pictures captured from videos made with drones. In the three exposed cases, the detection rate was 100% of detection rate (Fig. 6).

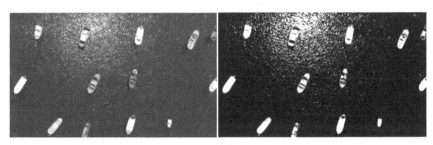

(a) Gray-scale and small-particles filter (b) Threshold in pure black and white
applied

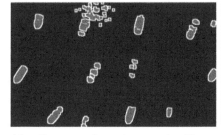 

(c) Contour detection and dilate (d) After some iterations a red box ap-
method applied   pears in detected objects

Fig. 5. PoC - 13 boats of different sizes

Table 2. Terminology and derivations from Table 1

$TP = 7$	$TN = 3$	$FP = 2$
$FN = 1$	$P = TP + FN = 8$	$N = TN + FP = 5$
$TPR = \frac{TP}{P} = 0.875$	$PPV = \frac{TP}{TP+FP} = 0.\overline{7} \approx 0.8$	$F1Score : 2 * \frac{PPV*TPR}{PPV+TPR} = 0.823$

(a) Container ship de- (b) Three ships detected (c) Small boat detected
tected (great quality)   (low quality)

Fig. 6. Different colour schemes, size and image quality PoC

4   Conclusions

In this paper has been presented a system for the detection of small boats (also known as pateras) to solve the irregular and illegal immigration. This study has been made based principally on the coasts of the Canary Islands where, in the last years, the immigration due to warlike conflicts and poverty has aroused. The project has combined emerging technologies such as drones to fly over an area taking photographs constantly making use of a smartphone to take these photos and communicates with the main server. Photographs are processed using an algorithm (following a number of steps) to detect patterns and shapes. As the main result, it is possible to conclude the profitability and efficiency of the detection of small boats using the system. In addition, the system is able to alert in case of detection by presenting information in real time in a web application that will be handled by an operator. As future work will include the possibility of including thermal vision in the system, which will greatly facilitate the actual detection of objects at sea. Besides, we will work on an improvement and inclusion of a complete analysis of the effectiveness of the classifier related to different flight altitudes.

References

1. Syrian refugees: Four million people forced to flee as crisis deepens. https://www.theguardian.com/global-development/2015/jul/09/syria-refugees-4-million-people-flee-crisis-deepens
2. The UN refugee agency. http://www.unhcr.org/
3. European border and coast guard agency. http://www.frontex.europa.eu/
4. Wastl-Walter, D., et al.: Borders, Fences and Walls: State of Insecurity? Ashgate Publishing Ltd., Farnham (2014)
5. Migratory routes map. http://frontex.europa.eu/trends-and-routes/migratory-routes-map/
6. Soddu, P.: Ceuta and Melilla. Security, human rights and frontier control, pp. 212–214. Institut Europeu de la Mediterrània (eds) IEMED Mediterranean Yearbook Med (2006)
7. Tello, M., López-Martnez, C., Mallorqu, J., Tares, T., Greidanus, H.: Advances in unsupervised ship detection with multiscale techniques, vol. 4, pp. IV979–IV982 (2009). Cited By 2
8. Díaz-Cabrera, M., Cabrera-Gámez, J., Aguasca-Colomo, R., Miatliuk, K.: Photogrammetric analysis of images acquired by an UAV. In: Moreno-Díaz, R., Pichler, F., Quesada-Arencibia, A. (eds.) EUROCAST 2013. LNCS, vol. 8112, pp. 109–116. Springer, Heidelberg (2013). https://doi.org/10.1007/978-3-642-53862-9_15
9. Margarit, G., Tabasco, A.: Ship classification in single-pol SAR images based on fuzzy logic. IEEE Trans. Geosci. Remote Sens. **49**(8), 3129–3138 (2011). Cited By 27
10. Huang, G., Wang, Y., Zhang, Y., Tian, Y.: Ship detection using texture statistics from optical satellite images, pp. 507–512 (2011). Cited By 2
11. Santos-González, I., Rivero-García, A., Caballero-Gil, P., Hernández-Goya, C.: Alternative communication system for emergency situations, vol. 2, pp. 397–402 (2014). Cited By 0

12. ZUC stream cipher. http://www.gsma.com/aboutus/wp-content/uploads/2014/12/eea3eia3zucv16.pdf
13. SNOW3G stream cipher. http://www.gsma.com/aboutus/wp-content/uploads/2014/12/snow3gspec.pdf
14. Wikipedia: Otsu's method – wikipedia, the free encyclopedia (2017). Accessed 10 May 2017
15. Wikipedia: Confusion matrix – wikipedia, the free encyclopedia (2017). Accessed 23 May 2017

Modifications of Model Free Control to FOTD Plants

Mikuláš Huba$^{(\boxtimes)}$ and Tomáš Huba

Slovak University of Technology in Bratislava,
Ilkovičova 3, 812 19 Bratislava, Slovakia
{mikulas.huba,tomas.huba}@stuba.sk
http://uamt.fei.stuba.sk

Abstract. Model free control (MFC) represents one of possible alternatives to traditional approaches as PID control, disturbance observer based control (DOBC), internal model control (IMC), etc. As one of its central features one could mention use of finite-impulse-response (FIR) filters in input disturbance reconstruction. The paper deals with characterizing links of MFC to these approaches and discussing the terminology used and simple procedures for tuning of such controllers for the plants approximated by the first order time delayed models.

Keywords: Model free control · Model based control · PID control
Disturbance observer based control · Internal model control

1 Introduction

With its product known as "intelligent" PID control, the model free control (MFC) [1] represents one popular alternative to such approaches as PID control [2], disturbance observer based control (DOBC) [3], internal model control (IMC) [4], or advanced disturbance rejection control (ADRC) [5–7]. Although it yields significant control design simplifications, the denotation "model free" may seem exaggerated. Despite its proclaimed simplicity, in doing a decision to use MFC one should still be aware of several important questions. Their list could include (but not be limited to) points as:

1. Does the method yield a continuous-, or a discrete-time solution?
2. What are the MFC special features with respect to other alternatives?
3. What is the algorithm complexity with respect to the alternative solutions?
4. What are the memory requirements with respect to the alternatives?
5. What is the achievable loop performance and robustness with respect to the alternatives?
6. Which methods may be used for an optimal-robust closed loop tuning?
7. What is the achievable noise attenuation with respect to the alternatives?
8. What are the numerical loop properties with respect to the alternatives?
9. Is the method appropriate for time-delays compensation? Etc.

By using the space available, the article attempts to discuss at least some of all these issues in more depth.

© Springer International Publishing AG 2018
R. Moreno-Díaz et al. (Eds.): EUROCAST 2017, Part I, LNCS 10671, pp. 175–182, 2018.
https://doi.org/10.1007/978-3-319-74718-7_21

2   Continuous- versus Discrete-Time Solution

Based on mathematically complex approaches of functional analysis and differential algebra, MFC has been formulated within the so called flatness-based control [1,8–11]. In order to bring the complex theory as close as possible to practice, proposed intelligent PID controllers offer great simplicity for their application and tuning. The simplest solution, the intelligent proportional (iP) "model free" controller [1] is based on an "ultra local" model written usually as

$$\dot{y} = F + \alpha u \tag{1}$$

Thereby, $\alpha \in \mathbb{R}$ is considered to be (in general) a non-physical constant parameter that may be interpreted as a gain with an estimate $\overline{\alpha}$. F is representing an equivalent disturbance acting on the integrator input. Thus, arround a fixed operating point with a constant (or zero) F, (1) obviously corresponds to an integral plant model. Approximations of the plants to be controlled by integral models may significantly simplify the plant description, which has been independently advertized also by ADRC.

When approximating F by a piecewise constant parameter ϕ, for a measured output derivative $\dot{y}_m$ one gets

$$\phi = \dot{y}_m - \overline{\alpha}u \tag{2}$$

By calculating a mean value of this parameter $\overline{\phi}$ by its integration over a time interval L it is possible to increase the loop robustness and to achieve an improved noise attenuation. In calculating this parameter the papers on MFC are usually using a special output differentiation, or ϕ is calculated by using an operational calculus according to

$$\overline{\phi} = -\frac{6}{L^3} \int_{t-L}^{t} [(L - 2\delta)y_m(\delta) + \overline{\alpha}\delta(L - \delta)u(\delta)]d\delta \tag{3}$$

Yet simpler alternatives may be derived by a manipulation of integral of (2)

$$\overline{\phi} = \frac{y_m(t) - y_m(t - L)}{L} - \frac{1}{L} \int_{t-L}^{t} \overline{\alpha}u d\tau \tag{4}$$

The integration may be expressed by a transfer function

$$Q_c(s) = (1 - e^{-sL})/(Ls) \tag{5}$$

Let us specify a required output reference trajectory by y^*, $\dot{y}^*$ and the tracking error trajectory as $e = y - y^*$, $\dot{e} = \dot{y} - \dot{y}^*$. By requiring an exponential tracking error decrease according to the differential equation

$$\dot{e} = \lambda e \tag{6}$$

with $\lambda < 0$ being the closed loop pole, the new control u_0 may be specified as

$$u_0 = \dot{y}^* - \overline{K}_P e \tag{7}$$

It means that the iP control algorithm may finally be written in the form

$$u = \frac{u_0 - \phi}{\overline{\alpha}} = \frac{\dot{y}^* - \overline{K}_P e - \phi}{\overline{\alpha}} \tag{8}$$

where $\overline{K}_P = -\lambda$ is a proportional gain defined, for example, by a required closed loop pole λ. Frequently, a closed loop time constant $T_c = -1/\lambda$ is used instead. Thus, the closed loop dynamics may be transformed to

$$\dot{y} = F - \phi + \frac{\alpha}{\overline{\alpha}} u_0 \tag{9}$$

In an ideal case with $\alpha = \overline{\alpha}$ and $F = \phi$ it leads to a single integrator control

$$\dot{y} = u_0 \tag{10}$$

By considering a simple nominal case with a piecewise constant reference trajectory specified by the setpoint $y^* = w = const$, when $\dot{y}^* = 0, \overline{\alpha} = \alpha, \overline{K}_P = K_P \overline{\alpha}$ the closed loop may be characterized by the transfer functions for the setpoint-to-output

$$F_{wy}(s) = \frac{Y(s)}{W(s)} = \frac{1}{1 + T_c s}; \; T_c = \frac{1}{K_P \alpha} \tag{11}$$

and for the input-disturbance-to-output dynamics

$$F_{iy}(s) = \frac{Y(s)}{D_i(s)} = \frac{\alpha(sL - 1 + e^{-sL})}{s(sL + K_P \alpha s L)} \approx \frac{L}{2K_P} \frac{s}{1 + T_c s} \tag{12}$$

However, since the integration (5) represents just a marginally stable operation, it is preferably accomplished in a discrete-time form by a FIR filter

$$Q_d(z) = \frac{1}{N} \sum_1^N z^{-i}; \; N = IP(\frac{L}{T_s}) \tag{13}$$

where $T_s < L$ is the sampling period, z^{-1} the shift operator and IP represents an integer part[1].

Thus, although the iP control is frequently introduced in the continuous-time domain, it represents a typically discrete-time control. Whereas the used FIR filters seem to represent the simplest possible alternative, there exist just very limited possibilities for their rigorous analytical tuning. From this point of view it represents an ideal field for application of the numerical performance portrait based loop analysis and optimization [12, 13].

---

[1] To be rigorous, in order to eliminate the algebraic loops, the integration in (4)–(5) has to be carried out over past signals, i.e. over the time interval $[t - L - \epsilon, t - \epsilon]$ with a small positive $\epsilon > 0$, in the discrete-time case with $\epsilon = T_s$.

3   Model-Free or Model-Based Control?

Similarly as in the disturbance-observer based control, where the combination of the parallel I-action with a disturbance observer leads to some unwanted performance (overshooting [14]), it is questionable, if an additional parallel I action of the iPI control brings some advantages. On the other hand, since the iP control inherently includes an integral action for reconstruction and compensation of possible external and internal disturbances, a more concise approach would be (similarly as in the disturbance-observer based control) to speak already in such a case about iPI control.

Another misleading moment may be related to the denotation "model free" control. As already mentioned above, (1) corresponds to an integral plant model. But, why to limit the advantages brought by the FIR low-pass filters just to integral plant models? Consider, for example, a "local" first order linear plant model

$$\dot{y} + ay = F + \alpha u \tag{14}$$

Then, with a denoting gain of the internal plant feedback, for $F = 0$ such system may be described by a first-order transfer function

$$S(s) = \left[\frac{Y(s)}{U(s)}\right]_{F=0} = \frac{\alpha}{s+a} \tag{15}$$

With $\bar{a}$ denoting the plant model parameter corresponding to a, the disturbance reconstruction may be calculated according to

$$\bar{\phi} = \frac{y_m(t) - y_m(t-L)}{L} - \frac{1}{L}\int_{t-L}^{t} [\bar{\alpha}u - \bar{a}y_m]\,d\tau \tag{16}$$

an a controller accomplished according to

$$u = \frac{u_0 - \phi}{\bar{\alpha}} = \frac{\dot{y}^* + \bar{a}y^* - \overline{K}_P e - \phi}{\bar{\alpha}} \tag{17}$$

Of course, with the aim of simplicity, one may use $\bar{a} = 0$ even in situations, in which obviously $a \neq 0$. On the other hand, there surelly exist situations, in which it is advantageous to work with $\bar{a} \neq 0$ Then, since the corresponding approaches obviously represent modification of a model based approach, the denotation "model-free" does not seem to yield the most eloquent terminology. With respect to the interesting results offered by MFC, core of its success should be described by using a correspondingly refined terminology that would establish clear relations to equivalent approaches.

In [15] the authors discuss an adverse impact of possible time delays on the closed loop dynamics of iP control. However, this adverse impact holds also for all the above mentioned alternative approches. Thereby, a short study by the performance portrait method [13] showed that the use of FIR filters of iP control may yield performance preferrable to the alternative approaches with disturbance observers based on infinite impulse response (IIR) filters. Thus, their effect should be studied more systematically and in combination with both integral and more complex plant models (14).

4   Performance Measures for iP Evaluation

Intuitively, one may expect that the reference trajectory y^*, the controller gain $\overline{K}_P$ and the length of the FIR filter N (together with the sampling period T_s) are going to play the key roles in implementation of the iP control. Thereby, an appropriate choice of the sampling period and the corresponding P controller tuning are far from being trivial. Although there exist huge amount of literature devoted to performance of FIR filters in the signal processing area (see e.g. Matlab Signal Processing Toolbox https://www.mathworks.com/help/signal/ug/fir-filter-design.html), less attention is given to its analysis form the control design perspective.

Therefore, let us examine some its properties by simulation. In carrying out control experiments, the speed of the transients achieved at the output will be evaluated by the integral of absolute error IAE_Σ Thereby, the IAE_s (Integral of Absolute Error) is defined as

$$IAE = \int_0^\infty |e(t)|\, dt; \quad e = w - y; \quad w = setpoint \tag{18}$$

In evaluating disturbance steps characterized by IAE_d $w = 0$.

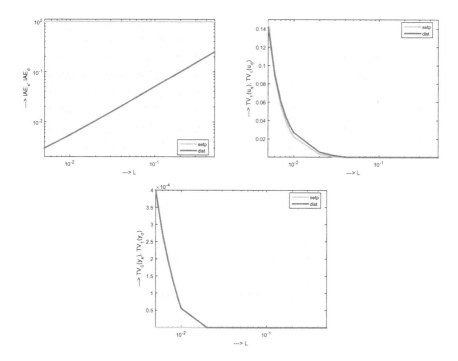

Fig. 1. Performance measures versus L for iP control based on (4), $L \in [.005, 0.5]$; $\alpha = 1$; $K_P = 1$; $a = \overline{a} = 0$; $T_s = 0.001$

Deviations of the output setpoint responses from a pure monotonicity will be evaluated by the modified Total Variance measure

$$TV_0(y_s) = \sum_i |y_{i+1} - y_i| - |y_\infty - y_0| \tag{19}$$

Similarly, deviations [12,16] of the output disturbance responses from 1P shapes with a monotonic output return to a new steady state value y_∞ after an extreme initial deviation y_{extr} caused by a disturbance setpoint change, the relative total variance measure

$$TV_1(y_d) = \sum_i |y_{i+1} - y_i| - |2y_{extr} - y_\infty - y_0| \tag{20}$$

will be used.

At the input in both output situations one has to expect 1P ideal shapes consisting of two monotonic intervals. Deviations from such ideal input characterizing an additional control effort will be evaluated by the relative total variance measure $TV_1(u)$.

5   Short Illustration

By increasing the FIR length $L = NT_s$ for fixed values K_P and T_s (Fig. 1) one may see that the ideal monotonic, or 1P shapes at the plant input and output are kept just for a relatively high value $N = L/T_s \geq 30$. Thus, the iP control brings usually higher requirements on the internal memory than the disturbance observer using usual IIR (for example, first order) low-pass filters. This result corresponds well with the FIR properties identified within the signal processing area.

As signalized already by the transfer functions (11)–(12), the setpoint responses are nearly invariant against the choice of L, whereas the disturbance responses (Fig. 2) become by increasing L slower.

For the relatively short L, the responses at the plant input and output become oscillatory (Fig. 3), which is signalized by increased shape-related deviations in Fig. 1.

Whereas the high length of FIR filters may contribute to an improved noise attenuation in noise-corrupted tasks, it may be excessive in loops with a high quality measurement. In such situations, the usual approaches based on a continuous-time design and an additional discretization have to be replaced by a fully discrete-time design carried out e.g. by the performance portrait method. Such a design may simultaneously respect also additional loop delays [13], or some parameters uncertainty [12].

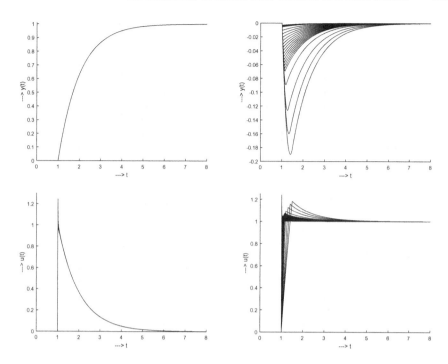

Fig. 2. Setpoint (left) and disturbance step responses (right) for $\alpha = 1$; $K_P = 1$; $a = \bar{a} = 0$; $T_s = 0.001$; $L = [0.005\ .006\ .007\ .008\ .009\ .01\ .02\ .03\ .04\ .05\ .06\ .07\ .08\ .09$ $.1\ .11\ .12\ .13\ .14\ .15\ .2\ .3\ .4\ .5]$

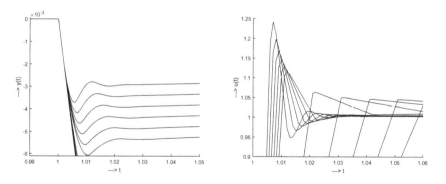

Fig. 3. Details of the disturbance step responses for the shortest L showing deviations from 1P shapes, $\alpha = 1$; $K_P = 1$; $a = \bar{a} = 0$; $T_s = 0.001$; $L \in [.005, 0.5]$;

6 Conclusions

In the paper, the iP controller has been shown as a discrete-time control augmenting a stabilizing P control by reconstruction and compensation of internal disturbances and a possible feedforward specified by the reference trajectory

$y^*, \dot{y}^*$. Up to now, it is not fully using possibilities of the disturbance observer based control based on FIR low-pass filters and thus it has been overlooked that it represents just a new type of the model based approach using simplified (integral) plant models.

Acknowledgment. Supported by the grants APVV-0343-12 and VEGA 1/0937/14.

References

1. Fliess, M., Join, C.: Model-free control. Int. J. Control **86**(12), 2228–2252 (2013)
2. Åström, K.J., Hägglund, T.: Advanced PID Control. ISA, Research Triangle Park (2006)
3. Schrijver, E., van Dijk, J.: Disturbance observers for rigid mechanical systems: equivalence, stability, and design. ASME J. Dyn. Sys. Measur. Control **124**(4), 539–548 (2002)
4. Morari, M., Zafiriou, E.: Robust Process Control. Prentice Hall, Englewood Cliffs (1989)
5. Han, J.: From PID to active disturbance rejection control. IEEE Trans. Ind. Electron. **56**(3), 900–906 (2009)
6. Gao, Z.: Active disturbance rejection control: a paradigm shift in feedback control system design. In: 2006 American Control Conference, pp. 2399–2405 (2006)
7. Gao, Z.: On the centrality of disturbance rejection in automatic control. ISA Trans. **53**(4), 850–857 (2014)
8. Fliess, M.: Noise: a nonstandard analysis [analyse non standard du bruit]. C.R. Math. **342**(10), 797–802 (2006)
9. Fliess, M., Join, C.: Intelligent PID controllers. In: Proceedings 16th Mediterrean Conference on Control and Automation, Ajaccio, France, pp. 326–331. IEEE (2008)
10. Fliess, M., Join, C.: Model-free control and intelligent PID controllers: towards a possible trivialization of nonlinear control? IFAC Proc. Vol. (IFAC-PapersOnline) **15**, 1531–1550 (2009)
11. Fliess, M., Join, C., Riachy, S.: Revisiting some practical issues in the implementation of model-free control. In: 18th IFAC World Congress, Milano, Italy, vol. 18, Part 1, pp. 8589–8594 (2011)
12. Huba, M.: Performance measures, performance limits and optimal PI control for the IPDT plant. J. Process Control **23**(4), 500–515 (2013)
13. Huba, M., Bisták, P.: Tuning of an iP controller for a time-delayed plant by a performance portrait method. In: 25th Mediterranean Conference on Control and Automation, Valletta, Malta, 2017
14. Huba, M.: Analyzing limits of one type of disturbance observer based PI control by the performance portrait method. In: IEEE Mediterranean Control Conference, MED 2015, Torre Molinos, Spain (2015)
15. Fliess, M., Join, C.: Stability margins and model-free control: a first look. In 2014 European Control Conference (ECC), pp. 454–459, June 2014
16. Huba, M.: Comparing 2DOF PI and predictive disturbance observer based filtered PI control. J. Process Control **23**(10), 1379–1400 (2013)

PIRX³ᴰ – Pilotless Reconfigurable Experimental UAV

Jens Altenburg⁽⊠⁾, Christopher Hilgert,
and Johannes von Eichel-Streiber

University of Applied Science, Bingen am Rhein, Germany
{j.altenburg,c.hilgert,
j.voneichelstreiber}@th-bingen.de

Abstract. The combination of smart control algorithms, miniaturized avionics and optimized 3D printing technology is the base of a powerful UAV. The mechanical construction of the sensor, payload and driving module allows an ideal adjustment to different mission scenarios. Using a test environment of a flight simulator and an external motion platform allows testing of PID control algorithms in the laboratory under realistic conditions. The implementation of the results is carried out on an ARM Cortex-M3 microcontroller. Within the UAV, the data are exchanged under hard real-time conditions on the Futaba SBUS.2 bus system. All mechanical modules have their own control processor. For an ideal adaptation to the aerodynamics, components of the airframe will be manufactured by 3D printers.

Keywords: UAV · Flight simulator · PID · 3D-printer · Hardware-in-the-loop
ARM Cortex M3 · Direct memory access (DMA) · Futaba SBUS.2
Realtime operation system · Cooperative multitasking

1 Basic Design and Simulation

At first glance, a "flying robot" looks like a usual embedded control project. Consisting of a few electronic components, some additional software and a common model aircraft, it doesn't look very special, does it?

Actually, there were lots of unknown parameters, like noise on sensor data, dynamic range of measured values, etc. Most of the parameters are linked to other ones. There is also a broad range of possible (model-) aircrafts existing. During earlier experiments with flight control systems we have got to the intricate dependences between electronics, software and mechanical characteristics of the used model aircrafts [1, 7]. Therefore, straight forward design steps wouldn't be a good idea. It is very important to get a view of all main parts of the flight controller, focusing on only a few things, i.e. the hardware platform, isn't enough for a successful development.

As long as no (flying) hardware exists, no real sensor values are given. Without data, no algorithms can be validated and tested. Huge companies or universities use more or less complex software models for developing their further products, i.e. manned aircrafts, etc. Obviously this way requires more time and money than we have. A shortcut will be the use of an existing flight simulator. It is not the same as a

© Springer International Publishing AG 2018
R. Moreno-Díaz et al. (Eds.): EUROCAST 2017, Part I, LNCS 10671, pp. 183–190, 2018.
https://doi.org/10.1007/978-3-319-74718-7_22

especially defined software simulation, but with some limitations, a flight simulator for model aircrafts will be very helpful. To get data as real as possible, the simulated data stream is not used as a direct input to the flight controller. The simulator has been expanded with an external moving platform. Controlled by two servos, two gyro axes can be moved simultaneously by the simulated model aircraft on the PC. In fact, it is not a piece of reality, but not so far away from it (Fig. 1).

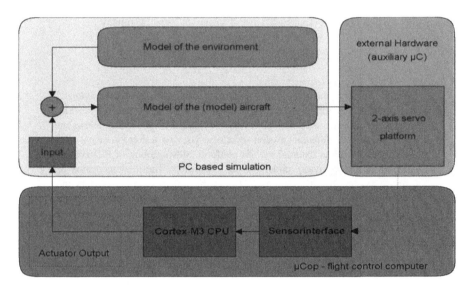

Fig. 1. Simulated data from the PC feed the two servos of the moving platform. With the sensorinterface (gyro, accelerometers, …) the CPU recons pulse with modulated actuator signals from those values. An input level shifter adapts these signals from the actuator output stage to the PC simulation.

The simulated aircraft on the PC is controlled by an external joystick. The Reflex simulator accepts the serial sequence of pulse with modulated signals from the control box as a control input. It is the way to connect the hardware later as a direct source of control signals for servos.

2 PID Control and Flight Simulator

In the first step it is possible to reduce the complexness of the onboard computer by using only four input sources: flight course, flying altitude, lateral control and pitch axis [6].

Figure 2 illustrates how the necessary input signals were connected to the control and decision block (CDB). Every input channel has its own PID controller. The main task of the CDB is the selecting of the control parameters of the different PID blocks and the oversteering of the equipment by a ground based pilot. Also it is possible to define specialised control parameters for several flight maneuvers. The control set in manual operation carried out with the 2-axis joystick. It can feed the specifications for the altitude and direction of flight as voltage values of two potentiometers in the controller.

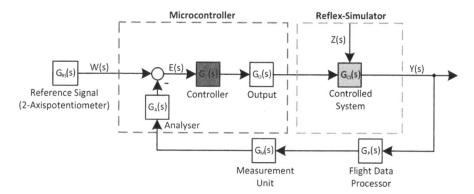

Fig. 2. PID controller for testing the control parameters in the flight simulator

The PID controller $G_R(s)$ processes the nominal variables and generates output signals $G_O(s)$. The output data are emitted in the well known pulse-pause-telegram that is needed in the standard remote control devices. The flight simulation can read this pulse telegram immediately and transfer it to control movements of the simulated aircraft. The control loop is closed by feeding back the flight attitude of the simulated model aircraft into block diagram $G_A(s)$ (Fig. 3).

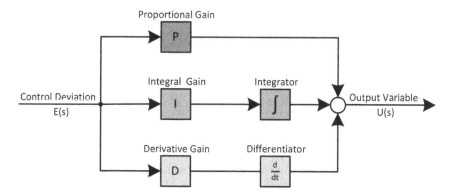

Fig. 3. MATLAB/Simulink model of the control section

The implementation of this control section is carried out on an ARM Cortex-M3 microcontroller. This closed loop allows further analysis of the process identification of the lateral and longitudinal axes. On the system snap-action and ramp functions bring answers that are recorded. Using these measurements and a MATLAB/Simulink model process behavior of the individual axes could be determined mathematically. A numerical optimization is calculated taking into account the energy consumption and the demanded reaction speed of control parameters. The interface to the simulator is also implemented on this CPU. The software modules are separated from each other in different tasks. To use the PID control algorithms can be used directly in the onboard computer of the UAV after testing in the simulator. In other words, the programming and testing of the PID controller may begin more quickly and almost without a field test.

A data stream is sent every 50 ms by the serial port of the PC with a baud rate of 38400 Bd. Every axis is represented by standard vectors $X = [x_x, x_y, x_z]$. In case of normal flight of the model in X axis the vectors are $X = [1_{(x)}, 0_{(y)}, 0_{(z)}]$, $Y = [0_{(x)}, 1_{(y)}, 0_{(z)}]$ and $Z = [0_{(x)}, 0_{(y)}, 1_{(z)}]$. Only the data of the X and Y axis are actually used for calculating PID values (Fig. 4).

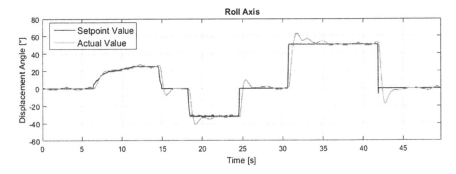

Fig. 4. Results of the simulation of the roll-axis using Reflex^{XTR} data

The frequency domain equation of the PID controller is given by:

$$u_k = u_{k-1} + e_k * \left(K_P + K_I * T_S + \frac{K_D}{T_S} \right) + e_{k-1} * \left(-K_P - 2 * \frac{K_D}{T_S} \right) + e_{k-2} * \left(\frac{K_D}{T_S} \right) \quad (1)$$

- u_k current command signal
- $u_{(k-1)}$ previous command signal
- e_k current control deviation (error signal)
- $e_{(k-1)}$ previous control deviation
- T_s sampling time

The C-implementation is given in [3]. The program duration of this computing function is approximately 2.3 μs on a ARM Cortex-M3 clocked with 84 MHz. The function is calling every 20 ms sampling time.

The Matlab/SIMULINK model calculates the following control parameter K_P, K_I and K_D:

- $K_P = 0.74$ (float fKP in the software example)
- $K_I = 0.52$ (float fKI)
- $K_D = 0.13$ (float fKD)
- $T_S = 20$ ms Timertick (sampling time)

The complete autopilot needs more than one PID controller and more input signals. Additional signals and all of the controllers have to be connected by a set of regulations.

3 Concept of the On-Board Flight Computer

The UAV consists of three electronic blocks, the board computer, the payload computer and the drive control. The onboard computer is the central part of the on-board electronics. Situational awareness, the flight control algorithms and the telemetry are concentrated in it. An additional RADAR sensor can be mounted.

For communication with the other electronic blocks the Futaba SBUS.2 serves primarily [2]. This bus transmits control commands to the steering gear and is used to query individual sensor values. The exact specifications Futaba provides after signing a LOI. More bus lines supplement the communications within the UAVs.

For position detection of the aircraft a 9-DOF sensor BNO055 (http://www.bosch-sensortec.com) is used. This sensor module provides the location data (acceleration, angular rate and magnetic field) as Euler angles for direct processing of the control algorithms. Alternatively, the use of quaternions for improved location calculation would be conceivable.

GPS values for height, course and speed over ground, and a sensor for the true airspeed (Pitot tube) supplement these data. The altitude is additionally barometric detected and measured in low altitudes by a radar sensor as true height above ground.

The main task of the on-board computer is the data acquisition and position control of the UAV. In addition, the communication channels to the ground station and the remote control are monitored by the on-board computer. In the computer network of the UAV the processed data is provided to the payload CPU afterwards.

It is impossible to predict all user applications in the payload. Therefore, the payload computer supports serial interfaces (UART, SPI, I²C), digital logic-I/O connections and power. The user does not need their own communication lines to the ground (Fig. 5).

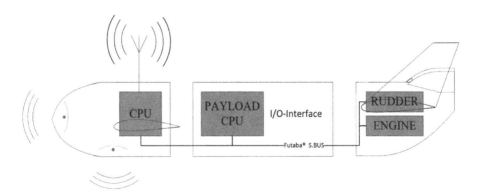

Fig. 5. Computer network and bus structure of the modular UAV PIRX³ᴰ

By means of the Futaba-SBUS.2 important real-time information is transmitted to the control surfaces. Roughly outlined it involves 26 bytes for the servos, which are updated on the bus every 15 ms. The data transfer rate is 100 kbps, which means a data volume of approx. 17 kBit/s. At first glance this doesn't seem to be particularly large, but the necessary real-time processing requires a worst-case response time of the CPU of max. 100 μs. That is the time, a changing event on the bus must be given an answer. In other words, the sum of all non-interruptible highest-priority system tasks (interrupts, DMA, etc.) must be less than 100 μs.

To meet the conflicting requirements, by on the one hand extremely fast response times and on the other hand a high level of abstraction in programming, cooperative multitasking is implemented with extremely low system overhead on the on-board computer. The total amount of the runtime of all tasks and interrupts for a tick must not exceed this time (sampling time, i.e. 20 ms). The sum of the running times of all nested interrupts is determined by the worst-case timing of bus communication (100 μs). The time management of all system functions (tasks, interrupts) requires care, but allows for the precise classification as deterministic software system with predictable response times. To optimize throughput times parts of the bus communication are configured as DMA (direct memory access).

4   Hardware Design of the Main Computer

Due to the very dedicated requirements regarding system optimization, the on-board computer has not been purchased as a "Component-of-the-shelf", but also redesigned. The pure material cost of the CPU fall in small or medium quantities of little consequence.

Fig. 6. The hardware of the flight computer with interfaces to the on-board, the payload, the extension module and the additional RADAR equipment

After careful consideration and some test trials, the hardware has been based on a STM32F446 by ST-Microelectronics. The CPU is clocked at 84 MHz and allows a compact electronics design in the result of the various integrated modules which the manufacturer has integrated on the silicon.

The entire electronic system is constructed as a stack of a few boards. This modularity makes a good adaptation to customer requirements possible. On top, the basic design of the on-board computer is also used as a payload computer and for controlling the engine in the aircraft. With slight modifications, the board computer serves as a centerpiece of the ground station and it can even be used in combination with the Reflex flight simulator to test the control algorithms in the laboratory. In Fig. 6 a detailed picture is shown. Details for the design of the electronics are described in [4, 5].

5   Airframe and System Design

Issues of transport and set-up time play an important role for the practical use. Therefore the airframe is constructed of three mechanical parts. The sensor module contains the on-board computer, including the position detectors, the barometer, the pitot tube, GPS, telemetry and remote control receiver as well as the optional radar equipment (Fig. 7).

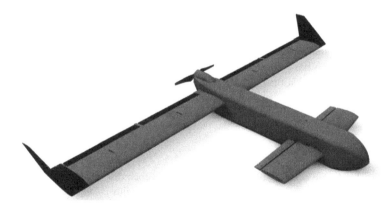

Fig. 7. The first prototype of the UAV PIRX^{3D} manufactured from 3D printed parts

In the mounting frame between these functional modules are the electrical connectors and the mating mechanical detent and stop points. Originally, the functional units of special foam (EPP) have been thermally cut. The thusproduced components are manufactured inexpensively and relatively quickly. Unfortunately, the quality of the components meet the requirements only insufficiently. The first flight tests were still successful, confirming the usefulness of the selected design.

Therefore the next step, the change to production of the modules, was tested by using a 3D printer. A major weakness of the technology quickly stepped to day - the great weight of the 3D printed parts. In the first experiment all 3D printed parts have been designed as a hollow body with an internal honeycomb structure. However, the computer-generated structures have not achieved the desired weight, it is still too much material in areas where no stability is required. The structure of the airframe in a CAD calculated honeycomb structure provides significant weight savings. Unfortunately, this way can only be gone up to a certain limit. The smaller the printed wall thickness, the greater is the risk of delamination under stress.

The greatest stresses occur within the payload module, between the connection set of the modules and in the wing attachment. That is why the end rib of the sensor module, payload box and the engine unit are printed in composite. Composite, in this context, a glass fiber plate is covered with 3D printer material. The outer shell is made by 3D printing.

6 Conclusion

Based on the requirements, it has been possible to develop a powerful design of a UAV. Only under consideration of the specific features of electronic hardware, customized software and specialized mechanics, an acceptable compromise for the overall system can be found. Besides the known CAD development tools for mechanical design, circuit design and software development, a simulation system has been created, which allows the validation of the flight control system under laboratory conditions. The results obtained with this simulator are accurate enough to perform successful field experiments.

The inclusion of 3D printers in the prototype was not initially provided. It was not until after the first (successful) test flights loomed the limits of the use of foam materials, that the step has ventured to redesign. This part of the design has not been complete yet, the optimization of the 3D printed parts is still in full swing. Nevertheless, it can be said that the use of 3D printers is a success.

References

1. Altenburg, J.: AONE - A highly sophisticated test bench for Flight Control Systems. Scientific Reports - 21st International Scientific Conference Mittweida, 26–27 October 2011, No. 4 (2011). ISSN 1437-7624
2. http://www.futabarc.com/sbus/index.html
3. von Eichel-Streiber, J.: Entwicklung und Inbetriebnahme einer eingebetteten Mehrachs-Regelung zur Flugstabilisierung eines UAV. Master thesis, TH Bingen (2016)
4. Hilgert, C.: Entwurf und Realisierung eines Bordcomputers für ein UAV. Bachelor thesis, FH Bingen (2014)
5. Hilgert, C.: Spezifizierung und Analyse von Radarsensoren zur Nutzung in unbemannten Flugsystemen. Master thesis, TH Bingen (2016)
6. Franke, H.-M.: Flugregler-Systeme. Oldenburg Verlag, München Wien (1968)
7. Altenburg, J.: INES – Intelligentes Elektroniksystem für Flugdrohnen, Scientific Reports – 23rd International Scientific Conference Mittweida, 05–06 October 2014, No. 3 (2011). ISSN 1437-7624

The Models that Can Be Matched by Feedback

Vladimír Kučera[(⊠)] [ID]

Czech Institute of Informatics, Robotics and Cybernetics,
Czech Technical University, Prague, Czech Republic
vladimir.kucera@cvut.cz

Abstract. Model matching (or exact model matching) is a problem of great interest in systems theory and applications. It consists of compensating a given system so as to achieve a specified target transfer function matrix. For linear time-invariant systems and regular static state feedback compensation, the solution is well known. The non-regular case, when the number of external inputs does not match the number of control inputs, the problem has not been solved until recently. The solution is not entirely algorithmic and requires some kind of trial and error. That is why it may be of interest to be able to decide on solvability prior to determining a solution. This is made possible by an efficient parametrization of all target models that can be matched by feedback applied to the given system.

Keywords: Linear systems · Model matching · Static state feedback

1 Model Matching

Given a linear, time-invariant system (A, B, C, D) with n states, m inputs, and p outputs described by the equations

$$\dot{x}(t) = Ax(t) + Bu(t), \quad y(t) = Cx(t) + Du(t) \tag{1}$$

that gives rise to the $p \times m$ transfer function matrix

$$T(s) = C(sI - A)^{-1}B + D,$$

we seek to determine a static state feedback law (F, G) of the form

$$u(t) = Fx(t) + Gv(t) \tag{2}$$

with r external inputs v such that the closed-loop transfer function matrix

$$T_{F,G}(s) = (C + DF)(sI - A - BF)^{-1}BG + DG \tag{3}$$

equals a specified, $p \times r$ proper rational matrix $T_m(s)$.

This work was supported by the Technology Agency of the Czech Republic under Project TE01020197 Center for Applied Cybernetics.

R. Moreno-Díaz et al. (Eds.): EUROCAST 2017, Part I, LNCS 10671, pp. 191–196, 2018.
https://doi.org/10.1007/978-3-319-74718-7_23

The problem is difficult to solve in full generality. In case $r = m$ and G is square and nonsingular the state feedback law is said to be regular and the problem greatly simplifies. The original regular solution was published in [1] whereas a recent overview of methods and results related to regular model matching is presented in [2].

The solvability conditions in the general, non-regular case are obtained as follows. Let $N_1(s)$, $D_1(s)$ be right coprime polynomial matrices such that

$$(sI - A)^{-1}B = N_1(s)D_1^{-1}(s) \tag{4}$$

and $D_1(s)$ is column reduced with column degrees $d_1, d_2, \ldots, d_m$. Denote

$$N(s) := CN_1(s) + DD_1(s), \tag{5}$$

so that

$$T(s) = N(s)D_1^{-1}(s). \tag{6}$$

Thus, the input-to-state transfer function matrix (4) and the (input-to-output) transfer function matrix (6) have the same denominator matrix $D_1(s)$ but different numerator matrices $N_1(s)$ and $N(s)$. Note that while the pair $N_1(s)$, $D_1(s)$ is right coprime by definition, the implied pair $N(s)$, $D_1(s)$ need not be right coprime.

Finally, let $N_m(s)$, $D_m(s)$ be right coprime polynomial matrices such that

$$T_m(s) = N_m(s)D_m^{-1}(s). \tag{7}$$

Theorem 1. There exists a solution of the model matching problem via static state feedback if and only if

(a) the equation

$$N(s)U(s) = N_m(s)V(s) \tag{8}$$

admits a polynomial matrix solution pair $U(s)$, $V(s)$ such that $V(s)$ is $r \times r$ and nonsingular;

(b) there exist a polynomial matrix $H(s)$ and a constant matrix Z, where $H(s)$ is $m \times m$ and nonsingular, column reduced and with column degrees $d_1, d_2, \ldots, d_m$, such that

$$H^{-1}(s)Z = U(s)[D_m(s)V(s)]^{-1}. \tag{9}$$

Proof. [3].   □

The synthesis of a matching feedback law is illustrated by a simple example.

Example 1. Consider a system (1) with

$$A = \begin{bmatrix} 0 & 0 \\ 0 & -1 \end{bmatrix}, \quad B = \begin{bmatrix} 1 & 0 & 0 \\ 0 & 1 & 0 \end{bmatrix}, \quad C = \begin{bmatrix} 1 & 0 \\ 0 & 1 \end{bmatrix}, \quad D = \begin{bmatrix} 0 & 0 & 1 \\ 0 & 0 & 0 \end{bmatrix}, \quad (10)$$

which gives rise to the transfer function matrix

$$T(s) = \begin{bmatrix} \frac{1}{s} & 0 & 1 \\ 0 & \frac{1}{s+1} & 0 \end{bmatrix}.$$

Then the input-to-state transfer function matrix (4) reads

$$(sI - A)^{-1}B = \begin{bmatrix} \frac{1}{s} & 0 & 0 \\ 0 & \frac{1}{s+1} & 0 \end{bmatrix} = \begin{bmatrix} 1 & 0 & 0 \\ 0 & 1 & 0 \end{bmatrix} \begin{bmatrix} s & 0 & 0 \\ 0 & s+1 & 0 \\ 0 & 0 & 1 \end{bmatrix}^{-1} = N_1(s)D_1^{-1}(s)$$

and a fractional description of $T(s)$ is given by (6)

$$T(s) = C(sI - A)^{-1}B + D = \begin{bmatrix} \frac{1}{s} & 0 & 1 \\ 0 & \frac{1}{s+1} & 0 \end{bmatrix} = \begin{bmatrix} 1 & 0 & 1 \\ 0 & 1 & 0 \end{bmatrix} \begin{bmatrix} s & 0 & 0 \\ 0 & s+1 & 0 \\ 0 & 0 & 1 \end{bmatrix}^{-1}$$

$$= N(s)D_1^{-1}(s)$$

on using (5).

Also specified is the model transfer function matrix

$$T_m(s) = \begin{bmatrix} \frac{s}{s+1} \\ \frac{1}{s+1} \end{bmatrix}$$

whose right coprime fractional description (7) is given by

$$N_m(s) - \begin{bmatrix} s \\ 1 \end{bmatrix}, \quad D_m(s) = s+1.$$

Since $m = 3$ and $r = 1$, we have to solve a non-regular model matching problem.
To determine $U(s)$ and $V(s)$ satisfying condition (a) of Theorem 1, one can determine a polynomial basis for the right kernel of the matrix

$$[N(s) \quad -N_m(s)] = \begin{bmatrix} 1 & 0 & 1 & -s \\ 0 & 1 & 0 & -1 \end{bmatrix},$$

which is

$$\begin{bmatrix} s & -1 \\ 1 & 0 \\ 0 & 1 \\ 1 & 0 \end{bmatrix}. \quad (11)$$

To determine $H(s)$ and Z that satisfy condition (b) of Theorem 1, one can determine a polynomial basis for the left kernel of the matrix

$$\begin{bmatrix} U(s) \\ -D_m(s)V(s) \end{bmatrix}.$$

By trial and error, we select a suitable linear polynomial combination of the basis vectors in (11) as

$$\begin{bmatrix} U(s) \\ V(s) \end{bmatrix} = \begin{bmatrix} s \\ 1 \\ 0 \\ \overline{1} \end{bmatrix} + s \begin{bmatrix} -1 \\ 0 \\ 1 \\ 0 \end{bmatrix} = \begin{bmatrix} 0 \\ 1 \\ s \\ \overline{1} \end{bmatrix}.$$

Then

$$[H(s) \quad Z] = \begin{bmatrix} s & 0 & 0 & 0 \\ 0 & s+1 & 0 & 1 \\ 0 & 1 & 1 & 1 \end{bmatrix}$$

and we note that the column degrees of $H(s)$ are indeed the same as those of $D_1(s)$, namely 1, 1, 0 and that Z is constant.

The matrix

$$W(s) := D_1(s)H^{-1}(s)Z$$

can be thought of as a compensator that achieves the matching, because

$$T(s)W(s) = N(s)H^{-1}(s)Z = [N(s)U(s)][D_m(s)V(s)]^{-1} = N_m(s)D_m^{-1}(s) = T_m(s). \tag{12}$$

In order to realize the action of $W(s)$ on the system by static state feedback, solve the equation

$$XD_1(s) + YN_1(s) = H(s) \tag{13}$$

for constant matrices X, Y such that X is nonsingular. Then we obtain a matching feedback law by setting

$$F = -X^{-1}Y = \begin{bmatrix} 0 & 0 \\ 0 & 0 \\ 0 & -1 \end{bmatrix}, \quad G = X^{-1}Z = \begin{bmatrix} 0 \\ 1 \\ 1 \end{bmatrix}.$$

The system and matching feedback is visualized in Fig. 1. The given system is shown in solid line and the feedback in dashed line. □

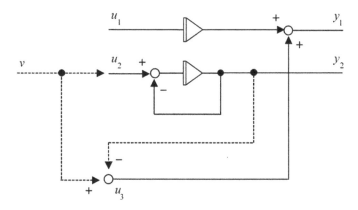

Fig. 1. A visualization of the system and matching feedback.

2   Parameterization of Matchable Models

The construction of matching feedback, however, is not entirely algorithmic and involves some kind of trial and error. This makes the task difficult.

It is therefore of interest to be able to check *a priori* whether a given model can be matched or not with the given system. A way to do it is to explicitly characterize, preferably in parametric form, the class of models that can be matched by compensating a given system, that is to say, to describe the set of the closed-loop transfer function matrices $T_{F,G}(s)$ obtainable by feedback. Such a parametrization is provided by all constant matrices F and G such that Eq. (3) holds. We shall find a more efficient parametrization, however.

Theorem 2. The set of all $p \times r$ model transfer function matrices that can be matched with a given system (1) by static state feedback (2) is given by

$$T_m(s) = N(s)[D_1(s) + Q(s)]^{-1}G \tag{14}$$

where $Q(s)$ is any $m \times m$ polynomial matrix such that $Q(s)D_1^{-1}(s)$ is strictly proper and G is any $m \times r$ real matrix.

Proof. The proof is based on the fact [4] that the set P of m-row polynomial vectors p (s) such that $p(s)D_1^{-1}(s)$ is strictly proper, is a real vector space of dimension $d :=$ $d_1 + d_2 + \ldots + d_m$, and the rows of $N_1(s)$ form a basis for P.

It follows from (12) and (13) that

$$T_m(s) = N(s)H^{-1}(s)Z = N(s)[XD_1(s) + YN_1(s)]^{-1}Z = N(s)[D_1(s) - FN_1(s)]^{-1}G.$$

Comparing with (14), we conclude that $Q(s) := -FN_1(s)$ is any polynomial matrix such that $Q(s)D_1^{-1}(s)$ is strictly proper.   □

Example 2. Consider the system of Example 1 and parameterize all 2×1 transfer function matrices $T_m(s)$ that can be obtained by static state feedback matching.

Since $r = 1$, we set

$$G = \begin{bmatrix} g_1 \\ g_2 \\ g_3 \end{bmatrix}$$

and all 3×3 polynomial matrices $Q(s)$ such that $Q(s)D_1^{-1}(s)$ is strictly proper are in fact constant, of the form

$$Q(s) = \begin{bmatrix} q_{11} & q_{12} & 0 \\ q_{21} & q_{22} & 0 \\ 0 & 0 & 0 \end{bmatrix}.$$

Thus,

$$T_m(s) = N(s)[D_1(s) + Q(s)]^{-1}\, G = \frac{1}{s^2 + \alpha_1 s + \alpha_0} \begin{bmatrix} \beta_2 s^2 + \beta_1 s + \beta_0 \\ \gamma_1 s + \gamma_0 \end{bmatrix}$$

where α_0, α_1, β_0, β_1, β_2, γ_0, and γ_1 is a set of free real parameters, which depend on g_1, g_2, g_3, q_{11}, q_{12}, q_{11}, q_{21}, and q_{22}. Note that we need 7 parameters, whereas (F, G) involves 9.

The model we have matched in Example 1, namely

$$T_{m1}(s) = \begin{bmatrix} \frac{s}{s+1} \\ \frac{1}{s+1} \end{bmatrix},$$

corresponds to $\alpha_0 = 0$, $\alpha_1 = 1$, $\beta_0 = \beta_1 = 0$, $\beta_2 = 1$, $\gamma_0 = 0$, and $\gamma_1 = 1$. On the other hand, the transfer function matrix

$$T_{m2}(s) = \begin{bmatrix} \frac{1}{s+1} \\ \frac{s}{s+1} \end{bmatrix}$$

cannot be matched using static state feedback.   □

References

1. Wolovich, W.A.: The use of state feedback for exact model matching. SIAM J. Control **10**, 512–523 (1972). https://doi.org/10.1137/0310039
2. Kučera, V., Castañeda Toledo, E.: A review of stable exact model matching by state feedback. In: Proceedings of the 22nd Mediterranean Conference on Control and Automation, Palermo, pp. 85–90 (2014). https://doi.org/10.1109/MED.2014.6961331
3. Kučera, V.: Stable model matching by non-regular static state feedback. IEEE Trans. Auitom. Control **61**, 4138–4142 (2016). https://doi.org/10.1109/TAC.2016.2547871
4. Hautus, M.L.J., Heymann, M.: Linear feedback – an algebraic approach. SIAM J. Control. Optim. **16**, 83–105 (1978). https://doi.org/10.1137/0316007

A Performance Assessment of Network Address Shuffling in IoT Systems

Aljosha Judmayer[1], Georg Merzdovnik[1,2], Johanna Ullrich[1],
Artemios G. Voyiatzis[1,2(✉)], and Edgar Weippl[1,2]

[1] SBA Research, Vienna, Austria
{ajudmayer,gmerzdovnik,jullrich,avoyiatzis,eweippl}@sba-research.org
[2] TU Wien, Vienna, Austria

Abstract. While the large scale distribution and unprecedented connectivity of embedded systems in the Internet of Things (IoT) has enabled various useful application scenarios, it also poses a risk to users and infrastructure alike. Recent incidents, like the Mirai botnet, have shown that these devices are often not sufficiently protected against attacks and can therefore be abused for malicious purposes, like distributed denial of service (DDoS) attacks. While it may be an impossible task to completely secure all systems against attacks, moving target defense (MTD) has been proposed as an alternative to prevent attackers from finding devices and endpoints and eventually launching their attacks against them. One of these approaches is network-based moving target defense which relies on the obfuscation and change of network level information, like IP addresses and ports. Since most of these approaches have been developed with desktop applications in mind, their usefulness in IoT applications has not been investigated.

In this paper we provide a study on the applicability of network-based MTD for low-power devices. We investigate their capabilities to regularly change addresses. We furthermore investigate their performance with multiple assigned IP addresses, for both IPv4 and IPv6. We show that although some functionality of these systems may be impeded by constantly changing addresses, network-based MTD might nonetheless be a viable option to protect Internet-connected embedded systems from attacks.

1 Introduction

Networked embedded systems of different forms and shapes are the building blocks for the Internet of Things (IoT). The security of these systems is already of paramount importance for the function of the Internet. Despite constraints in memory, processing, and power, network connectivity is in most cases more than ample. IoT devices often have network access interfaces of 100 Mbps or even 1 Gbps. This combined with the exploding number of Internet-connected devices gives rise to new forms of attacks.

The two most severe distributed denial of service (DDoS) attacks ever faced on the Internet, with an aggregate traffic volume of more than 1.1 Tbps, occurred

© Springer International Publishing AG 2018
R. Moreno-Díaz et al. (Eds.): EUROCAST 2017, Part I, LNCS 10671, pp. 197–204, 2018.
https://doi.org/10.1007/978-3-319-74718-7_24

in 2016. These attacks became feasible due to the availability of numerous vulnerable and subsequently compromised IoT systems, including devices such as digital video recorders (DVRs), IP cameras, and smart thermostats.

The resource constraints of these embedded systems make it difficult to integrate network security mechanisms that are commonplace in enterprise IT environments. Furthermore, the former are often part of smart homes and smart environments. In these settings, the consumers opt for a set-and-forget approach. In addition, software updates and upgrades are in many cases impossible to realize.

In this modus operandi, it is crucial to engineer defenses that are *preventive* in nature. Most of the IoT systems are, from a network perspective, *"sitting ducks"*, i.e., they are easily accessible from the network and passively receive all kind of attacks aiming to compromise them.

Moving target defense (MTD) is an approach to improve the standing of defending information systems in general by breaking the attacker-defender asymmetry [11]. The key assumption for MTD is that the attackers will first perform reconnaissance to identify possible targets. Then, at a next phase, they launch their (targeted) attack against the devices found before. Under this assumption, MTD dictates to mobilize available resources, so that attackers hit wrong or non-existent targets and the defenders thus succeed in protecting their systems; collect evidence of the attacker behavior; and provide enough time to deploy network-wide defenses (e.g., honeypots) for further studying their practices and delaying further attacks [19].

While host- and application-level MTDs are hard to realize in embedded systems, network-level ones (e.g., a time-varying topology) are considered feasible [3]. IPv6 and IPv4 network address shuffling, i.e., periodically changing the network addresses of the devices in a coordinated way, is an example of network-level MTD [2].

In this paper, we augment existing literature by exploring the capability of modern IoT systems to handle network address shuffling. More specifically, we study the performance overhead and the impact of periodically changing network addresses and ports in Linux-based IoT systems (namely, Raspberry Pi and Carambola2) under different probing and network scanning activity scenarios.

2   Network-Based MTD

Moving target defense aims to remove the attacker-defender asymmetry [11]. In a static system, an attacker has plenty of time to gain information about the victim, i.e., for performing reconnaissance, before eventually launching the attack.

MTD permanently adapts a system for reasons of defense. The system under protection regularly changes its appearance rendering the attacker's information from previous reconnaissance worthless when finally launching the attack against the system. MTD relies on the time gap between the reconnaissance phase and the actual attack phase. Protection following the principle of MTD might be applied at different abstraction levels, including memory [14], software [4, 10], and the network layer [13].

2.1 MTD in IPv4 and IPv6 Networks

In the network level, MTD relies on regularly changing addresses to reduce an attacker's chance of hitting a victim. MTD based on address shuffling have been proposed for both, IPv4- and IPv6-based networks.

Networks based on IPv4 exhibit already address scarcity, and in consequence, frequent address changes have remained a challenge. DHCP [5] assigns addresses in a centralized fashion, leading to a certain degree of slow-paced address shuffling. An IPv4-based approach for MTD was proposed in [15]. There, the sender decides for the correct destination address out of an address pool using a deterministic function. The latter chooses the respective address based on the timestamp field which implies that addresses change with a granularity of milliseconds. This approach however bears the drawback of requiring multiple routers spanning several administrative domains in order to gain a pool of IPv4 addresses.

IPv6 has been developed to overcome the address scarcity of IPv4 and has quadrupled the address length. This leads to a sparsely-occupied IPv6 address space for the time being. IPv6 address assignment is performed in a partially decentralized way [17]. A router announces the first part of the address (network prefix). The second part is chosen by the host itself providing the freedom to change the address whenever desired.

With the high amount of potential addresses, this new version also increases its applicability for MTD. Nevertheless, early approaches of IPv6 addresses defined static interface identifiers [9] or semi-static ones [1, 8]. The very first dynamic address format has been the privacy extension [16]; the address changes in an interval of 24 h per default to protect against passive correlation attacks. Privacy addresses have originally been intended to be used in addition to static ones as the consequences on networking were inestimable back then, even when running at a moderate interval of 24 h. Intended for clients initiating outgoing requests but not awaiting incoming ones, identifiers could be generated randomly – even though the specification defines an insecure pseudo-random algorithm [18].

Beyond, there are two approaches that generate addresses based on a shared key: MT6D calculates a receiver's addresses in an interval of ten seconds; due to its extra protocol header it however adds overhead [6, 7]. 6HOP defines a similar algorithm [12], but embeds itself natively into IPv6 networking, i.e., without an additional header.

The performance of dynamic address schemes is investigated only marginally: [7] reached a transmission speed of about 1.8 MB/s at a fixed interval of 10 seconds but neither analyzed the impact of shorter intervals nor of multiple simultaneously assigned addresses on networking.

2.2 Moving to an Internet of Things

Internet-of-Things (IoT) devices are more often than not resource-constrained. Thus, sophisticated security solutions cannot be deployed. Also, software tailored for IoT systems appears to be developed without security in mind. This results in exploitable applications and systems lacking a proper update mechanism for software patching.

From an IoT deployment perspective, there is often only a limited perimeter protection, if any, as for example a simple network firewall. This is more evident in private households positioning vulnerable IoT devices in an unprotected path on the Internet. Either way, IoT devices have to assign an address in order to fulfill their task. In this situation, it appears beneficial to create these addresses in a way providing most security. Including the principle of MTD into address assignment, address shuffling changes the device's address in regular intervals, constraining the attacker's time to perform an attack.

The faster the addresses alternate, the better in general is the level of protection. However, network stacks and network communication have been designed with long-living, almost static addresses in mind. Rapidly alternating addresses could potentially ruin network functionality and service.

The implication of fast-changing addresses on IoT networking are yet unclear. The paper at hand aims to shed light on these aspects.

3 Experimental Setup

To understand the impact of network-based MTD on IoT systems, we investigated different network level aspects on several low-power devices.

One aspect we are interested in is how fast devices are able to change IP addresses since this is a vital aspect for MTD. Depending on the specific scheme that is used, it might be possible that a device is not only reachable on a single address, but has to listen to several aliases. Therefore, we are interested in the performance impacts of having multiple addresses assigned to the same network interface. Furthermore, we want to measure the impact of constantly changing addresses on network transfers.

For our measurements, we selected five different Linux-based system on chips (SoC). The measurements were implemented in the form of bash scripts to make them portable and comparable across devices. IP address changes were performed with the *iproute2* tool suite. Measurements were performed on the devices listed in Table 1.

Our test script continuously adds and/or deletes addresses to/from a single interface. In particular, the script adds addresses to the interface without deleting them to evaluate the impact of multiple addresses. Thereby, the accumulated amount of addresses is assigned to the network interface. The performance impact on network operations is investigated by measuring the response time of the embedded device to five ping commands per second from another computer connected to the same network switch. To ping request were addressed toward the very first address that had been assigned to our system under test by DHCP.

Table 1. Test-devices

Device	SoC	CPUs	Clock speed	Memory
Raspberry Pi Model B	Broadcom BCM2835 (ARMv6Z)	1	700 MHz	512 MB
Raspberry Pi 2	Broadcom BCM2836 (Armv7-A)	4	900 MHz	1 GB
Raspberry Pi 3	Broadcom BCM2837 (ARMv8-A)	4	1.2 GHz	1 GB
Carambola2	Qualcomm/Atheros AR9331 SoC	1	400 MHz	64 MB

4 Results

Our findings indicate that network address shuffling is feasible in IoT environments. However, special care must be taken when implementing such techniques. The number of possible IP address changes per second varies significantly per device, as depicted in Fig. 1.

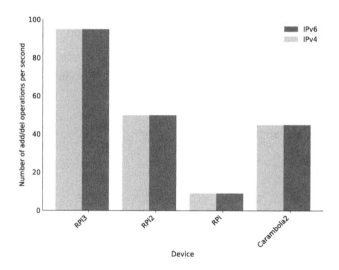

Fig. 1. Rate of address changes (add/delete operations) per second and device

The simultaneous use of multiple IP addresses at a single interface has an impact on the performance, especially in the case of IPv6, as depicted in Fig. 2. The more addresses already in use, the more time it needs to add new ones. Interestingly enough, Carambola2, running OpenWRT with a clock rate of 400 MHz clearly outperforms the *stronger* Raspberry Pi B+ clocked at 700 MHz, reaching the figures of Raspberry Pi 2, which is clocked at 900 MHz and has four cores.

Finally, continuous assignment of new address also manifests itself in an impact on network performance. Figure 3 provides the average round trip times for ping requests to the device under test during continuous assignment of new

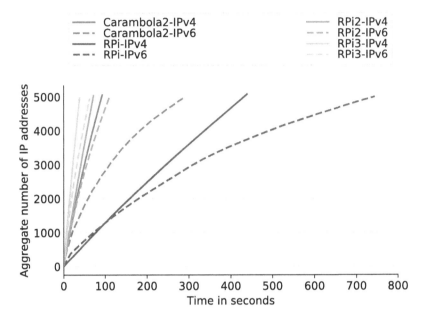

Fig. 2. Total number of IP addresses (add operations only)

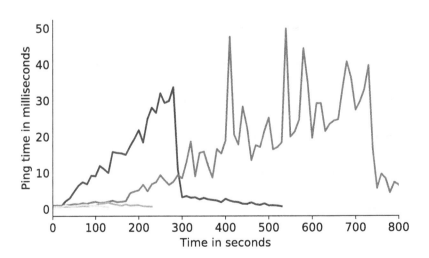

Fig. 3. Average ping times during IPv6 address add operations

IPv6 addresses. Cumulative assignment of addresses not only impacts the performance of adding new addresses, it furthermore influences the response time for ping requests. The more addresses are already added, the longer it takes for the device to respond. However, this is only while constantly adding new addresses. As soon as address assignment is stopped, and deletion of addresses is started again, response times immediately drop to a lower level again.

5   Conclusions and Future Work

Our results provide evidence that MTD based on network address shuffling is a viable method to defend IoT devices from attackers. Even the low-power devices that have been investigated in our study are capable of changing the address multiple times per second. Although network operation can be impeded if many addresses are assigned to the same interface simultaneously, this problem will not arise for most applications where only a single address needs to be assigned. A problem when permanently changing addresses is the possibility of collisions. While this might pose a problem for IPv4-based schemes, the huge address space of IPv6 allows much more freedom during address selection.

Acknowledgments. This work was supported partly by the Christian Doppler Forschungsgesellschaft (CDG) through Josef Ressel Center (JRC) projects TARGET and u'smile and the Austrian Research Promotion Agency (FFG) through projects SBA-K1, A2Bit, and CyPhySec.

References

1. Aura, T.: Cryptographically Generated Addresses (CGA). RFC 3972 (Proposed Standard), March 2005. http://www.ietf.org/rfc/rfc3972.txt. Updated by RFCs 4581, 4982
2. Cai, G., Wang, B., Wang, X., Yuan, Y., Li, S.: An introduction to network address shuffling. In: 18th International Conference on Advanced Communication Technology (ICACT), pp. 185–190. IEEE (2016)
3. Casola, V., De Benedictis, A., Albanese, M.: A moving target defense approach for protecting resource-constrained distributed devices. In: 14th International Conference on Information Reuse and Integration (IRI), pp. 22–29. IEEE (2013)
4. Christodorescu, M., Fredrikson, M., Jha, S., Giffin, J.: End-to-end software diversification of internet services. In: Jajodia, S., Ghosh, A., Swarup, V., Wang, C., Wang, X. (eds.) Moving Target Defense. Advances in Information Security, vol. 54, pp. 117–130. Springer, New York (2011). https://doi.org/10.1007/978-1-4614-0977-9_7
5. Droms, R.: Dynamic Host Configuration Protocol. RFC 2131 (Draft Standard), March 1997. http://www.ietf.org/rfc/rfc2131.txt. Updated by RFCs 3396, 4361, 5494, 6842
6. Dunlop, M., Groat, S., Urbanski, W., Marchany, R., Tront, J.: MT6D: a moving target IPv6 defense. In: 2011 Military Communications Conference (MILCOM 2011), pp. 1321–1326, November 2011

7. Dunlop, M., Groat, S., Urbanski, W., Marchany, R., Tront, J.: The blind man's bluff approach to security using IPv6. IEEE Secur. Privacy **10**(4), 35–43 (2012)
8. Gont, F.: A Method for Generating Semantically Opaque Interface Identifiers with IPv6 Stateless Address Autoconfiguration (SLAAC). RFC 7217 (Proposed Standard), April 2014. http://www.ietf.org/rfc/rfc7217.txt
9. Hinden, R., Deering, S.: IP Version 6 Addressing Architecture. RFC 4291 (Draft Standard), February 2006. http://www.ietf.org/rfc/rfc4291.txt. Updated by RFCs 5952, 6052, 7136, 7346, 7371
10. Huang, Y., Ghosh, A.K.: Introducing diversity and uncertainty to create moving attack surfaces for web services. In: Jajodia, S., Ghosh, A., Swarup, V., Wang, C., Wang, X. (eds.) Moving Target Defense. Advances in Information Security, vol. 54, pp. 131–151. Springer, New York (2011). https://doi.org/10.1007/978-1-4614-0977-9_8
11. Jajodia, S., Ghosh, A.K., Swarup, V., Wang, C., Wang, X.S. (eds.): Moving Target Defense: Creating Asymmetric Uncertainty for Cyber Threats, vol. 54. Springer Science & Business Media, Heidelberg (2011). https://doi.org/10.1007/978-1-4614-0977-9
12. Judmayer, A., Merzdovnik, G., Ullrich, J., Voyiatzis, A., Weippl, E.: Lightweight address hopping for defending the IPv6 IoT. In: International Conference on Availability, Reliability and Security (ARES) (2017)
13. Kampanakis, P., Perros, H., Beyene, T.: SDN-based solutions for moving target defense network protection. In: Proceeding of IEEE International Symposium on a World of Wireless, Mobile and Multimedia Networks, pp. 1–6, June 2014
14. Kil, C., Jun, J., Bookholt, C., Xu, J., Ning, P.: Address space layout permutation (ASLP): towards fine-grained randomization of commodity software. In: 2006 22nd Annual Computer Security Applications Conference (ACSAC 2006), pp. 339–348, December 2006
15. Krylov, V., Kravtsov, K.: IP fast hopping protocol design. In: 10th Central and Eastern European Software Engineering Conference in Russia, CEE-SECR 2014, pp. 11:1–11:5 (2014)
16. Narten, T., Draves, R., Krishnan, S.: Privacy Extensions for Stateless Address Autoconfiguration in IPv6. RFC 4941 (Draft Standard), September 2007. http://www.ietf.org/rfc/rfc4941.txt
17. Thomson, S., Narten, T., Jinmei, T.: IPv6 Stateless Address Autoconfiguration. RFC 4862 (Draft Standard), September 2007. http://www.ietf.org/rfc/rfc4862.txt. Updated by RFC 7527
18. Ullrich, J., Weippl, E.: Privacy is not an option: attacking the IPv6 privacy extension. In: Bos, H., Monrose, F., Blanc, G. (eds.) RAID 2015. LNCS, vol. 9404, pp. 448–468. Springer, Cham (2015). https://doi.org/10.1007/978-3-319-26362-5_21
19. Zhuang, R., DeLoach, S.A., Ou, X.: Towards a theory of moving target defense. In: Proceedings of the First ACM Workshop on Moving Target Defense, pp. 31–40. ACM (2014)

Mobile Wrist Vein Authentication Using SIFT Features

Pol Fernández Clotet[(✉)] and Rainhard Dieter Findling

Department of Mobile Computing, University of Applied Sciences Upper Austria,
Softwarepark 11, 4232 Hagenberg, Austria
pol.fernandez@students.fh-hagenberg.at, rainhard.findling@fh-hagenberg.at
http://www.fh-ooe.at/en/

Abstract. Biometrics have become important for authentication on modern mobile devices. Thereby, different biometrics are differently hard to observe by attackers: for example, veins used in vein pattern authentication are only revealed with specialized hardware. In this paper we propose a low cost mobile vein authentication system based on Scale-Invariant Feature Transform (SIFT). We implement our approach as vein recording and authentication prototype, evaluate it using a self recorded vein database, and compare results to other vein recognition approaches applied on the same data.

Keywords: Mobile authentication · Wrist veins · NIR · SIFT features

1 Introduction

Modern mobile devices have access to, store, and process much private information, including messaging (email, SMS), contacts, access to private networks (VPN, WiFi), or even mobile banking. Thus, many devices provide local device access protection mechanisms, such as PIN, password, or fingerprint authentication. With those the authentication secret could be observed by attackers and used in replay attacks. However, some biometrics are more difficult to observe by attackers, as they largely remain hidden without using special sensors. For example, observing biometric information is harder for vein than for face authentication. By combining multiple such biometrics, also including weak biometrics like gait, strong and reliable mobile authentication can be achieved [3].

One biometric authentication less explored with mobile devices is vein authentication, which has gained popularity outside the mobile environment for being contactless. As skin largely absorbs the visible spectrum of light, veins mostly remain hidden in normal conditions, which prevents vein from being reliably observed in this spectrum. Light in the (NIR) or infrared (IR) spectrum has maximum depth of penetration of skin tissue. Hence, veins are illuminated with NIR/IR light and captured using cameras with optical NIR/IR bandpass filters [11,14]. Most vein authentication approaches use finger, hand dorsal, palm, or wrist vein patterns [13,15], with vein capturing devices designed for medical

© Springer International Publishing AG 2018
R. Moreno-Díaz et al. (Eds.): EUROCAST 2017, Part I, LNCS 10671, pp. 205–213, 2018.
https://doi.org/10.1007/978-3-319-74718-7_25

and security fields of research [6]. For mobile users wrist veins have the advantage of being relatively easy to access – which could be used e.g. with smart-watches, thereby not requiring any additional effort or changes in user behavior.

In this paper we investigate mobile wrist authentication using SIFT features. Our main contributions are (a) development of a low cost wrist vein capturing device, (b) using SIFT features for vein authentication, and (c) evaluation of our approach using a new wrist vein database with 120 wrist vein images from 30 participants.

2   Related Work

Different approaches to vein visualization, recognition, and authentication have been proposed for medical and security purposes [2,8]. In this section we provide an overview of approaches most important to mobile vein authentication. In [13] NIR (800 nm) and far infrared (FIR, 800–1400 nm) light is used to acquire vein pattern images of the wrist, back, and palm of the hand. The capturing device is mounted on a board with a charge-coupled device (CCD) camera with IR filter and NIR lamp. After image acquisition, they use skeletonization to obtain vein patterns and perform matching by measuring the Hausdorff distance (LHD) between different patterns. Within a similar setup [10] uses fast spatial correlation for matching hand vein patterns in vein authentication.

In [9] wrist veins are captured using a physical structure and NIR illumination. They collect a database of 5 samples from the left and right hand for each of 50 subjects. Evaluation is done comparing nine different state-of-the-art vein matching techniques. Results indicate that Log-Gabor and Sparse Representation Classifier (LG-SRC) are the models with the best vein matching performance. [7] uses a local threshold for vein segmentation and 2D correlation coefficient for classification of obtained vein patterns. They evaluate on a self-recorded database of 1200 wrist images acquired from 50 volunteers for both left and right hands.

Finally [6] describes the design, development and initial evaluation of mVein-Vision, a mobile medical application for assisting and improving venipuncture. The application is implemented on a standard mobile device and intended to be a low cost alternative to commercial NIR devices. In contrast to our work, they only focus on vein detection and visualization as educational and clinical tool.

Summarizing, previous work does not yet combine wrist vein authentication with mobile environments. Therefore, our goal is to build and evaluate a mobile wrist vein authentication system, for which we propose to combine a low cost mobile vein capturing system with SIFT features for vein authentication.

3   Wrist Vein Authentication Based on SIFT Features

We present a low cost, mobile wrist vein authentication system, utilizing the visualization principle adopted from [6]. Our approach consists of three constituent parts (Fig. 1a) vein visualization and data collection using NIR illumination, (b) vein image enhancement to obtain clear vein patterns, and (c) vein authentication based on vein pattern matching using SIFT features.

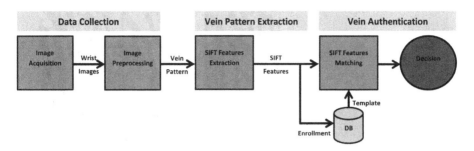

Fig. 1. Constituent parts of our approach to mobile vein pattern authentication.

3.1 Wrist Veins Capturing

Considering the good results of [7] we adopt their approach of using a low cost CCD camera with NIR illumination for vein capturing. As we operate in a mobile environment, wrists cannot be assumed to be placed in front of the sensor in a fixed or uniform position. This freedom of positioning implies three challenges that need to be addressed for successful vein authentication: non-uniformness in shift, rotation, and scale of sensor data. One could use hand pegs (cf. [9]) to address shift and rotation. However this would make capturing images in a mobile environment overly cumbersome. Consequently, we instead use a region of interest (ROI) of about 5.8×9.7 cm size when capturing vein images. Users position their wrist accordingly inside the ROI, then the image is cropped to only contain information within the ROI.

3.2 Image Preprocessing

After obtaining vein images (Fig. 2a) we adopt preprocessing from [7] to increase quality and visibility of vein patterns. We apply a 3×3 median and Gaussian Blur filter to reduce noise (Fig. 2b), image binarization and a 15×15 mean auto local threshold (Fig. 2c), morphological closing to reduce outliers and sharpen veins (Fig. 2d), and pixel inversion to obtain veins as white and background as black pixels (Fig. 2e).

3.3 Wrist Vein Feature Extraction and Matching

After preprocessing vein pattern images we derive features to distinguish individuals based on their vein patterns. Such vein patterns could be slightly scaled and rotated to each other, resulting from the freedom of wrist position during capturing vein images. Thus, we use a SIFT features based matching algorithm [1] to extract and match features in a scale and rotation invariance manner (Fig. 3). Moreover, SIFT features have not yet been used for wrist vein recognition, but for extracting image characteristics in object recognition, movement detection, and image registers, and have proven to work for finger vein and face recognition [4,5].

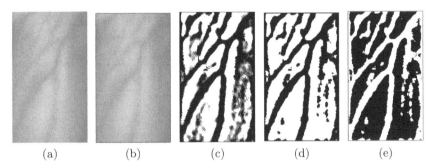

(a)   (b)   (c)   (d)   (e)

Fig. 2. Vein image preprocessing: sample after applying cropping (a), filtering (b), auto local threshold (c), morphological closing (d), and pixel inversion (e).

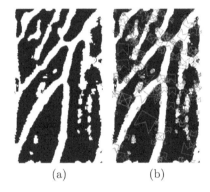

(a)   (b)

Fig. 3. Preprocessed vein pattern sample (a) and its extracted SIFT features (b).

Using the similarity between SIFT features of two vein patterns we derive if those are actually from the same person. For two samples I_A, I_B with corresponding SIFT features $S_a\{f_{A1}, f_{A2}, ..., f_{An}\}$ and $S_b\{f_{B1}, f_{B2}, ..., f_{Bm}\}$, our first step is to calculate a list of matching SIFT features L_{ab} between S_a and S_b: $L_{ab} = \{f_{A1} - f_{B3}, f_{A3} - f_{A2}, ...f_{An} - f_{Bm}\}$. L_{ab} already contains suitable matches between SIFT features of the two samples – based on which we propose to enhance the accuracy of feature matching by going one step further. We propose to use the Euclidean distance of all possible pairs of SIFT features of S_a and S_b, to ensure matched features in L_{ab} actually have the minimum distance compared to all other possible matches using the same features. Using the obtained L_{ab}, for each proposed matched pair of features ($f_{Ai} - f_{Bj}$ with $i \in [1, n]$ and $j \in [1, m]$), we calculate the Euclidean distance of these features $D(f_{Ai} - f_{Bj})$ to all other features $D(f_{Ai} - f_{B1}), D(f_{Ai} - f_{B2}), ..., D(f_{Ai} - f_{Bm})$. If thereby $D(f_{An} - f_{Bm})$ is the minimum distance we say that $f_{An} - f_{Bm}$ are a feature match (Eq. 1):

$$\forall x : D(f_{Ai} - f_{Bj}) < D(f_{Ai} - f_{Bx}) \Rightarrow \text{feature match} \tag{1}$$

After obtaining all matching features between two vein patterns, the number of matches C_{ab} is used together with a predefined threshold τ as similarity between those patterns. If $C_{ab} \geq \tau$ we conclude that those patterns are originated by the same person, and by different people otherwise.

So far our approach acts in a 1:1 vein pattern comparison manner: it requires one sample to enroll users and one further sample to perform authentication. To improve authentication accuracy we propose to instead use majority voting with N vein pattern samples for both enrollment and authentication. Thus, during authentication, comparisons between N enrollment samples $I_{A,n}$ and N authentication samples $I_{B,m}$ result in N^2 individual results $C_{ab,i}$. We apply a majority voting like approach over all $C_{ab,i}$ to obtain an overall authentication result. Such can be done using mean, median, standard deviation, or similar, based on individual results. In our approach we compare using the mean and median with $N = 4$ samples, thereby on $N^2 = 16$ individual comparison results. The obtained similarity $\overline{C_{ab}}$ of two vein pattern samples is used with a threshold τ to decide if they were originated by the same person. If $\overline{C_{ab}} \geq \tau$ we say the samples are from the same person, respectively from different people otherwise.

4   Evaluation

To evaluate our approach we built a wrist vein capturing device and recorded a wrist vein dataset. Our device consists of three main parts (Fig. 4): a cluster of 24 LEDs emitting NIR light (880 nm), a CCD camera with an optical NIR bandpass filter (700–1000 nm), and an open physical structure. The camera is placed about 15 cm above of the wrist, with the LED array being about 8 cm away from the camera and about 17 cm from the capturing point – emitting light with an angle of about 62° to the wrist. Using this setup we obtained a reasonable illumination of wrist veins (Fig. 4d). The aim of our physical structure is to ease recording and emulate a mobile device equipped with NIR hardware. Using an open physical structure thereby provides for more realistic data in the mobile environment than frequently used closed box recording approaches with absolute darkness except the NIR illumination. Using our recording setup we simulate users placing their wrists inside a ROI frame on mobile devices. We record 4 vein image samples of the right wrist from 30 participants, which results in a total of 120 vein images.

4.1   Evaluation Setup

We apply image preprocessing, wrist vein feature extraction, and matching as explained in Sect. 3 to our dataset. We then partition the preprocessed data using gallery independence to evaluate our approach independently of participants in training data. We select $50\% = 15$ users for training our model and use the remaining 50% as held-back test set for exclusively testing the final model. The threshold $\tau = 0.760$ to separate true matches (P) and false matches (N) is based on the model's equal error rate (EER) from the training partition (Fig. 5a): we

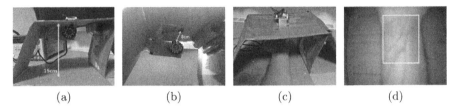

Fig. 4. Capturing device camera position (a), NIR LEDs position (b), hand position (c), camera view with ROI (d).

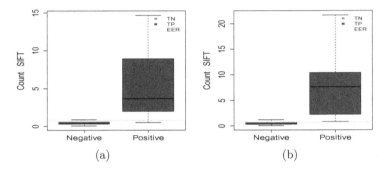

Fig. 5. Training Positive (blue) and Negative (red) classes distribution with derived EER threshold (orange) (a), testing Positive (blue) and Negative (red) classes distribution with evaluated EER threshold (orange). (Color figure online)

thereby obtain the same error for the P an N class during training. We then use τ with the test partition: as this data is originated by yet unseen participants we can derive how our system performs on new and unknown users.

4.2 Results and Model Comparison

We evaluate a number of different models using our dataset. The first three use our approach based on SIFT features: SIFT, SIFT mean, SIFT median. The first uses 1 sample for enrollment/authentication, the others 4 samples – and either mean or median with majority voting. For comparison, we further test three models based on 2D Cross Correlation: CC, CC mean, CC median [7]. The core difference to our SIFT based models is that instead of using SIFT feature similarity, those models use cross correlation similarity as underlying metric (Fig. 6).

Results indicate that using majority voting is in general preferable over using similarity measures of single samples for enrollment and authentication (Table 1). When comparing mean and median based majority voting, mean leads to better results for both SIFT and CC based models. When comparing SIFT and CC based models, SIFT models obtain overall better results (visible in both area under the ROC curve (AUC) and EER), which we assume originates from the SIFT features matching progress being more scale and rotation invariant.

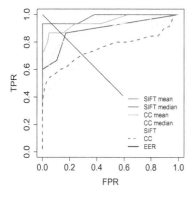

Fig. 6. Performance of our approach for different configurations.

Table 1. Performance of our approach including the decision threshold τ used with the EER.

Model	AUC	Acc	EER	τ
SIFT mean	0.980	0.858	0.072	0.760
SIFT median	0.890	0.742	0.153	0.010
SIFT	0.705	0.636	0.319	0.010
CC mean	0.950	0.775	0.143	0.100
CC median	0.890	0.758	0.175	0.100
CC	0.710	0.631	0.292	0.100

Thereby, the overall best performing approach is using a SIFT mean model, which resulted in an AUC of 0.98 and an EER of 0.072. However, we cannot conclude that our SIFT approach is completely rotation and scale invariant, as some strongly similar but different features still match. A possible approach to further increase the rotation invariance would be to compute different (de)rotations of captured vein images during authentication [7], which bears the drawback of being a computationally intensive task. Table 2 states a comparison of our results to approaches from previous and related approaches which used different dataset for evaluation.

Table 2. Performance of our approach in comparison to related approaches.

Model	EER
SIFT mean	0.072
SIFT median	0.143
LG-SRC [9]	0.016
Multiscale match filter [9]	0.134
2D correlation [7]	0.038

Our approach having a lower overall authentication accuracy may be caused by different reasons: our approach does not use a closed physical structure which would prevent ambient illumination from non-NIR light sources, and only use 4 vein pattern samples for enrollment and authentication (in contrast to 5 and 12 samples with [7,9]).

5   Conclusions and Future Work

Wrist vein authentication is promising for multiple fields of application, including mobile environments. In the future wrist vein authentication could be included in e.g. smart watches and wristbands and combined with other unobtrusive authentication approaches to obtain strong yet user friendly mobile authentication.

In this paper we focused on a mobile wrist vein authentication system based on a low cost capturing device, which can be adapted to suit mobile environments. We proposed a novel matching approach based on SIFT features that gave promising results (EER $= 0.072$) when using 4 vein images for enrollment and authentication. In the future our approach could be improved in terms of data recording capabilities: due to the freedom of the wrist, rotation needs to be addressed accordingly. Though this could be solved by trying different image rotations, such would cause a huge computational overhead. Further, addressing completely free illumination conditions [12] and human influence, such as different skin colors, light penetration, or vein behavior with sports could be investigated.

References

1. Burger, W., Burge, M.J.: Digital Image Processing: An Algorithmic Introduction Using Java. Texts in Computer Science, 2nd edn. Springer, Heidelberg (2016). https://doi.org/10.1007/978-1-4471-6684-9
2. Crisan, S., Crisan, T.E., Curta, C.: Near infrared vein pattern recognition for medical applications. Qualitative aspects and implementations. In: International Conference on Advancements of Medicine and Health Care through Technology, September 2007
3. Gad, R., El-Sayed, A., El-Fishawy, N., Zorkany, M.: Multi-biometric systems: a state of the art survey and research directions. (IJACSA) Int. J. Adv. Comput. Sci. Appl. **6**, 128–138 (2015)
4. Geng, C., Jiang, X.: Face recognition using sift features. In: 2009 16th IEEE International Conference on Image Processing (ICIP), pp. 3313–3316, November 2009
5. Huafeng, Q., Lan, Q., Lian, X., He, X., Chengbo, Y.: Finger-vein verification based on multi-features fusion. Sensors **13**, 11 (2013)
6. Juric, S., Zalik, B.: An innovative approach to near-infrared spectroscopy using a standard mobile device and its clinical application in the real-time visualization of peripheral veins. BMC Med. Inform. Decis. Making **14**(1), 100 (2014)
7. Kabaciński, R., Kowalski, M.: Vein pattern database and benchmark results. Electron. Lett. **47**(20), 1127–1128 (2011)
8. Luo, H., Yu, F.-X., Pan, J.-S., Chu, S.-C., Tsai, P.-W.: A survey of vein recognition techniques. Inf. Technol. J. **9**, 1142–1149 (2010)
9. Raghavendra, R.: A low cost wrist vein sensor for biometric authentication. In: IEEE-IST 2016 (2016)
10. Shahin, M., Badawi, A., Kamel, M.: Biometric authentication using fast correlation of near infrared hand vein patterns. Int. J. Biomed. Sci. **2**(3), 141–148 (2007)
11. Shrotri, A., Rethrekar, S., Patil, M., Kore, S.N.: IR-webcam imaging and vascular pattern analysis towards hand vein authentication (2010)

12. Suarez Pascual, J., Uriarte-Antonio, J., Sanchez-Reillo, R., Lorenz, M.: Capturing hand or wrist vein images for biometric authentication using low-cost devices. In: Intelligent Information Hiding and Multimedia Signal Processing (IIH-MSP) 2010, pp. 318–322, October 2010
13. Wang, L., Leedham, G., Cho, S.-Y.: Infrared imaging of hand vein patterns for biometric purposes. IET Comput. Vis. **1**, 113–122 (2007)
14. Xueyan, L., Shuxu, G.: Chapter 23 - the fourth biometric - vein recognition (2008)
15. Yu, C.-B., Qin, H.-F., Zhang, L., Cui, Y.-Z.: Finger-vein image recognition combining modified Hausdorff distance with minutiae feature matching (2009)

Modular 3D-Printed Robots for Education and Training for Industrie 4.0

Dirk Jacob[1(✉)], Patrick Haberstroh[2], Dominik Neidhardt[1],
and Benno Timmermann[1]

[1] Faculty of Electrical Engineering, University of Applied Sciences Kempten,
Kempten, Germany
dirk.jacob@hs-kempten.de
[2] B&R Industrie-Elektronik GmbH, Bad Homburg, Germany
patrick.haberstroh@br-automation.com

Abstract. Robotics as one of the main components of Industrie 4.0 has to be integrated in today's teachings. To provide educational institutions a low-cost but industry-oriented robot system, a modular 3D printed robot kit is developed which can be controlled by an industrial controller or a low-cost PC board.

Keywords: Robotics · 3D printing · Teaching

1 Introduction

Robotics is one of the major components of automation technologies within the framework of Industrie 4.0 [1, 2]. As robot are programmable handling systems, they are in principle able to adapt to changes of the positions resulting from flexible production tasks [3]. In order to be able to train future engineers on a high level, appropriate learning environments are required. The problem is often that robots on the one hand are very cost-intensive for the procurement of educational institutions such as professional schools or schools for technicians or universities. On the other hand, there is often not enough room in the facilities to be able to operate several robotic systems in parallel.

2 Teaching of Robotics

The teaching content in the field of robotics is extremely complex. One typical subject of robotics is mechatronics. Here, the development of mechanics, which can be stationary as well as mobile, and the corresponding control and regulation of the mechanics, are in the foreground. In addition to mechatronics, computer science is also a specialty that is intensively involved with robotics. In particular, application developments such as the evaluation of multimodal sensor networks as well as the integration of artificial intelligence to improve the performance of robot systems are currently the main focus. Another focus of the activities of computer science in robotics is the interaction between man and machine, to provide simplified programming environments or to improve the intuitive handling of robots.

R. Moreno-Díaz et al. (Eds.): EUROCAST 2017, Part I, LNCS 10671, pp. 214–219, 2018.
https://doi.org/10.1007/978-3-319-74718-7_26

Focused on the field of automation technology, the above-mentioned fields of interest in the teaching of industrial robots are to be found, too. In the field of mechatronic development of robots, topics such as coordinate transformation, path planning, dynamics, as well as control of single axes as well as the entire robot mechanics are taught. For the application of industrial robots different aspects may be considered in teaching, too. Starting with the task to program a robot and to discuss for example the effect of different path planning variants. Another topic is the off-line simulation of robots as well as the emerging challenges in the implementation of the simulated programs in reality. The programming of robot applications to solve dedicated tasks from the industrial environment and the integration of robots into applications, as well as the integration of one or more sensors into a robotic system can also be topics in teaching. Moreover the integration of one or more robots into a production environment can also be the subject of the teaching.

In order to be able to impart these different contents to the students, the best practical implementation of the teaching contents as exercises and lab exercises is to be done with real robots. However, the teaching institutions are confronted with various challenges, which are currently hindering the wider spread of robots in the teaching environment. For example, control systems from industrial robots are encrypted so that subjects from the field of mechatronic development cannot be tested on the real robot. Likewise, it is not possible to simply consider different versions of mechanical configurations. Even if a special mechanical configuration would be available on the market, it had to be procured. In the field of robots, high costs for the supply of a robot system as well as the laboratory space required for an industrial robot argue against the use of a larger number of robots in the teaching. In addition, the safety guidelines for the use of robots are very strict even in the laboratory environment due to the possible risk of injury.

Based on these limitations, training and lab exercises with robot systems for pupils and students are often limited to a very short period of time so that training for core technology cannot be deepened. Small cost-effective robotic models for training can be used as a makeshift, but they do not have an industrial control system. Thus the trainee is able to train a larger number of educators, but in the professional life the students need completely different programming environments and standards.

3 Approach

In order to find a solution to these limitations of the use of robots in teaching, a modular system has been developed, which can be used to implement cost-effectively different mechanical configurations on the one hand, and to program and apply robots in typical industrial environments on the other hand.

Mechanics
For the set-up of different robotic kinematics different modules have been developed out of basic functions, which are used in robotic mechanics.

Linear axes or rotary axes are necessary for building the mechanics of a serial robot. If a six-axis articulated robot arm is to be constructed with a central hand, rotary axes are required from axis 1 to axis 3, where mechanical levers like carrousel, linkage arm and arm are attached to. To realize the wrist additional 3 rotatory axes are installed, which are concentrated in a narrow space. This cannot be implemented as individual modules. Therefore hand modules with 2 or 3 axes are designed.

With this developed kinematic kit, it is possible to build different robotic mechanical configurations from simple 2- or 3-axis models to complex 6-axis articulated arm robots (Fig. 1). In addition to the classical articulated arm kinematics, the construction of a SCARA robot as well as a kinematics of a palletizing robot with 4 degrees of freedom is possible, too.

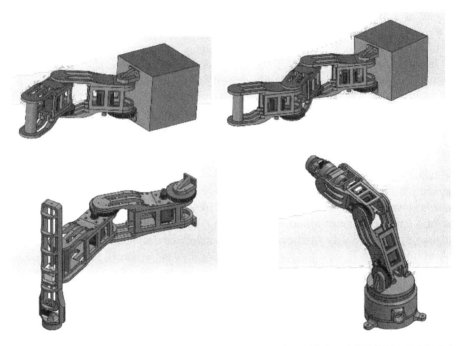

Fig. 1. Different mechanical configurations: plane kinematics with 2 and 3 DOF (top) SCARA kinematics and articulated 6-axis robot (bottom)

Rotary or linear axis units are also necessary for the construction of parallel kinematics. Furthermore, in parallel kinematics the joints, which must have up to 3 degrees of freedom, are the greatest challenge. Currently, the modules for the construction of parallel kinematics are planned but not realized yet.

Control System

In the field of control systems, appropriate control environments should be available depending on the application. If the robots are to be used in the field of mechatronic development, a control environment has to be available in which all the freedoms and possibilities of current development environments are given. The ROS development environment is an operating system developed as a freeware for research and development. Due to the modular design of ROS, this operating system is used for the development of robot control systems and the integration of sensors in robot control systems of many research facilities worldwide. In addition, ROS does not place high demands on the control hardware, so that a cost-efficient PC board like a RaspBerry Pi can be used.

If the robot is to be used in an industrial environment, such a control is not useful. In this case, a control system should be used, which already provides typical interfaces to bus systems, safety technology, the integration of sensors from the industrial environment and the connection to higher-level control systems. The challenge here is that the control must be able to support the different robotic kinematics that are possible due to the modular design.

4 Implementation

Mechanics

The design of the modules of the robot mechanics optimized to print the individual parts in a 3D printer for thermoplast. The parts are dimensioned to use a 3D printer with a work space with a length of edge of 200 mm. These printers are cost-effective and are thus frequently available in institutions of education. The implementation of the mechanics in plastic, which can be printed, also has the advantage that the mechanics can be manufactured cost-effectively and also simple user-specific changes to the mechanics can be implemented. The drive technology consists of stepping motors in three different performance and sizes, each of which is supplemented by a gearbox to increase the available torque at the output axle. The individual components of the kinematics are connected to one another by the drive units, thus completing the overall kinematics. The cable routing is implemented within the robot kinematics so that no interfering cables are located outside the mechanics. The CAD models of the parts are designed in such a way that the mechanical lever lengths of the individual kinematic components can be adapted due to individual requirements (Fig. 2).

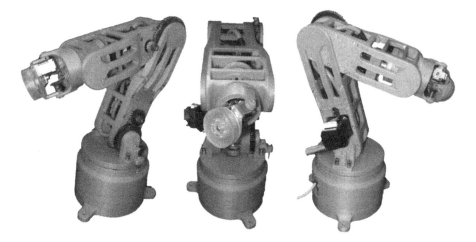

Fig. 2. Implementation of the development kit as 6 axis articulated arm

The parts of the robot mechanic were made by FDM process from PLA and after the deburring the parts are connected with screws. The gears of the gear stages are made of high-resistance synthetic resin using an SLA process. Alternatively, standard plastic gear wheels can be used, too.

Control System

An industrial controller from Bernecker + Rainer Industrie Elektronik Ges.mb.H (B & R) is used to control the robot models. B & R Software Automation Studio is used as a programming environment. This uses a so-called "mapp.Technology" to be able to program and control various modular automation systems. For the controlling of the robot "mapp.Robotics" is used, a sub-field of "mapp.Technology". This is a software that provides modular function blocks and data structures for robot programming. "Mapp.Robotics" offers already various predefined kinematic models as central objects, which contain all the necessary information for the inverse kinematics. These central objects command the basic tasks for robotic mechanics, such as the switching on of the arm powers or the referencing of the axes. In addition, the central object coordinates the automatic operation of the robot as well as the manual operation and provides diagnostic options for the motion control. With a corresponding configuration of the parameter lists, the function block executes, for example, "jog" commands for single axes, or moves the TCP along Cartesian coordinate systems. The respective object works internally with "PLCopen" blocks. If the predefined central object does not offer certain functionalities, it is possible to add these via "PLCopen" blocks. For standard models of the robotic construction kit, pre-defined configurations are available for the respective mechanical configurations and drives, so that the robot models can be put into operation and programmed without great effort. For individualization of the configuration, corresponding parameter lists and input aids are available in the Automation Studio.

In principle, it is also possible to operate the robotic models via a small PC such as a Raspberry-Pi with appropriate interface card. This can be useful, for example, when developing your own coordinate transformations or your own rule algorithms is at the forefront of teaching.

5   Conclusion

The modular robotic kit makes it possible to manufacture cost-effective robotic machines individually as printed models and to operate them with an industrial control system. The modular design of the development environment of the industrial control system also allows the control parameters to be adapted to individual robot configurations. In addition, the robot models can be controlled with available control solutions, such as Raspberry Pi to be individually operated. This provides a variety of possibilities for training and teaching in order to be able to integrate robots as one of the core components of Industry 4.0 intensively in today's universities and schools.

Acknowledgement. The project was donated by Bernecker + Rainer Industrie Elektronik Ges. mb.H and Nanotec Electronic GmbH & Co KG.

References

1. Huber, W.: Industrie 4.0 in der Automobilproduktion. Ein Praxisbuch, p. 21. Springer, Heidelberg (2016). https://doi.org/10.1007/978-3-658-12732-9
2. Gilchrist, A.: Industry 4.0. The Industrial Internet of Things, p. 42. Apress, New York (2016)
3. Bauernhansl, T., ten Hompel, M., Vogel-Heuser, B.: Industrie 4.0 in Produktion, Automatisierung und Logistik. Anwendung, Technologien, Migration, p. 13. Springer, Heidelberg (2014). https://doi.org/10.1007/978-3-658-04682-8

Anticipating the Unexpected: Simulating a Health Care System Showing Counterintuitive Behavior

Markus Schwaninger[(✉)]

University of St. Gallen, Dufourstrasse 40a, 9000 St. Gallen, Switzerland
markus.schwaninger@unisg.ch

Abstract. Complex systems often exhibit counterintuitive behavior. They confront us with the unexpected, and the idea of anticipating the unexpected is a challenge to commonsense. The purpose of this contribution is to demonstrate the power of modeling and simulation in discovering the structures that generate counterintuitive behavior in and of organizations. The research question here is if and how these generative "mechanisms" that produce unexpected behavior can be ascertained. If this can be achieved, then unexpected patterns of behavior become amenable to being anticipated as contingencies. If not, system behavior cannot be anticipated, and it remains in the dark. To answer our research question, we revert to a case study of a health-care system that showed unexpected behavior.

Keywords: Dynamic modeling and simulation · Mathematical modeling
System Dynamics · Counterintuitive system behavior · Case study
Health care

1 Introduction

In organizations, unexpected things happen all the time. It seems paradoxical to claim that one can anticipate the unexpected. Commonsense understanding tells us that we can not. Our claim is that unexpected behavior can be anticipated by good dynamic models with the help of simulation. We apply System Dynamics, a widely used methodology for the modeling and simulation of complex dynamic systems.

A case study is used to underpin our claim. The study was realized in the health care system of Carinthia, a nation-state of the Republic of Austria. The full case study covering 30+ years, is documented elsewhere (Schwaninger and Klocker 2017a). In this piece, we concentrate on the last phase of the case study (2011–2016), and elaborate on aspects pertinent to the purpose of this contribution: to use modeling and simulation for the anticipation of unexpected behavior of complex systems. We are focusing on organizations, but in principle our study is relevant for any kind of social system.

R. Moreno-Díaz et al. (Eds.): EUROCAST 2017, Part I, LNCS 10671, pp. 220–227, 2018.
https://doi.org/10.1007/978-3-319-74718-7_27

2   Case Study

This is a long-term, ongoing case from a health organization in which the author has been involved for 30+ years, in an advisory function. He collaborated with the head of that organization, which was developed under their conceptual leadership on the basis of systemic thinking. The scenery of our case study is the Oncological Care System (OCS) of Carinthia, Austria, with its hub at the central hospital of Klagenfurt, the capital. As our focus is on the OCS as a whole, we consider the overall system: it includes the hub and the network of medical services, with several peripheral hospitals as well as local registered doctors from all over the state. In 2010, the OCS showed a record of continual successes, which had been achieved in its history since 1985. However, 26 years after the foundation of the unit, in 2011, the administration of the central hospital announced that it would cut the budgets of all departments, "... to improve the economic situation". The leaders of the OCS expected that such a cut would have severe implications for the success of this unit.

To analyze the situation more closely, a simulation model was built cooperatively: the oncologists contributed the substantive knowledge about the issues under study, while the author (MS) furnished modeling and simulation know-how.

3   Model

In an initial phase, a qualitative analysis by means of causal-loop diagrams ("Qualitative System Dynamics") was carried out, in which a set of reinforcing and balancing feedback loops was identified (documented in detail in Schwaninger and Klocker 2017a). Then all the identified loops were synthesized into a quantitative simulation model. To this end we used the System Dynamics methodology (Forrester 1961, Sterman 2000). The overall picture of the model is shown in the Fig. 1, in the form of a Stock-and-Flow Diagram. That scheme is, to some extent, self-explanatory. For example, Loop B1, - the Finance-Personnel Loop. It shows a connection that is straightforward: the allocation of financial resources enables hiring people, which increases the workforce; the larger the workforce, the higher the cost, which in turn decreases earnings (indicated by the "-" sign) and financial resources available. This is a balancing loop, denoted as "B". The B-loops are normally controlled by a goal or limiting factor. In this case, the personnel budgeted delimits the quantity of personnel.

Compared to this first loop, the second one – R1 – highlights a countervailing relationship: the larger the workforce, the greater the experience and knowledge extant in the organization. The more experienced people are, the less susceptible to stress they become. "Stress" is a proxy for a working climate that enhances the number of exits, which reduce personnel. This Experience-Stress Loop is self-reinforcing (denoted with "R"), leading either to a virtuous or a vicious cycle.

The Stress-Quality Loops R3a and R3b show the causes and implications of both stress and quality of care. A lack of personnel leads to overload and stress, which is a major factor that jeopardizes quality of care. The success of care and the resulting number of cures alleviate the load of patients under treatment, consequently improving the personnel-patients ratio and alleviating stress. Lower stress means higher quality of

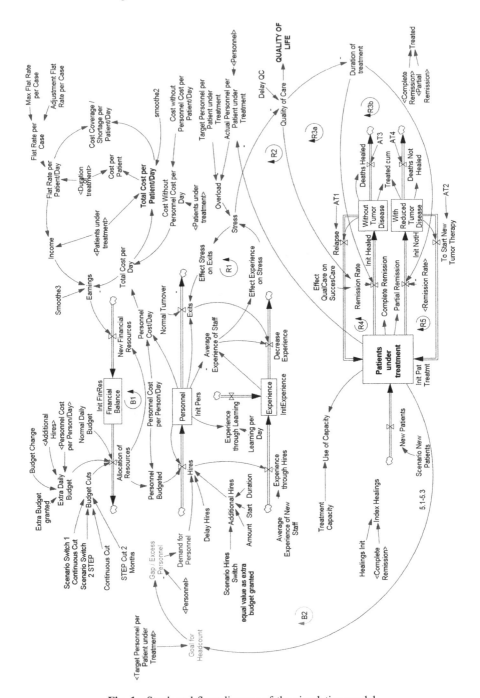

Fig. 1. Stock-and-flow diagram of the simulation model

care and then higher success of care. As loop R3b shows, quality of care – due to better dedication of staff and superior organization – reduces the duration of treatment, which affects the number of cures: shorter duration of treatment results in more healings. Both loops are of the reinforcing type.

The quantitative model is made up of equations, mostly differential equations. It covers a period of three years, from January 1, 2013 to December 31, 2015. It runs over 1095 days, with a small time step, sized 0.0625, to avoid rounding error. The equations have been documented earlier (Schwaninger and Klocker 2017b).

4 Simulations

The Budget Cut Scenario – which is the most important one here - examines the implications of the measures announced by the hospital administration. This scenario assumes a continuous curtailment of daily budgets by 15% over the whole simulation period. We make the simplifying assumption that all budget retrenchments are applicable to staff expenses only, leading to decreases in the workforce.

The graphs in Fig. 2 show that the personnel drops enormously as a consequence of the budget cut. That leads to counterintuitive behavior of the system under study. The decrease of personnel intensity induces overload and more stress, which accelerates the exit of employees, and leads to a lower quality of care. The next consequence is a longer duration of treatments, i.e., people stay in hospital longer, so that the number of patients under treatment and therewith cost are enlarged. What recedes, and dramatically, is the quality of care – a cornerstone and lead indicator of a health care system: growing stress and falling staff experience occur at the price of unsatisfactory treatment of patients.

A second and even more surprising result concerns the economic dimension. The budget cuts appear to be successful in that fewer resources need to be allocated. However, that impression is misleading. It turns out that the flow of earnings, which is positive in the base scenario for 15.7 months, becomes negative already after 8.6 months, in the budget cut scenario. With (−10.8) Mio Euros the cumulated earnings are negative, – twice the amount of the base scenario. In other words, not even the economic quantities respond to the interventions in a desirable way.

Much of the behavior of the model is counterintuitive from the viewpoint of the managers, while it makes sense from the stance of the medical staff. Even so, the working of the "mechanisms" just analyzed was fully understood by the doctors only in hindsight, when they saw the results of the simulations and had observable light-bulb moments.

Among many possible sensitivity analyses we only mention two. The first one to answer the question: "Could changes in the daily budget bring about positive cumulative earnings?" In the maximum budget cut scenario, cumulative earnings already turn negative after 14 months. On the other hand, an expansion of the budget by 15% leads to positive cumulative earnings over the whole period of three years.

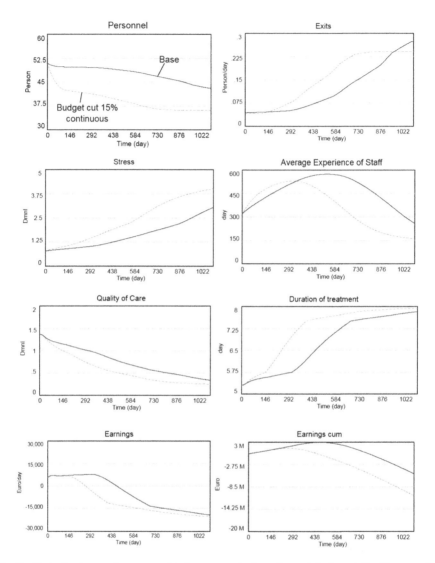

Fig. 2. Budget cut scenario compared to base scenario (based on original model)—Base run (no budget cut); — Budget Cut 15%, continuous

Second, hiring is a policy test among others examined. The sensitivity analysis shows that recruiting people with more professional experience entails positive consequences throughout: quality of care can be maintained, stress is mitigated, duration of treatment is hardly increased, and earnings remain positive (details are in Schwaninger and Klocker 2017b). The advantages of this policy are obvious, but it is difficult to implement it, as the market for hospital staff in the region has run dry.

5 Validation and Calibration

The model was submitted to a large set of validation tests from the "canon", which is the standard of practice in the System Dynamics Society (for details see: Schwaninger and Klocker 2017b).

In early 2016, as historical data became available, the model was submitted to behavior reproduction tests. The quantitative model was calibrated on the accessible data series of personnel headcount and patient healings (complete and partial remissions). A list with the parameter values is available in Schwaninger and Klocker 2017b. In Fig. 3, the simulation outcomes of the model are compared with the historical data ("real") obtained from the OCS.[1]

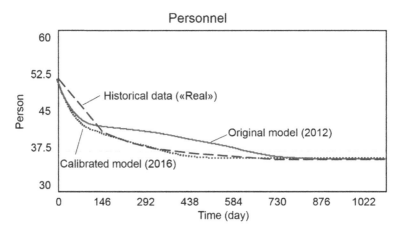

Fig. 3. Behavior reproduction of the model

The correlation between simulated and historical data is close to the maximum of 100%. Values for Mean Squared Error (0.06) and Mean Absolute Percentage Error (0.52) are very low. The analysis of the Mean Squared Error with the Theil Statistics (Theil 1966; Sterman 1984) confirms that the model captures the overall trend in historical behavior (details: Schwaninger and Klocker 2017b). The fit obtained is very high; strong confidence in the model is justified.

6 Results

We have explored dynamic simulation as an instrument, by means of which the counterintuitive, unexpected outcomes of certain policies or interventions can be anticipated. In line with the purpose of this contribution, the power of modeling and

---

[1] A cut of 15% of the budget is assumed.

simulation in discovering structures that generate counterintuitive system behavior, has been demonstrated.

The better the space of potential system behaviors is understood, the greater is the potential for successful moves. The design of a policy need not meet a discrete objective such as reaching or avoiding a precise outcome. It rather needs building the system for criteria such as robustness (tolerance or resistance to parameter variations) and resilience (response to perturbations that re-establishes equilibrium. In this context simulation can also help to unveil "new parameters" and their consequences. In our analysis, for example "experience" and "stress" were such novel parameters.

The results of the simulations disclosed clear patterns in the OCS's behavior, some of which were unexpected. The model presented amounts to more than a photograph of a system state. It captures the dynamics of both the short run, with the efficiency view, and the long term, with an effectiveness perspective. If the budget is cut, the consequence is a decrease in personnel costs; this is a short-term success. However, in the longer run, antagonistic forces emerge, and not only in economic terms. More important, the purpose of the hospital – quality of care, together with the healings quota - is affected. This difference between the short-term and long-term views is a clearly palpable instance of the distinction between efficiency and effectiveness.

It is not necessary to increase the budget, but at the same time it is imperative not to cut crucial resources painfully. Budget cutting offers the path of least resistance, promising the relief of economic concerns. In contrast, the longer-term consequences refute such short-termism, which paves the way to disaster. As quality erosion creeps in, the virtuous path of success is lost. Vicious circles establish themselves, even before their consequences become palpable. Once these vicious loops are established, it is difficult to find a path back to a virtuous trajectory.

Our study uncovers a structure that generates characteristic patterns of behavior. These conform to the expectations of the medical staff, but are counterintuitive in the logic of the administrators (Forrester 1971). *Our main finding of that kind* – detected ex ante and corroborated a posteriori - *is that the intervention of a budget cut, contrary to the* expectations of the administrators, *led to a decrease in earnings instead of an improvement of the economic situation of the organization.*

The structural features of our model are of a generic type. In other words, they are applicable to multiple contexts, representing a "wider class" of real-world situations (Forrester 1961, p. 208). Hence, their applicability is not limited to public organizations, for our model also covers private firms.

References

Forrester, J.W.: Industrial Dynamics. MIT Press, Cambridge (1961)
Forrester, J.W.: Counterintuitive behavior of social systems. Technol. Rev. **73**(3), 52–68 (1971)
Schwaninger, M., Klocker, J.: Systemic development of health organizations: an integrative systems methodology. In: Qudrat-Ullah, H., Tsasis, P. (eds.) Innovative Healthcare Systems for the 21st Century. UCS, pp. 87–139. Springer, Cham (2017a). https://doi.org/10.1007/978-3-319-55774-8_4

Schwaninger, M., Klocker, J.: Efficiency versus effectiveness in hospitals: a dynamic simulation approach. In: Borgonovi, E., Anessi-Pessina, E., Bianchi, C. (eds.) Outcome-Based Performance Management in the Public Sector. SDPM, vol. 2, pp. 397–424. Springer, Cham (2017b). https://doi.org/10.1007/978-3-319-57018-1_20

Sterman, J.D.: Appropriate summary statistics for evaluating the historical fit of system dynamics models. Dynamica **10**(2), 51–66 (1984)

Sterman, J.D.: Business Dynamics. Systems Thinking and Modeling for a Complex World. Irwin/Mc Graw-Hill, Boston (2000)

Theil, H.: Applied Economic Forecasting. Rand McNally, Chicago (1966)

Inscrutable Decision Makers: Knightian Uncertainty in Machine Learning

Rick Hangartner$^{(\boxtimes)}$ and Paul Cull

School of Electrical Engineer and Computer Science,
Oregon State University, Corvallis, OR 97331, USA
hangarr09@gmail.com, pc@eecs.oregonstate.edu

Abstract. In building models that causally explain observed data and future data, econometricians must grapple with quantifiable uncertainty, or risk, and unquantifiable Knightian uncertainty, or ambiguity. In contrast, machine learning practitioners work with statistical models for a data set that enable predictions about data items imputed to be in the data set. Recently these two distinct modeling concepts have become topics of mutual interest in economics and machine learning. We take the viewpoint here that a data set implicitly embodies the ambiguity of the generating processes from which it arises. We present a data model incorporating ambiguity that we dub the *Inscrutable Decision Maker* (IDM) derived from the Anscombe-Aumann model of subjective utility.

Keywords: Machine learning · Uncertainty · Ambiguity

1 Introduction

In 2010, economists Angrist and Pischke revived a 25-year-old debate with their essay *The Credibility Revolution in Empirical Economics* [11]. They opined that econometric tools had brought about "a credibility revolution, with a consequent increase in policy relevance and scientific impact." They were responding to Leamer's 1983 essay: *Let's Take the Con out of Economics* [15] in which he argued that real and natural experiments in data-driven econometrics failed to adequately explain economic behavior. Leamer asserted "it is therefore important that we study fragility (of inferences) in a much more systematic way."

Leamer obligingly countered Angrist and Pischke with *Tantalus on the Road to Asymptopia* [16]. Writing in 2010 as economists and policymakers were still working to fully explain the great recession of 2008, he argued that the real challenge econometricians face remained largely unaddressed: "I think the roots of the problem are deeper, calling for a change in the way we do business and calling for a book that might be titled: *The Myth of the Data Generating Process: A History of Delusion in Academia*." Unobservable factors pose a fundamental problem for economic model builders so that "the evolving, innovating, self-organizing, self-healing human system we call the economy is not well described by a fictional 'data-generating process'." To deal with unobservable

© Springer International Publishing AG 2018
R. Moreno-Díaz et al. (Eds.): EUROCAST 2017, Part I, LNCS 10671, pp. 228–236, 2018.
https://doi.org/10.1007/978-3-319-74718-7_28

factors, Leamer advocated a strategy incorporating what he termed "sensitivity analyses" that "begins with the admission that the historical data are compatible with countless alternative generating models."

Some in the machine learning community have a different perspective. Domingos [5] acknowledged that "it's true that some things are predictable and some aren't." He argued, however, that the 2008 recession was foreseen, just not by the models most banks used (many presumably built by economists). "But that was due to well-understood limitations of those models, not limitations of machine-learning (ML) in general." Referring to Taleb's probabilistic concept of "Black Swan" events [23], Domingos argues the goal of ML "is to learn everything that *can* be known, and that's a vastly wider domain than Taleb and others imagine." Even as this epistemological debate has developed, others have raised questions about inherent limitations of data-driven, statistical models for decision-making [10,12,17,19].

Leamer's focus on unobservable factors and Domingos emphasis on probabilistic machine learning traces to a distinction Knight first observed in 1921 between two qualitatively different types of uncertainty [13]. Knight reserved the term "uncertainty", now commonly referred to as *Knightian uncertainty* or *ambiguity*, for unquantifiable uncertainty. He distinguished this from probability, or quantifiable uncertainty, which he referred to as *risk*.

The difference between ambiguity and risk is the point of departure for this paper. Leamer's assertion that in some cases multiple alternative generating models can be compatible with historical economic data suggests the concept of ambiguity in a data set. Equivalently, any single instance of a specific model for a data set would be a compromise model in a sense we describe. We develop a framework for detecting when ambiguity exists in a data set, rather than deriving the best compromise model of the specified type for the data set.

2   Uncertainty

We begin by reviewing the formal concept of uncertainty. Table 1 summarizes the taxonomies of uncertainty from several disciplines. Knight [13] provided an early key insight that uncertainty involved multiple distinct notions when he distinguished between *risk* and *(Knightian) uncertainty*. He defined the former as "a quantity susceptible of measurement." and the latter as being "of the non-quantitative type." Smithson [21] placed uncertainty within a philosophical taxonomy for the more general notion of *ignorance*. Building on Arrow's [2] landmark work on choice theory, Hansen and Marinacci [9] introduced a categorization of uncertainty in the applicability of data models to observed behaviors. Walker *et al.* [24] have developed a formal taxonomy that focuses on future outcomes of decisions. Finally, Lane and Maxfield [14], and more recently Smith [20], have offered a categorization that arises from the notion of *prescriptive* (aids to decision-making) and *descriptive* (models of decision-making) AI.

In some cases only an approximate alignment of types of uncertainty exists, but the table is illustrative for our purposes here. All researchers identify unquantifiable uncertainty in their categorizations. Unlike their work, however, we focus

Table 1. Taxonomies of uncertainty

Knight [13] (economics)	Smithson [21] (philosophy)	Hansen and Marinacci [9] (econometrics)	Walker *et al.* [24] (decision theory)	Lane and Maxfield [14], Smith [20] (AI/ML)
Risk	Probability	Risk	Alternate probabilistic futures	Truth uncertainty
	Vagueness	Misspecification	Ranked futures	Semantic uncertainty
Unquantifiable uncertainty	Ambiguity	Ambiguity	Multiple plausible futures	Ontological uncertainty

on identifying ambiguity in the generating source(s) for a data set and embodying that ambiguity in a data model useful for ML applications.

3   Ambiguity in Decision Theory

In this section, we describe Anscombe and Aumann's (AA) model of subjective utility [1,8]. The AA model defines a *preference functional* as a decision maker's subjective expected utility over all state events in $\mathcal{A}$ for a capacity ν:

$$V(f) = \int_{\Psi} \left(\sum_{B \in \text{supp } f(\psi)} u(B)f(\psi) \right) d\nu = \int_{\Psi} U(f, \psi) \, d\nu \tag{1}$$

Letting $\mathcal{A}$ be an event algebra and $\mathcal{B}$ be a Borel field over *outcomes* Ω:

- An *act* $f : \Psi \to \Delta(\Omega)$ maps a *state* of the world ψ into a *lottery* μ over a set of *consequences* $\Delta(\Omega)$.
- A *lottery* $\mu : \mathcal{B} \to [0, 1]$ maps *alternatives* in $\mathcal{B}$ to probabilities.
- An *objective utility* $u : \mathcal{B} \to \mathbb{R}$ maps alternatives in $\mathcal{B}$ to numbers representing utility.
- A non-additive *capacity* $\nu : \mathcal{A} \to [0, 1]$ maps *events* in $\mathcal{A}$ to weights.
- $U(f, \psi)$ is the expected utility for an objective utility u and an act $f(\psi)$.

Because ν is a non-additive capacity, the outer integral averaging over the set of states Ψ is a Choquet integral [4]. This integral becomes a conventional Riemann integral over the states Ψ when ν is an additive probability measure.

3.1   A Data Model Inspired by Subjective Utility

The AA preference functional (1) is the starting point for a data model we dub an *Inscrutable Decision Maker* (IDM) that embodies ambiguity. The model imposes fewer constraints on the data generating processes and the relationships between them than any other model of which we are aware. Hidden Markov models [3,22], and perhaps others, are special cases with additional constraints.

Recalling that acts $f : \Psi \to \Delta(\Omega)$ map discrete world states into lotteries, which are discrete conditional distributions $\Pr(B|A, \psi, \theta)$ with finite support, without loss of generality we can assume that Ψ is ordered such that $U(f, \psi_0) = 0$ and $U(f, \psi_j) \leq U(f, \psi_{j+1})$. Let $\Psi_j = \{\psi_j, \ldots, \psi_{|\Psi|}\}$ and $\Psi_{|\Psi|+1} = \emptyset$. Finally, noting that the capacity $\nu : \mathcal{A} \to [0, 1]$ is positive valued, we can extend and rewrite the preference functional, where the first expression is a Choquet integral, as:

$$
\begin{aligned}
V(f) &= \int_\Psi \left(\sum_{B \in \text{supp } \Pr(B|A,\psi,\theta)} u(B) \Pr(B \mid A, \psi, \theta) \right) d\nu \\
&= \int_0^\infty \nu \left(\{\psi \mid U(f, \psi) \geq x\} \right) dx \\
&= \sum_{j=1}^{|\Psi|} \nu(\Psi_j) \left(U(f, \psi_j) - U(f, \psi_{j-1}) \right) \\
&= \sum_{j=1}^{|\Psi|} \left(\nu(\Psi_j) - \nu(\Psi_{j+1}) \right) U(f, \psi_j) \\
&= \sum_{B \in \text{supp } \Pr(B|A,\theta)} u(B) \left\{ \sum_{j=1}^{|\Psi|} \Pr(B \mid A, \psi_j, \theta) \, \nu'(\psi_j, \psi_{j+1}) \right\}
\end{aligned}
\tag{2}
$$

In summary, we assume an act f is a set of nondeterministic choices by the IDM of alternate lotteries. The choices are indexed by the state ψ. When we estimate the capacity ν of the IDM's preference functional we assume it only summarizes the observed data and has no intrinsic value for predicting the IDM's future choices. In essence, we can ignore whether ν is a non-additive capacity or an additive probability measure because the IDM replaces ν with ambiguity.

3.2 An Urn Problem Described by the AA Model

The AA model can be used to describe Ellsberg's well known balls and urn experiments that explore decision-making in the face of ambiguity and risk [6]. Figure 1 depicts a K-urn generalization of Ellsberg's one-urn model. Each of the K Risk Urns corresponds to a state ψ of the AA model. The number of black and white balls in each of the K urns embodies a lottery μ. The Uncertainty Urn represents the problem as seen by an outside observer, an urn that contains some minimum number N_w and N_b of white and black balls, respectively. The AA capacity ν represents knowledge about the likelihood of draws from each urn. Finally, the utility u is associated with the winnings for the gambles on each color. The IDM draws balls from different urns, but unpredictably selects the urn for each draw by preferences only the IDM knows.

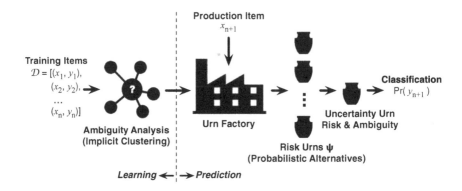

Fig. 1. Uncertainty identification process

4   A Statistical Perspective on Ambiguity

Referring again to Fig. 1, as the IDM model assumes that the capacity ν estimated along with the model parameters θ from the observed data $\mathcal{D}$ in the *Ambiguity Analysis* has no predictive value, we face a dilemma: We estimate the θ and ν from observed data $\mathcal{D}$. For a new item x_{n+1} (an independent event A), we can compute a new set of $\Pr(B \mid A, \psi, \theta)$ denoted by the *Random Urns*. But, as represented by the *Uncertainty Urn*, we have no principled way to determine which ψ applies for that item since we don't know y_{n+1} (the alternatives B). This is what we seek to predict. We address this question by looking at the IDM from a statistical perspective.

Mathematical statistics distinguishes between *marginal* statistical models for populations and *conditional* statistical models for individuals in a population [18]. Traditional conditional models incorporate *random effects* to model individuals. Frequently one can derive a marginal model for a population by marginalizing a conditional model over the random effects. The opposite does not hold, however, and deriving a conditional model from a marginal model can be difficult if not impossible unless one assumes generic random effects.

4.1   Individuals as Population Exemplars: The Marginal Model

Marginal models describe population statistics. We might reasonably assume in a population that individuals are not completely unique and that the model includes a state ψ for each possible group to which an individual represented by $x_{n+1} = A$ could belong. We could then view ν as an *ex post* estimate for the composition of the population and compute the marginal model:

$$\Pr(B \mid A, \theta) = \sum_{j=1}^{|\Psi|} \Pr(B \mid A, \psi_j, \theta)\, \nu'(\psi_j, \psi_{j+1}) \tag{3}$$

Given a new x_{n+1}, we would first find $\Pr(B \mid x_{n+1}, \theta)$. We might then also find $y_{n+1} = B$, such as by finding a maximum likelihood estimate for B, using techniques suited to the model involved. The marginal model essentially just averages away ambiguity by assuming individuals are exemplars of the population.

4.2 Individual Variation: The Conditional Model

Conditional models capture individual variation as a random effect. The IDM model inherently represents a conditional model as a collection of conditional probabilities $\Pr(B \mid A, \psi, \theta)$ derived from probability measures $\mathcal{M}_\psi$ across the states $\psi \in \Psi$. If we know the ψ we should use for a new $x_{n+1} = A$, we could compute $\Pr(B \mid A, \psi, \theta)$ and $y_{n+1} = B$. However, since we assume ν is derived *ex post* and has no predictive role, we don't *ex ante* know ψ.

To resolve this difference we develop a formalization for ambiguity from the AA model (1). Let the alternatives of a decision problem be defined over an event set $\mathcal{E} \subseteq \mathcal{B}$. Define the *unambiguous event set* $\mathcal{U} \subseteq \mathcal{E}$ to be the union of events in $\mathcal{E}$ each of which we can view as singleton event sets and assign some probability in a particular setting. Define an *ambiguity set* $\mathcal{X} \subseteq \mathcal{E}$ to be a subset of events for which we can only assign a probability to the whole subset, but not to all individual events as singleton sets. We can non-uniquely decompose $\mathcal{E}$ as:

$$\mathcal{E} = \mathcal{U} \cup \bigcup_{l=1}^{L} \mathcal{X}_l \tag{4}$$

Ambiguity sets in (4) give rise to three types of ambiguity:

- **Type 1:** Ambiguity between events in the non-singleton $\mathcal{X}_l$ sets.
- **Type 2:** Ambiguity between $\mathcal{U}$ and $\mathcal{X}_l$ if $\mathcal{U} \cap \mathcal{X}_l \neq \emptyset$.
- **Type 3:** Ambiguity between $\mathcal{X}_j$ and $\mathcal{X}_l$ if $\mathcal{X}_j \cap \mathcal{X}_l \neq \emptyset$.

If only Type 1 ambiguity holds, $\mathcal{U}$ and all of the $\mathcal{X}_l$ are disjoint subsets. We can compute a probability distribution over the individual events in $\mathcal{U}$ and the ambiguity sets $\mathcal{X}_l$ from the probability measures $\mathcal{M}_\psi$. If we have either Type 2 or Type 3 ambiguity, we cannot enforce the additivity of the probability measure. Absent additional constraints we can only specify a capacity over $\mathcal{U}$ and the $\mathcal{X}_l$.

We use this definition of ambiguity sets to formulate an alternate conditional model. Consider first the conditional probabilities $\Pr(B|A, \psi, \theta)$ in the AA model for each $\psi \in \Psi$. Using these probabilities and ambiguity sets we can define a new distribution over the unambiguous and ambiguous event sets:

$$1 = \Pr(B' \text{ is unambiguous}) + \Pr(B' \text{ is ambiguous})$$

$$= \sum_{B \in \bigcup_\psi \text{ supp } \Pr(B|A,\psi,\theta)} \Pr(B' = \{B\} \mid A, \theta) + \sum_{l=1}^{L} \Pr(\mathcal{X}_l \mid A, \theta) \tag{5}$$

As an example, for the two-color K-urn problem we can represent a draw from the final uncertainty urn as a ternary decision with probabilities:

$$\Pr(B' \mid A, \theta) = \begin{cases} p_{\text{Black}} = \min_\psi \Pr(\text{Black} \mid A, \psi, \theta) & B' = \{\text{Black}\} \\ p_{\text{White}} = \min_\psi \Pr(\text{White} \mid A, \psi, \theta) & B' = \{\text{White}\} \\ 1 - p_{\text{Black}} - p_{\text{White}} & B' = \mathcal{X} \end{cases} \quad (6)$$

4.3 The IDM Data Model as Meta Analysis

We can view the IDM data model from an additional perspective. Often we have alternative models for a prediction. Leamer [16] provides an example of two different models for predicting the chance of rain as the basis for making a decision whether to take an umbrella on an errand. Gilboa and Marinacci [8] discuss five different predictors for the risk of heart disease. The IDM data model combines multiple predictors as a form of meta-analysis that embodies the minimum and maximum probability of the certain outcomes—rain/no rain or heart disease/no heart disease—and the probability the outcome is ambiguous.

A key detail of combining models in this way is that the individual models can make predictions based on different subspaces of a combined state space represented by the unobservable states Ψ, and the observable state encoded by the predictors. Gilboa and Marinacci [8] comment that "(i)t is important to emphasize that in statistics and in computer science the state space, which is subject of prior and posterior beliefs, tends to be a restricted space that does not grow with the data." However, "the state space that is often assumed in economics is much larger than in other disciplines. Importantly, it increases with the size of the data." The IDM data model inherently embodies a notion of state space compatible with that of economics.

5   Concluding Remarks and Future Work

The Computing Research Association recently laid out a vision of the future of data science [7]. The CRA's Committee on Data Science observes that "many classic statistical assumptions and machine learning techniques do not fit current data science needs." The IDM addresses their concern that "(w)hile predictive modeling is important, many data science problems involve decision making, and the ability to reason about alternate courses of action is needed."

A few additional distinctions between the IDM data model and other statistical techniques are worth noting. A formal similarity exists between mixture models and an IDM data model where mixing probabilities are analogous to the capacity ν. The IDM distinguishes between the *ex post* nature of ν, and the *ex ante* character of ambiguity sets in the model. The probability distribution (5) of the IDM is over the (singleton) states and ambiguity sets of the states.

Finally, we postulate that patterns of ambiguity sets may reveal significant properties of phenomena described by IDM data models. The IDM data model can also be a component in more complex models. We can propagate the probability of ambiguity throughout a composite model incorporating one or more

IDM models. Although the probability of ambiguity could diverge to 100% in some cases, understanding and controlling this probability might be a useful approach for system theory, inference, and policy design.

References

1. Anscombe, F.J., Aumann, R.J.: A definition of subjective probability. Ann. Math. Stat. **34**(1), 199–205 (1963)
2. Arrow, K.J.: Alternative approaches to the theory of choice in risk-taking situations. Econometrica **19**(4), 404–437 (1951)
3. Baum, L., Petrie, T.: Statistical inference for probabilistic functions of finite state Markov chains. Ann. Math. Stat. **37**(6), 1554–1563 (1966)
4. Choquet, G.: Theory of capacities. Annales de l'institut Fourier **5**, 131–295 (1953)
5. Domingos, P.: The Master Algorithm: How the Quest for the Ultimate Learning Machine Will Remake Our World. Basic Books, New York (2015)
6. Ellsberg, D.: Risk, ambiguity, and the savage axioms. Q. J. Econ. **75**(4), 643–669 (1961)
7. Getoor, L., Culler, D., de Sturler, E., Ebert, D., Franklin, M., Jagadish, H.V.: Computing Research and the Emerging Field of Data Science (2016). http://cra. org/wp-content/uploads/2016/10/Computing-Research-and-the-Emerging-Field-of-Data-Science.pdf
8. Gilboa, I., Marinacci, M.: Ambiguity and the Bayesian paradigm. In: Arló-Costa, H., Hendricks, F.V., van Benthem, J. (eds.) Readings in Formal Epistemology. SGTP, vol. 1, pp. 385–439. Springer International Publishing, Cham (2016). https://doi.org/10.1007/978-3-319-20451-2_21
9. Hansen, L.P., Marinacci, M.: Ambiguity Aversion and Model Misspecification: An Economic Perspective (2016). http://didattica.unibocconi.it/mypage/dwload. php?nomefile=approximate-02-June-201620160608190839.pdf
10. Hvistendahl, M.: Crime forecasters. Science **353**, 1484–1487 (2016)
11. Angrist, J., Pischke, J.S.: The credibility revolution in empirical economics: how better research design is taking the con out of econometrics. J. Econ. Perspect. **24**(2), 3–30 (2010)
12. Kirkpatrick, K.: Battling algorithmic bias. Comm. ACM **59**, 16–17 (2016)
13. Knight, F.H.: Risk, Uncertainty, and Profit. Houghton Mifflin Co., New York (1921)
14. Lane, D.A., Maxfield, R.R.: Ontological uncertainty and innovation. J. Evol. Econ. **15**(1), 3–50 (2005)
15. Leamer, E.E.: Let's take the con out of econometrics. Am. Econ. Rev. **73**(1), 31–43 (1983)
16. Leamer, E.E.: Tantalus on the road to asymptopia. J. Econ. Perspect. **24**(2), 31–46 (2010)
17. Liptak, A.: Sent to prison by a software program's secret algorithms. New York Times, 1 May 2017. https://www.nytimes.com/2017/05/01/us/politics/sent-to-prison-by-a-software-programs-secret-algorithms.html
18. McCulloch, C.E., Searle, S.R., Neuhaus, J.M.: Generalized, Linear, and Mixed Models. Wiley Series in Probability and Statistics. Wiley, Hoboken (2008)
19. O'Neil, C.: Weapons of Math Destruction: How Big Data Increases Inequality and Threatens Democracy. Crown, New York (2016)
20. Smith, R.E.: Idealizations of uncertainty, and lessons from artificial intelligence. Econ.: Open-Access Open-Assess. E-J. **10**(2016-7), 1–40 (2016). https://dx.doi. org/10.5018/economics-ejournal.ja.2016-7

21. Smithson, M.: Ignorance and Uncertainty, Emerging Paradigms. Cognitive Science. Springer-Verlag, New York (1989). https://doi.org/10.1007/978-1-4612-3628-3
22. Stratonovich, R.: Conditional Markov processes. Theory Probab. Appl. **5**(2), 156–178 (1960)
23. Taleb, N.: The Black Swan: The Impact of the Highly Improbable, 2nd edn. Penguin Books, London (2010)
24. Walker, W., Lempert, R., Kwakkel, J.H.: Deep uncertainty. In: Gass, S., Fu, M. (eds.) Encyclopedia of Operations Research and Management Science. Springer, Berlin (2013). https://doi.org/10.1007/978-1-4419-1153-7

The Computer and the Calculator

Paul Cull[✉]

Computer Science, Kelley Engineering Center,
Oregon State University, Corvallis, OR 97331, USA
pc@cs.orst.edu

Abstract. Some recent discussions [1–5] have suggested that the concept of universal stored program computer is not useful in understanding the history of computing. In particular, there is the suggestion that this idea was so well known that all of the early computing devices already incorporated this concept. Here, we argue that all or almost all of the early digital machines were based on the idea of a *calculator* and that the *computer* was a real and significant new concept. We attempt to explain the differences between *calculator* and *computer*, and try to show that our contemporary computing is based on the *computer* rather than the *calculator*, and that the calculator model is inadequate to describe our current notions of computing.

1   Introduction

Below are two modern examples of a *computer* and a *calculator*

BIG QUESTION
Are modern computers and calculators really the same?
Can they compute the same things?

2   Ancient History

Human beings are counting animals. From earliest history, mankind has counted things, represented the counts with symbols, and computed with these symbols.

© Springer International Publishing AG 2018
R. Moreno-Díaz et al. (Eds.): EUROCAST 2017, Part I, LNCS 10671, pp. 237–244, 2018.
https://doi.org/10.1007/978-3-319-74718-7_29

While much of the calculation took place inside the head, external devices were also used, for example the clay tables of the Babylonians, or the sand-box of Archimedes (like an Etch-a-Sketch without the knobs). Eventually *machines* were designed to help with calculation. THE classic is, of course, the Abacus. This is an ancient device, but I saw it in practical use as little as 40 years ago. A quick history review shows that a variety of machines were designed over the centuries. It is our basic contention that all of these machines were designed as *calculators*. We will explain this in subsequent sections. It is often claimed that, in particular, Babbage [6] and Eckert-Mauchly [7] designed *computers*. We will try to explain why we believe that these machines were really calculators and try to elucidate the difference between computers and calculators.

Pascal [8] built an early calculator which was mechanical and could carry out the operations ADD and MULTIPLY. This machine is on the left. On the right is a replica of the Stepped Reckoner of Leibniz [9] which was designed as a Logic machine.

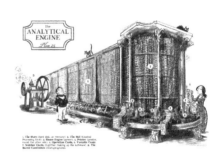

Babbage's designed two machines, the Difference Engine, a machine to calculate navigational tables by the method of differences, and the Analytical Engine. The Analytical Engine is sometimes called the first computer because it was programmable. Note that the PROGRAM is on punch cards (circled 7, next to Ada Lovelace in the picture) [10].

Burroughs built an early adding machine which was used to compute ballistics tables in WWI. Herman Hollerith built a punch card tabulator originally for the US Census. ENIAC [4] was a WWII era electronic machine to compute ballistics tables. Originally, this machine was programmed externally using plug-boards, there were no internal programs. I like the story that Nobert Wiener spent WWI in front of a calculator computing ballistics tables and since he did not want to spent the next war doing the same task, he convinced the US Army that a machine could be build to compute these tables. (Perhaps Wiener can be counted as one of the fathers of the ENIAC and one of the forefathers of the computer.)

3 Models

The function model is a very general model of computation:

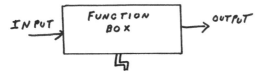

Function: For each **Input** there is **exactly one Output**.

Questions:

– For **Each Function** can we build a machine that computes **That Function**?
– Can we build **One Function** which will compute **ALL Functions**?
  (E.G. by pushing buttons on the machine to pick the function it computes.)

A **calculator** can compute a function by having the function built in and having a specific key pushed, or there can be an *external* program which would be a sequence of key pushes, or finally a program could be entered by a sequence of key pushes and then this *internal* program would specify which function is to be computed. A **computer**, by contrast, seems to need a program entered before it can compute a function.

In 1936, Turing [11] proposed a theoretical model of a computer, which we now call a Turing machine. The following is a schematic picture of a Turing machine.

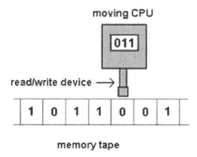

- UNBOUNDED memory (Tape)
- FINITE control box (which contains the program)

An instruction of a Turing machine has an *input* part and an *action* part as:

INPUT	ACTIONS (from among)
Contents of LOCAL memory	Change contents of tape cell
Contents of tape cell	Change contents of local memory
	Move the control box **RIGHT** or **LEFT**
	STOP

The most amazing part of Turing's analysis is the Universal Turing Machine [9], one machine that can simulate the calculation of ANY Turing machine. A schematic picture of a Universal Turing Machine 𝔘 is given below:

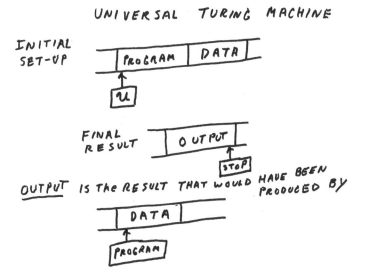

4   Computer vs Calculator

Now we want to compare these two types of machines

- the **calculator** following from historic to modern devices
- the **computer** as an embodiment of a Turing machine.

These devices have very different properties as described in the following lists.

Computer

- FINITE but unbounded (Potentially Infinite)
  NO bound on size of data
  NO bound on address space
- Operands
  Natural numbers (unbounded size)
  (or Strings of Symbols (unbounded size))
- Operations
  Change symbol
  Change local memory
  Move to new data location
- PROGRAM
  Repeat until a **STOP** instruction
  (Unbounded Iteration)

Calculator

- FINITE SIZE
  An upper limit on the number of digits in a operand
  A limited number of registers (Finite address space)
- Operands
  Fixed sized numbers
- Operators
  Arithmetic ops $(+, -, \div, ...)$ maybe others $(\sin, \exp, ...)$
- "PROGRAM"
  a sequence of ops,
  arithmetic formula

(Non) Limitations of Calculator

- Computes functions from a FINITE set to a FINITE set
- **ALL** functions are computable
- **ALL** functions are polynomials

Limitations of Computer

- Almost all functions are **NOT** computable
- There are **SPECIFIC** functions that can be shown to be **NON**-computable
- Computes PARTIAL functions
  i.e. sometimes it may **NOT** return an **ANSWER** (output)

An example of a specific function that CANNOT be computed is $D(n_p)$ given below: [12]

- Let $p(x_1, x_2 ..., x_n)$ be a polynomial with integer coefficients
- An integer vector $(v_1, v_2 ..., v_n)$ is a *root* of p **iff** $p(v_1, v_2 ..., v_n) = 0$ (evaluates to 0)
- Let n_p be the integer that ENCODES the polynomial $p(x)$
  (there is an easy-to-compute-and-invert coding of polynomials as integers)
-

$$D(n_p) = \begin{cases} 1 & \text{if } p(x) \text{ has an integer root} \\ 0 & \text{if } p(x) \text{ has no integer root} \end{cases}$$

- $D(n_p)$ is **NOT** a computable function

By restricting its Programming Language we can force a computer to compute only Functions (not Partial Functions). Gödel devised a syntax (programming language) for primitive recursion which can compute SOME but not ALL computable functions.

Theorem 1 [13,14]. *Any Programming Language which allows the computation of* **ALL** *computable functions, must also allow the computation of partial functions.*

5   Stored Programs?

John von Neumann was a consultant on the ENIAC project.

He made the following obervations [15] on how ENIAC could be changed to be more like the machine proposed by Turing:

– For general (full) computation, the computer must have a stored program which may be modified during execution.
– His example was a program to ADD an ARRAY of numbers.
  The program incremented one of its own instructions, so that it would address the next element in the array.
– Turing's model may SEEM to leave the program alone, but it must mark the instructions to find which instruction is being executed.

WWII Era Machines

– Bombe – special purpose (code breaking) [16]
– Colossus – special purpose (code breaking)
  Programmed using switches and plug panels [17]
– ATANASOFF-BERRY (ABC) – special purpose
  more a calculator than a computer [18]
– Zuse –original version was a calculator
  later versions were programmable [19]
– ENIAC (original version) – external program [4]
– Manchester Baby – Stored Program [20]
– ENIAC (redesigned version) – Stored Program [4]
– EDSAC – Stored Program – much larger memory than ENIAC [21]
– IAS machine –Stored Program [22]

From this table, it certainly looks like "Stored Program" forms the dividing line between early machines and computers. BUT, I think that this view is mistaken. Following Turing's analysis, stored program is implied by his model. As we saw above, it's possible to design a calculator with a stored program. So "Stored Program" is not the essential difference between a Computer and a Calculator.

The *ESSENTIAL DIFFERENCE* between Computer and Calculator is

– UNBOUNDEDNESS
(whether in address space or memory size)
– Whether or not the program is stored internally
(in the same space as data) is inessential

Minsky [13] has shown that the four basic operations *multiplication, division, addition* and *subtraction* are sufficient to carry out all computation. In fact, all of these operations can be limited so that one of its operands is a fixed number and the other operand is a arbitrary natural number, e.g. division by 2. The only extra stipulations are that two arbitrary natural numbers are needed in the computation, and that the operations are *conditional*, that is which number is to be multiplied and which is to be divided and what numbers to add or subtract depend on a small finite number of local conditions.

It seems that this demonstration shows that the calculator model is sufficient for ALL computation, but notice the two UNBOUNDED numbers are used. This then reinforces our claim that UNBOUNDEDness is the essential difference between the calculator and the computer.

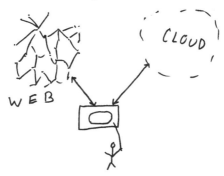

As the picture suggests Present Day Computing is not localized. The Web allows us to have a world-wide memory. The Cloud allows us to store arbitrary amounts of data and compute on these large masses of data. So for modern computing:

– No clear bound on memory. (except possibly, the universe is finite)
– The unbounded computer model is appropriate.

6   Conclusion

– For much "PRACTICAL" computing the distinction
  COMPUTER vs **CALCULATOR** is a side issue.
– Theoretically there is a **BIG** difference
  between **CALCULATOR** and **COMPUTER**.
– **COMPUTERS** are UNBOUNDED while
  CALCULATORS have FINITE BOUNDS.
– This distinction is important in designing systems,
  and ALL our large systems are designed based on the **COMPUTER**.

References

1. Vardi, M.: Who begat computing? Commun. ACM **56**, 5 (2013)
2. Copeland, J., Dresner, E., Proudfoot, D., Shagrir, O.: Time to reinspect the foundations? Commun. ACM **59**, 34–38 (2016)
3. Haigh, T.: Actually, Turing did not invent the computer. Commun. ACM **57**, 36–41 (2014)
4. Haugh, T., Priestly, M., Rope, C.: ENIAC in Action: Making and Remaking the Modern Computer. MIT Press, Cambridge (2016)
5. Sommaruga, G., Strahm, T. (eds.): Turing's Revolution. Springer, Cham (2015). https://doi.org/10.1007/978-3-319-22156-4
6. Swade, D.: The Difference Engine: Charles Babbage and the Quest to Build the First Computer. Viking, New York (2001)
7. Stern, N.B.: From ENIAC to UNIVAC: Appraisal of the Eckert-Mauchly Computers. Digital Press, Bedford (1981)
8. Goldstine, H.H.: The Computer from Pascal to von Neumann. Princeton University Press, Princeton (1980)
9. Davis, M.: The Universal Computer: The Road from Leibniz to Turing. W.W. Norton, New York (2000)
10. Padua, S.: The Thrilling Adventures of Lovelace and Babbage. Pantheon, New York (2015)
11. Turing, A.M.: On computable numbers, with an application to the entscheidungsproblem. Proc. London Math. Soc. **42**, 230–265 (1936)
12. Matiyasevich, Y.: Hilbert's Tenth Problem. MIT Press, Cambridge (1993)
13. Minsky, M.: Computation: Finite and Infinite Machines. Prentice-Hall, Englewood Cliffs (1967)
14. Cull, P.: Notes for CS 321. Computer Science, Oregon State University, Corvallis (2016)
15. von Neumann, J.: First Draft of a Report on the EDVAC. University of Pennsylvania, Philadelphia (1945)
16. Link, D.: Archaeology of Algorithmic Artefacts. Univocal, Minneapolis (2016)
17. Copeland, B.J., et al.: Colossus: The Secrets of Bletchley Park's Code-Breaking Computers. Oxford University Press, Oxford (2010)
18. Burks, A.R., Burks, A.W.: The First Electronic Computer: The Atanasoff Story. University of Michigan Press, Ann Arbor (1989)
19. Zuse, K.: The Computer, My Life. Springer, New York (1993). https://doi.org/10.1007/978-3-662-02931-2
20. Copeland, B.J., et al.: Alan Turing's Electronic Brain. Oxford University Press, Oxford (2005)
21. Wilkes, W.V., Renwick, W.: The EDSAC (electronic delay storage automatic calculator). Math. Comput. **4**, 61–65 (1950)
22. Dyson, G.: Turing's Cathedral. Pantheon, New York (2012)

Non-email Spam and Machine Learning-Based Anti-spam Filters: Trends and Some Remarks

Ylermi Cabrera-León[1], Patricio García Báez[2],
and Carmen Paz Suárez-Araujo[3(✉)]

[1] Universidad de Las Palmas de Gran Canaria (ULPGC),
Las Palmas de Gran Canaria, Spain
`ylermi.cabrera101@alu.ulpgc.es`
[2] Departamento de Ingeniería Informática y de Sistemas,
Universidad de La Laguna, San Cristóbal de La Laguna, Spain
`pgarcia@ull.es`
[3] Instituto Universitario de Ciencias y Tecnologías Cibernéticas, Universidad de Las
Palmas de Gran Canaria (ULPGC), Las Palmas de Gran Canaria, Spain
`carmenpaz.suarez@ulpgc.es`

Abstract. Electronic spam, or unsolicited and undesired messages sent massively, is one of the threats that affects email and other media. The high volume and ratio of email spam have generated enormous time and economic losses. Due to this, many different email anti-spam defenses have been used. This translated into more complex spams in order to surpass them. Moreover, the spamming business moved to the less protected yet quite profitable non-email media because of the numerous potential targets that results from their extensive usage. Since that moment, spams in these media have increased rapidly in quantity, sophistication and danger, especially in the most popular ones: Instant Messaging, SMS and social media. Therefore, in this paper some of the characteristics and statistics of instant spam, mobile spam and social spam are exposed. Then, an overview of anti-spam techniques developed during the last decade to fight these new spam trends is presented, focusing on hybrid and Machine Learning-based approaches. We conclude with some possible future evolutionary steps of both non-email spams and anti-spams.

Keywords: Spam filtering · Non-email spam · Social spam
Mobile spam · SMS spam · Instant spam · Spim · Social media
Machine Learning

1 Introduction

Spam, or electronic spam, is a word that has many different definitions [20]. Our proposal is precise yet straightforward: spam is any unrequested or unwanted message sent massively through any electronic media. It has been considered a synonym of "email spam" as this has been the most popular spamming media in the last decades, accounting for massive volumes and even spam ratios higher than 90% in 2008 [19].

© Springer International Publishing AG 2018
R. Moreno-Díaz et al. (Eds.): EUROCAST 2017, Part I, LNCS 10671, pp. 245–253, 2018.
https://doi.org/10.1007/978-3-319-74718-7_30

However, apart from email, unsolicited messages exist in social media, Voice over IP (VoIP), Massive Multiplayer Online Games (MMOGs), Instant Messaging (IM), Short Messaging Service (SMS), and many other electronic media [6]. As the complexity, number and hazard level of non-email spam raised, filters started to appear to counter them. At the outset, non-email spam filters were adapted from email ones, with variable but mostly positive performance. Later, more complex, specialized and novel approaches were deployed, more suitable to fight within these new environments. Adaptive, usually Machine Learning (ML), and modular methodologies came after that to fulfill the need to defend from different types of spams.

The objectives of this paper are to describe spam in SMS, IM and social media, and to briefly overview several filters based on ML, including combinations with other types of filters (hence called hybrid), that have appeared during the last decade to counter these non-email spam families. The rest of the paper is organized as follows. The most dangerous or popular non-email spam families - *i.e.* instant spam, SMS spam and social spam - are exposed in Sect. 2. Thereupon, some of the most prominent hybrid and ML-based anti-spams are presented in Sect. 3. Finally, the conclusions and how future non-email spam and anti-spam may further evolve can be read in Sect. 4.

2   Non-email Spam: Spim, Mobile Spam, and Social Spam

Due to the greater effectiveness of email spam filters and the subsequent reduction of spam profitability, spammers have adopted two strategies: evolve and improve spamming techniques, or move the business to other less protected and more cost-effective media. Indeed, many spammers got attracted by the unstoppable rise in popularity of these so-called non-email media, most notably the social media.

These three different non-email media are considered the most profitable for spammers, the most dangerous for the users and with the most potential targets:

- Instant Messaging: instant spam, a.k.a. spim, is spam found in IM systems, such as WhatsApp, Windows Live Messenger, XMPP and ICQ chat rooms, among others. Between 2002 and 2003 there were 500 million spim messages, and 7 out of 10 contained porn links [15]. Compared with email spam, spim has smaller size, gets delivered nearly in real-time, and consumes lower traffic and server storage [15]. Some of these characteristics complicate its filtering.
- Short Messaging Service: SMS spam is probably the most numerous threat that menaces mobile phones, albeit not necessarily the most dangerous one compared with viruses, worms, Trojans and spyware. As it has happened with emails, SMS messages have acted as vectors of transmission for other families of threats, so they are more dangerous than users commonly expect. Mobile spam has to overcome two types of barriers, economical and technological, that limit the quantity of sent SMS spam and also thwart its geographic widespread and propagation speed. As sending SMS spam is much

more expensive than sending email spam, spammers have lowered these costs by using inexpensive and unlimited prepaid SMS packages [9]. On the other hand, different regions may use diverse telecommunication technologies and character encoding. This can complicate the correct delivery of the SMS spam to the user or the adequate display of its content.

However, these handicaps were not enough. In 2003 there were more mobile spams than email spams, and 8 out of 10 European users have received at least one [11]. Around 8.3 trillion SMS were sent worldwide in 2013 [17]. The main topics found in SMS spams have been dating services, ringtones and premium-rate services (higher fee numbers), apart from software and drug sales frequent in email spam too [9,11].

- Social media: within this broad term there are several subgroups [6]: comments, product/service review pages, blogs, microblogs (mainly Twitter), Online Social Networks (OSNs) (for example, Facebook, Google+ and MySpace), social bookmarking sites (*e.g.* Reddit, Digg and Menéame), and location sites (such as Foursquare and HopOver). All of them are targeted by social spammers. Social spam has been widely distributed, with 200 million dangerous activities blocked daily in Facebook, and 1 out of 10 tweets in Twitter can be considered spam [22]. This can be explained by its high degree of economic efficiency, partly due to a higher click-through rate. For example, near a hundred times more visits are induced by spam tweets than by email spams [7]. Moreover, economic benefits and popularity of frequently spammed OSNs tend to decrease [1,6]. Social spam comes in different shapes: malicious links, false profiles, dishonest product reviews, threats, hate speeches, and bulk posting [6]. It is generated by real spammers, spambots and fake accounts. The latter are called "Sybils", which usually befriend extensive networks of real users, preferably celebrities, in order to mimic real users' accounts and hinder being detected prematurely. In August 2012 Facebook estimated that above 83 million active accounts were false or duplicates, of whom around 17.23% were potentially harmful ones [14].

3   Trends in Non-email Spam Filters

Non-email media, such as IM, SMS and social media, are quite different from email, with diverse infrastructure and technology. This fact entails that some email spam filters cannot be applied but, at the same time, permitted the introduction of new non-email-based ones. Henceforth, two big families of filters can be used against non-email spam: based on email methods, and novel approaches. Some of the email-based ones that could successfully adapt to these media were legislation, law enforcement agencies, whitelists, blacklists and some commercial products.

In the next subsections some of the most relevant filters for instant spam, SMS spam and social spam are introduced. Tables 1, 2 and 3 will only show results for the optimal classifier in each research.

3.1 Anti-spim

A problem that instant spam filters must deal with, and that email ones do not need to, is the historical preponderance of proprietary protocols until the development of open ones such as XMPP. As a result, diverse IM clients tend to be incompatible. Therefore, most anti-spim are constrained to one or a few IM networks, so not many have been deployed, and many IM users have migrated to Online Social Networks with Instant Messaging features.

Initial approaches involved the usage of personal, user-managed blacklists and whitelists. As it happened with email spam, more complex anti-spim appeared later, Table 1. Anti-spim methods can be classified based on their purpose or how the task is done. Spimmer detection has been the most common as it is considered more effective than filtering spims. This is so because creating a new IM account in some IM applications is more difficult and costly than email accounts. For instance, in WhatsApp, the most popular one nowadays, a valid and never used before SIM card or phone number is required.

Table 1. Instant spam filters. A = Accuracy, FNR = False Negative Rate, FPR = False Positive Rate, P = Precision, R = Recall

Purpose	Anti-spam family	Method	Results	Research
Spim, spimmer and Denial-of-Service attacks detection	Hybrid	2 blacklists, whitelist, challenge-response, collaborative user feedback, Bayesian, Longest Common Subsequence algorithm, Approximate Text Comparison, sending rate control & gateway/server-side filtering	FNR ≈ 0%, 2.3% FPR	[15]
Spimmer detection	List	"SpimRank": whitelist & blacklist	95% A	[4]
Spim and spimmer detection	Machine learning	J48 decision tree	89.49% A, 0.125% FPR	[16]
Packet classification		AdaBoost, J48, Naïve Bayes (NB), SVM	A ≈ 98%	[8]

Statistical and non-neural ML methods, such as Bayesian, Support Vector Machine (SVM) and J48 decision trees [8,15,16], have achieved good anti-spim performance. Still, simpler techniques, such as whitelists and blacklists, have demonstrated their adequateness to spim filtering, either on their own with automatic updates [4], or combined with other families of techniques in a layered fashion [15]. Probably the next generation of anti-spims will follow the latter path.

Researchers fulfilled the initial lack of spim messages in the public domain in two ways: using documents from other media, such as SMS and short email spams [8,15], or simulating IM networks [16].

3.2   Anti-mobile-spam

The content of SMSs and emails differ in terms of used language, maximum length, and type of content [9,11,17]. SMSs are mostly written with "SMS language", a mix of abbreviations, bad spelling, emoticons and slang, which is highly dependent on the sender's language, age and level of education. Up to 160 characters can be carried in a single SMS, with the default GSM 7-bit alphabet. Until the appearance of Enhanced Messaging Service (EMS) and Multimedia Messaging Service (MMS), a SMS message could only contain text, without images, text formatting or sounds.

Although these differences limit the usage and performance of some email-based filters, they also constrain mobile spam delivery. Additionally, they hamper the complete and correct deployment in SMS of several families of email spam and spamming methods such as image spam and some content obfuscation techniques. The latter might need an anti-obfuscation method, such as the hybrid and modular email anti-spam found in Cabrera [5], albeit after proper adaptation to this media.

Law enforcement agencies, legislation, prevention and collaborative user feedback have been often utilized to deal with mobile spam. More sophisticated and recent approaches can be grouped into the two following categories, SMS spamming botnets detectors and SMS spam detectors, Table 2.

Table 2. Mobile spam filters, all but the first one specialized in SMS spam detection. If comparisons, optimal classifier in bold. A = Accuracy, FPR = False Positive Rate, P = Precision, R = Recall

Anti-spam family	Method	Results	Research
Hybrid	Blacklist, rule-based, X-means (unsupervised clustering) & correlation rules	Multi-class problem No performance metrics published	[3]
Hybrid	Blacklist, whitelist & k-NN	With ham, 91.35% P, 84.50% R With spam, 83.72% P, 90% R	[10]
	"SMSAssassin" mobile app: blacklists, user feedbacks & classifier (Bayesian or SVM)	With Bayesian, 97% A for ham, 72.5% A for spam With SVM, 93% A for ham, 86% A for spam	[23]
Machine Learning	NB, C4.5, PART rule learner, **SVM**	With Spanish dataset, A > 95%, 0% FPR With mix of English datasets, A ≈ 89%	[11]
	Bayesian approaches, C4.5, k-NN, PART, **linear SVM**	97.64% A	[2]
	Linear SVM, Collaborative filter	With SVM, 94.63% A, 93.31% R, 4.05% FPR With collaborative, 79.09% A, 58,24% R, 0.06% FPR	[9]
	UCS, XCS, GAssist-ADI, Fuzzy AdaBoost; JRip rule learner, C4.5, SVM, k-NN, NB	93% R, 0% FPR	[13]
	SDA deep neural network	97.51% A, 95.47% P, 85.58% R	[17]

Most filters in these groups work on the server-side because cell phones, especially basic ones, are far less powerful than computers. The majority of analyzed mobile spam filters are specialized in SMS spam detection: in our case, only Alzahrani's proposal [3] is devoted to SMS spamming botnet detection. By far, SVM with linear kernel is the most common ML method to fight SMS spam, followed by Bayesian approaches (statistical) and C4.5 decision trees. So far, computational intelligence techniques have been less utilized. Among them, we can mention Deep Neural Networks, such as the Stacked Denoising Autoenconder (SDA) [17], and evolutionary algorithms, e.g. Fuzzy AdaBoost, Genetic clASSIfier sySTem-Adaptive Discretization Intervals (GAssist-ADI), eXtended Classifier System (XCS) and sUpervised Classifier System (UCS) [13]. As it happened with spim filters, hybrid techniques that unify ML approaches, such as k-NN or X-means clustering, with other types, principally blacklists and whitelists, exist too. Supervised methods are the most popular in both categories of filters.

Apart from personal datasets, which hinder the reproducibility of experiments, the most used SMS dataset has been the "SMS Spam Collection"[1] built by Almeida [2]. Websites where users can report SMS spams[2] is a less used source of them.

3.3   Anti-social-spam

Collaborative user feedback have been implemented in almost all OSNs so that any or a selected group of users are able to report spammy content. Other social media, such as Facebook, combine user feedback with some ML algorithms, which have reduced spamming rate to about 5% [6]. According to Al-Qurishi [1], the most analyzed OSN in the last decade has been Facebook due to its wider geographical usage and enormous user base.

We will focus on methods for Sybil detection [12, 24–26] and cross-media spam detection [21, 22], Table 3. Sybil detection has become popular for two reasons. First, Sybil activities are potentially highly dangerous, long-term and concealed, so it is important and urgent to avoid future damages. Second, stopping Sybils is seen as more efficacious than filtering social spam directly. Cross-media spams are unwanted messages that appear in a particular social media that sometimes can be later found in other media, something that might ease future spam filtering [21, 22]. Except the hybrid filter in Wang [21]; all of them are solely based on ML. It has been found that C4.5 decision trees and Random Forests are widely used against social spam in both Sybil and cross-media spam filtering. Neural networks, such as Multi-Layer Perceptron (MLP) with Back-propagation Learning and Extreme Learning Machine (ELM), and regression methodologies have been used too. Also, some hybrid methods that mix blacklists with ML modules, among others, have achieved good performance results. In this case, modules are mostly based on NB (statistical), C4.5 decision trees, and Sequential Minimal Optimization (SMO), a SVM variant.

---

[1] http://www.dt.fee.unicamp.br/~tiago/smsspamcollection/.
[2] http://www.grumbletext.co.uk/ (website unavailable at the time of writing).

Table 3. Social spam filters specialized in Sybil detection or cross-media spam detection. If comparisons, optimal classifier in bold. A = Accuracy, AUC = Area Under the Curve, FPR = False Positive Rate, P = Precision, R = Recall, TP = True Positives

Method	Dataset	Results	Research
C4.5 decision tree, AdaBoost, SMO, MLP, NB	MySpace	99.4% → 92.9% A, due to adversarial attacks	[12]
k-NN, Linear Regression, **Back-propagation**, NB		83% A	[26]
ELM, SVM, Bayesian, decision tree	Sina Weibo	With spammers, 99.9% P, 99% R With non-spammers, 99.4% P, 99.9% R	[24]
"ProGuard": **normalized Random Forest**, SVM, Gradient-Boosted Tree	Tencent QQ	0.9959 AUC, 96.67% TP rate, 0.3% FPR	[25]
Random Forest, Random Tree, J48, Logistic Regression (LR), Bagging, BayesNet, NB	Facebook & Twitter	A ≤ 98.9%, R ≤ 95.5%	[22]
Blacklist, similarity matching & supervised classifier (among others, NB, C4.5 and SMO)	TREC 2007 emails, MySpace & Twitter	95.9% A	[21]

On the other hand and due to the relevance of Twitter, two filtering approaches for this very popular microblogging site have been found, social spam campaign detection [7] and microblogging spam detection [18]. Hence, spam campaigns are currently present not only in email but also in some social media, thus indicating a similar evolution, which may be useful to anticipate future social spams.

In our opinion, future social filters will combine neural networks or Random Forests with non-ML modules, becoming more proficient hybrid solutions.

4 Conclusions

Non-email media have increasingly been attacked because of its high estimated profitability and lower level of defenses. Throughout this paper we have seen that new filters tend to be more sophisticated than the preceding generations. They frequently become modular solutions that combine different technologies, including ML methods, in order to defend from diverse spams and spamming methods in more optimal ways.

More complex defenses, possibly hybrid with ML, will be needed in the near future as the number of spam and other threats coming via non-email media keeps increasing and the spamming techniques continue improving and surpassing previous defenses. An end to this arms race induced by the coevolution between non-email spam and anti-spam techniques is highly improbable in the short term. We consider that social engineering will play an important role in next generations of attacks. Therefore, education and prevention will greatly help in countering it.

References

1. Al-Qurishi, M., Al-Rakhami, M., Alamri, A., AlRubaian, M., Rahman, M., Hossain, M.S.: Sybil defense techniques in online social networks: a survey. IEEE Access **5**, 1200–1219 (2017)
2. Almeida, T.A., Hidalgo, J.M.G., Yamakami, A.: Contributions to the study of SMS spam filtering: new collection and results. In: Proceedings of the 11th ACM Symposium on Document Engineering, pp. 259–262. ACM (2011)
3. Alzahrani, A.J., Ghorbani, A.A.: SMS-based mobile botnet detection framework using intelligent agents. J. Cyber Secur. Mobil. **5**(2), 47–74 (2016)
4. Bi, J., Wu, J., Zhang, W.: A trust and reputation based anti-SPIM method. In: IEEE INFOCOM 2008. The 27th Conference on Computer Communications, April 2008
5. Cabrera-León, Y., García Báez, P., Suárez-Araujo, C.P.: Self-organizing maps in the design of anti-spam filters. A proposal based on thematic categories. In: Proceedings of the 8th IJCCI 2016, NCTA, vol. 3, pp. 21–32. SCITEPRESS Digital Library, Porto, November 2016
6. Chakraborty, M., Pal, S., Pramanik, R., Ravindranath Chowdary, C.: Recent developments in social spam detection and combating techniques: a survey. Inf. Process. Manag. **52**(6), 1053–1073 (2016). https://doi.org/10.1016/j.ipm.2016.04.009
7. Chu, Z., Widjaja, I., Wang, H.: Detecting social spam campaigns on Twitter. In: Bao, F., Samarati, P., Zhou, J. (eds.) ACNS 2012. LNCS, vol. 7341, pp. 455–472. Springer, Heidelberg (2012). https://doi.org/10.1007/978-3-642-31284-7_27
8. Das, S., Pourzandi, M., Debbabi, M.: On SPIM detection in LTE networks. In: 2012 25th IEEE Canadian Conference on Electrical and Computer Engineering (CCECE), pp. 1–4, April 2012
9. Delany, S.J., Buckley, M., Greene, D.: SMS spam filtering: methods and data. Expert Syst. Appl. **39**(10), 9899–9908 (2012)
10. Duan, L., Li, N., Huang, L.: A new spam short message classification. In: 2009 First International Workshop on Education Technology and Computer Science, vol. 2, pp. 168–171. IEEE, March 2009
11. Gómez Hidalgo, J.M., Bringas, G.C., Sánz, E.P., García, F.C.: Content based SMS spam filtering. In: Proceedings of the 2006 ACM Symposium on Document Engineering, pp. 107–114. ACM Press, October 2006
12. Irani, D., Webb, S., Pu, C.: Study of static classification of social spam profiles in MySpace. In: ICWSM (2010)
13. Junaid, M.B., Farooq, M.: Using evolutionary learning classifiers to do Mobile Spam (SMS) filtering. In: Proceedings of the 13th Annual Conference on Genetic and Evolutionary Computation, pp. 1795–1802. ACM (2011)
14. Kelly, H.: 83 million Facebook accounts are fakes and dupes, Auguset 2012. http://www.cnn.com/2012/08/02/tech/social-media/facebook-fake-accounts/index.html

15. Liu, Z., Lin, W., Li, N., Lee, D.: Detecting and filtering instant messaging spam - a global and personalized approach. In: 1st IEEE ICNP Workshop on Secure Network Protocols, 2005 (NPSec), pp. 19–24. IEEE, November 2005
16. Maroof, U.: Analysis and detection of SPIM using message statistics. In: 2010 6th International Conference on Emerging Technologies (ICET), pp. 246–249, October 2010
17. Moubayed, N.A., Breckon, T., Matthews, P., McGough, A.S.: SMS spam filtering using probabilistic topic modelling and stacked denoising autoencoder. In: Villa, A.E.P., Masulli, P., Pons Rivero, A.J. (eds.) ICANN 2016. LNCS, vol. 9887, pp. 423–430. Springer, Cham (2016). https://doi.org/10.1007/978-3-319-44781-0_50
18. Sedhai, S., Sun, A.: Semi-Supervised Spam Detection in Twitter Stream. IEEE Trans. Comput. Soc. Syst. **PP**(99) (2017). https://doi.org/10.1109/TCSS.2017.2773581
19. Statista: Global spam volume as percentage of total e-mail traffic from 2007 to 2015, April 2016. http://www.statista.com/statistics/420400/spam-email-traffic-share-annual/
20. Subramaniam, T., Jalab, H.A., Taqa, A.Y.: Overview of textual anti-spam filtering techniques. Int. J. Phys. Sci. **5**(12), 1869–1882 (2010)
21. Wang, D., Irani, D., Pu, C.: A social-spam detection framework. In: Proceedings of the 8th Annual Collaboration, Electronic Messaging, Anti-Abuse and Spam Conference, pp. 46–54. ACM (2011)
22. Xu, H., Sun, W., Javaid, A.: Efficient spam detection across online social networks. In: 2016 IEEE International Conference on Big Data Analysis (ICBDA), pp. 1–6. IEEE, March 2016
23. Yadav, K., Kumaraguru, P., Goyal, A., Gupta, A., Naik, V.: SMSAssassin: crowd-sourcing driven mobile-based system for SMS spam filtering. In: Proceedings of the 12th Workshop on Mobile Computing Systems and Applications, pp. 1–6. ACM (2011)
24. Zheng, X., Zhang, X., Yu, Y., Kechadi, T., Rong, C.: ELM-based spammer detection in social networks. J. Supercomput. **72**(8), 2991–3005 (2016)
25. Zhou, Y., Kim, D.W., Zhang, J., Liu, L., Jin, H., Jin, H., Liu, T.: ProGuard: detecting malicious accounts in social-network-based online promotions. IEEE Access **5**, 1990–1999 (2017). https://doi.org/10.1109/ACCESS.2017.2654272
26. Zinman, A., Donath, J.S.: Is Britney Spears spam? In: CEAS, Mountain View, California (USA), pp. 1 10, August 2007

Theory and Applications of
Metaheuristic Algorithms

A General Solution Approach for the Location Routing Problem

Viktoria A. Hauder[1,2(✉)], Johannes Karder[1,3], Andreas Beham[1,3], Stefan Wagner[1], and Michael Affenzeller[1,3]

[1] Heuristic and Evolutionary Algorithms Laboratory, School of Informatics, Communications and Media, University of Applied Sciences Upper Austria, Hagenberg Campus, Hagenberg, Austria
{viktoria.hauder,johannes.karder,andreas.beham,stefan.wagner, michael.affenzeller}@fh-hagenberg.at
[2] Institute for Production and Logistics Management, Johannes Kepler University Linz, Linz, Austria
[3] Institute for Formal Models and Verification, Johannes Kepler University Linz, Linz, Austria

Abstract. Conventional solution methods for logistics optimization problems often have to be adapted when objectives or restrictions of organizations in logistics environments are changing. In this paper, a new, generic solution approach called optimization network (ON) is developed and applied to a logistics optimization problem, the Location Routing Problem (LRP). With this approach, required flexibility in terms of fast changing data within the advancement of industry 4.0 is addressed. In an ON, existing solution methods are applied to the basic problems of the LRP. A meta solver optimizes the overall result of the network with black box optimization. Based on this, an orchestrator is responsible for the introduction of new optimization runs. The developed approach guarantees that changing external influences only involve the adaption of affected optimization nodes within the ON and not of the whole solution approach. Results are compared with an already existing generic solver and show the potential of the new solution method.

Keywords: Optimization networks · Location Routing Problem
Integration · Synergy effects

1 Introduction

Within the fourth industrial revolution, also called industry 4.0, high adaptability and resource efficiency of organizations are essential key points [5]. In so-called smart enterprises, which are a major reference point within industry 4.0, information is exchanged in real time and all involved agents are connected [5]. Therefore, also in the field of optimization in production and logistics, a combination of interrelated problem models is gaining growing importance [1]. With

© Springer International Publishing AG 2018
R. Moreno-Díaz et al. (Eds.): EUROCAST 2017, Part I, LNCS 10671, pp. 257–265, 2018.
https://doi.org/10.1007/978-3-319-74718-7_31

the introduction of the simultaneous optimization of more than one production and logistics optimization problem, the ability to react and respond quickly to changing external and internal influences should be improved.

One already existing combination of two logistics optimization problems is the Location Routing Problem (LRP) [2,7]. It consists of the Facility Location Problem (FLP), a strategic logistics optimization problem, and the Vehicle Routing Problem (VRP), an operational optimization problem. The FLP determines whether new depots are opened or not and which customers are supplied by which depot [7]. The VRP is based on the FLP's solution. It decides in which sequence customers are delivered from one depot [8]. Since the determination of facility locations influences routing costs to other customers, there is an interdependence between the FLP and the VRP. By combining both problems, potential synergies can be exploited [2,7], as illustrated in Fig. 1.

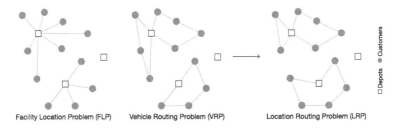

Facility Location Problem (FLP)   Vehicle Routing Problem (VRP)   Location Routing Problem (LRP)

Fig. 1. Difference between the FLP, the VRP and the LRP.

In the context of industry 4.0, not only an interconnection between problem models should exist. There should also be a continous exchange of information in order to support decision-making processes regarding temporary changes. Therefore, a system is needed which provides information for more than one problem instance and continuously supports decision makers. As a consequence, a methodology considering the high adaptability requirements of smart factories, called optimization network (ON), is developed. The LRP is split into its basic models and solved with existing solution approaches. After that, the models' solutions are united again. The overall result is improved by introducing a meta level black box optimization and several optimization runs, which lead to an improving solution quality. With the developed method, multiple problem models get connected and solved simultaneously. By considering and integrating several problems, synergy effects are utilized and opportunity costs are avoided. Moreover, due to the separate consideration of single problem models within the ON, real-world changing requirements of single departments only concern one problem model and not the whole LRP solution approach.

The article is organized as follows. In Sect. 2, related literature concerning the LRP is discussed. Section 3 contains the presentation of the new methodology of an optimization network for the LRP. Results are shown in Sect. 4. Finally, in Sect. 5 conclusions and directions for further research are drawn.

2 Literature Review

The Capacitated Location Routing Problem (CLRP) combines the Single-Source Capacitated Facility Location Problem (SS-CFLP) and the Multi-Depot Vehicle Routing Problem (MDVRP) [7]. The SS-CFLP decides if new depots are opened or not and assigns every customer to one opened depot under consideration of depot opening costs and delivery costs [7]. The MDVRP finds routes between customers and depots under consideration of distance costs and vehicle fixed costs [8]. The combination of the SS-CFLP and the MDVRP or the Capacitated Vehicle Routing Problem (CVRP) if only one depot is considered, leads to the following described mathematical model of the CLRP [6,7].

Let $G = (V, E)$ be a directed graph. V is a set of nodes with a subset I of m potential depots and a subset $J = V \setminus I$ of n customers, which need to be served. E is a set of edges which connects all nodes. Each edge $(i, j) \in E$ has distance costs C_{ij}. W_i represents the capacity and O_i the opening costs of each depot $i \in I$. Every customer $j \in J$ has a demand d_j. Q_k is the capacity of each vehicle $k \in K$. F are vehicle fixed costs which occur once for each route. Binary decision variables are $x_{ijk} = 1$ if node j is served after i with vehicle k, $y_i = 1$ if a depot i is opened, and $f_{ij} = 1$ if a customer j is assigned to a depot i.

$$\sum_{i \in I} O_i y_i + \sum_{i \in V} \sum_{j \in V} \sum_{k \in K} C_{ij} x_{ijk} + \sum_{i \in I} \sum_{j \in V} \sum_{k \in K} F x_{ijk} \tag{1}$$

subject to

$$\sum_{i \in V} \sum_{k \in K} x_{ijk} = 1 \quad \forall\, j \in J, \tag{2}$$

$$\sum_{i \in V} \sum_{j \in J} x_{ijk} d_j \leq Q_k \quad \forall\, k \in K, \tag{3}$$

$$\sum_{j \in J} d_j f_{ij} \leq W_i y_i \quad \forall\, i \in I, \tag{4}$$

$$\sum_{j \in V} x_{ijk} - \sum_{j \in V} x_{jik} = 0 \quad \forall\, i \in V, k \in K, \tag{5}$$

$$\sum_{i \in I} \sum_{j \in J} x_{ijk} \leq 1 \quad \forall\, k \in K, \tag{6}$$

$$\sum_{i \in S} \sum_{j \in S} x_{ijk} \leq |S| - 1 \quad \forall\, S \subseteq J, k \in K, \tag{7}$$

$$- f_{ij} + \sum_{u \in V} (x_{iuk} + x_{ujk}) \leq 1 \quad \forall\, i \in I, j \in J, k \in K, \tag{8}$$

$$f_{ij} \in \{0,1\} \quad \forall\, i \in I, j \in J, \quad x_{ijk} \in \{0,1\} \quad \forall\, i \in V, j \in V, k \in K, \tag{9}$$
$$y_i \in \{0,1\} \quad \forall\, i \in I.$$

With the objective function (1), depot opening costs, distance costs, and vehicle fixed costs are minimized. Constraints (2) define that every customer has to be served exactly once. With restrictions (3) and (4), capacities of vehicles and depots are satisfied. Conditions (5) and (6) guarantee the continuity of each route, and every vehicle is allowed to start and end at the same depot once. With restrictions (7), subtours are eliminated. Constraints (8) assign every customer to a depot if the customer is included in a route from that depot. Conditions (9) define boolean variables.

The CLRP is NP-hard, as the NP-hard proven MDVRP or CVRP and the SS-CFLP are included in this problem formulation [7,8]. Therefore, most of the algorithms for the CLRP are heuristic ones, such as metaheuristics, which are problem-independent solution approaches [7]. Concerning population based algorithms, there have already been successful approaches by splitting up the LRP into a FLP and a VRP [2]. With the integrated solution approach of matheuristics, mathematical programming is combined with metaheuristics [7].

An analysis of solution techniques which consider more than one problem shows that there is a new integrated optimization approach, where algorithmic strategies are proposed for the combination of existing heuristics to solve several problems within one optimization [1]. Multilevel decision-making optimization is another possibility to deal with various problem models [4]. Hierarchical, dezentralized management problems are combined within a decision-making model [4]. The idea originates from Stackelberg's game theory [9]. The upper level (leader) makes a decision and the lower level (follower) decides based on the leader's decision. An example for an already existing multilevel solution approach is the three-echelon supply chain network [10].

3   An Optimization Network for the Capacitated Location Routing Problem

The approach of this paper is based on the idea of embedding the CLRP into a network of optimizations. The optimization network consists of problem solvers for the subproblems of the CLRP (SS-CFLP and CVRP). An orchestrator is responsible for the necessary data transfer between the problem solvers and for the calculation of the overall CLRP result out of the single problem solver solutions. A meta solver is in charge of the improvement of the overall result of the single problem solvers. Several optimization runs are allowed within the ON, which lead to improving CLRP solutions. Subsequently, the operating mode of an ON is presented in Sect. 3.1 and a detailed explanation of the solution procedure and information on the computational evaluation are provided in Sect. 3.2.

3.1   Operating Mode

Within the ON, there is a model splitting and a solution merging phase. For the *model splitting*, the CLRP is divided into its basic models. Within a second stage, the *solution merging* process, the two single optimization results are united. The two phases are implemented by the introduction of three different parts within the ON. The *problem solvers*, the *meta solver* and the *orchestrator*.

- **Problem solvers.** Two problem solvers are applied, one for the optimization of the SS-CFLP and one for the CVRP. The CVRP is used instead of the MDVRP, as for every single opened depot out of the solution of the SS-CFLP optimization, one route optimization is started. For the SS-CFLP, the exact solver IBM CPLEX Optimizer[1] is used. The CVRP is solved by applying an already implemented genetic algorithm (GA) in the framework HeuristicLab[2]. The GA has been chosen since with such a heuristic approach, a more acceptable optimization run time can be guaranteed for big instances for practical applications than with exact solution methods.
- **Orchestrator.** The orchestrator is responsible for the data transfer between all optimization nodes and for the merging of the problem solver solutions.
- **Meta solver.** After the overall result is calculated, parts of the problem instance are optimized by the meta solver. One or more parts of the problem input have to be chosen for this optimization. In this work, one part of the SS-CFLP input is selected. In the first ON (ON1), depot coordinates dco_i are optimized, in the second one (ON2), customer coordinates cco_j. This black box optimization is applied by taking the overall result into account and leads to a better overall solution within the next optimization run. For this purpose, an evolutionary solution approach, the Covariance Matrix Adaptation Evolution Strategy (CMA/ES) [3], is applied.

After the meta solver optimization, a new optimization run is initiated within the ON. Different outcomes of both problem models are compared and the best ones are selected. The exact mechanism of the network, especially the change of input data and its reconversion into original data to ensure a valid real-world result after every optimization run, is described in the following Sect. 3.2.

3.2 Solution Procedure

The solution procedure of the ON is presented in the following seven steps with a related pseudo code shown in Algorithm 1. Moreover, parameter settings for the applied algorithms, selected benchmark instances and the chosen computational performance comparison are described.

Step 1. The meta solver generates a random real vector Vec_j^*. Its size corresponds to one part $Orig_j$ of the original input data of the SS-CFLP which has been chosen for the meta solver optimization (dco_i or cco_j).

Step 2. The orchestrator stores the original data $Orig_j$ and variegated one Vec_j^* and replaces $Orig_j$ by Vec_j^* in the SS-CFLP input data. Example SS-CFLP input data: depots $i \in I$, depot coordinates dco_i^* (instead of original dco_i), depot capacities W_i, depot opening costs O_i, customers $j \in J$, customer coordinates cco_j, customer demands d_j, distances C_{ij}.

---

[1] www.ibm.com/software/products/en/ibmilogcpleoptistud.
[2] http://dev.heuristiclab.com/.

Algorithm 1. Pseudo Code for the Optimization Network of a CLRP

```
1: for Vec_j^* generated by meta solver do
2:      SS - CFLP' ← Adapt(SS - CFLP, Vec_j^*)
3:      Depots, CustomerAssignment ← CPLEX(SS - CFLP')
4:      for i ∈ depots do
5:          CVRP ← InitializeCVRP(depot, CustomerAssignment[depot])
6:          Tour ← HeuristicLab(CVRP)
7:          Tours.Add(Tour)
8:      end for
9:      SOL* ← Evaluate(Depots, Tours)
10: end for
```

Step 3. The first problem solver is responsible for the optimization of the adapted SS-CFLP. The objective is the minimization of depot opening costs and delivery costs from depots to customers. If a depot is opened or not, $y_i = 1$ or 0 and if a customer is supplied by a depot or not, $x_{ij} = 1$ or 0.

Step 4. The orchestrator takes the solution of *Step 3*. For every opened depot $y_i = 1$ as a starting point for the CVRP, the changed input data Vec_j^* is substituted by inserting the original $Orig_j$ out of the stored correlation of *Step 2*. Assigned customers $x_{ij} = 1$ are given to the CVRP solver per opened depot with original data (*Example: if $y_2 = 1$, for the CVRP, dco_2 is taken instead of dco_2^**). Example CVRP input data: opened depot $i = 0 \in n$, original depot coordinates dco_i, vehicles K, vehicle capacity Q_k, vehicle fixed costs F, customers $j \in n \setminus 0$ with coordinates cco_j, customer demands d_j, and distances C_{ij} of all edges.

Step 5. The problem solver optimizes one CVRP for every opened depot $y_i = 1$ out of *Step 3* with original input data. The objective is to minimize distance costs and vehicle fixed costs. If one edge is served by one vehicle, $x_{ijk} = 1$.

Step 6. The orchestrator takes necessary solution parts out of both problem solvers' solutions and calculates the overall result SOL^* of the CLRP. From the SS-CFLP solver, O_i if $y_i = 1$ and from the CVRP solver, F and C_{ij} if $x_{ijk} = 1$.

Step 7. The meta solver takes the overall result SOL^* and optimizes the real vector Vec_j^*. The result is a new real vector Vec_j^*. If the current generation of the CMA/ES of the meta solver $G \leq G_{max}$, take the new optimization result Vec_j^* and go to *Step 2*, otherwise output the best overall result SOL^*.

Parameters for the GA and the CMA/ES are defined as follows. *GA:* popoluation size $P = 100$, maximum generations $MG_{max1} = 100$ if customer nodes < 100 else $MG_{max2} = 350$, mutation probability $m = 0.05$, MultiVRPSolutionCrossover and Manipulator, TournamentSelector. *CMA/ES:* maximum generations $G_{max} = 80$, population size $P = 20$, $\sigma = 0.5$, $\mu = 0$.

The developed generic ON are tested with established benchmark instances proposed by Prins et al. [6]. Their performance is compared with an already existing generic solver, the so-called LocalSolver[3], which combines various optimization techniques, from local search to nonlinear programming techniques.

4 Results

Since the used algorithms are evolutionary and genetic approaches, they are of random nature. Therefore, presented results in Table 1 are average solutions of ten repetitions. Instances are tested for the developed ON and for the Local Solver with a maximum run time of eight hours. Instance names, for example *20-5-1a*, represent *customer nodes-depot nodes-vehicle and depot capacities* [6]. Besides lower bounds (LB) and best known results (BKR) for the twelve presented instances, ON1 and ON2 represent the described method in Sect. 3.1 and LS the Local Solver.

When comparing average results of ON1, ON2, and LS in Table 1, it can be seen that with an average gap of 1.95% to BKR, with ON1, best solutions can be achieved. A comparison of ON1 and ON2 demonstrates a significant difference in solution quality of over 21%. The results indicate that the meta solver's solution strategy is responsible for this difference. Whereas ON1 focuses on depot coordinates, where a maximum of ten nodes have to be optimized, in ON2, the substantially higher quantity of 20 to 200 customer nodes seems to be a lot more difficult to optimize for the CVRP solver and the meta solver. A consideration of ON1 and LS shows that for instances up to 100 customer nodes, ON1 brings similar good results as LS. However, the more depot and customer nodes instances have got, the better ON1's results are compared to LS ones.

Table 1. Comparison of optimization networks and local solver results.

| | Instance | LB | BKR | Average Results | | | | | |
| | | | | ON 1 | | ON 2 | | LS | |
				Result	Gap [%]	Result	Gap [%]	Result	Gap [%]
1	20-5-1a	**54,793**	**54,793**	**54,793**	0.00	66,130	17.14	**54,793**	0.00
2	20-5-1b	**39,104**	**39,104**	**39,104**	0.00	45,866	14.74	**39,104**	0.00
3	50-5-1a	**90,111**	**90,111**	92,043	2.10	102,473	12.06	**90,111**	0.00
4	50-5-1b	**63,242**	**63,242**	**63,242**	0.00	79,848	20.80	**63,242**	0.00
5	100-5-1a	**274,814**	**274,814**	276,827	0.73	366,118	24.94	277,840	1.09
6	100-5-1b	207,037	213,615	215,672	4.00	255,091	18.84	216,854	4.41
7	200-10-1a		475,294	494,319	3.85	629,624	24.51	521,156	8.80
8	200-10-1b		377,043	392,701	3.99	485,136	22.28	444,944	15.26
9	200-10-2a		449,006	458,691	2.11	772,991	41.91	522,473	14.06
10	200-10-2b		374,280	380,765	1.70	626,225	40.23	432,299	13.42
11	200-10-3a		469,433	483,115	2.83	611,885	23.28	484,144	3.04
12	200-10-3b		362,653	370,400	2.09	478,934	24.28	457,772	20.78
				Average	**1.95**		**23.75**		**6.74**

---

[3] http://www.localsolver.com/.

5   Conclusion and Outlook

In this paper, a new, generic solution approach for the Capacitated Location Routing Problem has been proposed. With this approach, existing solvers are used for the optimization of the basic problems of the CLRP. By the introduction of a meta solver, parts of the problem instance are optimized with a black box optimization method. Several optimization runs lead to the utilization of synergy effects and therefore to improving overall results. An orchestrator ensures a valid transfer and conversion of input data.

Results show that depending on the optimization strategy of the input data within the meta solver, compared to a generic, already exising solver, the developed method is promising. Therefore, further optimizations with different adaptation strategies, for example the optimization of depot opening costs, have to be tested. The coupling of several strategies, for example depot coordinates and opening costs, also has to be examined. Moreover, another ON, consisting of a MDVRP instead of starting several CVRP should be tested to explore if further synergy effects are exploitable. Besides, for this new, generic solution approach, the optimization run time was defined with a maximum of eight hours. To be able to come closer to optimize real-world instances, the new solution approach has to be refined concerning the reduction of optimization run time.

Acknowledgments. The work described in this paper was done within the COMET Project Heuristic Optimization in Production and Logistics (HOPL), #843532 funded by the Austrian Research Promotion Agency (FFG) and the Government of Upper Austria.

References

1. Beham, A., Fechter, J., Kommenda, M., Wagner, S., Winkler, S.M., Affenzeller, M.: Optimization strategies for integrated knapsack and traveling salesman problems. In: Moreno-Díaz, R., Pichler, F., Quesada-Arencibia, A. (eds.) EUROCAST 2015. LNCS, vol. 9520, pp. 359–366. Springer, Cham (2015). https://doi.org/10.1007/978-3-319-27340-2_45
2. Drexl, M., Schneider, M.: A survey of variants and extensions of the location-routing problem. Eur. J. Oper. Res. **241**(2), 283–308 (2015)
3. Hansen, N.: The CMA evolution strategy: a comparing review. In: Lozano, J.A., Larrañaga, P., Inza, I., Bengoetxea, E. (eds.) Towards a New Evolutionary Computation, vol. 192, pp. 75–102. Springer, Heidelberg (2006). https://doi.org/10.1007/3-540-32494-1_4
4. Lu, J., Han, J., Hu, Y., Zhang, G.: Multilevel decision-making: a survey. Inf. Sci. **346**, 463–487 (2016)
5. MacDougall, W.: Industrie 4.0: Smart Manufacturing for the Future. Germany Trade & Invest, Berlin (2014)
6. Prins, C., Prodhon, C., Calvo, R.W.: Solving the capacitated location-routing problem by a grasp complemented by a learning process and a path relinking. 4OR: Q. J. Oper. Res. **4**(3), 221–238 (2006)
7. Prodhon, C., Prins, C.: A survey of recent research on location-routing problems. Eur. J. Oper. Res. **238**(1), 1–17 (2014)

8. Toth, P., Vigo, D.: Vehicle Routing: Problems, Methods, and Applications. SIAM (2014)
9. Von Stackelberg, H.: The Theory of the Market Economy. Oxford University Press, Oxford (1952)
10. Xu, X., Meng, Z., Shen, R.: A tri-level programming model based on conditional value-at-risk for three-stage supply chain management. Comput. Ind. Eng. **66**(2), 470–475 (2013)

A Matheuristic to Solve a Competitive Location Problem

Dolores R. Santos-Peñate[1(✉)], Clara M. Campos-Rodríguez[2],
and José A. Moreno-Pérez[2]

[1] Dpto de Métodos Cuantitativos en Economía y Gestión/TIDES,
Universidad de Las Palmas de G.C., 35017 Las Palmas de Gran Canaria, Spain
dr.santos@ulpgc.es
[2] Instituto Universitario de Desarrollo Regional, Universidad de La Laguna,
38271 La Laguna, Spain
{ccampos,jamoreno}@ull.es

Abstract. We consider the leader-follower problem in a discrete space.
We apply an algorithm which integrates the linear programming formulation of the problems for the leader and the follower and, in an iterative
process, finds a solution by solving a sequence of these linear problems.
We propose a matheuristic procedure where the problem of the leader is
solved via a kernel search algorithm.

Keywords: Competitive location · Leader-follower problem ·
$(r|p)$-centroid · $(r|X_p)$-medianoid · Linear programming · Kernel search

1 Introduction

In a competitive location problem two or more firms try to attract customers
making decisions about locations. The most usual objective is to maximize the
own market share but other objectives can be considered [6]. We study a discrete
leader-follower problem. Initially no firm is operating in the market. The leader
enters the market first with p facilities and seeks to minimize the maximum market share captured by a future competitor. Then, the follower opens r facilities
at the locations that maximize its market share. Given the set of location for
the leader, X_p, the problem of the follower is to find an $(r|X_p)$-medianoid. The
problem of the leader is to find an $(r|p)$-centroid. We assume that goods are
essential which means that all demand must be satisfied. The behaviour of the
customers is modeled using different functions. Besides the binary rule, according to which a customer uses the closest facility, we consider other customer
choice rules defined by a continuous, non-negative and non-decreasing S-shaped
function which takes values between zero and one.

Partially financed by Ministerio de Economía y Competitividad (Spanish Government) with FEDER funds, grant ECO2014-59067-P and TIN2015-70226-R, and by
Fundación Cajacanarias (grant 2016TUR19).

R. Moreno-Díaz et al. (Eds.): EUROCAST 2017, Part I, LNCS 10671, pp. 266–274, 2018.
https://doi.org/10.1007/978-3-319-74718-7_32

In this work we propose a matheuristic where, using the linear formulations of the leader's and follower's problems, the solution is obtained by solving a sequence of these linear problems. The problem of the follower is solved applying an exact algorithm. To solve the leader's problem we apply a kernel search procedure [5]. Different solution procedures for the leader-follower problem, exact and heuristic, for different choice rules and demand assumptions can be found in [1–4,7].

The rest of the paper is organized as follows. Section 2 contains the problem statement and linear programming formulations. The solution procedure is described in Sect. 3. Section 4 includes some computational results. Finally, Sect. 5 contains some conclusions.

2   Problem Statement and Linear Formulations

Let $C = \{c_k : k \in [1..n]\}$ be a finite set of demand points or clients and $L = \{l_i : i \in [1..m]\}$ be a finite set of potential locations for facilities. Every point $c_k \in C$ has a weight $w_k = w(c_k)$ which represents the demand at c_k. Let $W_T = \sum\limits_{k=1}^{n} w_k$ be the total demand. Let $d_{ki} = d(c_k, l_i)$ be the distance between point $c_k \in C$ and point $l_i \in L$, and let $d_{kX} = \min_{x \in X} d(c_k, x)$ be the distance between the point $c_k \in C$ and the subset $X \subseteq L$.

Let $f_k(\delta)$ be the function defined by

$$
f_k(\delta) = \begin{cases}
0 & \text{if} & \delta \leq a_k \\
2\left(\dfrac{\delta - a_k}{b_k - a_k}\right)^2 & \text{if } a_k < \delta \leq \dfrac{a_k + b_k}{2} \\
1 - 2\left(\dfrac{\delta - b_k}{b_k - a_k}\right)^2 & \text{if } \dfrac{a_k + b_k}{2} < \delta \leq b_k \\
1 & \text{if} & \delta > b_k
\end{cases}
$$

where $a_k \leq 0 < b_k$.

The leader (firm A) enters the market first with p facilities located at $X_p \in L^p$. Then, the follower (firm B) opens r facilities located at $Y_r \in L^r$. The demand at point c_k captured by the firms depends on the difference $\delta_k = d_{kX_p} - d_{kY_r} = d_{kA} - d_{kB}$. The market share for firms A and B are given, respectively, by $W_A = W_T - W_B$ and

$$
W_B = W_B(X_p, Y_r) = \sum_{k=1}^{n} w_k f_k(\delta_k) \tag{1}
$$

2.1   Follower's Problem

If the leader has p facilities at X_p, then the problem of the follower is to determine the set Y_r such that

$$
W_B(X_p, Y_r) = \max_{Y \in L^r} W_B(X_p, Y). \tag{2}
$$

For $i \in [1..m]$ and $k \in [1..n]$ we define the following variables,

$$y_i = \begin{cases} 1 \text{ if the follower opens a facility at point } l_i \\ 0 \text{ otherwise} \end{cases}$$

$$z_{ki} = \begin{cases} 1 \text{ if customer at } c_k \text{ visits a facility at point } l_i \\ 0 \text{ otherwise.} \end{cases}$$

Moreover, for $i \in [1..m]$ and $k \in [1..n]$ we have

$$h_{ki} = w_k f_k(\delta_{ki}) \text{ where } \delta_{ki} = d_{kX_p} - d_{ki}.$$

The problem can be formulated as follows,

$$\max \sum_{i=1}^{m} \sum_{k=1}^{n} h_{ki} z_{ki}$$

$$\sum_{i=1}^{m} y_i = r$$

$$\sum_{i=1}^{m} z_{ki} \leq 1 \qquad k \in [1..n] \qquad (3)$$

$$z_{ki} \leq y_i \qquad i \in [1..m],\ k \in [1..n]$$

$$z_{ki},\ y_i \in \{0,1\} \qquad i \in [1..m],\ k \in [1..n].$$

2.2 Leader's Problem

The problem of the leader is to determine the set X_p that minimizes the maximum market share that the follower could achieve:

$$\min_{X \in L^p} \max_{Y \in L^r} W_B(X, Y). \qquad (4)$$

The score of $X \in L^p$ is $S(X) = \max_{Y \in L^r} W_B(X, Y)$.

For $i \in [1..m]$ and $k \in [1..n]$ we define the following variables,

$$x_i = \begin{cases} 1 \text{ if the leader opens a facility at point } l_i \\ 0 \text{ otherwise} \end{cases}$$

$$u_{ki} = \begin{cases} 1 \text{ if customer at } c_k \text{ visits a facility at point } l_i \\ 0 \text{ otherwise.} \end{cases}$$

For $i \in [1..m]$, $k \in [1..n]$ and $J = [1..\binom{m}{r}]$, we define

$$h_{ki}^j = w_k f_k(\delta_{ki}^j) \qquad\qquad \delta_{ki}^j = d_{ki} - d_{kY_j}.$$

The notation h_{ki}^j represents the demand at point c_k captured by the follower when the follower has its facilities located at Y_j and the customer at c_k visits a leader's facility at point l_i (when $u_{ki} = 1$).

The problem of the leader is formulated as follows,

$$\min W$$

$$\sum_{i=1}^{m} x_i = p$$

$$\sum_{i=1}^{m} \sum_{k=1}^{n} h_{ki}^j u_{ki} \leq W \quad j \in J$$

$$\sum_{i=1}^{m} u_{ki} = 1 \qquad k \in [1..n]$$

$$u_{ki} \leq x_i \qquad i \in [1..m], \ k \in [1..n]$$

$$u_{ki}, \ x_i \in \{0,1\} \qquad i \in [1..m], \ k \in [1..n].$$

(5)

3   Solution Procedure

We adapt the scheme of the exact algorithm described in Fig. 1 to incorporate the heuristic procedure which solves the leader's problem in Step 2.2. The solution is obtained by solving a sequence of problems for the leader and the follower. In this sequence the leader's model is modified by replacing the total set of follower's feasible solutions, set J in Problem (5), by a family of good follower's solutions $\mathcal{F}$. The optimum for the leader's problem restricted to the family $\mathcal{F}$ provides a lower bound of the optimum W^* while the optimum for the follower's problem gives an upper bound. The optimum W^* is achieved when both bounds, the lower and upper bounds, coincide.

We apply a basic heuristic and a kernel search algorithm (Fig. 2). The kernel search procedure consists of solving the problem using a reduced set of variables, the kernel, which is updated during the process by the incorporation of one of the buckets made of the rest of variables and the elimination of the less promising variables in the kernel [5]. The basic heuristic (BH) consists of steps 1, 2 and 4 of the kernel search (KS) algorithm, it is the KS when only the initial kernel is considered. For the leader's problem, in Step 3 of the KS algorithm, the location variables which are null in the solution to the relaxed problem are sorted in non-decreasing order of the average of the reduced costs obtained for the associated allocation variables.

4   Computational Experience

We consider several scenarios defined by the customer choice rule, the number of facilities, p and r, and the demand distribution. For the binary case, ties are

---

Algorithm 1. Exact procedure to solve the leader-follower problem

1. *Initialization.*
    1.1 Select s feasible leader's solutions X_i, $i = 1, ..., s$.
    1.2 Solve the follower's problem for X_i, $i = 1, ..., s$.
    1.3 Calculate an upper bound of $\overline{W}$ of the optimum W^*
        $\overline{W} = \min_i S(X_i)$.
    1.4 Let $X^* = X$ with $S(X) = \overline{W}$.
    1.5 Let $\mathcal{F} = \{Y_i\}_{i=1}^s$ be the selected family of good follower's feasible solutions.
    1.6 Set a lower bound $\underline{W}$.

2. *Iterations.* Repeat, until a stop rule condition is satisfied.
    2.1 Do $i = i + 1$.
    2.2 Solve the leader's problem using the family $\mathcal{F}$ of follower's solutions.
        Let X be the optimal solution obtained.
        i. If the optimal value obtained $S_{\mathcal{F}}(X)$ verifies $S_{\mathcal{F}}(X) > \underline{W}$ then do $\underline{W} = S_{\mathcal{F}}(X)$.
        ii. If $\underline{W} = \overline{W}$, then $W^* = \underline{W} = \overline{W}$ is the optimal value and X^* is the optimal location set for the leader.
    2.3 Solve the follower's problem for X.
        i. If $S(X) < \overline{W}$ then set $\overline{W} = S(X)$ and $X^* = X$.
        ii. If $\underline{W} = \overline{W}$, then $W^* = \underline{W} = \overline{W}$ is the optimal value and X^* is the optimal location set for the leader.
    Set $\mathcal{F} = \mathcal{F} \cup \{Y(X)\}$, where $Y(X)$ is the solution to the follower's problem.

Fig. 1. The exact procedure

solved by assigning half of the demand to each player. For the S-shaped functions we consider five scenarios defined by different values of a_k, b_k. The extremes of the interval, a_k and b_k, have been chosen doing $a_k = -\alpha_k \times \rho$ and $b_k = \beta_k \times \rho$, where α_k, $\beta_k > 0$ and ρ is the average of the absolute values $|d_{ki} - d_{kj}|$ with $k \in K$, $i, j \in I$ and $i \neq j$. We consider five scenarios: (1) $\alpha_k = \beta_k = 0.01$ $\forall k$; (2) $\alpha_k = \beta_k = 0.05$ $\forall k$; (3) $\alpha_k = \beta_k = 0.10$ $\forall k$; (4) $\alpha_k = \beta_k = 0.25$ $\forall k$; and (5) α_k and β_k randomly generated using a uniform distribution with values between 0.01 and 0.25. For the number of facilities we consider four scenarios: (1) $p = r = 2$; (2) $p = 2, r = 3$; (3) $p = 3, r = 2$; and (4) $p = r = 5$. For each scenario (S-function, p, r) we have the twenty instances used in [1], where $C = L$, $n = m = 100$. For instances 1 to 10 the demand is $w_k = 1$ for all k, for instances 11 to 20 the demand is generated by a uniform distribution in $[0, 200]$. We introduce a stop rule which fixes the maximum number of iterations (100 iterations).

In the kernel search algorithm, the initial kernel size is q where q is the number of nonzero variables in the solution to the relaxed problem. The bucket size is also q. The maximum number of investigated buckets is $\min\{nbuck, 3\}$ where $nbuck$ is the number of buckets. The location variables are eliminated from the kernel if these variables are zero for the last 2 solved problems.

Algorithm 3. Kernel search heuristic to solve the leader-follower problem

1. Solve the LP-relaxation of the leader's problem restricted to family $\mathcal{F}$.
2. Build the initial kernel $(K, U(K))$ being

$$K = \{x_i : x_i \neq 0 \text{ for the LP-relaxation solution }\} \qquad U(K) = \{u_{ki} : x_i \neq 0\}.$$

3. Sort the other variables according to a criterion based on their reduced costs and build a sequence of disjoint variable buckets with the same fixed length, except the last bucket that can have a smaller length.
4. Solve the MILP problem on the initial kernel.
5. Repeat until a certain number of buckets have been analysed.
   (a) Let $(\hat{K}, U(\hat{K}))$ be the current kernel plus a bucket.
   (b) Solve the MILP problem on $(\hat{K}, U(\hat{K}))$ with two constraints:
       i. Set an upper bound on the objective function value.
       ii. At least one facility of he current bucket must be selected.
       Let K_h^+ be the set of facilities of the current bucket that are selected in the feasible solution.
       Let K_h^- be the set of facilities belonging to the current kernel that havent been selected in s of the previously solved restricted problems since they have been added.
   (c) Update the current kernel: $\hat{K} \leftarrow \hat{K} \cup K_h^+ \setminus K_h^-$.

Fig. 2. Kernel search algorithm

The computational results presented in this section were obtained using a PC Intel(R)Core(TM) i7-2700K CPU 3.50 GHz, RAM 16 GB. The solutions were obtained using the CPLEX solver in GAMS.

The results are summarized in tables. W^* denotes the best objective value found, $ER = 100 \times \dfrac{\overline{W} - W}{W_T}$, IT is the total number of iterations until the algorithm stops, ITO is the number of iterations until the best objective value is found, T is the total time (seconds) consumed, and TO is the time (seconds) employed until the best solution is achieved. For the matheuristic, $\Delta W^* = 100 \times \dfrac{W^*_{heu} - W^*_{ex}}{W_T}$ where W^*_{heu} and W^*_{ex} are the best objective value obtained by the matheuristic and the exact procedure, respectively.

The results for the binary rule are summarized in Table 1. This table shows the average values for the two groups of instances 1–10 and 11–20. The exact procedure provides the optimum $(ER = 0)$ for all cases except for $p = r = 5$. Last column shows that the KS gives the same value as the exact method in three cases. For $p = r = 5$ and scenarios 1–10, the value W^* provided by the KS is slightly better than the best value obtained by the exact procedure after 100 iterations. Times $(T$ and $TO)$ consumed by the heuristic are normally much lower than the time required by the exact procedure, specially for $p = r = 5$.

Table 2 shows the average values obtained for the exact and KS procedures for $S = 5$ (a_k, b_k randomly generated). For the exact procedure, column 5 (ER) shows that only for one case the optimum is not reached for all the instances into the group before 100 iterations, it is the case $p = r = 5$ for the group of instances 1–10. The highest time values correspond to $p = r = 5$, the lowest times are obtained for $p = r = 2$. Normally, times consumed by the exact algorithm are higher than times used by the heuristic, the differences are more significant

Table 1. Binary rule. Average values for different procedures and scenarios. Minimum and maximum time values are indicated in bold

Scenario			W^*	ER (%)	IT	ITO	T (s)	TO (s)	ΔW^* (%)
p	r	Instances							
Exact									
2	2	1–10	52.30	0	10.00	4.20	**77.69**	**28.74**	
2	2	11–20	5370.35	0	14.70	9.20	86.92	44.63	
2	3	1–10	72.05	0	41.90	20.70	1374.91	620.55	
2	3	11–20	7273.95	0	51.40	16.20	1562.71	170.01	
3	2	1–10	37.20	0	12.00	6.60	95.30	59.17	
3	2	11–20	3756.30	0	15.20	10.40	134.86	143.32	
5	5	1–10	53.45	0.20	53.10	15.80	3844.15	225.09	
5	5	11–20	5386.90	0.11	75.20	52.10	**13434.71**	**5963.88**	
BH									
2	2	1–10	52.30		8.00	4.10	**12.59**	6.48	0
2	2	11–20	5402.90		18.50	6.50	32.99	10.22	0.34
2	3	1–10	72.60		29.50	3.80	46.80	**5.82**	0.55
2	3	11–20	7375.10		17.30	5.60	31.69	8.26	0.97
3	2	1–10	37.45		10.20	7.40	16.78	12.36	0.25
3	2	11–20	3771.35		29.60	6.80	59.12	11.98	0.15
5	5	1–10	53.45		33.80	15.30	91.04	36.31	0
5	5	11–20	5403.00		62.70	38.30	**277.39**	**161.07**	0.16
KS									
2	2	1–10	52.30		10.40	4.50	**44.06**	**14.70**	0
2	2	11–20	5378.00		14.00	9.10	67.07	37.75	0.09
2	3	1–10	72.35		34.10	15.20	196.47	68.62	0.30
2	3	11–20	7301.95		60.80	18.10	491.80	83.73	0.27
3	2	1–10	37.20		12.60	6.90	65.45	35.81	0
3	2	11–20	3756.30		24.20	8.90	138.29	56.09	0
5	5	1–10	53.35		49.80	19.10	1132.37	162.79	−0.10
5	5	11–20	5394.35		77.20	39.20	**3721.35**	**1308.11**	0.07

in the cases with the highest time consumption for the exact procedure. The average values provided by the BK, no presented in the table, are close to the values obtained by the exact procedure and the time consumption is much lower for the BK algorithm.

Table 2. Average values for $S = 5$. Maximum and minimum time values are indicated in bold

p	r	Inst	Exact						KS				
			W^*	ER	IT	ITO	T	TO	W^*	IT	ITO	T	TO
2	2	1–10	51.72	0	9.00	3.63	59.33	**19.09**	51.72	7.40	3.10	**31.54**	**10.39**
2	2	11–20	5232.43	0	9.90	6.89	**50.62**	31.06	5232.43	9.30	6.30	37.11	24.46
2	3	1–10	70.96	0	33.60	17.00	862.11	239.05	71.00	49.70	11.60	343.77	49.48
2	3	11–20	7083.66	0	33.70	18.57	668.86	304.83	7116.21	48.70	7.90	341.11	34.89
3	2	1–10	37.39	0	13.30	11.25	135.07	123.14	37.39	21.70	8.30	119.28	49.16
3	2	11–20	3715.43	0	11.30	9.22	86.36	67.61	3715.85	20.00	8.40	102.30	50.32
5	5	1–10	52.91	0.02	51.30	40.30	3763.70	**1965.34**	52.91	50.60	40.20	1189.93	667.31
5	5	11 20	5300.14	0	59.60	33.90	**6268.37**	1478.33	5300.14	60.00	36.70	**1919.55**	**712.13**

5 Conclusions

The main contribution of this paper is the proposal of a kernel search heuristic to solve the leader-follower problem in locations. We solve the leader-follower problem for 480 scenarios using three procedures, an exact algorithm, a basic heuristic (BH) and a kernel search heuristic (KS). We consider that the results provided by the BH and KS are promising but a deeper analysis would give more information about the performance of the algorithms and possible improvement strategies, for example in relation to different sorting criteria to select the variables in the buckets construction, among other issues.

References

1. Alekseeva, E., Kochetova, N., Kochetov, Y., Plyasunov, A.: Heuristic and exact methods for the discrete $(r|p)$-centroid problem. In: Cowling, P., Merz, P. (eds.) EvoCOP 2010. LNCS, vol. 6022, pp. 11–22. Springer, Heidelberg (2010). https://doi.org/10.1007/978-3-642-12139-5_2
2. Biersinger, B., Hu, B., Raidl, G.: Models and algorithms for competitive facility location problems with different customer behaviour. Ann. Math. Artif. Intell. **76**, 93–119 (2015)
3. Campos-Rodríguez, C.M., Santos-Peñate, D.R., Moreno-Pérez, J.A.: An exact procedure and LP formulations for the leader-follower location problem. TOP **18**(1), 97–121 (2010)
4. Davydov, I.A., Kochetov, Y.A., Mladenovic, N., Urosevic, D.: Fast metaheuristics for the discrete $(r|p)$-centroid problem. Autom. Remote Control **75**(4), 677–687 (2014)

5. Guastaroba, G., Speranza, M.G.: Kernel search for the capacitated facility location problem. J. Heuristics **18**(6), 877–917 (2012)
6. Hakimi, S.L.: Location with spatial interactions: competitive locations and games. In: Mirchandani, P.B., Francis, R.L. (eds.) Discrete Location Theory, pp. 439–478. Wiley, New York (1990)
7. Roboredo, M.C., Pessoa, A.A.: A branch-and-bound algorithm for the discrte $(r|p)$-centroid problem. EJOR **224**, 101–109 (2013)

Station Planning by Simulating User Behavior for Electric Car-Sharing Systems

Benjamin Biesinger$^{(\boxtimes)}$ (iD), Bin Hu, Martin Stubenschrott, Ulrike Ritzinger, and Matthias Prandtstetter (iD)

Center for Mobility Systems – Dynamic Transportation Systems,
AIT Austrian Institute of Technology, Giefinggasse 2, 1210 Vienna, Austria
{benjamin.biesinger,bin.hu,martin.stubenschrott,ulrike.ritzinger,
matthias.prandtstetter}@ait.ac.at

Abstract. The planing of a full battery electric car sharing system involves several strategic decisions. These decisions include the placement of recharging stations, the number of recharging slots per station, and the total number of cars. The evaluation of such decisions clearly depends on the demand that is to be expected within the operational area as well as the user behavior. In this work we model this as combinatorial optimization problem and solve it heuristically using a variable neighborhood search approach. For the solution evaluation we use a probability model for the user behavior and approximate the expected profit with a Monte-Carlo method. The proposed algorithm is evaluated on a set of benchmark instances based on real world data of Vienna, Austria. Computational results show that by simulating user behavior the expected profit can increase significantly and that other methods assuming the best case for user behavior are likely to overestimate the profit.

1 Introduction

Over the last years the increased air pollution and the awareness for sustainability has lead to a steady growth of the market for privately owned electric vehicles. While so far the high acquisition costs and the limited battery range of these vehicles hinder the wide-spread use, car-sharing systems with electric cars could potentially decrease the use for conventional vehicles in urban areas [9]. Such car-sharing systems offer cars in a pre-defined area which can be rented by customers to perform their desired trips. Compared to systems using conventional cars, in electric car-sharing systems charging stations have to be available within the operational area to recharge the battery of the vehicles. In this work, we consider station-based (in contrast to free-floating) systems in which cars can only be rented and returned at specific stations. The most important strategic decisions when introducing such a system in a new area are where to place

This work has been partially funded by the Austrian Federal Ministry for Transport, Innovation and Technology (bmvit) in the JPI Urban Europe programme under grant number 847350 (e4share).

R. Moreno-Díaz et al. (Eds.): EUROCAST 2017, Part I, LNCS 10671, pp. 275–282, 2018.
https://doi.org/10.1007/978-3-319-74718-7_33

the stations, how many charging slots to install, and how many electrical cars to deploy. For being able to make a statement about the viability a demand model is needed which gives a forecast of the customer requests, i.e., when the potential customers want to use a shared car and where they want to go. As the customers are usually willing to walk a short distance to or from a station which is close to their desired starting or ending point, each customer request has a set of potential starting and ending stations. In this work, we model the strategic decisions on the locations of stations, the number of charging slots per station, and the total number of deployed cars with respect to a limited budget as a combinatorial optimization problem and solve it heuristically using a variable neighborhood search approach. For evaluating of these strategic decisions we simulate user behavior by using a probability model and thereby model how the cars are used over time. The user behavior determines which trips are fulfilled, resulting in an expected profit value that is used to assess the quality of the station and car decisions. Figure 1 shows an example of a real-world instance of this problem for the inner part of Vienna, Austria. Possible station locations are shown along with origin and destination points for all requested trips in this area. This example indicates that real-world instances tend to involve a lot of decisions and a large solution space, which makes the use of fast heuristics appealing.

Fig. 1. Example of a real-world instance of Vienna, Austria. The red dots are the start and end points of the requested trips and the blue rectangles denote possible locations for recharging stations. On the right side a solution candidate is shown with the chosen station locations, their approximated area of attraction, and the acceptable trip requests. (Color figure online)

In Sect. 2 we formally define the station planning problem and in Sect. 3 we give an overview of related work. Then, the solution approach is presented including the description of the modeling of the user behavior in Sect. 4. The evaluation of the algorithm is shown in Sect. 5, in which also the benchmark instances are described. Finally, conclusions are drawn in Sect. 6 where also a view on possible future work is given.

2   Problem Definition

The formal problem definition, which is based on [2,6], is as follows. The charging station location problem with user decisions (CSLP-UD) is defined on a road network $G = (V, A)$, where the set A represents road segments, and V the crossings. Each arc $a = (i, j) \in A$, $i, j \in V$ with length l_{ij} has an associated travel time δ_{ij} needed to travel from vertex i to j. Possible station locations $S \subseteq V$ are given by a subset of the vertices and each potential station $i \in S$ has an associated opening cost $F_i \geq 0$, a capacity $C_i \in \mathbb{N}$, and a cost per slot $Q_i \geq 0$. The maximum number of cars is given by H, and each car has the same acquisition cost F_c, battery capacity B^{max}, and charging rate per time unit ρ.

The demand model is given by a set of trip requests K, where each trip $k \in K$ has a starting s_k and ending time e_k, where $s_k, e_k \in T = \{0, \dots, T_{\mathrm{max}}\}$ with $e_k > s_k$, an origin $o_k \in V$, and a destination $d_k \in V$. Furthermore, a duration δ_k, an estimated battery consumption b_k, and a profit p_k is given which is proportional to the trip duration. A parameter for maximum walking distance β^w determines the set of possible starting $N(o_k)$ and ending stations $N(d_k)$ for a request $k \in K$. If one of the sets of a request k is empty or none of these stations are opened, then k is not fulfillable and not considered anymore. The right part of Fig. 1 shows a selection of the stations and the resulting fullfilable trip requests.

The goal of the CSLP-UD is to find the set of stations to open $S' \subseteq S$, the number of slots z_s to use for each open station $s \in S'$, and the total number of cars $H' \leq H$ in the system with a limited budget W such that the total expected profit under the given user decision model is maximized. The user decision model defines probabilities how the users behave, i.e., which user gets the car in case of concurrent demand and which ending station is chosen for returning the car. This randomness can cause strongly different sets of fulfilled trip requests and the goal is to maximize the average profit over all possible scenarios. One scenario can be described as a set of fulfilled trips K'_c for each car $c = 1 \dots, H'$. These sets have to fulfill several constraints to represent a feasible solution. *Capacity feasibility* is given when at each time-step $t \in T$ there are no more cars in station $s \in S$ than the available number of slots. *Battery feasibility* is given if the battery capacity of the car is sufficient for performing the requested trip taking potential preceding battery charging into account. More formally, the solution is battery feasible if between two consecutive trips $k^1, k^2 \in K$ starting/ending at station i of a car $\min\{(s_{k^2} - e_{k^1})\rho + B^{k^1}, B^{\mathrm{max}}\} \geq b_{k^2}$ is valid, where B^{k^1} is the remaining battery capacity of the car after performing trip k^1. *Connectivity* is given when the ending station of a trip is equal to the starting station of the next trip.

3   Related Work

Although the literature about optimization problems in the domain of car-sharing is huge, when considering battery electric vehicles within such a system the literature is scarce. Brandstätter et al. [4] give an overview of this and several other optimization problems arising in the domain of e-car sharing systems and suggest possible research directions. The problem described in this article without consideration of user behavior has already been approached with exact algorithms in the form of mixed integer linear programming formulations by Brandstätter et al. [6,7] and with metaheuristic methods in our previous work by Biesinger et al. [2]. A variant of this problem focusing on the stochastic aspects of the CSLP is presented by Brandstätter et al. [5]. Considering relocation decisions for moving cars from areas of low density to high density regions within the location problem is described by [3]. Weikl and Bogenberger [11] investigated variants of relocation strategies for free-floating car sharing system with conventional vehicles. A related problem of choosing locations for recharging stations for electric taxis is described by Asamer et al. [1] who suggest regions for placing stations.

4   Solution Approach

Similar to Biesinger et al. [2], the algorithm uses the vector $z = (z_0, \ldots, z_{|S'|-1})$ and H' as solution representation. After an initial solution is generated using a method described shortly, a variable neighborhood search (VNS) [8] approach is employed, which uses neighborhood structures (NBs) that only operate on z, whereas H' is determined by the remaining budget.

4.1   Initial Solution and Variable Neighborhood Search

For generating an initial solution, each station is assigned a value representing an attractiveness factor which is computed by counting the number of requests that can either start or end at this station. Then, in descending order, a new station is iteratively opened with a randomly chosen number of slots until the budget limit is reached. This initial solution is taken as input by the VNS, which uses four NBs in the following order: The *close station* NB closes a previously opened station and thereby increases H'. The *open station* NB opens a previously closed station while respecting the budget constraint. The *change slots* NB changes the number of slots of an open station, and the *swap* NB swaps the number of slots of two open stations. For the last NB we use a repair method, which iteratively reduces the number of slots of either station, to ensure budget feasibility.

4.2   Solution Evaluation

The solution evaluation is an essential part of the algorithm and involves the decisions which trip requests can be fulfilled. As this problem is itself a difficult optimization problem, in previous work we developed several heuristics based on a greedy criterion [2]. They all use a time discretization and generate a time-expanded location network in which the vehicle paths through space and time are iteratively computed. In this work we do the same, however, as already mentioned in Sect. 1, we do not assume the best case in which the profit is maximized as all state-of-the-art approaches (e.g., [3]) but we simulate user behavior.

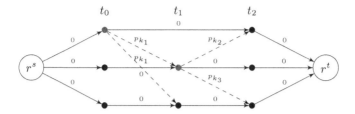

Fig. 2. Time-expanded location network used for solution evaluation.

Figure 2 shows an example of the generated time expanded location network (TELN). The TELN consists of a node for each open station and time slot, an artificial source r^s and target r^t node. There are three different sets of arcs: *Initialization arcs* exist between r^s and all station nodes at time instant t_0 and between all station nodes at time instant t^{max} (t_2 in this example) and r^t. *Waiting arcs* exist between any two consecutive time slots of each station but a node can possibly be skipped if it is not relevant (i.e., degree of two). They correspond to a vehicle waiting at a station. *Travel arcs* exist for each trip $k \in K$ between all start and end station combination. All waiting and travel arcs have two types of weight; an energy consumption value and a profit value. Waiting arcs have a negative energy consumption value indicating the loading of the battery and a profit value of zero. Travel arcs have a positive energy consumption value and a profit value corresponding to the profit p_k of the corresponding trip k.

During solution evaluation, for each H' cars a path from r^s to r^t is computed and the TELN is updated. The user decision is modeled in the choice of the path: Whenever there are two or more outgoing travel arcs at the current node, each of these arcs gets assigned a probability as follows. Assume that there are $K' \subset K$ trips that can start at the current node. First, a $k \in K'$ is chosen uniformly at random, and then the destination station $s' \in N(d_k)$ is chosen uniformly at random as well. When r^t is reached a path of a car has been successfully found, and the TELN is updated by deleting all trip arcs of the performed trips and for all stations at full capacity the incoming arcs are deleted. Then, another path for the next car is computed in the updated TELN. A Monte-Carlo simulation (see, e.g., [10]) is used by repeating this process of finding H' paths for a number of times, which is determined by the sample size parameter of the overall algorithm. The average of the achieved profit during these repetitions is the expected profit and the objective value of the solution candidate.

5   Computational Results

The proposed algorithm is evaluated on a set of benchmark instances based on real-world data from Vienna, Austria. OpenStreetMap data is used for the underlying road network and we assume potential locations for stations at supermarkets, parking lots, and areas next to subway stations. The number of slots for each station is between 1 and 10, chosen uniformly at random, and the costs are $F_i \in \{9000, \ldots, 64000\}$ and $Q_i \in \{22000, \ldots, 32000\}$ Euro. The customer demand model is based on real taxi data while only trips longer than 500 m are considered. We set the maximum walking distance to/from stations to 5 min, consider a time horizon of 8 h, and use a time discretization of 15 min. The car data is based on the real data from a Smart ED, which is a small full battery electric vehicle and we choose a maximum budget W between 1 and 5 million Euro. The overall instance contains 693 potential station locations and 37965 trip requests but we only use 4 subsets $I_1, \ldots, I_4$ of increasing size of the inner part of Vienna corresponding to viable business areas. These subsets contain between 105 and 280 station locations and between 108 and 1347 trip requests.

We compare the proposed simulation-based evaluation with a greedy-based evaluation, which also uses the TELN as described in Sect. 4.2, but finds the paths through the TELN deterministically using a greedy criterion based on a potential profit value (see [2] for a more detailed description). For the simulation-based evaluations a sample size of 10 is used and the time limit of both algorithm variants is 1 h. The following questions are approached by the computational study: How large is the error if we assume the best case of the user behavior when planing recharging stations? Is it viable to use a more realistic but time-consuming method? We answer these questions in Table 1 which shows the above described comparison between the simulation (Sim. Eval.) and the greedy evaluation (Greedy Eval.) for the different instances and maximum number of cars H aggregated over the budget sizes of 1 to 5 million. The final solution of these methods is evaluated both with the simulation and the greedy method and their geometric means are shown in the column $\overline{obj^*_{\text{sim}}}$ and $\overline{obj^*_g}$, respectively. Furthermore, the relative differences are shown in column *diff*.

Table 1. Results of the simulation-based compared to the greedy-based evaluation.

Instance		Sim. Eval.			Greedy Eval.		
I	H	$\overline{obj^*_{\text{sim}}}$	$\overline{obj^*_g}$	diff.	$\overline{obj^*_{\text{sim}}}$	$\overline{obj^*_g}$	diff.
I_1	10	7493.2	12633.0	39.08%	5973.0	18755.1	67.61%
	25	11577.6	16842.1	30.49%	5368.0	24434.4	76.19%
	50	12616.7	17360.7	26.90%	5170.8	25724.4	77.11%
I_2	10	10238.6	17411.3	39.42%	10870.4	31720.7	65.36%
	25	17064.9	24587.9	29.92%	10811.7	41394.5	73.85%
	50	17402.1	24815.9	29.28%	10256.5	43920.6	75.94%
I_3	10	23472.2	40739.0	40.07%	16575.2	63650.0	73.70%
	25	31040.2	47649.2	33.49%	17814.4	87202.0	79.56%
	50	31981.6	48075.5	32.34%	18025.0	93342.3	80.57%
I_4	10	24624.2	41758.9	39.22%	21054.6	67233.5	68.40%
	25	33083.8	49996.7	32.76%	20047.4	93056.8	78.35%
	50	32413.1	49254.6	33.26%	19072.8	101595.1	80.77%

For answering the first question, we take a look at the final profits of the greedy evaluation and compare them to the final simulation profits. The difference between these values vary between about 65% and 80% and seems to be getting larger for bigger instances. This shows that the error introduced by always choosing the best case for trip-acceptance and user behavior is large and does not correspond to the more realistic value. Only because these values have a large gap does not necessarily mean that the greedy evaluation is not a good approximation to the real profit; the expected profit could just have a large variance. When we inspect the final simulation profits of the simulation evaluation, however, we see a higher value than the final simulation profits of the greedy evaluation for all instances which shows that in the same amount of time, a better solution is found when the simulation evaluation is used, which also answers the second question.

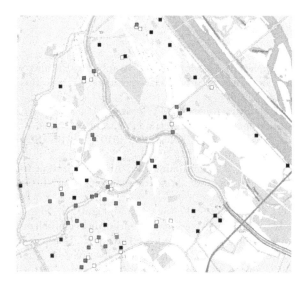

Fig. 3. Graphical comparison of a solution from the simulation (□) and the greedy (■) evaluation. Locations chosen in both solutions are marked with (▣).

Finally, Fig. 3 shows a comparison of station locations of two solutions from the greedy and simulation pathfinder, respectively. The locations obtained by the simulation evaluation seem to be more clustered in area with a high trip density, while the locations obtained by the greedy evaluation are more dispersed over the whole operational area. This corresponds to the intuition that in the latter solution the potentially accepted trips are longer and thus more profitable. This only works out, however, in a best case scenario where all trips take place in an optimal way. Since this cannot be guaranteed in practice, the simulation evaluation leads to solutions that are more robust with respect to expected profit.

6   Conclusions and Future Work

In this work we propose a more realistic evaluation of strategic decisions for a full electric car sharing system. User decisions are modeled as random variables and Monte-Carlo simulations are used for computing the expected profit. For the station locations and design a variable neighborhood search with several neighborhood structures is used. The results show that by using so-far used evaluation methods large errors can be made by assuming the best case of the user behavior which can negatively influence the strategic decisions. Furthermore, it pays off to invest more time into the more realistic approximation of the expected profit. For future work we plan to also consider relocation of vehicles which can influence the strategic decisions and is in many practical systems a major factor of the operational costs. Another future research direction is to model free-floating car sharing systems in which the users can rent and return cars anywhere within the operational area.

References

1. Asamer, J., Reinthaler, M., Ruthmair, M., Straub, M., Puchinger, J.: Optimizing charging station locations for urban taxi providers. Transp. Res. Part A: Policy Pract. **85**, 233–246 (2016)
2. Biesinger, B., Hu, B., Stubenschrott, M., Ritzinger, U., Prandtstetter, M.: Optimizing charging station locations for electric car-sharing systems. In: Hu, B., López-Ibáñez, M. (eds.) EvoCOP 2017. LNCS, vol. 10197, pp. 157–172. Springer, Cham (2017). https://doi.org/10.1007/978-3-319-55453-2_11
3. Boyacı, B., Zografos, K.G., Geroliminis, N.: An optimization framework for the development of efficient one-way car-sharing systems. Eur. J. Oper. Res. **240**(3), 718–733 (2015)
4. Brandstätter, G., Gambella, C., Leitner, M., Malaguti, E., Masini, F., Puchinger, J., Ruthmair, M., Vigo, D.: Overview of optimization problems in electric car-sharing system design and management. In: Dawid, H., Doerner, K.F., Feichtinger, G., Kort, P.M., Seidl, A. (eds.) Dynamic Perspectives on Managerial Decision Making. DMEEF, vol. 22, pp. 441–471. Springer, Cham (2016). https://doi.org/10.1007/978-3-319-39120-5_24
5. Brandstätter, G., Kahr, M., Leitner, M.: Determining optimal locations for charging stations of electric car-sharing under stochastic demand (2016, submitted)
6. Brandstätter, G., Leitner, M., Ljubić, I.: Location of charging stations in electric car sharing systems (2016, submitted)
7. Brandstätter, G., Leitner, M., Ljubic, I.: On finding optimal charging station locations in an electric car sharing system. In: Annual Workshop of the EURO Working Group on Vehicle Routing and Logistics Optimization (VeRoLog 2016), p. 13 (2016)
8. Mladenović, N., Hansen, P.: Variable neighborhood search. Comput. Oper. Res. **24**(11), 1097–1100 (1997)
9. Pelletier, S., Jabali, O., Laporte, G.: Goods distribution with electric vehicles: review and research perspectives. Transp. Sci. **50**(1), 3–22 (2016)
10. Rubinstein, R.Y., Kroese, D.P.: Simulation and the Monte Carlo method. Wiley, Hoboken (2016)
11. Weikl, S., Bogenberger, K.: Relocation strategies and algorithms for free-floating car sharing systems. IEEE Intell. Transp. Syst. Mag. **5**(4), 100–111 (2013)

Measures for the Evaluation and Comparison of Graphical Model Structures

Gabriel Kronberger[1]([✉])([iD]), Bogdan Burlacu[1,2], Michael Kommenda[1,2],
Stephan Winkler[1], and Michael Affenzeller[1,2]

[1] Heuristic and Evolutionary Algorithms Laboratory,
University of Applied Sciences Upper Austria,
Softwarepark 11, 4232 Hagenberg, Austria
gabriel.kronberger@fh-hagenberg.at
[2] Institute for Formal Models and Verification, Johannes Kepler University,
Altenbergerstr. 69, 4040 Linz, Austria

Abstract. Structure learning is the identification of the structure of graphical models based solely on observational data and is NP-hard. An important component of many structure learning algorithms are heuristics or bounds to reduce the size of the search space. We argue that variable relevance rankings that can be easily calculated for many standard regression models can be used to improve the efficiency of structure learning algorithms. In this contribution, we describe measures that can be used to evaluate the quality of variable relevance rankings, especially the well-known normalized discounted cumulative gain (NDCG). We evaluate and compare different regression methods using the proposed measures and a set of linear and non-linear benchmark problems.

Keywords: Graphical models · Structure learning · Regression

1 Introduction

We aim to define an efficient algorithm for learning the structure of a graphical model solely from observational data. The algorithm should work for processes with continuous variables, non-linear dependencies, and noisy measurements.

Graphical models are not only useful for visualization purposes but can be facilitated for online control. Consider for example a system, which has multiple continuous inputs and dependent outputs as well as internal variables. In such applications, the system parameters are often interrelated. Therefore, a controller cannot set parameter values independently of each other. An explicit and complete model of all variable dependencies allows a controller to set valid values for all parameters and achieve a stable process.

Many variants of structure learning algorithms have been proposed, including exact as well as approximate algorithms. However, most of the algorithms work only for discrete variables or impose constraints on the type of dependencies (e.g. only linear dependencies). Our intention is to use standard regression algorithms

© Springer International Publishing AG 2018
R. Moreno-Díaz et al. (Eds.): EUROCAST 2017, Part I, LNCS 10671, pp. 283–290, 2018.
https://doi.org/10.1007/978-3-319-74718-7_34

to learn models for each variable and calculate a variable relevance ranking for each model. The relevance ranking can later be used to guide a heuristic structure learning algorithm.

In this paper, we focus on the sub-problem of evaluating the quality of variable relevance rankings, i.e. we try to answer the research question: *How can we quantify the accuracy of variable relevance rankings produced by regression algorithms?*

2   Related Work

Structure learning for Bayesian networks is NP-hard [2]. In the past, a large number of algorithms for structure learning have been formulated including exact methods as well as approximate algorithms (cf. [11]). Exact methods such as the algorithms described in [9,16] use dynamic programming and work for problems up to 30 variables [1]. Approximate algorithms can handle much larger networks [1]. The PC algorithm [17] is a classical exact algorithm for structure learning which assumes an oracle for determining whether pairs are (conditionally) independent. An improved variant is the MMPC algorithm [18]. Many of the published algorithms work only for discrete variables. A simple approach for continuous variables and linear systems is to fit a lasso regression using each variable as the response and the others as predictors [14]. A more systematic approach is the graphical lasso [4] which optimizes a global penalized likelihood and can solve sparse problems with 1000 nodes in less than a minute [4]. The "ideal-parent" algorithm [3] can be used for structure learning for non-linear systems with continuous variables. However, the dependencies between variables are limited to generalized linear models with non-linear link functions. Artificial neural networks as well as Gaussian process networks have been used for structure learning for non-linear and continuous Bayesian Networks [5,7].

Variable interaction networks can be used to visualize dependencies between variables of a system [12,15]. Variable interaction networks are specific types of graphical models [10] in which nodes represent variables of the system and directed edges represent dependencies between variables. So far, variable interaction networks have been used primarily for visualization with the aim to gain a better understanding of complex systems (see e.g. [12,13,15,19]). In these applications, empirical models (e.g. symbolic regression models) have been used to identify and describe statistical dependencies between continuous variables.

3   Methods

We use standard regression modeling methods to generate a ranking of input variables by relevance for each dependent variable e.g. using the explained variance measure.

Assuming multiple different methods are available for generating graphical models from observational data, we want to determine which of those produces the best structure. This can be determined only if the optimal system structure is known. Therefore, we use a set of synthetic problem instances where the data generating process is known.

3.1 Measures for the Quality of Variable Relevance Rankings

We propose to use one of three indicators to evaluate and compare variable rankings: (1) Gini coefficient [6], (2) Spearman's rank correlation, and (3) normalized discounted cumulative gain (NDCG) [8]. The Gini coefficient can be used to quantify how well a modeling method is able to discriminate between the actually necessary inputs and unnecessary inputs and is closely related to the AUC measure for classification problems [6]. The Gini coefficient can be calculated even if no information on the actual importance ranking of variables is available, as it only depends on how well the modeling method discriminates between necessary and irrelevant variables.

Spearman's rank correlation coefficient and NDCG allow comparison of the estimated ranks with an ideal ranking. The former weights all elements of the ranking equally and is therefore strongly influenced by the relative ranking of irrelevant variables if there are only a few actually relevant variables. The later uses an exponential weighting scheme to assign more weight on the most important variables. Therefore, NDCG is an ideal measure for the evaluation of variable relevance rankings for structure learning where it is most important that the top-most variables are correctly identified while the ordering of irrelevant variables should not have a strong impact.

3.2 Empirical Evaluation of Variable Relevance Measures

For the empirical evaluation we use four different regression algorithms: linear regression (LR), Gaussian process regression (GPR), random forest regression (RF), and symbolic regression (SR) with genetic programming. Each of those four methods is used to estimate regression models for all dependent variables and the relevance of all input variables is determined. The resulting rankings are compared to the actual variable relevance rankings and the quality of the ranking is calculated. Finally, the arithmetic mean of all ranking qualities is determined to produce an overall quality for each method and problem instance.

3.3 Problem Instances

We have generated data by sampling from random processes where all dependencies between variables are known. Linear as well as non-linear systems are considered. For the linear instances all dependent variables are described by randomly generated linear models. For the non-linear systems all dependent variables are described by randomly sampled Gaussian processes. We have generated problem instances with different dimensionality $d \in \{10, 20, 50, 100\}$ and different noise levels $\in \{0\%, 1\%, 5\%, 20\%\}$. All instances represent graphical models in which the variables can be assigned to one of four levels. The relative number of variables in each level is fixed. The first and second level contain 33% of the variables, the third level contains 20% of the variables and the fourth level contains the remaining 14%. Variables in the first level are sampled independently from a zero-mean unit-variance Gaussian distribution. The variables in the other levels are sampled from generative models which use input variables randomly chosen from all lower levels.

The effective dimensionality d_{eff} is the number of input variables which are actually used in each model and is sampled according to the following expressions:

$$u \sim \text{uniform}(0, 1) \tag{1}$$
$$r = -2 \log(1 - u) \tag{2}$$
$$d_{\text{eff}} = \lfloor 1.5 + r \rfloor \tag{3}$$

Variable values are randomly sampled starting with the variables in the first level. After the values in the first level have been assigned, the sampling procedure continues with the variables in the next higher level. Considering only one dependent variable y, the values y_i $(1 \leq i \leq N)$ are generated using the following expression, where x_i contains the values of the randomly selected input variables for this dependent variable.

$$y_i \overset{i.i.d}{\sim} N\left(\mu = f(x_i), \sigma = \sqrt{\frac{\text{ratio}_{\text{noise}}}{1.0 - \text{ratio}_{\text{noise}}}}\right) \tag{4}$$

We used $N = 250$ samples for each problem instance. Correspondingly, the high-dimensional problems are more difficult as it is more likely to observe spurious strong correlations of irrelevant variables.

For the linear instances we sampled random models $f(x_i)$ using the following:

$$f(x_i) = g(x_i, w) = x_i^T w \tag{5}$$
$$w_j = \frac{\lambda_j^2}{\text{Var}(x_j)}, \, 1 < j \leq d_{\text{eff}} \tag{6}$$
$$\lambda_j \overset{i.i.d}{\sim} N(\mu = 0, \sigma = 1) \tag{7}$$

For the non-linear instances we sampled random models $f(x)$ from a zero-mean Gaussian process prior with squared exponential covariance function with randomly sampled length scales ℓ for each dimension to determine the relevance of each variable.

$$f(x) \sim \mathcal{GP}(0, k(x, x')) \tag{8}$$
$$k(x_p, x_q) = \exp\left(-(x_p - x_q)^T (\ell I)^{-1} (x_p - x_q)\right) \tag{9}$$
$$\ell_j \overset{i.i.d}{\sim} \text{uniform}(0.5, 2.5) \tag{10}$$

4   Results

Table 1 shows a comparison of the average accuracies of the variable rankings calculated with the three proposed measures for all benchmark instances with ratio$_{noise}$ = 5%. As expected, linear regression works better for linear systems and is not able to identify relevant variables correctly for the non-linear systems. For the non-linear instances, RF produces the best variable rankings.

Table 1. Comparison of variable relevance rankings. Values are averages over the values for all dependent variables as well as the ranks of the values in each row. Noise level is 5% for all problem instances. All three measures often produce the same rankings of methods. The best NDCG values for each problem are highlighted using bold font. SR works best for the linear problem instances and RF for non-linear instances.

Problem type	d	Measure	Average value				Rank			
			LR	SR	RF	GPR	LR	SR	RF	GPR
Linear	10	Gini	0.98	0.83	0.69	0.96	1	3	4	2
		NDCG	**0.99**	0.93	0.85	0.97	1	3	4	2
		Spearman	0.74	0.57	0.37	0.72	1	3	4	2
	20	Gini	0.95	0.93	0.89	0.93	1	3	4	2
		NDCG	0.94	**0.95**	0.90	0.91	2	1	4	3
		Spearman	0.49	0.47	0.43	0.48	1	3	4	2
	50	Gini	0.98	0.93	0.97	0.98	2	4	3	1
		NDCG	**0.96**	**0.96**	**0.96**	0.97	3	4	2	1
		Spearman	0.31	0.33	0.30	0.32	3	1	4	2
	100	Gini	0.98	0.98	0.93	0.98	3	2	4	1
		NDCG	0.93	**0.96**	0.91	0.92	2	1	4	3
		Spearman	0.24	0.32	0.23	0.26	3	1	4	2
Gaussian process	10	Gini	0.34	0.88	0.83	0.85	4	1	3	2
		NDCG	0.71	**0.95**	**0.95**	0.94	4	1	2	3
		Spearman	0.29	0.70	0.65	0.67	4	1	3	2
	20	Gini	0.77	0.85	0.94	0.80	4	2	1	3
		NDCG	0.79	0.94	**0.95**	0.88	4	2	1	3
		Spearman	0.39	0.47	0.49	0.41	4	2	1	3
	50	Gini	0.71	0.74	0.92	0.76	4	3	1	2
		NDCG	0.64	0.85	**0.89**	0.73	4	2	1	3
		Spearman	0.23	0.32	0.30	0.26	4	1	2	3
	100	Gini	0.50	0.69	0.80	0.58	4	2	1	3
		NDCG	0.53	0.74	**0.79**	0.63	4	2	1	3
		Spearman	0.12	0.26	0.20	0.14	4	1	2	3

Interestingly, the ordering of the methods similar, regardless of the measure, which is used to compare the methods. For example, for the Gaussian process system with 20 variables the ordering of methods is the same for all three measures.

For a sensitivity analysis we generated problem instances with different noise levels. Table 2 shows the NDCG values for all problem instances and noise ratios. The linear instances without noise (0%) are ill-conditioned because indirect and direct dependencies cannot be distinguished by the standard regression algorithms. Therefore, all methods produce better variable rankings when at least a small amount of noise is present. RF produces the best NDCG values for the non-linear instances and is able to identify to top-ranked variables even for high dimensionality.

Table 2. NDCG values for the variable relevance rankings reached for all instances. The best values for each problem instance are highlighted using bold font. SR works best for the linear instances, RF works best for the non-linear instances. The values for 5% noise are shown in Table 1.

d	LR	SR	RF	GPR	LR	SR	RF	GPR
Linear instances								
	Noise = 0%				Noise = 1%			
10	0.68	**0.96**	0.93	0.70	**1.00**	**1.00**	0.97	**1.00**
20	0.64	**0.91**	0.87	0.64	**0.96**	**0.96**	0.88	**0.96**
50	0.54	0.80	**0.85**	0.55	0.92	**0.95**	0.84	0.91
100	0.59	**0.87**	0.85	0.57	0.88	**0.89**	0.84	0.84
	Noise = 10%				Noise = 20%			
10	**1.00**	**1.00**	0.99	**1.00**	**1.00**	**1.00**	0.99	**1.00**
20	0.93	0.89	**0.94**	0.93	0.97	**0.98**	0.95	0.97
50	0.93	0.93	**0.94**	**0.94**	**0.97**	**0.97**	0.96	**0.97**
100	**0.97**	0.95	0.93	0.96	0.94	**0.96**	0.94	0.93
Non-linear instances								
	Noise = 0%				Noise = 1%			
10	0.90	0.93	**0.98**	0.91	0.73	0.93	**0.99**	0.96
20	0.70	0.89	**0.90**	0.88	0.73	0.84	**0.95**	0.90
50	0.74	0.86	**0.90**	0.78	0.58	0.77	**0.82**	0.71
100	0.55	0.81	**0.83**	0.64	0.55	0.74	**0.76**	0.64
	Noise = 10%				Noise = 20%			
10	0.81	**0.93**	0.92	0.90	0.75	0.93	**0.96**	0.94
20	0.68	0.80	**0.86**	0.82	0.78	0.86	0.85	**0.88**
50	0.63	0.80	**0.83**	0.74	0.61	**0.81**	0.79	0.67
100	0.58	0.78	**0.81**	0.66	0.55	0.74	**0.80**	0.59

5   Discussion and Conclusions

Our overall aim is to use estimated variable relevance rankings for guiding structure learning algorithms for graphical models with continuous variables and non-linear dependencies. In previous work we have used a similar approach to generate so-called variable interaction networks [12]. However, variable interaction networks are useful only for visualization purposes. Instead, we would like to use graphical models also for example for online predictive control where multiple interrelated process variables must be controlled through multiple parameters which might also be interrelated.

We have described how the quality of variable relevance rankings can be measured relative to a gold standard and have used three different measures (i) Spearman's rank correlation coefficient, (ii) the Gini coefficient, and (iii) the normalized discounted cumulative gain (NDCG). We have demonstrated how the measures can be used to compare the quality of the variable relevance rankings for different regression algorithms. For an empirical evaluation, we have used linear as well as non-linear benchmark problems.

Analysis of the results shows that the order of regression methods is frequently similar for all three measures. For the linear problem instances, symbolic regression most often produced the best variable relevance ranking whereas for the non-linear instances random forest regression consistently produced the best relevance rankings.

An interesting result is that Gaussian process regression performed worse than random forest regression on the non-linear instances even though the data for these instances have been generated by sampling from Gaussian processes. The effect is especially visible for the high-dimensional problem instances. This indicates that our approach of maximum likelihood learning with automatic relevance determination does not work well for the identification of the relevant input variables in this case.

Acknowledgements. The authors gratefully acknowledge financial support by the Austrian Research Promotion Agency (FFG) and the Government of Upper Austria within the COMET Project #843532 Heuristic Optimization in Production and Logistics (HOPL).

References

1. Campos, C.P., Ji, Q.: Efficient structure learning of Bayesian networks using constraints. J. Mach. Learn. Res. **12**, 663–689 (2011)
2. Chickering, D.M.: Learning Bayesian Networks is NP-Complete, pp. 121–130. Springer, Heidelberg (1996). https://doi.org/10.1007/978-1-4612-2404-4_12
3. Elidan, G., Nachman, I., Friedman, N.: "Ideal Parent" structure learning for continuous variable Bayesian networks. J. Mach. Learn. Res. **8**(8), 1799–1833 (2007)
4. Friedman, J., Hastie, T., Tibshirani, R.: Sparse inverse covariance estimation with the graphical lasso. Biostatistics **9**(3), 432–441 (2008)

5. Friedman, N., Nachman, I.: Gaussian process networks. In: Proceedings of the Sixteenth Conference on Uncertainty in Artificial Intelligence (UAI), pp. 211–219. Morgan Kaufmann Publishers (2000)
6. Hand, D.J., Till, R.J.: A simple generalisation of the area under the ROC curve for multiple class classification problems. Mach. Learn. **45**(2), 171–186 (2001). https://doi.org/10.1023/A:1010920819831
7. Hofmann, R., Tresp, V.: Discovering structure in continuous variables using Bayesian networks. In: Advances in Neural Information Processing Systems (NIPS), pp. 500–506 (1996)
8. Järvelin, K., Kekäläinen, J.: Cumulated gain-based evaluation of IR techniques. ACM Trans. Inf. Syst. (TOIS) **20**(4), 422–446 (2002)
9. Koivisto, M., Sood, K.: Exact Bayesian structure discovery in Bayesian networks. J. Mach. Learn. Res. **5**(May), 549–573 (2004)
10. Koller, D., Friedman, N.: Probabilistic Graphical Models: Principles and Techniques - Adaptive Computation and Machine Learning. MIT Press, Cambridge (2009)
11. Koski, T.J., Noble, J.: A review of Bayesian networks and structure learning. Math. Applicanda **40**(1), 51–103 (2012)
12. Kronberger, G.: Symbolic Regression for Knowledge Discovery - Bloat, Overfitting, and Variable Interaction Networks. Reihe C: Technik und Naturwissenschaften, Trauner Verlag (2011)
13. Kronberger, G., Fink, S., Kommenda, M., Affenzeller, M.: Macro-economic time series modeling and interaction networks. In: Di Chio, C., Brabazon, A., Di Caro, G.A., Drechsler, R., Farooq, M., Grahl, J., Greenfield, G., Prins, C., Romero, J., Squillero, G., Tarantino, E., Tettamanzi, A.G.B., Urquhart, N., Uyar, A.Ş. (eds.) EvoApplications 2011. LNCS, vol. 6625, pp. 101–110. Springer, Heidelberg (2011). https://doi.org/10.1007/978-3-642-20520-0_11
14. Meinshausen, N., Bühlmann, P.: High-dimensional graphs and variable selection with the lasso. Annal. Stat. **34**(3), 1436–1462 (2006)
15. Rao, R., Lakshminarayanan, S.: Variable interaction network based variable selection for multivariate calibration. Anal. Chim. Acta **599**(1), 24–35 (2007)
16. Singh, A.P., Moore, A.W.: Finding optimal Bayesian networks by dynamic programming. Technical report CMU-CALD-05-1062, School of Computer Science, Carnegie Mellon University, June 2005
17. Spirtes, P., Glymour, C., Scheines, R.: Causation, Prediction, and Search. Springer, New York (1993). https://doi.org/10.1007/978-1-4612-2748-9
18. Tsamardinos, I., Brown, L.E., Aliferis, C.F.: The max-min hill-climbing Bayesian network structure learning algorithm. Mach. Learn. **65**(1), 31–78 (2006)
19. Winker, S., Affenzeller, M., Kronberger, G., Kommenda, M., Wagner, S., Jacak, W., Stekel, H.: Variable interaction networks in medical data. In: Proceedings of the 24th European Modeling and Simulation Symposium EMSS 2012, pp. 265–270. Dime Universitá di Genova (2012)

Towards System-Aware Routes

Matthias Prandtstetter[(✉)] [iD] and Clovis Seragiotto

AIT Austrian Institute of Technology,
Center for Mobility Systems – Dynamic Transportation Systems,
Giefinggasse 2, 1210 Vienna, Austria
{matthias.prandtstetter,clovis.seragiotto}@ait.ac.at
http://www.ait.ac.at/

Abstract. We introduce solution methods for finding a system-optimal or at least system-aware set of routes for multiple concurrently travelers on a road network such that congestion is minimized, i.e. traffic flow is maximized. Beside a dynamic programming approach, we present a basic heuristic mimicking the current state-of-the-art resulting in a user equilibrium and a multistart local search based approach approximating the system-optimum. Experimental results show that the chosen approaches are promising and that heavy congestion can be avoided when applying such a planning approach.

Keywords: Routing · System-aware optimization

1 Introduction

The way of our living changed significantly in the last decade. With the introduction of smartphones and appropriate (mobile) internet technologies we started to be online all around the clock—including when we travel. This implies, that we started to use our mobile phones—among others—for tasks as navigation. However, we are rather selfish and uncooperative when it comes down to route planning. In other words, we do not (yet) connect with other travelers in order to gain the maximum positive effect with respect to load distribution in (road) traffic which results in congested streets, annoyance and dissatisfaction.

We, therefore, present within this paper a first approach towards a system-aware route planning algorithm which is capable of coordinating routes for concurrently traveling people leading to an even utilization of the street network. The results are reduced congestion, maximized traffic flow and therefore reduced (average) travel times and reduced emissions. The overall goal is to reach a system optimum, i.e., a system state which cannot be improved with respect to system-wide optimization objectives [5]. First investigations of system optima with respect to individual routes are shown in [3], but that work focuses more

This project has received funding from the European Union's Horizon 2020 research and innovation programme under grant agreement No. 636160 (OPTIMUM).

R. Moreno-Díaz et al. (Eds.): EUROCAST 2017, Part I, LNCS 10671, pp. 291–298, 2018.
https://doi.org/10.1007/978-3-319-74718-7_35

on the theoretical traffic assignment rather than on the computation of individual routes. To the best of our knowledge, no further works exist which address the problem as stated within our work.

We first give a more formal description of the problem statement and then propose a first basic heuristic which, however, results in a user equilibrium [5], i.e., the current situation observed on a daily basis. We further propose a dynamic programming approach capable of optimally solving small instances and come up with a (meta)heuristic solution approach in order to tackle larger (real-world) instances. We end our paper with (preliminary) experimental results and conclude with a short section on future works.

2   Problem Description

We are given a (directed) graph $G(V, E)$ with V denoting the set of nodes (i.e. crossings) in a street network and $E \subseteq V^2$ denoting the arcs (i.e. roads). We are given a set of travelers (or users) U and for each traveler $i \in U$ we are given her origin $o(i) \in O \subseteq V$ and her destination $d(i) \in D \subseteq V$. Further, we are given for each traveler a departure time $s(i) \in S$. Furthermore, we are given a mutual influence function which is defined by the fundamental diagram of traffic flow [1]. The fundamental diagram of traffic flow, as sketched in Fig. 1 for highways, represents the mutual influence of multiple cars when traveling along the same road at the same time. The resulting variance in traffic flow and therefore travel speed is easily visible in Fig. 1.

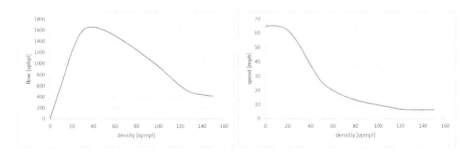

Fig. 1. A schematic representation of the fundamental diagram of traffic flow. On the right, we see speed of vehicles over density (in vehicles per kilometer per lane). On the left, we see the corresponding flow (vehicles per hour per lane) over density.

We now search a set R of individual routes r_i, for $i \in U$, such that the overall travel time $\sum_{i \in U} l(r_i)$ is minimized, with $l(r_i)$ denoting the travel time for route r_i which is computed via the fundamental diagram of traffic flow. The resulting formal formulation is

$$\min \sum_{i \in U} l(r_i) \tag{1a}$$

$$r_i \text{ is a path from } o(i) \text{ to } d(i), \qquad \forall i \in U \tag{1b}$$

Algorithm 1. USEREQUILIBRIUM

---

begin
    foreach $i \in U$ **do**
        compute r_i based on expected traffic situation;
        $R \leftarrow R \cup \{r_i\}$;
    return R;

Status Quo. In order to have a baseline for further evaluations, we additionally present the formulation related to the current situation as we can find it on the road. In general, each traveler optimizes her route independently from the other users. In most cases, the individual fastest route is chosen. However, travel time is estimated on an uncongested road network or—in some cases—based on an expected traffic situation (e.g. based on historic data). The resulting problem formulation is

$$\min l'(r_i), \qquad\qquad\qquad \forall i \in U \qquad\qquad (2a)$$
$$r_i \text{ is a path from } o(i) \text{ to } d(i), \qquad \forall i \in U \qquad\qquad (2b)$$

Obviously, this approach results in a user equilibrium where no user can improve her own route resulting in congestion if multiple users plan to travel the same—or a similar—path. Please note, that $l'(r_i)$ denotes the length (or travel time) of route r_i without any influences of other travelers.

3   Basic User Equilibrium Approach

Algorithm 1 represents the algorithmic solution for the formulation represented by Eq. 2. Obviously, for each traveler a route is computed independently of all other (to be) planned routes. The resulting set of computed routes is then returned. In order to later compare the resulting situation with a system optimum approach, we want to note that a fair evaluation is necessary. This means, that for the evaluation formula $\sum_{i \in U} l(r_i)$ is used, i.e. the sum of all travel times respecting mutual influence among the planned routes.

4   Dynamic Programming Approach for System Optimum

In order to find a system optimum, we present a dynamic programming based approach which is strongly related to the algorithm by Dijkstra for finding shortest paths [2]. The basic idea is identical, i.e., for each state we generate all possible follow-up states and determine then, which of these states shall be further evaluated and which of these states can be omitted. Therefore, we define a state $\lambda = (\rho, t, A, I, F, P)$ as a m-tupel of a reference to the previous state ρ, time stamp t, the set of active users A, the set of inactive users I, the set of finished

---

Algorithm 2. GenerateNewStates($\lambda = (t, A, I, F, P)$)

---

begin
    /* generate all possible edge selections (based on λ.   */
    $\Lambda = \{\}$;
    $\Pi \leftarrow$ all new possible combination of "follow-up" edges;
    foreach $\pi \in \Pi$ **do**
        $\rho' \leftarrow \lambda$, $t' \leftarrow t$, $A' \leftarrow A$, $I' \leftarrow I$, $F' \leftarrow F$, $P' \leftarrow P$;
        compute for each user $i \in A$ at which time $t(i)$ she will reach the
        end-node of the chosen edge;
        /* choose the earliest arrival time among all   */
        $t' \leftarrow \min \{t(i) : i \in A\}$;
        /* choose whether an inactive user $i \in I$ has started   */
        foreach $i \in I'$ **do**
            if $t(i) \leq t'$ **then**
                $t' \leftarrow t(i)$; $I' \leftarrow I' \setminus \{i\}$; $A' \leftarrow A' \cup \{i\}$;

        /* compute for all users $i \in U'$ their current positions   */
        $P' \leftarrow$ set of current positions $\forall i \in U$ at time stamp t';
        /* check if some users already reached their destinations   */
        foreach $i \in A'$ **do**
            if $p(i) == d(i)$ **then**
                $A' \leftarrow A' \setminus \{i\}$; $F' \leftarrow F' \cup \{i\}$;

        $\lambda' \leftarrow (\rho', t', A', I', F', P')$;
        $\Lambda \leftarrow \Lambda \cup \{\lambda'\}$;
    return Λ;

---

users F, and the set of current positions P of all users. We refer to active users as travelers who are already traveling, inactive users as travelers who did not start their trips yet and finished users as those travelers who already reached their destination. Positions as stored in set P are either crossings (i.e. nodes $v \in V$) or are positions along the roads. If the current position of a user is a crossing, possible follow-up edges are—obviously—all roads leaving the current crossing. If the current position is along a road, the only possible follow-up edge is the edge referring to the current road. We refer to Algorithm 2 for method of generating new states based on a given input state λ. The output of this algorithm is the set of possible follow-up states. Algorithm 3 builds the main routine of the DP approach. Algorithm 4 is used to remove duplicate states.

5   Iterated Local Search Based Approach

Iterated Local Search [4] is a multistart metaheuristic built on basic local search (LS). The LS procedure is iterated multiple times with each start beginning from a (possibly) different start solution. It is therefore important to have a construction heuristic capable of providing varying solutions.

Algorithm 3. DP()

---

begin

    $\lambda \leftarrow (\emptyset, 0, \{\}, U, \{\}, O);\ \Lambda \leftarrow \{\lambda\};$

    while $\Lambda \neq \{\}$ **do**

        /* select the state with the minimal time stamp   */

        $\lambda \leftarrow \arg\min\{t : \lambda = (\rho, t, A, I, F, P) \in \Lambda\};\ \Lambda \leftarrow \Lambda \setminus \{\lambda\};$

        $\Lambda' \leftarrow$ GenerateNewStates$(\lambda);\ \Lambda \leftarrow \Lambda \cup \Lambda';$

        /* remove duplicates in Λ -- keep only the better ones   */

        foreach $(\lambda, \lambda') \in \Lambda^2$ **do**

            $\Lambda \leftarrow$ RemoveDuplicate$(\Lambda,\ \lambda,\ \lambda');$

        /* end, if all users are finished in at least one state */

        if $\exists \lambda \in \Lambda : A \cup I = \emptyset$ **then**

            return $\lambda;$

Algorithm 4. RemoveDuplicates($\Lambda,\ \lambda,\ \lambda'$)

---

begin

    if $(A = A')\&(I = I')\&(F = F')\&(P = P')$ **then**

        if $t \leq t'$ **then return** $\Lambda \setminus \{\lambda'\}$;

        else return $\Lambda \setminus \{\lambda\}$;

    return $\Lambda;$

5.1 Construction Heuristic

The straight-forward construction heuristic intended for our work is based on the idea to mimic the behavior of "informed" travelers, i.e. travelers who consider all the other already planned routes when planning their own route. In order to achieve some kind of variation, we build on a random order when planning routes. See Algorithm 5 for more details.

5.2 Local Search Based Approach

While the construction heuristic provides a first start solution R^{init}, it can be assumed that optimization potential is existing. Therefore, a basic local search is applied afterwards. This local search consists of one move type which is iteratively applied to the current (best) solution until no further improvement can be gained. In this move, we remove a randomly chosen route and re-calculate the route for the corresponding user regarding all the other routes (for all the other users). Whenever a new best solution can be reached when applying the current move, the new solution is used for further computations. See Algorithm 6 for details. In order to further improve the obtained results, we restart the local search with randomly generated initial solutions.

Algorithm 5. CONSTRUCTION

---

begin

    $U' \leftarrow U$;

    randomly mix elements in U';

    while $U' \neq \emptyset$ **do**

        $i \leftarrow \mathit{first}(U')$;

        $U' \leftarrow U' \setminus \{i\}$;

        $r_i \leftarrow$ route from $o(i)$ to $d(i)$ based on expected traffic situation and already planned routes;

        $R \leftarrow R \cup \{r_i\}$;

    return R;

Algorithm 6. LOCALSEARCH

---

begin

    $R^* \leftarrow R^{init}$;

    while *abortion criterion is not met* **do**

        $i \leftarrow$ randomly chosen element in U;

        $R' \leftarrow R^* \setminus \{r_i\}$;

        $r_i \leftarrow$ route from $o(i)$ to $d(i)$ based on expected traffic situation and routes in R';

        $R' \leftarrow R' \cup \{r_i\}$;

        if $l(R') < l(R^*)$ **then**

            $R^* \leftarrow R'$;

    return R^*;

---

6 Preliminary Results

In order to assess whether or not the proposed methods result in promising solutions, we conducted first experiments. It turned relatively fast out that the dynamic programming approach is not capable of solving instances larger than a view (i.e. below ten) concurrent travelers. Therefore, no further detailed results are presented within this section.

For testing the (meta)heuristic approach, we generated six random instances based on the street network of Vienna, Austria, with x randomly chosen pairs of start and end positions. For variance, x was chosen among $\{3, 5, 10\}$. Further, we varied the departure time of the travelers between "all start at 7am" and "all start in a time window of ten minutes between 7am and 7:10am" (shown as 7am+ in Table 1). Finally, we decided to have 24 and 240 concurrent travelers. Therefore, we ended up with twelve instances which were all solved with the basic user equilibrium approach and the multistart local search based approach— labeled UE and SO in Table 1, respectively. Table 1 further lists the total travel time (tot.tt), the travel time of the maximum longest route (max.tt) and the computation time needed (comp.time). As the SO approach is randomized we

Table 1. Results obtained for instances of size 3×3, 5×5 and 10×10. The total travel time (tot.tt), the length of the longest route (max.tt) and the computation time (comp.tt) are given in seconds. For SO averages over 30 runs are presented.

	7am				7am+			
	24		240		24		240	
	UE	SO	UE	SO	UE	SO	UE	SO
3×3								
tot.tt [s]	63251	18356	8690876	326907	18542	18248	771094	174304
max.tt [s]	5958	1057	63974	2777	1090	1090	9919	1111
comp.time [s]	6	63	37	2451	7	21	16	1946
5×5								
tot.tt [s]	15639	12681	3232376	266144	13097	12412	354929	145194
max.tt [s]	1499	972	61670	2947	1135	972	11568	1120
comp.time [s]	5	26	21	2133	5	13	15	1318
10×10								
tot.tt [s]	14897	13711	2115568	168065	13742	13676	320224	151138
max.tt [s]	1024	867	58363	1411	868	868	9871	1074
comp.time [s]	5	33	21	1229	6	16	15	753

list the average over 30 runs. We decided to stop the local search part of the SO approach after at most 50000 iterations or 1000 iterations without improvements. The multistart was repeated 100 times. We want to highlight that of those theoretically possible five millions of iterations no more than 23000 iterations were needed through all tests.

It can be easily seen that the UE approach is much faster while the overall travel times for UE are much longer than for the SO approach. Furthermore, we see that for many instances the duration of the longest route in the UE solution is significantly longer than the duration for the longest route in the SO solution indicating that heavy congestion is observed. With respect to computation times we can conclude that the UE approach is rather fast while the computation times SO are noticeable. The main reason for the rather slow performance is reasoned by the fact that the evaluation procedure of this problem formulation is rather complex as for each change (e.g. removal of one route) all routes have to be re-evaluated according to the fundamental diagram of traffic flow.

7   Conclusions

Within this paper, we formulated a routing problem where the goal is to find a set of system-aware routes, i.e. a set of routes where influences of the routes on other (concurrently planned) routes are minimized. This is equivalent to the goal is to minimize the overall travel time in the system. Although currently not a real-world problem statement, modern technologies like smartphones and mobile internet connections support developments in this direction.

We presented a dynamic programming based approach which was unfortunately only capable of solving (very) small sized instances with below 10 concurrent users. We also presented a heuristic mimicking the way traffic currently develops meaning that non-collaborative routes are computed. We later show in comparison with the third presented approach based on multistart local search that the non-collaborative approach results in significant congestion while the multistart local search based approach is capable of providing route sets where the mutual negative effects are not as significant.

We conclude that although the first results are rather promising further research is necessary as the performance of the proposed algorithms is not applicable for real-world settings. We further have to highlight that the fundamental diagram as used within this work is not the best estimation for congestion in cities as this diagram originates from highway data. Nevertheless, we think that the chosen approach (based on fundamental diagrams) is promising and more sophisticated optimization approaches are capable of significantly reducing the computation times.

References

1. Ashton, W.D.: The theory of road traffic flow (1966)
2. Dijkstra, E.W.: A note on two problems in connexion with graphs. Numer. Math. **1**(1), 269–271 (1959)
3. Köhler, E., Möhring, R.H., Skutella, M.: Traffic networks and flows over time. In: Lerner, J., Wagner, D., Zweig, K.A. (eds.) Algorithmics of Large and Complex Networks. LNCS, vol. 5515, pp. 166–196. Springer, Heidelberg (2009). https://doi.org/10.1007/978-3-642-02094-0_9
4. Lourenço, H.R., Martin, O.C., Stützle, T.: Iterated local search: framework and applications. In: Gendreau, M., Potvin, J.Y. (eds.) Handbook of Metaheuristics, vol. 146, pp. 363–397. Springer, Boston (2010). https://doi.org/10.1007/978-1-4419-1665-5_12
5. Wardrop, J.G.: Some theoretical aspects of road traffic research. Proc. Inst. Civil Eng. Part II **1**, 325–378 (1952)

GRASP-VNS for a Periodic VRP with Time Windows to Deal with Milk Collection

Airam Expósito[1(✉)], Günther R. Raidl[2], Julio Brito[1],
and José A. Moreno-Pérez[1]

[1] Department of Computer and Systems Engineering, Universidad de La Laguna,
38271 La Laguna, Canary Islands, Spain
{aexposito,jbrito,jamoreno}@ull.es
[2] Institute of Computer Graphics and Algorithms, TU Wien,
Favoritenstraße 9–11/1861, 1040 Vienna, Austria
raidl@ac.tuwien.ac.at

Abstract. This paper considers the planning of the collection of fresh milk from local farms with a fleet of refrigerated vehicles. The problem is formulated as a version of the Periodic Vehicle Routing Problem with Time Windows. The objective function is oriented to the quality of service by minimizing the service times to the customers within their time windows. We developed a hybrid metaheuristic that combines GRASP and VNS to find solutions. In order to help the hybrid GRASP-VNS find high-quality and feasible solutions, we consider infeasible solutions during the search using different penalty functions.

Keywords: Periodic Vehicle Routing Problem with Time Windows
Quality of service · Milk collection · GRASP · VNS · Penalty functions

1 Introduction

Logistics and transport management systems for perishable products have operational specificities associated with demands, handling, storage equipment and transport infrastructure. Models to solve the problems of collecting, sharing and distributing these products must adapt to new objectives and constraints. The minimization of total travel cost is an important logistics and transport objective and is the main criterion for the optimization of supply and distribution chains. Nevertheless, there are further important aspects to consider than just the special importance of the costs in perishable products. Quality assurance of service of perishable products constitute the main criteria for the optimization of supply and distribution chains for this kind of goods.

In this work we specifically address specifically the problem of planning the collection of fresh milk from local farms through a fleet of refrigerated trucks. The scattered small-scale family farms type have limited isothermal facilities for storing milk. In these circumstances the collection by the industry demands a precise temporal organization to preserve the quality of the product [4]. Milk

© Springer International Publishing AG 2018
R. Moreno-Díaz et al. (Eds.): EUROCAST 2017, Part I, LNCS 10671, pp. 299–306, 2018.
https://doi.org/10.1007/978-3-319-74718-7_36

collection needs not be daily because the farms have facilities to store milk for one to three days. The collection planning is done in weekly periods [2].

The problem to determine the most appropriate routes for collecting milk from a set of known farms in a given planning period of several days, including a time window for each pick up, is modelled as a Periodic Vehicle Routing Problem with Time Windows (PVRPTW).

The rest of the paper is organized as follows. The next section describes the PVRPTW model and the objective function. Section 3 explains the proposed solution approach to solve the problem. In Sect. 4 computational experiments and results are described and analyzed. Finally, some conclusions and future works are included in the last section.

2   Periodic Vehicle Routing Problem with Time Windows

The PVRPTW, first mentioned in [3], ask for a number of routes for each day over a given planning horizon with the aim of minimizing the total travel cost while satisfying the constraints on vehicle capacity, route duration, customer service time windows, and customer visit requirements [5,6].

The PVRPTW is defined on a complete directed graph $G = (V, A)$, where $V = \{v_0, v_1, \ldots, v_n\}$ is the vertex set and $A = \{(v_i, v_j) : v_i, v_j \in V, i \neq j\}$ is the arc set. The planning horizon considers t days, also referred to as set $T = \{1, \ldots, t\}$. Vertex v_0 represents the depot with time window $[e_0, l_0]$ at which are based m vehicles that have capacity limited to Q and maximum working time D. Each vertex $v_i \in V$, $i \neq 0$, corresponds to a customer and has an associated demand $q_i \geq 0$, a service duration $d_i \geq 0$, a time window $[e_i, l_i]$, a service frequency f_i and a set $C_i \subseteq T$ of allowable combinations of visit days. For each arc $(v_i, v_j) \in A$ there is a cost $c_{ij} \geq 0$. The problem then consists in selecting a single visit combination per customer and designing (at most) m feasible vehicle routes for each of the t days on G [5,6].

With respect to our application in milk collection, we consider a special version of the PVRPTW with an objective function focused on quality of service, since it is in practice hard to meet the farms ideal milk collection time windows. The quality of service is improved by reducing the time that farms have to wait to be served within their time windows. This new objective is based on variables s_{ik} representing the time when vehicle k arrives at farm i, and e_i and l_i corresponding to the the earliest start time of service and the latest start time of service at the farm i, and n is the number of farms.

The consecutive values of the variable s_{ik} are computed iteratively in each route by $s_{jk} = \max(e_i, s_{ik}) + u_i + c_{ij}$, if vehicle k goes from v_i to v_j, with u_i as the time it takes to perform the service on the farm i. The objective function of the S solution is defined as follows:

$$f(S) = \min \frac{1}{n} \sum_k \sum_j \frac{\max\{(s_{jk} - e_j), 0\}}{l_j - e_j} \tag{1}$$

The objective function thus aims at maximizing milk quality by minimizing the lateness of the collection at each farm.

3   Solution Method

VRPs in general are known to be also difficult to solve in practice. PVRPTW and obviously also our variant of the problem are NP-hard. Accordingly, meta-heuristic methods are appropriate to optimize our model for the milk collection problem and a real-world process.

We propose a hybrid metaheuristic that combines GRASP (Greedy Randomized Adaptive Search Procedure) [7] and VNS (Variable Neighborhood Search) [8]. GRASP is an iterative two-phase metaheuristic made up of a construction phase, in which a feasible solution is produced, and a post-optimization phase, in which this feasible solution is improved. The GRASP solution construction mechanism builds a solution step-by-step by adding at random a new node from a restricted candidate list (RCL). We use a variant of VNS, VND (Variable Neighborhood Descent).

VND consists in changing the neighbourhoods each time the local search is trapped in a local optimum with respect to current neighbourhoods. VND is basically iteratively determining a better solution from the current solution by some transformation or movement. Standard VND considers several neighborhood structures of solution S as $N_k(S)$ for $k = 1, \ldots, k_{max}$, being k_{max} the number of neighborhood structures. Nevertheless, in our method we use the value of k for control the size of movements that will be described later.

This hybrid approach uses GRASP as an outer framework for diversification and VND for intensification, i.e., for locally improving and post-processing constructed solutions as shown in Fig. 1.

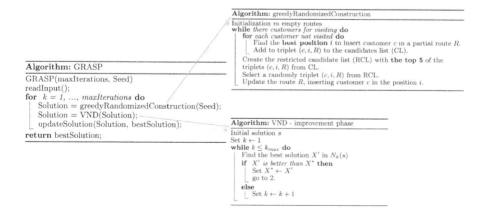

Fig. 1. General solution approach: GRASP-VND hybrid

Generally two approaches which deal with PVRPTW are offered in the literature. The first one begins by assigning days to dairy farms and in a second step the routing problem for every single day is solved using classical techniques for solving the VRP [9]. In the second approach, routes are developed and then assigned to days of the week. The method we present here follows the second approach and consists of the following two steps as shown in Fig. 2.

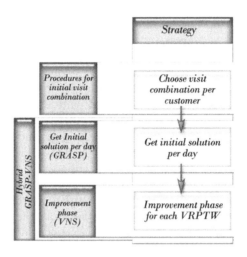

Fig. 2. General solution approach

The first step of our method consists in assigning a single visit combination to each dairy farm. Customers are then assigned to the corresponding days of the planning horizon. A list of customers is created in descending order relative to the time window size. Customers are assigned to a single visit combination alternatively in descending and ascending order with respect to this list.

In the second step, a hybrid GRASP-VND is used to solve the Vehicle Routing Problem with Time Windows (VRPTW) for each day of the planning horizon. In the first step we allow infeasible VRPTW due to violation of constraints as total duration of the routes, farm time windows or vehicle capacity. In order to help the hybrid GRASP-VND find high-quality and feasible solutions, we consider infeasible solutions during the search. Capacity, duration, and time window constraints can be violated and are penalized by including proportional penalty terms in the objective function. As a specific focus of our work, we experimentally compare different kinds of penalty functions. The penalty functions are described in the next subsection.

Following the second step, the GRASP is used to obtain an initial solution to each day of the planning horizon. This GRASP tries to satisfy the constraints. If this is not possible the GRASP procedure assigns customers that do not satisfy constraints to the last route. The initial solutions obtained by GRASP are improved by using a VND with three different movements:

- *Change visit combinations.* Change the visit combination of a farm with a new combination. The farm then has to be removed from routes of the days in the first combination that are not in the second one and inserted in the routes of the days that are in the new combination and not in the previous one.
- *k-chain moves.* Take a chain of k consecutive farms in a route of the solution and move it to another part of the same route or in other route.
- *k-swap moves.* Interchange the position of two chains with length k in the solution. Both chains can be in the same route or in different routes.

We consider a dynamic neighbourhood order to obtain high-quality solutions. We use a composition of h neighborhood structures, where $h = 9$, $N_1, \ldots, N_9$. The neighborhood structures *k-swap chain*, *k-move chain* and *change visit combination*. The nine neighborhood structures are the different combinations of the movements described above. A weight is assigned to each neighbourhood structure, this weight reflects the performance during the search.

This weight considers two measures; the improvement over time it_h^j and the utilization u_h^j of the neighbourhood structure. Given that a solution S^j at iteration j, let S^* be the solution obtained by $N_h(S^j)$ and t_h^j the CPU time spent, the measures are defined as follows:

- Improvement over time: $it_h^j = \frac{f(S^j) - f(S^*)}{t_h^j}$
- Utilization: u_h^j of N_h, the number of times neighborhood structure N_h has been applied.

In this way, when a neighborhood N_h is applied at iteration j, we calculate:

- $in_h^{j+1} = \delta \cdot in_h^j + (1 - \delta) \cdot it_h^j, \delta \in [0, 1)$ being a strategy parameter.
- $u_h^{j+1} = u_h^j + 1$.

The corresponding weight of neighbourhood structure N_h is calculated as

$$r_h = \frac{in_h^j}{u_h^j}. \tag{2}$$

And the probability of selecting N_h as neighborhood structure to be applied is

$$p_h = \frac{r_h}{\sum_{h=1}^k r_h} \tag{3}$$

3.1   Penalized Cost Functions

To guide the search and help the hybrid GRASP-VNS find high-quality and feasible solutions we explicitly allow infeasible solutions during the search process. We relax the constraints related with vehicle capacity Q, maximum working time D, and time windows.

For a solution S, we denote the quality of service objective function as

$$qos(S) = \frac{1}{n} \sum_{k=1}^{m} \sum_{i=0}^{n} \frac{s_{ik} - e_i}{l_i - e_i}, \tag{4}$$

total violation of load constraints as $q(S)$ calculated considering the maximum Q_k, total violation of time windows constraints $tw(S)$ calculated as

$$\sum_{k=1}^{m} \sum_{i=1}^{n} \max\{0, s_{ik} - l_i\}, \tag{5}$$

where s_{ik} is the time when vehicle k arrives to farm i, and total violation of duration constraints $rlt(S)$ calculated considering D_k. The objective function is defined as $f(S) = qos(S) + \alpha \cdot q(S) + \beta \cdot tw(S) + \gamma \cdot rlt(S)$, where α, β and γ are positive weight factors that depend on the kind of penalty function.

We propose two different kinds of penalty functions. The first one is a static penalty function where the penalty terms do not depend on the current iteration of the search process, therefore, remain constant during the entire search. Secondly a dynamic penalty function is proposed where the penalty term depends on the solutions obtained during the search. In both kind of penalty functions the values of α, β, γ are defined as follow:

- $\alpha = (q(S)_{max} - q(S)_{min})/q(S)_{avg}$
- $\beta = (tw(S)_{max} - tw(S)_{min})/tw(S)_{avg}$
- $\gamma = (rlt(S)_{max} - rlt(S)_{min})/rlt(S)_{avg}$.

In the case of the static penalty function the maximum, minimum and average bounds of the violation of the constraints were obtained by preliminary computational experiments. In the dynamic penalty function the bounds are updated with the values obtained during the search.

4   Experimentation and Results

This section describes the results from the computational experiments that were carried out in our study. The aim of the experiment is to test the practical feasibility of the proposed hybrid procedure GRASP-VNS to solve the milk collection problem and compare the hybrid GRASP-VNS that considers infeasible solutions to the GRASP-VNS that discard infeasible solutions.

Only some characteristics from daily milk collection real-world data are known to us so far but no concrete instancdes are presently available. Therefore, we adapted benchmark instances for PVRPTW [3]. Specifically we use the instances p01, p07, p11 and p17 because the characteristics are similar to the real data. The data provides the position of a set of farms with service duration, demand and time windows, and we changed the visit combination for each dairy farm by setting it as company data. The number of days of the planning horizon and the maximum number of routes for each day of the planning horizon is also included. For more details concerning the used instances, see Table 1.

Table 1. Characteristics of instances, taken from Cordeau et al. [3]

Instances	Farms	Routes per day	Max. time per day	Max. load of truck	Days
$p01$	48	3	500	200	4
$p07$	72	5	500	200	6
$p11$	48	3	500	200	4
$p17$	72	4	500	200	6

Regarding the parameter of the hybrid GRASP-VNS, the size of the restricted candidate list is fixed to 5 and k_{max} for VND is set to 3. The solution approach was run 100 times for each of the instances and parameters used in experimentation. The results of the computational experiments can be seen in Table 2, where the hybrid GRASP-VNS with two kinds of penalty functions and the hybrid GRASP-VNS that discards infeasible solutions are compared. It can be seen that the results of the hybrid GRASP-VNS with dynamic penalized function are better.

Table 2. Results on benchmark instances from Cordeau et al. [3]

		Penalty function	p01	p07	p11	p17
GRASP-VND	Average	Static	1.19	1.13	1.06	1.05
	Best		1.10	0.93	0.87	0.91
GRASP-VND	Average	Dynamic	1.17	1.05	0.98	0.95
	Best		1.06	0.89	0.81	0.72
GRASP-VND	Average	None	1.43	1.48	1.19	1.12
	Best		1.25	1.13	0.95	0.89

5   Conclusions and Further Research

In this study, we presented a heuristic solution approach for the planning of the collection of fresh milk from local farms with a fleet of refrigerated vehicles, modeled as a variant of the PVRPTW. The proposed objective function is oriented towards the quality of service in order to preserve the quality of fresh milk. In order to solve the problem to get high quality solutions in reasonable time a hybrid GRASP-VNS metaheuristic has been used. The approach considers infeasible solutions during the search relaxing constraints and smoothing the search space, using two kind of penalty functions. The computational experiments confirm that the proposed approach is reasonable to practically solve this model. Future work will extend experimentation with other instances, among which some will be real cases. The behavior of other metaheuristics, other neighbourhood structures in VND procedure and other procedures for choosing initial visit combination per customer will also be studied. A special future line is related to using other kinds of penalty functions and other techniques to deal with infeasible solutions.

Acknowledgment. This work has been partially funded by the Spanish Ministry of Economía y Competitividad with FEDER funds (TIN2015-70226-R) and by Fundación Cajacanarias (2016TUR19).

Contributions from Airam Expósito-Márquez are supported by la Agencia Canaria de Investigación, Innovación y Sociedad de la Información de la Consejería de Economía, Industria, Comercio y Conocimiento and by the Fondo Social Europeo (FSE).

References

1. De Armas, J., Lalla, E., Expósito, C., Landa, D., Melián, B.: A hybrid GRASP-VNS for ship routing and scheduling problem with discretized time windows. Eng. Appl. Artif. Intell. **45**, 350–360 (2015)
2. Claassen, G., Hendriks, T.: An application of special ordered sets to a periodic milk collection problem. Eur. J. Oper. Res. **180**(2), 754–769 (2007)
3. Cordeau, J.F., Laporte, G., Mercier, A.: A unified tabu search heuristic for vehicle routing problems with time windows. J. Oper. Res. Soc. **52**(8), 928–936 (2001)
4. Lahrichi, N., Gabriel Crainic, T., Gendreau, M., Rei, W., Rousseau, L.M.: Strategic analysis of the dairy transportation problem. J. Oper. Res. Soc. **66**(1), 44–56 (2015)
5. Pirkwieser, S., Raidl., G.: A variable neighborhood search for the periodic vehicle routing problem with time windows. In: Proceedings of the 9th EU/MEeting on Metaheuristics for Logistics and Vehicle Routing, pp. 23–24 (2008)
6. Pirkwieser, S., Raidl, G.R.: Multiple variable neighborhood search enriched with ILP techniques for the periodic vehicle routing problem with time windows. In: Blesa, M.J., Blum, C., Di Gaspero, L., Roli, A., Sampels, M., Schaerf, A. (eds.) HM 2009. LNCS, vol. 5818, pp. 45–59. Springer, Heidelberg (2009). https://doi.org/10.1007/978-3-642-04918-7_4
7. Feo, T.A., Resende, M.G.C.: Greedy randomized adaptive search procedures. J. Global Optim. **6**, 109–133 (1995)
8. Hansen, P., Mladenović, N., Moreno, J.A.: Variable neighbourhood search: methods and applications. Ann. Oper. Res. **175**(1), 367–407 (2010)
9. Beltrami, E.J., Bodin, L.D.: Networks and vehicle routing for municipal waste collection. Networks **4**(1), 65–94 (1974)

Solving the Traveling Thief Problem Using Orchestration in Optimization Networks

Johannes Karder[1,2(✉)], Andreas Beham[1,2], Stefan Wagner[1], and Michael Affenzeller[1,2]

[1] Heuristic and Evolutionary Algorithms Laboratory, University of Applied Sciences Upper Austria, Softwarepark 11, 4232 Hagenberg, Austria
{jkarder,abeham,swagner,maffenze}@heuristiclab.com
[2] Institute for Formal Models and Verification, Johannes Kepler University Linz, Altenberger Straße 69, 4040 Linz, Austria

Abstract. Optimization problems can sometimes be divided into multiple subproblems. Working on these subproblems instead of the actual master problem can have some advantages, e.g. if they are standard problems, it is possible to use already existing algorithms, whereas specialized algorithms would have to be implemented for the master problem. In this paper we approach the NP-hard Traveling Thief Problem by implementing different cooperative approaches using optimization networks. Orchestration is used to guide the algorithms that solve the respective subproblems. We conduct experiments on some instances of a larger benchmark set to compare the different network approaches to best known results, as well as a sophisticated, monolithic approach. Using optimization networks, we are able to find new best solutions for all of the selected problem instances.

Keywords: Optimization network · Traveling Thief Problem
Metaheuristic · Evolutionary algorithm · HeuristicLab

1 Introduction

In this paper, we use the concepts of optimization networks [9] and orchestration to find good solutions for instances of the NP-hard Traveling Thief Problem (TTP) [5]. The problem definition states that a thief wants to steal items that are stored at different locations. Each item has a certain value and weight. The thief can only carry a maximum amount of weight in his backpack for which a renting rate applies. The more weight he carries, the longer he is buccaneering and the more renting costs arise. He must visit all locations in a round-trip which starts and ends at his hideout. The goal is to steal those items in that round-trip so that the overall profit gets maximized. This *master problem* can be divided into a Knapsack Problem (KSP) [10] and a problem similar to the Traveling Salesman Problem (TSP) [3]. In the context of optimization networks, a KSP optimizer is used to yield good item selections, whereas a TSP optimizer

ⓒ Springer International Publishing AG 2018
R. Moreno-Díaz et al. (Eds.): EUROCAST 2017, Part I, LNCS 10671, pp. 307–315, 2018.
https://doi.org/10.1007/978-3-319-74718-7_37

is used to generate profitable tours. Therefore, optimization networks are quite applicable to solve the TTP since cooperative approaches can be implemented to optimize both subproblems in collaboration, as shown in [4].

The presented work is motivated by different aspects. One is to reuse already existing algorithms in a cooperative fashion, together with available problem implementations and solution encodings, instead of adapting these algorithms or even create new problem implementations or solution encodings specifically for the TTP. For one, this means that many different algorithms can be applied to optimize the subproblems and users can experiment with different algorithm combinations. Secondly, it encourages the reuse of existing implementations of profound, well-known and tested algorithms and problems. Furthermore, we want to try novel approaches for solving the TTP by implementing different network variants in order to optimize the aforementioned subproblems in a cooperative manner. When using multiple algorithms, one can e.g. generate parts of a solution and thus is able to reduce the problem for another cooperative algorithm.

In Sect. 2, the concepts of optimization networks and orchestration are explained. The implemented network variants are described in Sect. 3. Finally, the conducted experiments, as well as the achieved results are shown in Sect. 4, followed by conclusions and suggestions for future developments in Sect. 5.

2   Optimization Networks and Orchestration

Connecting different subproblem optimizers leads to the creation of so called optimization networks. Such networks are composed of multiple nodes that are able to communicate with each other. Each node has its own purpose:

- *Parameter nodes* store required network and problem parameters. Other nodes are able to query parameter nodes for these parameters.
- *Algorithm nodes* are used to execute optimization algorithms. They are able to query other nodes to reevaluate solution qualities and report optimization results after their algorithms finish.
- *Control nodes* can manage the communication between different nodes. For example, they can query parameter nodes for problem data, create respective problem instances and instruct algorithm nodes to optimize these.

To manage a collaborative optimization approach, the concept of *orchestrators* is used, which can be seen in Fig. 1. Orchestrators are control nodes that are able to steer subproblem optimizers in two different ways:

- They can recalculate actual qualities of solutions for the subproblems. The quality as determined by a subproblem optimizer might not reflect the true quality of the (partial) solution in the original master problem.

– They can adapt the subproblem parameters to yield better (partial) solutions for the master problem. By changing a subproblem's fitness landscape, the respective optimizer converges to other areas of the search space that it would have ignored in the original subproblem instance, because solutions in this area would be evaluated as weak right away.

Depending on the chosen orchestration strategy, the orchestrator will interact with the optimizers on a quality adaptation or problem adaptation basis, for which it is necessary to define according routines. These routines are problem-specific, therefore specialized orchestrators have to be provided for each problem that will be divided into subproblems. We implemented specialized orchestrators for the TTP to be used within optimization networks in the HeuristicLab optimization environment [11].

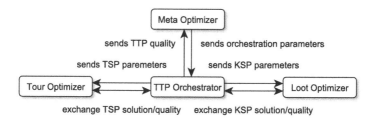

Fig. 1. The orchestrator steers the tour and loot optimizers. A meta-optimizer is used to optimize orchestration parameters e.g. adaptation parameters for the KSP item weights or TSP city coordinates.

The goal is to design specific *orchestrator nodes* for different network variants. A well-conceived architecture makes it possible to implement problem-specific orchestrator nodes with little effort. Specific TTP orchestrators are used to steer the overall optimization approach. We will experiment with both the quality adapting, as well as the problem adapting orchestration techniques and try different orchestration routines.

3   TTP Networks

Solutions for the TTP constitute of a binary vector representing the item selection, as well as an integer permutation representing the round-trip. All network implementations utilize a meta-optimizer that is used to evolve different orchestration parameters depending on the network variant. The parameters are evaluated by the orchestrator, who for this purpose manages the execution of the subproblem optimizers and uses the parameters for e.g. problem adaptation.

3.1 Loot First Network

The approach of this network is to first find a good item selection and then a good round-trip for this item selection in order to create a good TTP solution. An orchestrator is used for problem instance adaptation, as well as recalculation of solution qualities. The generation and evolution of the problem adaptation parameters is done by a meta-optimizer (i.e. an evolution strategy with covariance matrix adaptation, or CMA-ES [8]). When it needs to evaluate a parameter vector, the meta-optimizer sends it to the orchestrator. The orchestrator's task is to calculate the parameter vector's quality and report it back to the meta-optimizer. Therefore, it first creates an adapted KSP instance by multiplying each original item value with a real-valued number between $[-1.0, 1.0]$ from a provided parameter vector, making the specific item more or less valuable. The orchestrator then sends the adapted instance to the loot optimizer (i.e. a Parameter-less Population Pyramid, or P3 [7]), which tries to find good item selections and sends the best found solution back to the orchestrator. Then the tour optimizer is tasked by the orchestrator to find the best tour for this item selection. Therefore, the original TSP instance is sent to the tour optimizer (i.e. a local search algorithm). For each tour that needs to be evaluated, the tour optimizer will send it to the orchestrator, which will return the overall profit (that the thief generates by using the aforementioned item selection and traveling this particular tour) as quality. After the tour optimizer finishes, the best tour is sent back to the orchestrator, which is now able to calculate the quality of the parameter vector (i.e. the TTP quality of the best found item selection in combination with the best found round-trip) and report it back to the meta-optimizer.

3.2 Tour First Network

The Tour First network uses the same concept as the Loot First network, except that the parameter vector is used to adapt the TSP city coordinates, then the tour optimizer works on the adapted TSP instance and finally the loot optimizer tries to find the best item selection for the best found tour.

3.3 Loot and Tour Network

The Loot and Tour network tries to combine both aforementioned approaches by changing the fitness landscapes of both subproblems (i.e. adapting the KSP item values and TSP city coordinates) using the meta-optimizer's parameter vector. Each adapted subproblem is then given to the respective optimizers and the best found solutions are returned to the orchestrator in order to report a (TTP) quality to the meta-optimizer.

3.4 Switchover Network

The Switchover network uses a quite different approach. Its meta-optimizer (i.e. a genetic algorithm with offspring selection, or OSGA [2]) evolves binary vectors, which serve as initial item selections in the orchestrator. When the meta-optimizer needs to evaluate a binary vector, it will be sent to the orchestrator. For each binary vector evaluation, the orchestrator stores three different variables, being the current item selection, the current tour and the best found TTP quality. The orchestrator sets the sent binary vector as the current item selection and tasks the tour optimizer to find a good tour for it. After the tour optimizer finishes, it will report back the best found round-trip to the orchestrator, who will set it as the current tour. Then the TTP quality of the current item selection and tour is calculated. If the evaluated quality is better than the best found so far, the original item selection will be replaced by the current item selection and the best found quality will be updated. After the quality evaluation, the loot optimizer is tasked by the orchestrator to find the best item selection for the current tour. When the loot optimizer finishes, the best item selection is reported back to the orchestrator and quality evaluation is started again. After a certain number of switchovers have been executed, the orchestrator reports back the best found item selection and its quality to the meta-optimizer, where they replace the original item selection and quality.

4 Experiments and Results

Focus has been laid upon the Loot First and Switchover networks, since the other implementations did not perform well in preliminary tests. We therefore decided to not perform further testing with these implementations for the time being. The used benchmark set has been defined by [6] and contains 9720 instances. We selected 60 instances out of the berlin52 subset, 30 instances with 1 item per city and 30 instances with 3 items per city. To create another base line for comparison besides the already available best known qualities, we also conducted experiments with SASEGASA [1], a sophisticated self-adaptive, segregative genetic algorithm with simulated annealing aspects. Detailed algorithm configurations can be found on our additional material page[1]. The tested networks, as well as SASEGASA, have been given a computational budget of 2 h, after which the execution was stopped. Table 1 lists the qualities of the best known and best found solutions. Figure 2 shows the relative distances from the best known qualities in [6] to the best found for each tested instance, where positive percentages mean that our best found quality is better than the best known.

---

[1] https://dev.heuristiclab.com/wiki/AdditionalMaterial.

Table 1. Best achieved results (rounded) from the Loot First network, the Switchover network and SASEGASA, compared to best knowns (BK). Bold numbers indicate the best quality within a row.

#	Instance	BK [6]	Loot First	Switchover	SASEGASA
01	berlin52_n51_bounded-strongly-corr_01.ttp	4257.75	4454.54	4437.46	**4454.54**
02	berlin52_n51_bounded-strongly-corr_02.ttp	6775.21	7149.07	7126.95	**7149.07**
03	berlin52_n51_bounded-strongly-corr_03.ttp	7624.95	7847.19	7797.82	**7872.25**
04	berlin52_n51_bounded-strongly-corr_04.ttp	10607.40	10891.55	**10892.07**	10886.16
05	berlin52_n51_bounded-strongly-corr_05.ttp	13963.40	**14271.24**	14204.70	14244.42
06	berlin52_n51_bounded-strongly-corr_06.ttp	13401.80	13672.28	**13889.56**	**13889.56**
07	berlin52_n51_bounded-strongly-corr_07.ttp	16203.00	16675.66	16844.64	**16875.81**
08	berlin52_n51_bounded-strongly-corr_08.ttp	15845.57	16921.50	**17095.90**	**17095.90**
09	berlin52_n51_bounded-strongly-corr_09.ttp	18081.60	19239.39	19512.51	**19563.50**
10	berlin52_n51_bounded-strongly-corr_10.ttp	15270.80	16179.01	**16671.81**	**16671.81**
11	berlin52_n51_uncorr-similar-weights_01.ttp	1614.03	1656.06	1652.27	**1656.06**
12	berlin52_n51_uncorr-similar-weights_02.ttp	3720.01	3781.12	**3812.76**	**3812.76**
13	berlin52_n51_uncorr-similar-weights_03.ttp	4975.06	5127.88	5126.47	**5127.88**
14	berlin52_n51_uncorr-similar-weights_04.ttp	5114.96	5198.41	5140.19	**5198.41**
15	berlin52_n51_uncorr-similar-weights_05.ttp	6636.45	**6749.52**	**6749.52**	**6749.52**
16	berlin52_n51_uncorr-similar-weights_06.ttp	6603.46	6585.63	**6672.90**	**6672.90**
17	berlin52_n51_uncorr-similar-weights_07.ttp	7400.42	7387.14	**7464.48**	**7464.48**
18	berlin52_n51_uncorr-similar-weights_08.ttp	8585.31	9064.72	**9110.12**	**9110.12**
19	berlin52_n51_uncorr-similar-weights_09.ttp	8815.79	8693.92	**8854.33**	**8854.33**
10	berlin52_n51_uncorr-similar-weights_10.ttp	8956.09	**9087.86**	**9087.86**	**9087.86**
21	berlin52_n51_uncorr_01.ttp	2916.25	**3110.93**	3104.03	**3110.93**
22	berlin52_n51_uncorr_02.ttp	4216.75	4477.85	**4495.73**	**4495.73**
23	berlin52_n51_uncorr_03.ttp	5914.82	6218.07	**6317.48**	**6317.48**
24	berlin52_n51_uncorr_04.ttp	5808.78	5780.00	**6007.87**	**6007.87**
25	berlin52_n51_uncorr_05.ttp	6311.51	6074.16	6458.53	**6458.96**
26	berlin52_n51_uncorr_06.ttp	7586.51	7602.78	**7704.43**	**7704.43**
27	berlin52_n51_uncorr_07.ttp	8529.13	8444.88	**8647.17**	**8647.17**
28	berlin52_n51_uncorr_08.ttp	9208.45	9239.04	**9350.67**	**9350.67**
29	berlin52_n51_uncorr_09.ttp	9534.32	**9847.05**	**9847.05**	**9847.05**
30	berlin52_n51_uncorr_10.ttp	9465.35	9598.02	**9649.10**	**9649.10**
31	berlin52_n153_bounded-strongly-corr_01.ttp	10048.20	9732.95	10273.32	**10431.23**
32	berlin52_n153_bounded-strongly-corr_02.ttp	18260.50	17984.24	18802.40	**18851.32**
33	berlin52_n153_bounded-strongly-corr_03.ttp	23835.80	23560.56	24092.99	**24217.51**
34	berlin52_n153_bounded-strongly-corr_04.ttp	34704.30	34070.14	34709.87	**34859.18**
35	berlin52_n153_bounded-strongly-corr_05.ttp	37420.30	35060.06	**37519.66**	37126.16
36	berlin52_n153_bounded-strongly-corr_06.ttp	43951.00	42888.89	45094.56	**45139.88**
37	berlin52_n153_bounded-strongly-corr_07.ttp	44542.50	44275.07	**46969.48**	**46969.48**
38	berlin52_n153_bounded-strongly-corr_08.ttp	48446.20	48766.27	**52380.08**	52371.29
39	berlin52_n153_bounded-strongly-corr_09.ttp	47568.80	46109.26	49763.99	**50009.58**
40	berlin52_n153_bounded-strongly-corr_10.ttp	47221.90	46803.72	**51334.31**	**51334.31**
41	berlin52_n153_uncorr-similar-weights_01.ttp	5792.81	5806.24	5771.91	**5831.64**
42	berlin52_n153_uncorr-similar-weights_02.ttp	10126.20	10229.20	10292.22	**10319.97**
43	berlin52_n153_uncorr-similar-weights_03.ttp	11851.90	11550.20	**12162.66**	**12162.66**
44	berlin52_n153_uncorr-similar-weights_04.ttp	12053.60	10911.52	12478.04	**12564.68**
45	berlin52_n153_uncorr-similar-weights_05.ttp	16624.70	16533.84	**17324.56**	**17324.56**
46	berlin52_n153_uncorr-similar-weights_06.ttp	17614.70	17227.68	**19065.69**	**19065.69**
47	berlin52_n153_uncorr-similar-weights_07.ttp	19817.70	18380.48	**21488.95**	**21488.95**
48	berlin52_n153_uncorr-similar-weights_08.ttp	22331.00	22586.79	**23832.82**	**23832.82**
49	berlin52_n153_uncorr-similar-weights_09.ttp	24281.10	23941.09	25459.39	**25465.91**
50	berlin52_n153_uncorr-similar-weights_10.ttp	26711.10	25846.08	**27625.89**	**27625.89**
51	berlin52_n153_uncorr_01.ttp	10578.10	11061.85	10971.45	**11201.05**
52	berlin52_n153_uncorr_02.ttp	12696.70	12023.48	**12839.20**	12830.83
53	berlin52_n153_uncorr_03.ttp	18634.20	17847.23	18671.90	**18799.50**
54	berlin52_n153_uncorr_04.ttp	20317.30	18687.65	**20610.88**	20605.62
55	berlin52_n153_uncorr_05.ttp	20343.40	18807.33	20516.48	**20521.00**
56	berlin52_n153_uncorr_06.ttp	18463.20	16634.99	**18678.68**	18676.22
57	berlin52_n153_uncorr_07.ttp	20612.40	19441.19	**20766.88**	**20766.88**
58	berlin52_n153_uncorr_08.ttp	21943.80	20343.66	22291.06	**22431.38**
59	berlin52_n153_uncorr_09.ttp	21914.00	20693.68	22100.06	**22145.81**
60	berlin52_n153_uncorr_10.ttp	23821.60	22948.90	24070.31	**24160.36**

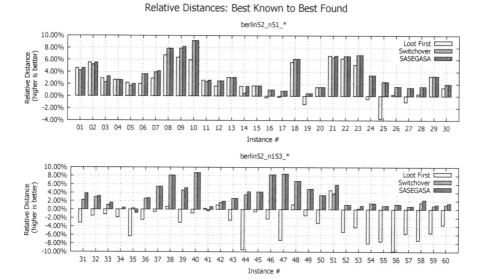

Fig. 2. The relative distances from the best known qualities in [6] to the best found qualities for all berlin52_n51 and berlin52_n153 instances.

5   Conclusion and Outlook

Three network variants have been implemented to adapt the subproblem instances (KSP item values, TSP city coordinates, both), one variant repeatedly switches between optimizing loot and tour. The performance greatly depends on the network variant. Furthermore, problem instance adaptation does not scale well with dimensionality. In the Tour First network, the meta-optimizer presumably has difficulties evolving good adaptation parameters for the city coordinates. Additionally, in the Loot and Tour network, adaptation parameters for the item values have to be found, making it even more difficult for the meta-optimizer to evolve good parameter vectors.

For 29 out of 60 instances (48.33%), the Loot First network was able to find new best solutions. The Switchover network and SASEGASA both found even more best solutions for 59 out of 60 instances (98.33%). For a few instances, some of our approaches were not able to find solutions with better quality than the published ones. For particular instances, the Switchover network achieved the best results and was even better than SASEGASA. It is crucial to note that even though the chosen problem instances had rather low dimensionality, we were able to find even better best solutions. In our opinion, this shows how hard these instances and the problem itself actually are.

Optimization networks combined with the orchestration approaches, as presented here, are still, to the best of our knowledge, a very unexplored subject in the field of metaheuristic optimization. First steps in the direction of orchestration have already been taken. Future development could include more sophisticated orchestration strategies. Furthermore, it should be possible to dynamically adjust algorithm parameters for subproblem optimizers depending on their performance. Orchestrators should be able to exchange algorithms or reconfigure their parameters on-the-fly. Depending on the algorithm, parameter adaptation could even be possible during its execution. Another enhancement is dynamic allocation of computational resources for each node within a network. Finally, creating optimization networks using multiple computers is also conceivable.

Acknowledgments. The work described in this paper was done within the COMET Project Heuristic Optimization in Production and Logistics (HOPL), #843532 funded by the Austrian Research Promotion Agency (FFG) and the Government of Upper Austria.

References

1. Affenzeller, M., Wagner, S.: SASEGASA: a new generic parallel evolutionary algorithm for achieving highest quality results. J. Heuristics - Spec. Issue New Adv. Parallel Meta-Heuristics Complex Probl. **10**, 239–263 (2004)
2. Affenzeller, M., Wagner, S.: Offspring selection: a new self-adaptive selection scheme for genetic algorithms. In: Ribeiro, B., Albrecht, R.F., Dobnikar, A., Pearson, D.W., Steele, N.C. (eds.) Adaptive and Natural Computing Algorithms, pp. 218–221. Springer, Heidelberg (2005). https://doi.org/10.1007/3-211-27389-1_52. Springer Computer Series
3. Applegate, D.L., Bixby, R.E., Chvátal, V., Cook, W.J.: The Traveling Salesman Problem: A Computational Study. Princeton University Press, Princeton (2006)
4. Beham, A., Fechter, J., Kommenda, M., Wagner, S., Winkler, S.M., Affenzeller, M.: Optimization strategies for integrated knapsack and traveling salesman problems. In: Moreno-Díaz, R., Pichler, F., Quesada-Arencibia, A. (eds.) EUROCAST 2015. LNCS, vol. 9520, pp. 359–366. Springer, Cham (2015). https://doi.org/10.1007/978-3-319-27340-2_45
5. Bonyadi, M.R., Michalewicz, Z., Barone, L.: The travelling thief problem: the first step in the transition from theoretical problems to realistic problems. In: 2013 IEEE Congress on Evolutionary Computation, pp. 1037–1044. IEEE, June 2013
6. Faulkner, H., Polyakovskiy, S., Schultz, T., Wagner, M.: Approximate approaches to the traveling thief problem. In: Proceedings of the 2015 Annual Conference on Genetic and Evolutionary Computation, GECCO 2015, pp. 385–392. ACM, July 2015
7. Goldman, B.W., Punch, W.F.: Parameter-less population pyramid. In: Proceedings of the 2014 Annual Conference on Genetic and Evolutionary Computation, GECCO 2014, pp. 785–792. ACM, July 2014
8. Hansen, N., Müller, S.D., Koumoutsakos, P.: Reducing the time complexity of the derandomized evolution strategy with covariance matrix adaptation (CMA-ES). Evol. Comput. **11**(1), 1–18 (2003)
9. Karder, J., Wagner, S., Beham, A., Kommenda, M., Affenzeller, M.: Towards the design and implementation of optimization networks in HeuristicLab. In: Proceedings of the 2017 Annual Conference on Genetic and Evolutionary Computation, GECCO 2017. ACM, July 2017. https://doi.org/10.1145/3067695.3082475

10. Sahni, S.: Approximate algorithms for the 0/1 Knapsack problem. J. ACM (JACM) **22**(1), 115–124 (1975)
11. Wagner, S., et al.: Advanced methods and applications in computational intelligence, topics in intelligent engineering and informatics. In: Klempous, R., Nikodem, J., Jacak, W., Chaczko, Z. (eds.) Advanced Methods and Applications in Computational Intelligence, vol. 6, pp. 197–261. Springer International Publishing, Heidelberg (2014). https://doi.org/10.1007/978-3-319-01436-4_10

Optimizing the Movement of Containers on the Yard of a Maritime Container Terminal

Israel López-Plata[✉], Christopher Expósito-Izquierdo, Belén Melián-Batista,
and J. Marcos Moreno-Vega

Department of Computer Engineering and Systems, Universidad de La Laguna,
38271 La Laguna, Spain
{ilopezpl,cexposit,mbmelian,jmmoreno}@ull.edu.es

Abstract. The present paper presents the Internal Delivery Vehicle Synchronization Problem which seeks to manage the available fleet of internal delivery vehicles on a yard of a maritime container terminal in its container movement operations. The objective of this problem is to obtain the planning of the vehicles optimized in terms of working time. This optimization problem must be done in scenarios where synchronization when accessing to specific positions on the yard is required. With the goal of solving this problem from an approximate point of view, a Variable Neighbourhood Search is here proposed. The computational experiments disclose the suitable performance of this algorithmic approach, which allows to reach high-quality solutions within reasonable computational times.

Keywords: Variable Neighbourhood Search · Internal Delivery
Vehicle Synchronization Problem · Maritime container terminal

1 Introduction

Maritime container terminals are logistic interfaces between sea and land means of transport [5], which provide temporary storage, consolidation, and freight transshipment. These terminals are split into three different functional areas: sea-side, hinterland, and yard. Optimizing the logistic operations of terminals is a highlighted research area due to the fact that these have high level of complexity and must be then performed with high level of reliability. There is a huge amount of papers in the scientific literature which seek to optimize the operations of the three areas in a maritime container terminal. The interested reader is referred to [8,9] to obtain comprehensive literature reviews in this regard.

The containers are moved between the three functional areas around the terminal by a fleet of internal delivery vehicles [2]. The management of these vehicles in a terminal can be divided into three different operations: (i) assigning a requested container movement to a determined vehicle [10], (ii) setting the appropriate routes of each vehicle, and (iii) determining the sequence in which each vehicle has to perform all its assigned tasks.

© Springer International Publishing AG 2018
R. Moreno-Díaz et al. (Eds.): EUROCAST 2017, Part I, LNCS 10671, pp. 316–322, 2018.
https://doi.org/10.1007/978-3-319-74718-7_38

The present paper introduces the Internal Delivery Vehicle Synchronization Problem. This problem seeks to optimize the usage of internal delivery vehicles in terms of working time. The main goal of the problem is to plan the vehicle movement in such a way that the makespan of its undelying schedule is minimized.

An optimization approach based upon the general framework of the well-known Variable Neighbourhood Search [4], in short VNS, is here proposed to solve the optimization problem under analysis. The computational results disclose the suitability of this algorithm when addressing the Internal Delivery Vehicle Synchronization Problem and open several promising lines for further research.

The remainder of this work is organized as follows. Firstly, Sect. 2 describes the Internal Delivery Vehicle Synchronization Problem. Section 3 presents the algorithmic technique aimed at solving the proposed problem. The computational experiments carried out in the work are presented and discussed in Sect. 4. Finally, Sect. 5 extracts the main conclusions from the work and indicates several lines for further research.

2   Problem Description

In this work, the Internal Delivery Vehicle Synchronization Problem (hereafter termed IDVSP) is proposed. In the IDVSP a set of jobs $J = \{1, 2, \ldots, n\}$ which has to be performed by a set of internal delivery vehicles $V = \{1, 2, \ldots, m\}$ are given. Performing a job consists in (i) picking up a determined container from a source position, (ii) moving the container from this source position $s(j)$ to a target position $t(j)$, and (iii) performing the container drop-off operation. All positions are denoted as $P = \{1, 2, \ldots, p\}$. In addition, a vehicle $v \in V$ is available only after its ready time $\alpha(v) \geq 0$.

A feasible solution of the IDVSP consists in a set of routes $\phi = (r_1, r_2, \ldots, r_m)$, where r_v is the route to be followed by the internal delivery vehicle $v \in V$, to serve all jobs in J. A particular route is represented by $r_v = (r_v^0, r_v^1, r_v^2, \ldots, r_v^{n(v)})$, where $n(v)$ is the number of jobs to be performed by the vehicle, and r_v^i is the i-th job performed in the route. It is important to know the time required to perform a job $j \in J$ in a determined route $r_v \in \phi$ when evaluating a route. This is denoted as $time(r_v, j)$ and is defined as the addition of three different times:

- Time to reach the picking up position of the job
- Time to move the container from pick-up to drop-off positions
- Time spent to pick up and drop off operations.

With these three times we can represent the time spent by a vehicle $v \in V$ when following its route r_v. This is as follows:

$$time(r_v) = \alpha(v) + \sum_{j \in r_v} time(r_v, j) \tag{1}$$

It is assumed that all the presented times satisfy the triangle inequality [3]. The optimization objective of the IDVSP is to minimize the makespan of the underlying schedule for all the vehicles. This objective can be formally expressed as follows:

$$\min \max_{v \in V} time(r_v) \tag{2}$$

It is important to remark that the problem considers the synchronization when the vehicles try to access to position where have to perform operations to pick up or drop off containers. This means that two vehicles cannot share positions on the yard in the same time instant. In these cases, the vehicles are queued in such a way that only one of them can perform its operations. To decide which of two vehicles has to wait in this situation, every vehicle has assigned a priority $\delta(v)$ that indicates the preference of the terminal manager.

In scenarios where synchronization is not required, the IDVSP can be reduced to the well-known Vehicle Routing Problem with pick-ups and deliveries [1]. This means that the IDVSP is also an $\mathcal{NP}$-hard class of problem in the strong sense.

3   Algorithmic Approach

This work proposes an algorithm based in the well-known framework of Variable Neighbourhood Search [4], in short VNS, for solving the Internal Delivery Vehicle Synchronization Problem introduced in Sect. 2 from an approximate point of view. The VNS is an algorithm that has the capacity to explore the solution space by an efficient way, changing systematically between different neighbourhood structures associated with the current solution. The execution of a basic VNS algorithm is divided into three different phases: (i) generating an initial solution to start with the exploration and, for every exploration, (ii) a shaking process, and (iii) a local search.

To generate the initial solution, a constructive method which provides feasible solutions is used. This solution is created using a greedy constructive method with a semi-random behaviour.

Before exploring a determined neighbourhood, the perturbation phase is executed. This phase consists in a shaking process that gets a random solution inside the current neighbourhood and allows to get out of possible valleys reached in previous explorations. After this phase, a local search is executed to get a local optimum in the current neighbourhood. In this particular cases, the set of neighbourhoods explored is denoted as N_k, where ($k = 1, 2, \ldots, k_{max}$). This set is composed of the following movements:

- *Swap*. Movement that interchanges two arbitrary elements in the same sequence. This neighbourhood function never changes the number of elements in the sequence [6]. In the context of the presented problem, this movement generates neighbours exchanging the performing order of two different jobs assigned to the same vehicle.

Algorithm 1. Pseudo-code of the Variable Neighbourhood Search to solve the Internal Delivery Vehicle Synchronization Problem

---

 1: **while** (stop criterion is not met) **do**
 2:     Creating solution s using constructive method
 3:     $k \leftarrow 1$
 4:     **repeat**
 5:         $s_{shaken} \leftarrow Shake(s, k)$
 6:         $s_{LS} \leftarrow LocalSearch(s_{shaken}, k)$
 7:         **if** $(f(s_{LS}) < f(s))$ **then**
 8:             $s \leftarrow s_{LS}$
 9:             $k \leftarrow 1$
10:         **else**
11:             $k \leftarrow k + 1$
12:         **end if**
13:     **until** $k = k_{max}$
14:     **if** $(f(s) < f(s_{VNS}))$ **then**
15:         $s_{VNS} \leftarrow s$
16:     **end if**
17: **end while**

- *Insertion.* Neighbourhood structure that deletes an element from a sequence and reinserts it in another sequence [6]. In the IDVSP, the insertion removes the assignment of a job to its vehicle and reassigns it to another one.
- *Two-opt.* Movement that breaks two arcs from one tour and reconnects the obtained subtours to create a new tour [7]. In this context, the tour where the two-opt movement is applied is the performing route of a determined vehicle.

The details of the proposed VNS to solve the IDVSP are shown in the pseudo-code depicted in Algorithm 1. This follows the schema of Basic VNS explained in [4]. The first step of the algorithm is to create the initial solution where starts the neighbourhood exploration explained above (line 2). Once the initial solution is created, the exploration of each neighbourhood structure in N_k is performed. The algorithm executes a shaking process (line 5) in each exploration and thereafter a local search (line 6). Once obtained the local optimum solution, this solution is compared to the best solution obtained from all previous explorations. If this local optimum solution is better than the best solution (line 7), it is considered the new best solution and the neighbourhood exploration is restarted from the first one (line 9). Otherwise, the algorithm continues with the exploration of the next neighbourhood structure (line 11).

Once all neighbourhood structures are explored without improving the current solution (line 13), this solution is compared to the best solution obtained during the search (line 14). If the current solution improves the best one, it is considered the new best solution of the search (line 15).

4   Analysis

In the following, the suitability of the Variable Neighbourhood Search (VNS) introduced in Sect. 3 to solve the Internal Delivery Vehicle Synchronization Problem (IDVSP) described in Sect. 2 is adequately assessed. To know the performance of the proposed algorithm and due to neither the heuristic model to obtain the optimal solution nor another optimization approach to solve this particular problem exist in the literature, the results obtained by means of the proposed VNS have to be compared with the solutions returned by other well-known algorithms:

- The first of the algorithms included in the comparison is a multistart search with multiple neighbourhoods and a systematic change of neighbourhood structure when the solution cannot be improved. This algorithm can be seen as a simplified version of the Variable Neighbourhood Search proposed in this paper. Due to represents the nearest solution to the optimal one, it is executed 100 times more than other algorithms in the analysis. The solutions reported by this multistart method are known as Best Known Solutions (BKS). Also, due to the fact that this is the algorithm of reference in the comparison its execution time is not included in the computational experiments.
- Local Search (LS). It reports a local optimum solution using only one of the available neighbourhood structures explained above.
- The last algorithm included in the analysis is a generation of a feasible solution randomly (Rnd).

The previous algorithms used in the analysis have been implemented in Java Standard Edition 8 and executed on a computer equipped with an Intel Core i7-6.50 GHz and 8 GB of RAM. Table 1 reports the comparison between all the algorithms explained above when executed with the same set of problem instances. These instances represent a wide range of realistic problem scenarios. First three columns represents the details of every problem instance included in the experiments. Particularly, the number of jobs that have to be performed (n), the number of available positions in the terminal (p), and the number of vehicles used to perform all jobs (m) are respectively shown. It is important to remark that each line in the table represents the execution of 5 different problem instances with the same characteristics. The subsequent column (f_{BKS}) shows the objective function value of the Best Known Solution. Columns *Gap.* show the deviation on percent of the average objective function value returned by each algorithm in comparison with f_{BKS}. Finally, columns t. report the computational times measured in seconds of all algorithms under analysis.

The computational results indicate that the performance of the *VNS* has negative Gap in all the computational experiments under analysis. This means that the proposed algorithm returns better solutions in all the situations than the Best Known Solution, with an average Gap of around 26.9%. The results are obtained even with 100 times less executions. With these results we can conclude than the proposed VNS returns better results than expected. The algorithms *LS*

Table 1. Computational comparison of algorithm performance

n	p	m	f_{BKS}	t_{VNS} (s.)	Gap_{VNS}	t_{LS} (s.)	Gap_{LS}	t_{Rnd} (s.)	Gap_{Rnd}
20	50	2	1778.2	2.403	−0.956	0.038	21.021	0.006	25.205
		5	800.0	4.234	−3.725	0.108	32.975	0.007	39.600
		10	494.4	7.703	−7.686	0.225	41.950	0.002	59.385
	100	2	1859.8	2.905	−1.043	0.044	13.249	0.004	15.389
		5	822.0	4.227	−3.139	0.102	21.290	0.007	28.686
		10	477.0	6.871	−1.426	0.176	37.610	0.004	43.354
40	50	2	3531.2	12.624	−4.474	0.090	68.668	0.011	65.808
		5	1835.2	18.054	−18.668	0.493	77.278	0.011	93.025
		10	1101.2	27.767	−26.862	0.716	61.733	0.014	78.641
	100	2	3441.6	11.627	−0.703	0.123	36.239	0.011	43.119
		5	1565.8	18.428	−6.872	0.526	49.495	0.011	75.131
		10	899.2	28.107	−8.096	1.174	67.905	0.012	79.471
60	100	2	5286.6	36.864	−3.397	0.186	67.355	0.018	67.253
		5	2700.6	51.325	−17.507	1.184	80.241	0.030	94.883
		10	1577.6	69.827	−22.642	3.458	67.748	0.028	91.899
	200	2	5054.6	38.866	−0.372	0.556	40.442	0.028	47.240
		5	2374.6	55.026	−6.182	2.333	57.837	0.030	68.559
		10	1337.4	71.610	−9.137	5.093	71.362	0.022	90.339
-	−	−	**2052.4**	**26.026**	**−7.938**	**0.924**	**50.800**	**0.014**	**61.499**

and *Rnd* return solutions in less time than the *VNS*, but with much worse quality. As final conclusion, we can deduce that the algorithm proposed in Sect. 3 returns high-quality solutions in a reasonable computational time.

5   Conclusions and Future Work

The Internal Delivery Vehicle Synchronization Problem (IDVSP) is an optimization problem whose objective is to create the set of routes that have to follow the internal delivery vehicles to perform a determined set of jobs. The available vehicles, the set of jobs to perform and all positions in the terminal are known in advance. To evaluate the quality of the returned solutions, the synchronization of the vehicles when they try to access to the same position in the same time instant has to be taken into account. Solving the IDVSP involves to make two decisions: (i) assigning each job to the vehicle that has to perform it and (ii) determining the route of each vehicle to perform all its assigned jobs.

In this paper, a Variable Neighbourhood Search (VNS) is proposed to solve the IDVSP from an approximate point of view. This algorithm performs an efficient exploration of the solution space. To start with the exploration, the VNS creates an initial solution and after that explores the environment of the current solution. It sequentially executes a shaking process and a local search. The computational experiments conducted in this paper indicate the suitable performance of the proposed algorithmic approach, which allows to reach high-quality solutions within reasonable computational times.

Several promising lines are still open for further research. One of them is to create the optimization model of the problem so the optimal solutions can be obtained and used to check the quality of the algorithmic approaches. Also, the algorithm can be studied to create a decision making tool, which allows to apply the obtained solutions directly in real environments. Lastly, the problem can be studied in dynamic situations. These situations could be, for instance, the addition of a new job to be performed and the removal of an internal delivery vehicle, among others.

Acknowledgements. This work has been partially funded by the Spanish Ministry of Economy and Competitiveness with FEDER funds (project TIN2015-70226-R).

References

1. Berbeglia, G., Cordeau, J.-F., Gribkovskaia, I., Laporte, G.: Static pickup and delivery problems: a classification scheme and survey. TOP **15**(1), 1–31 (2007)
2. Bish, E.K., Chen, F.Y., Leong, Y.T., Nelson, B.L., Ng, J.W.C., Simchi-Levi, D.: Dispatching vehicles in a mega container terminal. OR Spectr. **27**(4), 491–506 (2005)
3. Fleming, C.L., Griffis, S.E., Bell, J.E.: The effects of triangle inequality on the vehicle routing problem. Eur. J. Oper. Res. **224**(1), 1–7 (2013)
4. Hansen, P., Mladenović, N., Brimberg, J., Moreno Pérez, J.A.: Variable neighborhood search. In: Gendreau, M., Potvin, J.Y. (eds.) Handbook of Metaheuristics. International Series in Operations Research & Management Science, vol. 146, pp. 61–86. Springer, US (2010). https://doi.org/10.1007/978-1-4419-1665-5_3
5. Meisel, F.: Seaside Operations Planning in Container Terminals, vol. 1. Springer, Heidelberg (2009). https://doi.org/10.1007/978-3-7908-2191-8
6. Michiels, W., Korst, J., Aarts, E.: Theoretical Aspects of Local Search. Monographs in Theoretical Computer Science. An EATCS Series. Springer, Heidelberg (2007). https://doi.org/10.1007/978-3-540-35854-1
7. Mjirda, A., Todosijević, R., Hanafi, S., Hansen, P., Mladenović, N.: Sequential variable neighborhood descent variants: an empirical study on the traveling salesman problem. Int. Trans. Oper. Res. **24**, 1–19 (2016)
8. Stahlbock, R., Voß, S.: Operations research at container terminals: a literature update. OR Spectr. **30**, 1–52 (2008)
9. Steenken, D., Voß, S., Stahlbock, R.: Container terminal operation and operations research - a classification and literature review. OR Spectr. **26**, 3–49 (2004)
10. Tao, J., Qiu, Y.: A simulation optimization method for vehicles dispatching among multiple container terminals. Expert Syst. Appl. **42**(7), 3742–3750 (2015)

A Meta-heuristic Approach for the Transshipment of Containers in Maritime Container Terminals

Kevin Robayna-Hernández, Christopher Expósito-Izquierdo[✉],
Belén Melián-Batista, and J. Marcos Moreno-Vega

Department of Computer Engineering and Systems, Universidad de La Laguna,
38271 La Laguna, Spain
{krobayna,cexposit,mbmelian,jmmoreno}@ull.edu.es

Abstract. The present paper introduces a new multi-objective variant of the Berth Allocation Problem in the context of maritime container terminals. Specifically, this optimization problem seeks to minimize the waiting times of the incoming container vessels to serve, the costs derived from the movement of containers around the terminal, and the time the containers are at the terminal. This optimization problem is solved by means of an Adaptive Large Neighborhood Search, which uses a dynamic parameter to destroy part of the solutions while a building method is designed to later rebuild them. The computational performance of this technique is assessed over a wide range of realistic scenarios. The results indicate its high efficiency and effectiveness, reporting high-quality solutions in all the cases within short computational times.

Keywords: Maritime container terminal · Berth Allocation Problem
Meta-heuristic

1 Introduction

Over the last few decades, maritime container terminals have consolidated as highlighted infrastructures within global supply chains. The relevance of the terminals in the international trade arises from the huge socio-economic impact they have on the regions in which they are located. As an example, they are nowadays handling about 10 billion tons of goods worldwide, according to the United Nations Conference on Trade And Development[1]. In general terms, they are essential components of multi-modal transportation networks due to the fact that they are aimed at enabling the transshipment of freights between different transportation means. The most common means of transportation brought together in maritime container terminals are container vessels, trucks, and trains.

A maritime container terminal is usually split into the following functional areas [5]:

This work has been partially funded by the Spanish Ministry of Economy and Competitiveness with FEDER funds (project TIN2015-70226-R).

[1] http://www.unctad.org.

© Springer International Publishing AG 2018
R. Moreno-Díaz et al. (Eds.): EUROCAST 2017, Part I, LNCS 10671, pp. 323–330, 2018.
https://doi.org/10.1007/978-3-319-74718-7_39

- *Sea-side.* It serves the incoming container vessels. For this purpose, the sea-side of a terminal is usually divided into several berths, which are provided with technical equipment to perform the transshipment operations of containers. That is, loading and unloading containers onto/from the incoming vessels. The interested reader is referred to the book [9] to obtain an in-depth review of the main sea-side operations planning in container terminals.
- *Yard.* It is an open-air surface aimed at storing container until their later retrieval. With this goal in mind, the yard is usually organized into similar blocks. That is, three-dimensional physical structures composed of a predefined number of bays [4].
- *Land-side.* It is the part of the container terminal aimed at connecting the infrastructure and the hinterland transportation means [3].

The ever-increasing pressure on terminal managers highlights the need for improving the overall performance, especially when carrying out transshipment container operations. With this goal in mind, the main contributions of this paper are described in the following:

- Introducing the Berth Allocation Problem with transshipment flows, which seeks to determine how to schedule the arrival of the incoming container vessels and handle the containers included into their stowage plans.
- Proposing an Adaptive Large Neighborhood Search to solve the proposed optimization problem efficiently.
- Presenting a set of benchmark problem instances to gather the main features associated with the transshipment of containers between several container vessels in a maritime container terminal.

The remainder of the paper at hand is organized as follows. Section 2 describes the main flows of containers found in a maritime container terminal and formalizes their management as an optimization problem. Afterwards, Sect. 3 describes a meta-heuristic optimization technique aimed at solving the problem previously introduced. Section 4 presents several computational experiments to assess the performance of the meta-heuristic approach proposed in this paper. Finally, Sect. 5 discusses the main conclusions extracted from the paper and indicates several lines for further research.

2   Transshipment of Containers

The transshipment operations at maritime container terminals refer to the movement of containers carried out between the sea-side and the yard of the infrastructure [1], denoted as Y. This way, every time a vessel arrives to port, a subset of its containers are unloaded from it and later stored on the yard. Similarly, a set of containers is loaded into the vessel to be transported to the next port along its shipping route.

In this work, we seek to solve a variant of the Berth Allocation Problem [9], in short BAP, in which the objective is to determine a berth and a suitable berthing time for each incoming vessel that satisfies its particular requirements (*i.e.*, dimensions, time windows, service constraints, etc.).

Input data of the BAP under analysis are composed of a set of n incoming container vessels, denoted as V, and a set of m berths, denoted as B, in which the incoming container vessels must be assigned over the planning horizon, $H > 0$. Each incoming container vessel has to be assigned to a berth and each berth is only available to serve vessels within a given time window. The time window of vessel $b \in B$ is defined as $[tw_b, tw'_b]$, where $tw'_b > tw_b$. Specifically, tw_b and tw'_b represent the earliest and the latest times in which berth $b \in B$ is available, respectively.

Each incoming vessel $v \in V$ has a time window to visit the terminal. This window is denoted as $[t_v, t'_v]$, where $t'_v > t_v$. In this environment, t_v and t'_v represent the earliest arrival and the latest departure time of the vessel v, respectively. The berthing time of each incoming vessel $v \in V$ is denoted as bt_v, where $bt_v \geq t_v$. The waiting time of the vessel $v \in V$ is the time between the beginning of its time window until its berthing time. This waiting time is defined as follows:

$$wt_v = bt_v - t_v. \tag{1}$$

Also, the service time of a vessel depends on the volume of containers carried out and the characteristics of the berth assigned (*e.g.*, number of quay cranes, productivity of the technical equipment, number and features of internal delivery vehicles, etc.). It is worth mentioning that the service time of a given container vessel $v \in V$ in the berth $b \in B$ is denoted as $s_{vb} > 0$. This way, the departure time of vessel $v \in V$ when assigned to berth $b \in B$ is $dt_v = bt_v + s_{vb}$, where $dt_v \leq t'_v$. Lastly, it is assumed that each container vessel can be berthed in any berth:

$$t_v + s_{vb} \leq t'_v, \forall v \in V, b \in B. \tag{2}$$

The containers in a terminal are organized into groups, denoted as C. A group $c \in C$ is a set of $n(c) > 0$ containers which share some characteristics (*e.g.*, destination port, weight, dimensions, etc.). The movement of the containers around the terminal requires of a set of internal delivery vehicles. The cost associated with moving a container from a source, $s \in T$, toward a destination, $d \in T$, is defined as $f_{sd} > 0$, where $T = \{B \cup Y\}$. This way, the cost derived from moving a group of containers, $c \in C$, is computed as $n(c) \cdot f_{s(c)d(c)}$. The time a group of containers, $c \in C$, is at the terminal is denoted as $t(c)$.

The following are the decisions to make when solving the BAP under analysis:

- Assigning a berth to each incoming container vessel.
- Determining the berthing time of each incoming container vessel.
- Selecting the source and target of each group of containers when these are unknown.

The objective function of the BAP seeks to minimize (i) the waiting time of the incoming container vessels, (ii) the cost derived from moving the containers between the berths and the yard, and (iii) the time the containers are at the terminal. This is formally expressed as follows:

$$\min \alpha_1 \cdot \sum_{v \in V} wt_v + \alpha_2 \cdot \sum_{c \in C} n(c) \cdot f_{s(c)d(c)} + \alpha_3 \cdot \sum_{c \in C} n(c) \cdot t(c), \qquad (3)$$

where α_1, α_2, and α_3 are parameters of the problem whose values are selected by the decision maker.

Finally, it is worth mentioning that the BAP with discrete berths is known to be $\mathcal{NP}$-hard [13]. At the same time, the BAP can be easily reduced to the well-known Vehicle Routing Problem when removing the costs derived from moving containers and the flows. Consequently, it can be claimed that the BAP is also $\mathcal{NP}$-hard.

3 Adaptive Large Neighborhood Search

This section introduces an optimization technique based on the general framework termed Adaptive Large Neighborhood Search, in short ALNS, with the aim of solving the variant of the Berth Allocation Problem introduced in the previous section.

The general framework of the ALNS was initially introduced in [12] to address the well-known Vehicle Routing Problem with pick ups and deliveries. The rational behind this approach is based on exploring several large neighborhoods in an adaptive fashion. Specifically, a large number of robust heuristics aimed at destroying and rebuilding solutions are applied iteratively and according to a classification function, which indicates the success of these heuristics. The ALNS has been successfully applied when solving a multitude of optimization problems found in the scientific literature. Some representative examples can be found in [6, 10, 11].

The ALNS proposed in this paper uses a set of parameters to solve the BAP under analysis. Some of these are described as follows:

- max. It represents the maximum number of iterations of the main procedure.
- $maxWI$. It indicates the maximum number of iterations without improvement in terms of objective function value in the main procedure.
- $\lambda_{initial}$. It represents an initial destruction factor.
- $maxPL$. It is the maximum number of iterations of the rebuilding procedure.
- Δ. It represents how much the destruction factor is increased per iteration in the rebuilding method.

The pseudocode of the proposed ALNS is shown is Algorithm 1. The execution of the algorithm starts by generating a solution by means of a greedy procedure. This procedure assigns the incoming container vessels to the available berths at the terminal. Also, this procedure assigns sources and targets to the transshipment container flows satisfying a topological order, as described in Sect. 3.1. Each solution is destroyed partially, and then it is rebuilt to be used as starting point for a local search. This process is repeated until a maximum number of iterations is achieved. Finally, the best solution found by the optimization technique is reported.

Algorithm 1. Adaptive Large Neighborhood Search

---

1 **Data**: *Problem P, max, maxWI, $\lambda_{initial}$, maxPL, Δ*
2 **Result**: Best solution
3 *bestSolution $\leftarrow$ Greedy(P)*
4 *count $\leftarrow$ 0*
5 **for** $i = 0; i < max; i + +$ **do**
6   *current $\leftarrow$ GreedyRebuilding(bestSolution, λ_{start}, maxPL, Δ)*
7   **if** *!current.isFeasible()* **then**
8   *current $\leftarrow$ Greedy(P)*
9   *actual $\leftarrow$ LocalSearch(P, current)*
10   **if** *f(current) < f(bestSolution)* **then**
11   *bestSolution $\leftarrow$ current*
12   *count $\leftarrow$ 0*
13   **else**
14   *count + +*
15   **if** *count = maxWI* **then**
16   finish loop
17 **return** *bestSolution*

Algorithm 2 depicts the pseudocode of the greedy rebuilding process used in the previous optimization technique. In this case, a solution is rebuilt to better explore the search space and thus increasing the probability of finding the global optimum solution of the optimization problem to solve. In an iterative and adaptive way, $\lambda_{start} + \Delta$ % of solution is destroyed, as indicated in line 6. This factor is denoted as λ' and is always lower than or equals to 100%, line 4. Once a part of the solution is destroyed, this should be rebuilt. A new topological order must be generated with this goal in mind. Those container vessels that are not included into the solution are iteratively included. The objective function value of the resulting solution is computed, line 8, and its feasibility is checked, line 9. The loop is finished if the solution is feasible, line 10, and the solution is reported, line 11. If the resulting solution is not feasible, an empty solution is reported.

3.1 Topological Order

The topological order is one of the most important terms in Graph Theory due to its innumerable applications. Some of the most representative can be found in [2,7,8]. In general terms, a topological order of a graph is a lineal order of its vertices in such a way that, if (i, j) is one of its arcs, then i appears before j in the order.

In the context of the Berth Allocation Problem introduced in Sect. 2, selecting a feasible set of source vessels and target vessel for each container group is one of the decisions to make. For this purpose, a suitable source and target vessel is selected. Specifically, a source vessel and target vessel must be determined for each movement of a container group between berths and the yard.

Algorithm 2. Greedy Rebuilding Method

---

1 **Data**: *solution* $S, \lambda_{start}, maxPL, lambdaI$
2 **Result**: *newSolution*
3 $newSolution \leftarrow emptySolution$
4 **for** $(\lambda' = \lambda_{start}; \lambda' \le 100; \lambda' + = \Delta)$ **do**
5   **for** $i = 0; i < maxPL; i + +$ **do**
6         $newSolution \leftarrow Destroy(S, \lambda')$
7         $newSolution \leftarrow Rebuild(newSolution, topologicalOrder(S))$
8         $newSolution.evaluate()$
9   **if** $newSolution.isFeasible()$ **then**
10   finish loop
11 **return** $newSolution$

---

4 Computational Experiments

The optimization approach proposed in this work has been coded in Java SE 8. It has been tuned and tested on a computer equipped with an Intel 3.16 GHz and 4 GB of RAM. The test problem instances have been generated according to realistic features of large maritime container terminals.

Table 1 shows a subset of representative computational results reported by a multi-start method, which generates a feasible solution by means of a constructive procedure and then applies a local search until the stop criterion is fulfilled, and the Adaptive Large Neighborhood Search when solving scenarios with $m = 10$ berths. The algorithm has been executed with a maximum number of iterations of 100 or 20 iterations without improvement in terms of best objective function value. Each row represents 10 problem instances, for which each optimization technique has been applied 10 times per instance. Column n represents the number of incoming container vessels, whose ranges from 10 up to 40. *Tran. (%)* represents the percentage of container groups to transship between pairs of vessels.

As can be checked in the results provided in Table 1, the ALNS is very effective and efficient when solving the problem instances under analysis. In fact, this optimization technique reports better computational results than the multistart method in all the cases when $\lambda = 25\%$ and $\Delta = 5\%$. The results show that the approach is able converge to (near-)optimal solutions in a reduced number of iterations while keeping a high level of diversification. The improvements obtained in the computational experiments reach near 40% in the largest problem instances within computational times below 23 s. It is worth mentioning that the execution of the proposal finishes requiring lesser than the iterations set in the stop criterion. Finally, the results are robust when the varying the characteristics of the problem instances to solve, which indicates that the approach is suitable for being applied in realistic environments.

Table 1. Average computational results when setting $\lambda = 25\%$ and $\Delta = 5\%$

n	Tran. (%)	Multi-start			ALNS		
		$f(x)$	Time (s.)	Iter.	Time (s.)	Iter.	Gap. (%)
10	25	10682675.71	0.25	35.81	0.34	46.11	−5.81
	50	10846399.71	0.31	37.92	0.51	47.53	−10.17
20	25	23094985.44	1.15	34.71	2.09	55.34	−21.22
	50	26602826.99	1.49	36.32	2.76	62.82	−21.23
30	25	44775497.05	2.58	38.42	7.97	65.02	−30.65
	50	42962754.69	2.66	39.09	6.04	67.44	−37.41
40	25	57603445.22	7.60	38.19	18.18	66.85	−32.98
	50	60064336.14	8.76	35.01	22.72	72.30	−37.26

5 Conclusions and Further Research

This work introduces a new multi-objective variant of the Berth Allocation Problem in the context of maritime container terminals. This optimization problem has a high impact in economic terms and the competitiveness of container trade.

Several optimization techniques has been here proposed to solve this problem. In particular, an efficient Adaptive Large Neighborhood Search has been designed. The computational results indicate the high performance of the proposed Adaptive Large Neighborhood Search. Specifically, this meta-heuristic provides solutions with high quality within short computational times.

Finally, it is worth mentioning that several open lines for further research. Some of these lines are described in the following:

- Studying how the destination positions of the incoming containers can be determined on the basis of its technical and/or physical features.
- Studying the dynamic Berth Allocation Problem, where unforeseen events (*e.g.*, arrival of new container vessels, delays of the arrivals, among others) appear during the planning horizon.
- Developing a decision-making tool to manage the processes brought together in maritime container terminals using the optimization techniques proposed to solve the problem under analysis.

References

1. Böse, J.W.: General considerations on container terminal planning. In: Böse, J.W., Sharda, R., Voß, S. (eds.) Handbook of Terminal Planning. Operations Research/Computer Science Interfaces Series, vol. 49, pp. 3–22. Springer, New York (2011). https://doi.org/10.1007/978-1-4419-8408-1_1
2. Brucker, P., Knust, S.: Complex job-shop scheduling. In: Brucker, P., Knust, S. (eds.) Complex Scheduling. GOR-Publications, pp. 239–317. Springer, Heidelberg (2012). https://doi.org/10.1007/978-3-642-23929-8

3. Carlo, H.J., Vis, I.F.A., Roodbergen, K.J.: Transport operations in container terminals: literature overview, trends, research directions and classification scheme. Eur. J. Oper. Res. **236**(1), 1–13 (2014)
4. Carlo, H.J., Vis, I.F.A., Roodbergen, K.J.: Storage yard operations in container terminals: literature overview, trends, and research directions. Eur. J. Oper. Res. **235**(2), 412–430 (2014). Maritime Logistics
5. Günther, H.O., Kim, K.H.: Container terminals and terminal operations. OR Spectr. **28**(4), 437–445 (2006)
6. Kovacs, A.A., Parragh, S.N., Doerner, K.F., Hartl, R.F.: Adaptive large neighborhood search for service technician routing and scheduling problems. J. Sched. **15**(5), 579–600 (2012)
7. Li, J., Pan, Y., Shen, H.: More efficient topological sort using reconfigurable optical buses. J. Supercomput. **24**(3), 251–258 (2003)
8. Liu, S.Q., Kozan, E.: New graph-based algorithms to efficiently solve large scale open pit mining optimisation problems. Expert Syst. Appl. **43**, 59–65 (2015)
9. Meisel, F.: Seaside Operations Planning in Container Terminals. Contributions to Management Science. Springer, Heidelberg (2009). https://doi.org/10.1007/978-3-7908-2191-8
10. Muller, L.F.: An adaptive large neighborhood search algorithm for the resource-constrained project scheduling problem. In: MIC 2009: The VIII Metaheuristics International Conference (2009)
11. Ribeiro, G.M., Laporte, G.: An adaptive large neighborhood search heuristic for the cumulative capacitated vehicle routing problem. Comput. Oper. Res. **39**(3), 728–735 (2012)
12. Ropke, S., Pisinger, D.: An adaptive large neighborhood search heuristic for the pickup and delivery problem with time windows. Transp. Sci. **40**(4), 455–472 (2006)
13. Umang, N., Bierlaire, M., Vacca, I.: Exact and heuristic methods to solve the berth allocation problem in bulk ports. Transp. Res. Part E: Logist. Transp. Rev. **54**, 14–31 (2013)

Multi-objective Topology Optimization of Electrical Machine Designs Using Evolutionary Algorithms with Discrete and Real Encodings

Alexandru-Ciprian Zăvoianu[1(✉)], Gerd Bramerdorfer[2,3], Edwin Lughofer[1], and Susanne Saminger-Platz[1]

[1] Department of Knowledge-Based Mathematical Systems,
Johannes Kepler University Linz, Linz, Austria
ciprian.zavoianu@jku.at
[2] Institute for Electrical Drives and Power Electronics,
Johannes Kepler University Linz, Linz, Austria
[3] Linz Center of Mechatronics, LCM, Linz, Austria

Abstract. We describe initial results obtained when applying different multi-objective evolutionary algorithms (MOEAs) to direct topology optimization (DTO) scenarios that are relevant in the field of electrical machine design. Our analysis is particularly concerned with investigating if the use of discrete or real-value encodings combined with a preference for a particular population initialization strategy can have a severe impact on the performance of MOEAs applied for DTO.

Keywords: Evolutionary algorithms · Multi-objective optimization
Discrete encoding · Real encoding · Topology optimization
Electrical machine design

1 Introduction and Motivation

When designing electrical motors, one generally aims to discover machines that are simultaneously optimal with regard to (at least a few of) several criteria like energy efficiency, manufacturing costs, fault tolerance and operating characteristics. The standard approach for tackling these real-life multi-objective optimization problems (MOOPs) is structured as a two-step procedure. In the first step, a domain expert (i.e., an electrical engineer) defines the complete geometric specifications of the future design. This actually means that the human expert creates or chooses (and likely adapts) a parametric model that will act as a generic template for any subsequent electrical drive design that aims to solve the given task (see Fig. 1a). In the second step, a multi-objective optimization algorithm (MOOA) is employed to discover those sets of parameter combinations that, when applied to the preselected generic template, will produce Pareto-optimal design solutions. The final choice for one (or more) of the

© Springer International Publishing AG 2018
R. Moreno-Díaz et al. (Eds.): EUROCAST 2017, Part I, LNCS 10671, pp. 331–338, 2018.
https://doi.org/10.1007/978-3-319-74718-7_40

Pareto-optimal designs rests with the domain expert or with a third party decision maker (i.e., a customer).

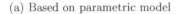

(a) Based on parametric model   (b) Based on discrete grid

Fig. 1. Example of two cross-sections of rotor designs. Light green parts denote iron elements and dark blue parts denote air gaps. (Color figure online)

Even when carrying out the optimization part via state-of-the-art population-based meta-heuristic solvers, like hybrid multi-objective evolutionary algorithms (MOEAs) [12] and particle swarm optimization strategies [1], one can easily argue that the truly creative part of the design process remains with the domain expert. When the wrong parametric model for the task at hand is chosen, no amount of optimization will be able to deliver good results. Thus, by imposing hard constraints in variable space, a choice for a parametric model actually entails restrictions on the shape of possible designs.

Direct topology optimization (DTO) [5] is an alternative approach that, when applicable, seems better suited to fully benefit from the explorative strength of modern MOEAs and recent advances in simulation software and computation power [7]. In this case, the domain expert only needs to define the boundaries of the design region and to choose a discretization factor. This results in a grid in which each cell can be from a limited set of values (see Fig. 1b). The simplest of such sets contains only two elements: iron and air. The task of the MOOA is to find those grid configurations (i.e., discrete matrices) that encode Pareto-optimal solutions. Thus, since the optimization problem is formulated in a manner that imposes virtually no restrictions on attainable geometries, the MOOA also "becomes responsible" for the more advanced/innovative part of the design automation process.

2   Research Focus and Approach

Our current aim is to gain some insights regarding the expected performance of multi-objective evolutionary algorithms (MOEAs) used on direct topology optimization scenarios. As such, we performed different types of numerical experiments on artificial and industrial problems. In particular, we investigated if the MOEAs currently used for template-based multi-objective optimization scenarios [8] are also suitable for DTO after minimal modifications. Across all DTO experiments, we have chosen to concatenate the rows of the topology matrix and to use a uni-dimensional (i.e., vector) encoding for all the tested MOEAs.

Our first idea was to adapt the very well known NSGA-II [3] to a DTO context by simply fitting it with genetic operators suitable for a *discrete-value encoding*: single point crossover and bit-flip mutation. This discrete encoding is the natural codification for a topology matrix/vector as a value of 0 can be used to denote air (i.e., cavities in the rotor design) and increasing non-negative values can encode various construction materials (e.g., iron, copper, magnets with different magnetization directions, etc.). In the present work we focus on the simplest (binary) DTO scenarios that consider only iron and air elements.

Secondly, we tested if the good convergence behavior exhibited by more advanced hybrid MOEAs – like DECMO [14] and DECMO2 [15] – on template-based scenarios also translates to DTO. Since a key feature of the two hybrid MOEAs is the integration of differential evolution (DE) [9] – a continuous optimization paradigm – applying them to DTO scenarios also requires a rather counter-intuitive *real-value encoding* of the topology matrices/vectors.

In order to get a better overview of the MOEA performance, we also tested different initialization strategies for both types of encodings. Thus, at the individual (i.e., topology vector) level we considered four initialization options: *1* – all cells are initialized with the value 1 (iron), *0* – all cells are initialized with the value 0 (air), *B* – each cell is initialized randomly with either 0 or 1, *R* – each cell is initialized randomly with a uniformly sampled value from $[0, 1]$. The *R* option is specific to the real-value encoding.

At the (start) population level we considered 6 initialization options. In four of them (marked by the prefix "all-") every member of the population was initialized using the same individual strategy: *all-1, all-0, all-B, all-R*. In the case of two population initialization strategies, namely *0/1/B* and *0/1/R*, the overall start population of the MOEA was divided into three subgroups of equal size and each subgroup was initialized with a different strategy.

3   Experimental Setup

Multi-objective Solvers

NSGA-II [3] is probably the best known metaheuristic multi-objective optimization strategy and can now be regarded as a (classical) go-to MOEA. Its main feature is a two-tier selection for survival operator based on a primary non-dominated sorting criterion and on a secondary objective-space crowding distance (i.e., niching) quality discriminant. NSGA-II also popularized the usage of two genetic operators: simulated binary crossover and polynomial mutation [2].

DECMO [14] is a proof-of-concept hybrid MOEA based on cooperative coevolution that uses two subpopulations of equal size in order to integrate two different search strategies during a single optimization run. Thus, one subpopulation relies on an evolutionary model that is similar to the one in NSGA-II while the other subpopulation is evolved via a differential evolution strategy resembling the one proposed in GDE3 [6].

DECMO2 [15] is an improvement over its predecessor as the former was specially designed for rapid convergence on a wide class of problems. The main novelty with respect to DECMO is the integration of an external archive of elite solutions that is maintained according to a decomposition-based principle similar to the one proposed by MOEA/D [11]. In order to smoothly accommodate all three multi-objective search space exploration paradigms, the fitness sharing mechanism in DECMO2 has also been redesigned to dynamically pivot towards the best performing strategy by allowing the latter to generate more offspring during certain parts of the optimization run.

Test Problems and Fitness Assessment

Even though various efficiency-related enhancements have enabled MOEAs to become state-of-the-art solvers in electrical machine design [13], a single optimization run can still take a few days even when distributing the computations over a high-throughput computing cluster of 50–100 nodes. Since MOEAs are stochastic methods, several repeats of an optimization experiment are required in order to estimate the average performance of these solvers. In light of these considerations, for this preliminary study of MOEA performance on DTOs, we have chosen to conduct extensive numerical experiments on two (self-defined) artificial benchmark problems. In the last part of Sect. 4 we also present results obtained by applying a MOEA to a realistic DTO scenario.

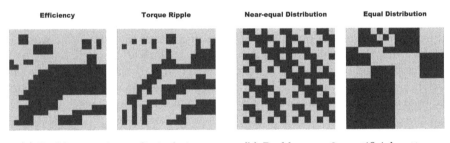

(a) Problem no. 1 – realistic design.   (b) Problem no. 2 – artificial pattern.

Fig. 2. Artificial benchmark problems. (Color figure online)

Each of the two artificial DTO benchmark MOOPs contains two objectives and each objective is defined as a binary template matrix (see Fig. 2). The idea is that, ideally, at the end of an optimization run, the solutions discovered by the MOEA at one extremum of the Pareto front should resemble the first template matrix while the solutions at the other extremum should resemble the second one. In-between Pareto non-dominated solutions are expected to: (1) contain all the subsections that are common in both templates and (2) cover the various trade-offs between the two objective matrices.

The binary (air and iron) templates that define the first benchmark DTO are shown in Fig. 2a and their are based on two realistic rotor design patterns.

The templates have a size of 20×20 elements (yielding a binary vector encoding of size 400) and, when considering both of them, the air (dark blue) to total elements ratio is 38.25%. The second benchmark DTO MOOP is defined by the 16×16 binary templates from Fig. 2b. The artificial patterns that define the second benchmark problem are more balanced (combined air to total elements ratio of 48.44%) but they describe fundamentally different designs (checkered vs continuous).

For both discrete and real value encodings, the internal MOEA fitness of an individual design vector $\mathbf{x}$ of size n with regard to the binary objective template $\mathbf{p}$ was computed as: $f(\mathbf{x}) = \sum_{0 \leq i < n} |x_i - p_i|$. In the case of the real value encodings, the presented results were recalculated/post-processed via a threshold-based modification: $f_R(\mathbf{x}) = \sum_{0 \leq i < n} |Round(x_i) - p_i|$.

MOEA Parameterization and Assessment of Solution Quality

For the numerical experiments on the two artificial MOOPs, we used the literature recommended genetic operators and standard parameter settings for all 3 MOEAs. The (total) population size was set at 200 and each optimization run was stopped after 100,000 fitness evaluations (i.e., 500 generations).

In order to estimate the convergence performance of MOEAs, we repeated each optimization run 50 times and, at every generation, we recorded the average relative hypervolume [4] of the MOEA population. The choice for the hypervolume metric is motivated by the monotonic behavior of this unary multi-objective quality indicator and the theoretical relative upper bound of 1 which is especially useful for MOOPs that, like our 2 benchmark problems, have an unknown Pareto-optimal set (i.e., solution).

4   Results - Comparative Performance

In Fig. 3 we present the comparative convergence performance of NSGA-II with four different initialization options suitable for discrete encodings. The difference from Fig. 3a between the *all-0* and *all-1* NSGA-II variants indicates that coupling the initialization strategy with the expected imbalance between air and iron elements can increase the overall convergence speed of the MOEA. The *0/1/B* initialization strategy delivers robust performance on both test MOOPs.

Figure 4 contains the convergence performance of real-value variants of NSGA-II, DECMO and DECMO2 for *all-R* and *0/1/R* – the best performing initialization strategies for real-value encodings. The results of the best discrete NSGA-II variant (*D NSGA-II 0/1/B*) are plotted to ease the comparison. As a general observation, real-value MOEA variants tend to converge slower than their discrete counterparts in the early parts of the run. Nevertheless, the real-value encoding seems to enable MOEAs to maintain a better population diversity and this directly translates into better results towards the end of the optimization.

Among the real-value solvers, DECMO2 coupled with an *all-R* initialization strategy delivers rather good results that are comparable to the ones obtained by the fastest converging discrete variants in the early stages of the run.

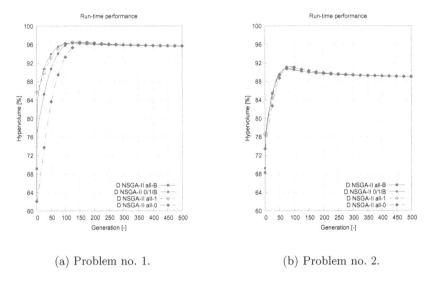

(a) Problem no. 1.   (b) Problem no. 2.

Fig. 3. Performance of NSGA-II with discrete encodings and different initialization strategies.

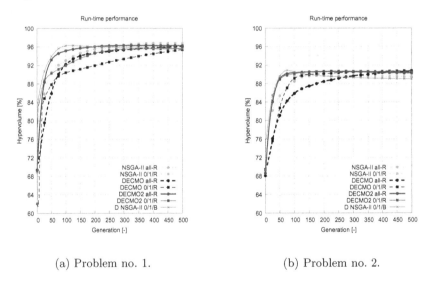

(a) Problem no. 1.   (b) Problem no. 2.

Fig. 4. Comparative convergence performance of the three MOEAs with real encodings versus the best performing version of NSGA-II with discrete encodings.

Although the performance of DECMO seems to be slightly inferior even to that of the real-value variants of NSGA-II, the former solver obtained very good preliminary results in a realistic DTO scenario [10] in which fitness assessment was performed via finite element (FE) simulations. This realistic multi-objective DTO task is related to a rotor design where one wises to simultaneously optimize output power and torque ripple. The design template is binary, as it allows for only stainless steel and air elements, and has a size of 15×15 (when considering symmetries). Figure 5 contains two rotor designs selected from the extrema of a Pareto front that was obtained after letting DECMO *all-R* evolve 20,000 designs (i.e., run for 100 generations).

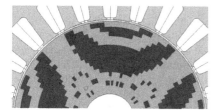
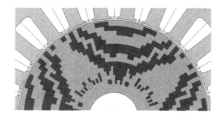

(a) Design that maximizes output power.   (b) Design that minimizes torque ripple.

Fig. 5. Rotor designs obtained by DECMO (with a real-value encoding) for the realistic DTO scenario.

5   Conclusions and Future Work

The results obtained on artificial and realistic multi-objective optimization problems indicate that applying existing MOEAs with discrete and real-value encodings to direct topology optimization (DTO) scenarios in the filed of electrical machine design can yield very promising solutions. As the preliminary innovative designs delivered by coupling MOEAs and DTO have been validated by domain experts (i.e., electrical engineers), the present study can be seen as a first step towards a more symbiotic relation between human experts and automated global search strategies inside the product design cycle.

Future work will revolve around two issues. Firstly, we plan to apply MOEAs to multi-material DTO problems. Secondly, we would like to perform the very computationally-intensive (but extremely useful) comparison between the advantages of using discrete vs. real encodings on multiple FE-based DTO scenarios.

Acknowledgments. This work was supported by the K-Project "Advanced Engineering Design Automation" (AEDA) that is financed under the COMET (COMpetence centers for Excellent Technologies) funding scheme of the Austrian Research Promotion Agency.

This work was partially conducted within LCM GmbH as a part of the COMET K2 program of the Austrian government. The COMET K2 projects at LCM are kindly supported by the Austrian and Upper Austrian governments and the participating scientific partners. The authors thank all involved partners for their support.

References

1. Bittner, F., Hahn, I.: Kriging-assisted multi-objective particle swarm optimization of permanent magnet synchronous machine for hybrid and electric cars. In: IEEE International Electric Machines and Drives Conference (IEMDC 2013), pp. 15–22. IEEE (2013)
2. Deb, K.: Multi-Objective Optimization Using Evolutionary Algorithms. Wiley, Hoboken (2001)
3. Deb, K., Pratap, A., Agarwal, S., Meyarivan, T.: A fast and elitist multiobjective genetic algorithm: NSGA-II. IEEE Trans. Evol. Comput. 6(2), 182–197 (2002)
4. Fleischer, M.: The measure of Pareto optima: applications to multi-objective meta-heuristics. In: Fonseca, C.M., Fleming, P.J., Zitzler, E., Thiele, L., Deb, K. (eds.) EMO 2003. LNCS, vol. 2632, pp. 519–533. Springer, Heidelberg (2003). https://doi.org/10.1007/3-540-36970-8_37
5. Im, C.H., Jung, H.K., Kim, Y.J.: Hybrid genetic algorithm for electromagnetic topology optimization. IEEE Trans. Mag. 39(5), 2163–2169 (2003)
6. Kukkonen, S., Lampinen, J.: GDE3: the third evolution step of generalized differential evolution. In: IEEE Congress on Evolutionary Computation (CEC 2005), pp. 443–450. IEEE Press (2005)
7. Silber, S., Koppelstätter, W., Weidenholzer, G., Bramerdorfer, G.: Magopt-optimization tool for mechatronic components. In: Proceedings of the ISMB14-14th International Symposium on Magnetic Bearings (2014)
8. Silber, S., Bramerdorfer, G., Dorninger, A., Fohler, A., Gerstmayr, J., Koppelstätter, W., Reischl, D., Weidenholzer, G., Weitzhofer, S.: Coupled optimization in MagOpt. Proc. Inst. Mech. Eng. Part I: J. Syst. Control Eng. 230(4), 291–299 (2016)
9. Storn, R., Price, K.V.: Differential evolution - a simple and effcient heuristic for global optimization over continuous spaces. J. Glob. Optim. 11(4), 341–359 (1997)
10. Straßl, M.: Topologische Optimierung von Synchronreluktanzmaschinen. Master's thesis, Johannes Kepler University Linz, Austria (2016)
11. Zhang, Q., Li, H.: MOEA/D: a multi-objective evolutionary algorithm based on decomposition. IEEE Trans. Evol. Comput. 11(6), 712–731 (2007)
12. Zăvoianu, A.C., Bramerdorfer, G., Lughofer, E., Silber, S., Amrhein, W., Klement, E.: Hybridization of multi-objective evolutionary algorithms and artificial neural networks for optimizing the performance of electrical drives. Eng. Appl. Artif. Intell. 26(8), 1781–1794 (2013)
13. Zăvoianu, A.C.: Enhanced evolutionary algorithms for solving computationally-intensive multi-objective optimization problems. Ph.D. thesis, Johannes Kepler University Linz, Austria (2015)
14. Zăvoianu, A.-C., Lughofer, E., Amrhein, W., Klement, E.P.: Efficient multi-objective optimization using 2-population cooperative coevolution. In: Moreno-Díaz, R., Pichler, F., Quesada-Arencibia, A. (eds.) EUROCAST 2013. LNCS, vol. 8111, pp. 251–258. Springer, Heidelberg (2013). https://doi.org/10.1007/978-3-642-53856-8_32
15. Zăvoianu, A.C., Lughofer, E., Bramerdorfer, G., Amrhein, W., Klement, E.P.: DECMO2: a robust hybrid and adaptive multi-objective evolutionary algorithm. Soft Comput. 19(12), 3551–3569 (2014)

Meta-Learning-Based System for Solving Logistic Optimization Problems

Alan Dávila de León[1]([✉]), Eduardo Lalla-Ruiz[2], Belén Melián-Batista[1],
and J. Marcos Moreno-Vega[1]

[1] Departamento de Ingeniería Informática y de Sistemas,
Universidad de La Laguna, La Laguna, Spain
{adavilal,mbmelian,jmmoreno}@ull.es
[2] Institute of Information Systems,
University of Hamburg, Hamburg, Germany
eduardo.lalla-ruiz@uni-hamburg.de

Abstract. The Algorithm Selection Problem seeks to select the most suitable algorithm for a given problem. For solving it, the algorithm selection systems have to face the so-called cold start. It concerns the disadvantage that arises in those cases where the system involved in the selection of the algorithm has not enough information to give an appropriate recommendation. Bearing that in mind, the main goal of this work is two-fold. On the one hand, a novel meta-learning-based approach that allows selecting a suitable algorithm for solving a given logistic problem is proposed. On the other hand, the proposed approach is enabled to work within cold start situations where a tree-structured hierarchy that enables to compare different metric dataset to identify a particular problem or variation is presented.

Keywords: Meta-learning · Logistic problems · Meta-heuristics

1 Introduction

The Algorithm Selection Problem (ASP) is introduced in [1] seeking to answer the research question *"Which algorithm is the best option to solve my problem?"* under those cases where the decision-maker or solver counts with more than one algorithm for a given problem. The importance of tackling the ASP is provided by: (*i*) No Free Lunch ([2], NFL) theorem, (*ii*) the big number of available algorithms, and (*iii*) the need of trying to obtain the best possible solution, and not only a correct one.

In the related literature, some systems have been proposed to solve the ASP. The machine learning toolbox project [3], continued in Statlog [4] and METAL [5], aims to select the best algorithm for a given dataset. Furthermore, in [5] a helping system for aiding the selection of machine learning algorithms, according to the dataset is proposed. They obtain meta-features that allow to compare different datasets and, by means of that, obtain a reduced group of datasets

© Springer International Publishing AG 2018
R. Moreno-Díaz et al. (Eds.): EUROCAST 2017, Part I, LNCS 10671, pp. 339–346, 2018.
https://doi.org/10.1007/978-3-319-74718-7_41

similar to the one at hand. Those reduced groups are later used to give a recommendation. In [6], a multilayer perceptron network is used to select the best
optimization algorithm to solve the quadratic assignment problem. The above-
mentioned contributions and systems are focused on recommending or choosing
algorithms for a particular problem. That is why a meta-learning [7] based system may be appropriate for those scenarios where a ranking of algorithms sorted
according to a provided criterion for any supported input problem is necessary.

In the context of algorithm selection systems, an usual appearing drawback
is the so-called *cold start* (see [8]). It defines the disadvantage that arises in those
cases where the system involved in the selection of the algorithm for providing
a solution has not enough information to give an appropriate recommendation
or selection. An extreme case of this problem happens when the system has no
previous information for comparing the input stream. Therefore, in this work,
we aim to propose, on the one hand, a meta-learning-based approach that allows
selecting suitable algorithms for solving given logistic problems, and on the other
hand, a tree-structured hierarchy that enables to address cold start situations.

The structure of this paper is as follows. Section 2 describes the meta-learning
based system and its main features. The proposed approach for solving the cold
start shortcoming appearing in algorithm selection is described in Sect. 3. The
results obtained by our proposed system are discussed in Sect. 4. Finally, Sect. 5
presents the conclusions together with future research lines.

2   Meta-Learning

The selection of the best algorithm for a given problem is studied by a sub-
field of Machine Learning known as *meta-learning* [7,9]. It aims to improve the
recommendation of algorithms for a given problem by applying machine learning
methods that take into account collected data from past problems with similar
features.

2.1   Meta-Learning-Based System

The algorithm selection problem (ASP) can be formally defined as follows: having
a problem instance $x \in P$, with given features $f(x) \in F$, the objective is to
perform a selection mapping $S(f(x))$ into the algorithms space A, with the
goal of selecting the algorithm $\alpha \in A$ that maximizes the performance mapping
$y(\alpha, x) \in Y$ such that $y(\alpha, x) \geq y(a, x), \forall a \in A$.

In order to start the algorithm selection process by means of a meta-learning-
based system (MLS), firstly is necessary to train the system. This is conducted
by running the set of available algorithms $A = \{a_1, ..., a_m\}$ on a set of training
instances $P_t = \{x_1, ..., x_n\}$. Each algorithm $a \in A$ is executed a certain number of iterations on each training instance in order to obtain the data related
to $Y = \{y(a_1, x_1), ..., (a_m, x_n)\}$. In this work, $y(\alpha, x)$ corresponds to the objective function value obtained by running the algorithm α on the instance x.
On the other hand, each instance $x \in P_t$ is analyzed to extract the features

$F = \{f(x_1), ..., f(x_n)\}$. Once the system is trained, a table whose rows correspond to each possible combination instance-algorithm is generated. This table associates to each row the features extracted from each instance $f(x)$ as well as the algorithm employed a_j. Moreover, each row contains the metrics obtained by $y(a, x)$ upon the execution of the experiment as, for example, the objective value. A machine learning algorithm whose input is the above-mentioned table is then applied. Its output provides a knowledge base used for providing the ranking of preferred algorithms based on the input instance. Once the system is prepared, when a new instance x' is provided, it firstly extracts its features $f(x')$ in order to start the algorithm selection process. The features that define the instance are subsequently included within a vector to be compared with the collected data using a machine learning algorithm. Based on that, a ranking $\tau_{x'}$ of recommended algorithms for the instance x' is provided.

The performance space Y is used to determine the preference of selecting an algorithm a instead of another for a particular instance x. This preference is expressed through a ranking of algorithms τ_x for each instance $x \in P_t$, where each algorithm gets a position according to their performance. This knowledge base is built offline from the set of training instances allowing to make recommendations with the premise that if two instances have similar features, the algorithms should exhibit a similar performance in both.

2.2 Problem and Algorithm Space

In this subsection, the problem and algorithm space are defined. The set P of problem instances is composed by 3 problems: (i) Travelling Salesman Problem (TSP, [10]), (ii) Capacitated Vehicle Routing Problem (CVRP, [11]), and (iii) Capacitated Vehicle Routing Problem with Time Windows (CVRPTW, [12]). Moreover, the algorithm space A used in the computational experiment is composed of 6 algorithms:

- Greedy Heuristic (GH): GH selects an unused vehicle v. Then, assign feasible nodes to vehicle v while its capacity is not exceeded. This process is repeated until a feasible solution has been created.
- Greedy Randomized Heuristic (GRH): GH but using a restricted list of candidates.
- Local Search (LS): swap environment, exhaustive sampling, and GRH.
- Greedy Randomized Adaptive Search Procedure ([13], GRASP): constructive phase = GRH, and improvement phase = LS.
- Simulated Annealing ([14], SA): GRH, initial temperature = 5, final temperature = 0.01, cooling rate = 0.01, and 100 iterations.
- Large Neighbourhood Search ([15], LNS): GRH, random destruction, degree of destruction = 0.1, and 10000 iterations.

2.3 KNN Supervised Classification

When the set of features F is available, it is possible to compare the features of the instances with each other to determine the most appropriate ranking, thus the use of a supervised classification algorithm such as k-nearest neighbors algorithm (KNN, [16]) is proposed. Because it works with rankings instead of labels, it is necessary to use a ranking aggregation methods in order to make a suitable prediction. In this work, the Borda's method [17] is used, where a score is assigned to each candidate in function of the position in which it appears in the ranking of each voter and is ordered by the total score obtained.

During the training phase, the system obtains the rankings for each problem instance. This ranking is categorized as ideal because it has all the information to properly build it. In the validation phase, it is necessary to determine the degree of success of the recommendations made. After the recommendation has been conducted, the ideal ranking is determined and compared to the recommended one in order to measure the degree of success of the system. This degree of success is measured by the distance between the rankings, for which we use the Spearman's rank correlation coefficient, where $\tau_j(i)$ is the position occupied by the candidate i in the ranking τ_j. It measures the correlation of the rankings and its value ranges between -1 and 1. The closer to 1 the higher the similarity between the rankings, while a value of -1 indicates that the rankings are inverse.

$$\rho(\tau_1, \tau_2) = 1 - \frac{6 \sum_{i=1}^{|A|} (\tau_1(i) - \tau_2(i))^2}{|A| \cdot (|A|^2 - 1)} \tag{1}$$

3   Cold Start

The cold start commonly arises when tackling real-world problems because it is very difficult to train the system to solve the ASP for any optimization problem and its variants. This problem worsens in real environments where each problem has its own peculiarities and features. To address this issue, we propose a tree-structured hierarchy approach which enables to work within cold start situations, where although the system does not have previous information of an introduced logistic problem, it may count with information from a similar problem or from a generalization of it. It allows, thus, to compare different metric dataset to identify a particular problem or variation.

In Fig. 1, the tree-structured hierarchy is depicted for the considered routing problems (i.e. TSP, VRP, CVRP, and CVRPTW). Each node corresponds to a set of features for a given problem or variation. Besides, each problem is defined by its features and those of its parents. The leaf nodes correspond to more specific problems, and as it rises in the hierarchy, the problems become more abstract or general. This means a loss of information about the specific given problem, but it allows to start the algorithm selection process based on generalizations of this problem. That is, if there are no data on the input problem, instead of not recommending a ranking, the MLS ascends in the hierarchy until it detects a problem for which it has enough information. Then, the features of that problem are extracted and the recommendation is made.

For example, suppose that the MLS is trained with TSP instances. Then, the MLS receives a VRP problem in order to start the algorithm selection process.

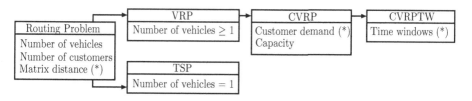

Fig. 1. Tree hierarchy of features for Rounting Problems. The features marked with an asterisk correspond to the statistics: minimum, average, maximum, standard deviation, and median

At this stage, the MLS is not able to provide a recommendation, since the TSP features do not match those extracted for the VRP. To solve it, MLS ascends in the hierarchy until detecting a problem for which it is available, (i.e. Routing Problem) and makes the recommendation based on its features.

4   Computational Results

In this experiment, we proceed to evaluate our solution approach to solve the cold start problem appearing when we proceed to start the algorithm selection process using input instance problems of the CVRPTW, but varying the problem with which the system is trained: (i) 37 TSP instances [18], (ii) 37 CVRP instances [19], and (iii) 56 CVRPTW instances [20].

In Tables 1 and 2 the Spearman's rank correlation coefficient ρ for three trained MLSs is depicted. The column 'Instance' show the CVRPTW instance identifier, while columns CVRPTW, CVRP, and TSP reflect the ρ value when MLS is trained with each problem. Thus, in columns CVRPTW there is no cold start problem. In this regard, the computational results report that the system can accurately predict rankings of metaheuristics, although the system has not been trained with instances of the same problem as the input instance. This information is summarized in Fig. 2, where three sets of columns are depicted. Each set of columns corresponds to a trained MLS for a single problem. For example, the 2nd set of columns is a trained MLS with CVRP instances. On the y-axis, the Spearman's rank correlation coefficient ρ is represented.

In the first set of columns, there is no cold start problem, since the system has been trained with CVRPTW instances. Therefore, this is our reference set. Note that the system is able to build the ideal ranking according to the previous data collected for similar problems, but it is not always possible although the system has been trained for the same problem. On the average case, ρ equals 0.91 which indicates a high correlation between the recommended and the ideal ranking. Finally, in the worst case, a relatively high correlation greater than zero (no correlation) is obtained, ρ equals 0.6.

In the second and third set, the system has been trained with CVRP and TSP instances, respectively. These two sets provide similar results, so they are set out together. Firstly, note that it is possible to obtain the ideal ranking even though

Table 1. Spearman's rank correlation coefficient ρ for three trained MLS and instances for CVRPTW with 25 and 50 customers.

25 customers				50 customers			
Instance	MLS_{CVRPTW}	MLS_{CVRP}	MLS_{TSP}	Instance	MLS_{CVRPTW}	MLS_{CVRP}	MLS_{TSP}
C101	0.94	0.60	0.66	C103	0.89	0.94	0.94
C105	1.00	0.66	0.77	C107	0.94	0.77	0.77
C109	0.77	0.66	0.26	C202	0.77	0.26	0.22
C204	0.66	0.83	0.83	C206	0.94	0.77	0.77
C208	0.77	0.37	0.37	R102	0.94	0.94	0.89
R104	0.83	0.60	0.37	R106	0.94	0.83	0.77
R108	0.94	0.83	0.66	R110	1.00	0.77	0.66
R112	0.94	0.77	0.60	R202	1.00	0.66	0.66
R204	0.94	0.94	0.94	R206	0.83	0.89	0.89
R208	0.94	0.94	0.94	R210	1.00	0.89	0.89
RC101	0.69	0.89	0.31	RC103	0.94	0.89	0.49
RC106	1.00	0.54	0.66	RC107	1.00	0.60	0.49
RC202	0.77	1.00	0.83	RC204	0.60	0.49	0.60
RC206	0.66	0.54	0.49	RC208	0.94	0.66	0.49
Average	0.85	0.73	0.62	Average	0.90	0.74	0.68

Table 2. Spearman's rank correlation coefficient ρ for three trained MLS and instances for CVRPTW with 100 and 200 customers.

100 customers				200 customers			
Instance	MLS_{CVRPTW}	MLS_{CVRP}	MLS_{TSP}	Instance	MLS_{CVRPTW}	MLS_{CVRP}	MLS_{TSP}
C103	0.94	0.94	0.94	C1_210	0.89	0.54	0.66
C107	0.94	0.60	0.66	C1_2_6	0.94	0.60	0.77
C202	0.94	0.89	0.89	C1_2_8	0.89	0.60	0.77
C206	0.94	0.77	0.77	C2_2_2	0.94	0.71	0.89
R102	0.94	0.89	0.77	C2_2_6	0.89	0.60	0.66
R106	0.94	0.83	0.77	R1_210	1.00	0.60	0.66
R110	0.94	0.83	0.77	R1_2_5	1.00	0.60	0.66
R202	0.89	0.89	0.89	R1_2_9	1.00	0.60	0.66
R206	0.89	0.66	0.66	R2_2_3	1.00	0.77	0.94
R210	0.94	0.77	0.77	R2_2_7	0.94	0.83	0.83
RC103	0.94	0.83	0.94	RC1_2_1	0.94	0.60	0.66
RC107	0.94	0.60	0.77	RC1_2_5	1.00	0.54	0.60
RC204	0.89	1.00	1.00	RC1_2_9	1.00	0.54	0.60
RC208	0.94	0.77	0.77	RC2_2_3	1.00	0.83	0.83
Average	0.93	0.80	0.81	Average	0.96	0.64	0.72

the system has not been explicitly trained to solve such problems, see Maximum columns. In addition, on the average case a high correlation is obtained, 0.73 and 0.71 when the system has been trained with CVRP and TSP instances, respectively, compared to the 0.91 obtained in the first set. This indicates a

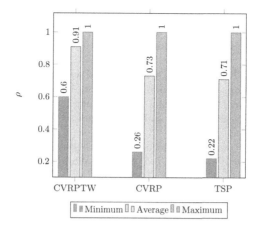

Fig. 2. Spearman's rank correlation coefficient ρ for three trained MLS.

strong relationship between the performance of the algorithms for problems of the same family where, as can be observed, there is a considerable decrease in the ρ value. Note how the values of the average and worst case decrease as the training problem differs from the input problem.

5   Conclusions

A novel way to select, from a pool of algorithms, a suitable algorithm for solving a given logistic problem has been proposed. Moreover, a new way for addressing the cold start is provided. It allows training the system with general problems and making recommendations for specific problems. We have investigated meta-features for TSP, CVRP, CVRPTW to be applied to the meta-learning process. Finally, we have shown that meta-learning models can accurately predict rankings of metaheuristics.

Additional future work includes extensively studying our approach on other well-known combinatorial problems such as those coming from the family of knapsack problems.

Acknowledgements. This work has been partially funded by the European Regional Development Fund, the Spanish Ministry of Economy and Competitiveness (project TIN2015-70226-R).

References

1. Rice, J.R.: The algorithm selection problem. Adv. Comput. **15**, 65–118 (1976)
2. Wolpert, D.H., Macready, W.G., et al.: No free lunch theorems for search. Technical report SFI-TR-95-02-010, Santa Fe Institute (1995)

3. Kodratoff, Y., Sleeman, D., Uszynski, M., Causse, K., Craw, S.: Building a machine learning toolbox. In: Steels, L., Lepape, B. (eds.) Enhancing the Knowledge-Engineering Process - Contributions from Esprit, pp. 81–108. Elsevier publishing company (1992)

4. Michie, D., Spiegelhalter, D.J., Taylor, C.C.: Machine Learning, Neural and Statistical Classification. Ellis Horwood, Chichester (1994)

5. Brazdil, P.B., Soares, C., Da Costa, J.P.: Ranking learning algorithms: using IBL and meta-learning on accuracy and time results. Mach. Learn. **50**(3), 251–277 (2003)

6. Smith-Miles, K.: Towards insightful algorithm selection for optimisation using meta-learning concepts. In: WCCI 2008: IEEE World Congress on Computational Intelligence, pp. 4118–4124. IEEE (2008)

7. Pappa, G.L., Ochoa, G., Hyde, M.R., Freitas, A.A., Woodward, J., Swan, J.: Contrasting meta-learning and hyper-heuristic research: the role of evolutionary algorithms. Genet. Program. Evolvable Mach. **15**(1), 3–35 (2014)

8. Schein, A.I., Popescul, A., Ungar, L.H., Pennock, D.M.: Methods and metrics for cold-start recommendations. In: Proceedings of the 25th Annual International ACM SIGIR Conference on Research and Development in Information Retrieval, pp. 253–260. ACM (2002)

9. Brazdil, P., Carrier, C.G., Soares, C., Vilalta, R.: Metalearning: Applications to Data Mining. Springer Science & Business Media, Heidelberg (2008). https://doi.org/10.1007/978-3-540-73263-1

10. Gutin, G., Punnen, A.P. (eds.): The Traveling Salesman Problem and its Variations, vol. 12. Springer Science & Business Media, Heidelberg (2006). https://doi.org/10.1007/b101971

11. Ralphs, T.K., Kopman, L., Pulleyblank, W.R., Trotter, L.E.: On the capacitated vehicle routing problem. Math. Program. **94**(2), 343–359 (2003)

12. Cordeau, J.-F., Groupe d'études et de recherche en analyse des décisions (Montréal, Québec): The VRP with Time Windows. Groupe d'études et de recherche en analyse des décisions, Montréal (2000)

13. Feo, T., Resende, M.: Greedy randomized adaptive search procedures. J. Glob. Optim. **6**(2), 109–133 (1995)

14. Kirkpatrick, S., Gelatt, C.D., Vecchi, M.P., et al.: Optimization by simulated annealing. Science **220**(4598), 671–680 (1983)

15. Pisinger, D., Ropke, S.: Large neighborhood search. In: Gendreau, M., Potvin, J.Y. (eds.) Handbook of Metaheuristics. International Series in Operations Research & Management Science, vol. 146, pp. 399–419. Springer, Boston (2010). https://doi.org/10.1007/978-1-4419-1665-5_13

16. Cover, T., Hart, P.: Nearest neighbor pattern classification. IEEE Trans. Inf. Theory **13**(1), 21–27 (1967)

17. Dwork, C., Kumar, R., Naor, M., Sivakumar, D.: Rank aggregation methods for the web. In: Proceedings of the 10th International Conference on World Wide Web, pp. 613–622. ACM (2001)

18. Reinelt, G.: TSPLIB - a traveling salesman problem library. ORSA J. Comput. **3**(4), 376–384 (1991)

19. Augerat, P., Belenguer, J.M., Benavent, E., Corberán, A., Naddef, D., Rinaldi, G.: Computational results with a branch-and-cut code for the capacitated vehicle routing problem (1998)

20. Solomon, M.M.: Algorithms for the vehicle routing and scheduling problems with time window constraints. Oper. Res. **35**(2), 254–265 (1987)

Analysing a Hybrid Model-Based Evolutionary Algorithm for a Hard Grouping Problem

Sebastian Raggl[1(✉)], Andreas Beham[1,2], Stefan Wagner[1],
and Michael Affenzeller[1,2]

[1] Heuristic and Evolutionary Algorithms Laboratory,
University of Applied Sciences Upper Austria, Hagenberg,
Softwarepark 11, 4232 Hagenberg, Austria
sebastian.raggl@fh-hagenberg.at
[2] Institute for Formal Models and Verification, Johannes Kepler University Linz,
Altenberger Straße 69, 4040 Linz, Austria

Abstract. We present a new hybrid model-based algorithm called Memetic Path Relinking (MemPR). MemPR incorporates ideas of memetic, evolutionary, model-based algorithms and path relinking. It uses different operators that compete to fill a small population of high quality solutions. We present a new hard grouping problem derived from a real world transport lot building problem. In order to better understand the algorithm as well as the problem we analyse the impact of the different operators on solution quality and which operators perform best at which stage of optimisation. Finally we compare MemPR to other state-of-the-art algorithms and find that MemPR outperforms them on real-world problem instances.

Keywords: Hybrid algorithm · Memetic algorithm
Estimation of distribution algorithm · Grouping problem

1 Introduction

Many modern optimisation algorithms are actually hybrids between different techniques. Memetic algorithms are the combination of population-based and trajectory based optimisation techniques. They have attracted a significant amount of research interest in recent years [1]. Memetic algorithms are especially successful at solving combinatorial optimisation problems [2]. Estimation of distribution algorithms are another class of algorithms that have been studied extensively [3]. They work by iteratively learning a statistical model of good solutions followed by the sampling of new solutions from that model. Path relinking (PR) was originally employed as an intensification strategy for tabu search [4]. It was also successfully used to explore paths between elite solutions in GRASP and scatter search [5,6].

The no free lunch theorem states that no algorithm can outperform all others on every problem [7]. The main idea behind the algorithm presented in this paper

© Springer International Publishing AG 2018
R. Moreno-Díaz et al. (Eds.): EUROCAST 2017, Part I, LNCS 10671, pp. 347–354, 2018.
https://doi.org/10.1007/978-3-319-74718-7_42

is that by incorporating all of the previously mentioned approaches we can get a very robust algorithm that can tackle a wide variety of problem instances with different characteristics. An algorithm that treats problems as a black box and can solve them reasonable well is very valuable, especially when dealing with real-world problems.

1.1 Real-World Problems

When optimising real-world problems it is very likely that the problem definition, i.e. constraints, objective, is going to be adapted over time. When presenting optimisation results of real world problems to practitioners it is very important that the presented solutions are at least local optima. Otherwise even very good results might be dismissed when a human can find, inspired by the presented solution, a very similar but better solution.

The rest of this article is structured as follows. In Sect. 2 we describe the real-world transport lot building problem we want to solve. In Sect. 3 we describe our new memetic path relinking algorithm. In Sect. 4 we analyse how much the different operators contribute to the population and how this contributions change over time. Additionally we investigate the influence of the operators on the quality of solutions found. We compare our algorithm against four different algorithms on real-world problem instances in Sect. 5. Finally, there is a short discussion and outlook.

2 Problem

The problem is concerned with grouping items into transport lots while respecting restrictions on which items can be transported together as well as dependencies between items. There is a set of n items $\mathcal{I} = \{i_1, ..., i_n\}$ that need to be grouped together into an ordered set of groups $\mathcal{S}$. The groups have a maximum size of N. The relationships between the items are described by an undirected weighted graph $G_p = (\mathcal{I}, \mathcal{R})$. The vertices are items and the weight of the edges are the costs of putting two items into the same group. There is a directed graph $G_d = (\mathcal{I}, \mathcal{D})$ with the items as vertices and edges describing dependencies between items. If an item a depends on another item b, item a must be either in the same group as b, or in a group with a higher index in $\mathcal{S}$. Such a dependency is modelled as an edge in the dependency graph $(a, b) \in D$.

$$\min |S| + \sum_{s \in \mathcal{S}} C(s) \tag{1}$$

$$s.t. \ (a, b) \in R \qquad \forall_{s \in \mathcal{S}} \ \forall_{a \in s, \ b \in s} \tag{2}$$

$$S(a) \leq S(b) \qquad \forall_{(a,b) \in \mathcal{D}} \tag{3}$$

$$|s| \leq N \qquad \forall_{s \in \mathcal{S}} \tag{4}$$

We want to find the assignment to the smallest number of groups with the lowest total cost. The function $C(\mathcal{S}) \rightarrow \mathbb{R}$ calculates the costs of a group using the weights on G_p. If an edge in G_p has a high weight placing the two items connected by that edge in the same group is unfavourable. Two items that do not share an edge in G_p must not be in the same group which is ensured by constraint 2. Constraint 3 guarantees that every item that depends on another one is either in the same group or in a group with a bigger index than the item it depends on. The function $S(\mathcal{I}) \rightarrow \mathbb{Z}$ maps an item to the index of the group it belongs to. The order of groups that do not depend on each other is irrelevant. A good way to check constraint 3 is to translate the item dependency graph into a group dependency graph for a given grouping and then check that this graph is acyclic. If this is the case, any topological ordering of the groups in the group dependency graph fulfils the constraint.

In order to be able to use single objective optimisation algorithms we model the constraints as soft constraints by adding a penalty for every constraint that is violated.

3   Memetic Path Relinking

Memetic Path Relinking (MemPR) is a new hybrid model-based algorithm based on the observations made above. The MemPR framework can be applied to a wide variety of problems thanks to the three variants using different solution encodings. The first variant optimises binary vectors, the second operates on permutations and the third can solve grouping problems by using the linear linkage encoding [12]. In this paper we are interested in the latter variant.

MemPR is designed to be used for black box optimisation and in order to be able to tackle problems with wildly different characteristics it does not rely on a single set of operators but instead uses six competing approaches for generating new individuals. MemPR uses the following operators:

- *Creation heuristic*
  MemPR can incorporate arbitrary creation heuristics but in the black box optimisation use case random solutions are sampled from the entire solution space.
- *Breeding*
  The breeding operator randomly selects two parents from the population and produces N children using any of the available crossover operators, where N is the number of dimensions the problem has. The best individual among these children is further improved by local search and reported as the result of the operator.
- *Path-relinking*
  This operator searches a path in the solution space between two randomly selected solutions and reports the best solution found on the path [4].
- *Path-delinking*
  The delinking procedure works the same as relinking but instead of a path starting from one solution and getting increasingly similar to a target solution it searches a path that leads away from the target solution.

- *Sampling from a bivariate model*
  The sampling operator builds a model of the probability of two items belonging to the same group using the current population. This model can subsequently be used to sample new solutions. Similarly to the breeding operator multiple samples are produced and the best one is returned.
- *Hill climbing*
  Different types of local search can be used, but we found that first improvement local search is well suited because it uses fewer evaluations.
- *Adaptive walk*
  A local search that may also accept steps that reduce quality is used to escape local optima.

Algorithm 1 shows how MemPR combines the operators described above. The population initially only consists of two individuals and can grow up to a predetermined size, usually twenty individuals. Those first two individuals are constructed by the creation heuristics and then improved using local search. In every iteration the breed, sample, relink and delink operators create new individuals and try to insert them into the population. Only if they are all unsuccessful is an adaptive walk used, to try to escape one of the local optima the population is comprised of.

Algorithm 1. Memetic Path Relinking

$pop \leftarrow$ [HillClimb(Create()), HillClimb(Create())]
while not termination criteria reached **do**
    $improved \leftarrow$ Replace(pop, Breed())
    $improved \leftarrow$ Replace(pop, Relink()) **or** $improved$
    $improved \leftarrow$ Replace(pop, Delink()) **or** $improved$
    $sample \leftarrow$ Sample(pop)
    $improved \leftarrow$ Replace(pop, $sample$) **or** $improved$
    if not $improved$ **and not** Replace(pop, HillClimb(sample)) **then**
        Replace(pop, AdaptiveWalk())
    end if
end while
return best(pop)

Due to the small population size it is very important to maintain enough diversity so that the population does not prematurely converge. The replacement procedure is a central piece of MemPR in that regard. It aims to keep the population as diverse as possible by using the following rules. As long as the population has not reached its maximum size new individuals are simply added to the population. If the maximum population size is reached, and the new solution is on a plateau, meaning there are two ore more solutions with the same fitness, the solution to be replaced is chosen such that the average distance within the plateau is maximised. If the new individual is not on a plateau, it replaces the member of the population that has lower or equal fitness and is most similar to the new one. If no suitable candidate could be found replacement fails.

4   Analysing the Operators

Given that we have multiple operators that all produce individuals competing to be part of a small population. It is interesting to compare the success rates of the operators. In all our tests we use real-world problem instances with 53, 54, 64, 85, 101, and 145 items.

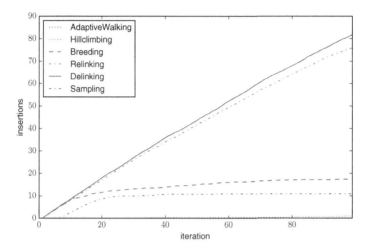

Fig. 1. Average number of individuals successfully inserted per operator into the population in the first 100 iterations

Figure 1 shows how often the different operators on average manage to create a solution that is accepted into the population. The delinking operator has the highest acceptance rate. This is explained by the fact that the replacement procedure favours a diverse population. Relinking has a slightly lower success rate. Breeding and sampling initially are as successful but fall behind as the optimisation run progresses. Both these operators essentially try to combine the groups found in individuals in the population into new individuals. Because of the nature of the problem a solution produced this way has a high chance of violating one of the constraints described in Sect. 2. The hill climber that is applied to the best individual has a chance to fix the violations. Unfortunately there is no guarantee that another solution would not be a better starting point for the hill climber. The sampling operator is only applied once the population has reached its full capacity which is why there are no insertions in the first few iterations. Adaptive walking and hill climbing are only applied when none of the other operators were managed to add a new individual to the population and therefore are only active in later stages of an optimisation run.

In order to asses the impact the operators have on the solution quality we experiment with disabling different combinations of operators. The algorithm was executed ten times for every configuration tested with a time limit of ten

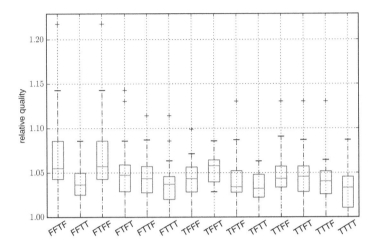

Fig. 2. Best relative quality achieved with different operator combinations

minutes per run. Figure 2 shows the best quality found for every combination relative to the best known quality. Four letter strings are used to signify which operator is enabled. The first letter is T, for true, if breeding is enabled and F, for false, if it is not. The second letter stands for delinking, the third for relinking and the last one for sampling.

The variant that uses only the sampling operator (FFFT) is excluded from the graph because it performs much worse. Between the variants only using a single operator the one using breeding performs best. Which is interesting considering that we showed earlier that the re/delinking are more likely to produce solutions that are integrated into the population. There are clearly interactions between the different operators, for example both relinking and sampling alone are not particularly good, but when combined into the FFTT variant they are quite competitive. On the other hand adding sampling to the breeding only variant decreases solution quality. The variants using all operators (TTTT) and the one only disabling delinking (TFTT) have similar and very good average solution qualities. The variant using delinking does however have the best median quality, because delinking helps preserve population diversity.

5   Algorithm Comparison

We compare MemPR with four algorithms available in the HeuristicLab optimisation environment [8] on real-world problem instances of various sizes. The algorithms we compare against are Variable Neighbourhood Search (VNS) [11], Evolution Strategy (ES), Offspring Selection Genetic Algorithm (OSGA) [10] and Age Layered Population Structure Genetic algorithm (ALPS) [9]. All algorithms use linear linkage encoding and randomly choose between greedy partition crossover, group crossover, lowest index max crossover and lowest index first crossover [12,13]. MemPR and VNS both need a definition of neighbourhood and the both use the same definition in our tests. The neighbourhood of a solution are all solutions reached by applying any move out of four different types of moves. The first two types are extracting an item from a group and either inserted into an existing or a new group. The remaining moves either split a group into two smaller groups, or merge two groups into one.

Table 1. Best quality found by different algorithms

Items	MemPR		ALPS		ES		OSGA		VNS	
	avg	std	avg	std	avg	std	avg	std	avg	std
53	15.00	0.00	15.00	0.00	16.10	0.00	16.80	2.25	19.80	92.25
54	**15.90**	0.32	16.00	0.00	16.00	0.32	16.00	0.63	20.00	2.39
64	**16.48**	0.33	16.51	0.10	17.06	0.11	16.67	0.39	17.19	0.29
85	**21.70**	0.29	22.40	0.82	24.30	0.13	22.46	0.34	27.67	2.35
101	**23.10**	0.32	24.30	0.48	34.10	31.94	28.20	2.35	95.10	53.33
145	**35.30**	0.48	36.20	0.42	39.50	0.71	52.70	32.10	100.10	34.49

Table 1 shows the average best quality reached in ten repetitions with an evaluation budget of 10^7. The parameters of the algorithms were tuned by hand. On the smallest instances ALPS and ES achieves comparable solutions, but MemPR consistently outperforms the other algorithms on the bigger instances. While the differences between MemPR and ALPS are numerically not very big, in the real world application saving a single transport is significant. The three population based evolutionary algorithms manage find better solutions than VNS. ALPS and OSGA perform roughly the same on the first four instances, on the bigger instances OSGA struggles with premature convergence while ALPS, which is designed to avoid this problem, does not.

6   Conclusion

We present a new memetic algorithm that incorporates crossover operators and local search with path-relinking and sampling from a model. From both the comparison with other algorithms as well as the analysis of the operators we conclude that combinations of the different operators performs better than every individual operator. We further observed that preserving diversity during the search is vital for finding good solutions for our real world grouping problem. The delinking operator together with the replacement strategy manage to preserve diversity. This leads to good performance compared to other algorithms tested.

Acknowledgements. The work described in this paper was done within the COMET Project Heuristic Optimization in Production and Logistics (HOPL), #843532 funded by the Austrian Research Promotion Agency (FFG).

References

1. Amaya, J.E., Cotta Porras, C., Fernández Leiva, A.J.: Memetic and hybrid evolutionary algorithms. In: Kacprzyk, J., Pedrycz, W. (eds.) Springer Handbook of Computational Intelligence, pp. 1047–1060. Springer, Heidelberg (2015). https://doi.org/10.1007/978-3-662-43505-2_52

2. Merz, P.: Memetic algorithms and fitness landscapes in combinatorial optimization. In: Neri, F., Cotta, C., Moscato, P. (eds.) Handbook of Memetic Algorithms. Studies in Computational Intelligence, vol. 379, pp. 95–119. Springer, Heidelberg (2012). https://doi.org/10.1007/978-3-642-23247-3_7

3. Pelikan, M., Hauschild, M.W., Lobo, F.G.: Estimation of distribution algorithms. In: Kacprzyk, J., Pedrycz, W. (eds.) Springer Handbook of Computational Intelligence, pp. 899–928. Springer, Heidelberg (2015). https://doi.org/10.1007/978-3-662-43505-2_45

4. Glover, F., Laguna, M.: Tabu Search. Kluwer Academic Publishers, Norwell (1997)

5. Glover, F., Laguna, M., Martí, R.: Fundamentals of scatter search and path relinking. Control cybern. 29, 653–684 (2000)

6. Resende, M.G.C., Ribeiro, C.C.: Greedy randomized adaptive search procedures. In: Glover, F., Kochenberger, G.A. (eds.) Handbook of Metaheuristics. International Series in Operations Research & Management Science, vol. 57, pp. 219–249. Springer, Boston (2003). https://doi.org/10.1007/0-306-48056-5_8

7. Wolpert, D.H., Macready, W.G.: No free lunch theorems for optimization. IEEE Trans. Evol. Comput. 1, 67–82 (1997)

8. Wagner, S., Kronberger, G., Beham, A., Kommenda, M., Scheibenpflug, A., Pitzer, E., Vonolfen, S., Kofler, M., Winkler, S., Dorfer, V., Affenzeller, M.: Architecture and design of the heuristiclab optimization environment. In: Klempous, R., Nikodem, J., Jacak, W., Chaczko, Z. (eds.) Advanced Methods and Applications in Computational Intelligence. Topics in Intelligent Engineering and Informatics, vol. 6, pp. 197–261. Springer, Heidelberg (2014). https://doi.org/10.1007/978-3-319-01436-4_10

9. Hornby, G.S.: ALPS: the age-layered population structure for reducing the problem of premature convergence. In: Proceedings of the 8th Annual Conference on Genetic and Evolutionary Computation, GECCO 2006, pp. 815–822. ACM, New York (2006)

10. Affenzeller, M., Wagner, S.: Offspring selection: a new self-adaptive selection scheme for genetic algorithms. In: Ribeiro, B., Albrecht, R.F., Dobnikar, A., Pearson, D.W., Steele, N.C. (eds.) Adaptive and Natural Computing Algorithms, pp. 218–221. Springer, Vienna (2005). https://doi.org/10.1007/3-211-27389-1_52

11. Mladenović, N., Hansen, P.: Variable neighborhood search. Comput. Oper. Res. 24, 1097–1100 (1997)

12. Ülker, Ö., Özcan, E., Korkmaz, E.E.: Linear linkage encoding in grouping problems: applications on graph coloring and timetabling. In: Burke, E.K., Rudová, H. (eds.) PATAT 2006. LNCS, vol. 3867, pp. 347–363. Springer, Heidelberg (2007). https://doi.org/10.1007/978-3-540-77345-0_22

13. Korkmaz, E.E.: Multi-objective genetic algorithms for grouping problems. Appl. Intell. 33, 179–192 (2010)

Solving a Weighted Set Covering Problem for Improving Algorithms for Cutting Stock Problems with Setup Costs by Solution Merging

Benedikt Klocker[✉] and Günther R. Raidl

Institute of Computer Graphics and Algorithms, TU Wien, Vienna, Austria
{klocker,raidl}@ac.tuwien.ac.at

Abstract. Many practical applications of the cutting stock problem (CSP) have additional costs for setting up machine configurations. In this paper we describe a post-processing method which can improve solutions in general, but works especially well if additional setup costs are considered. We formalize a general cutting stock problem and a solution merging problem which can be used as a post-processing step. To solve the solution merging problem we propose an integer linear programming (ILP) model, a greedy approach, a PILOT method and a beam search. We apply the approaches to different real-world problems and compare their results. They show that in up to 50% of the instances the post-processing could improve the previous best solution.

Keywords: Cutting stock problem · Discrete optimization · PILOT
Beam search · Solution merging

1 Introduction

There are many different kinds of cutting stock problems (CSPs) occurring in practice and in theory. They have in common that they ask for a set of patterns, where each pattern is a collection of elements, to satisfy given element demands while minimizing the total costs of the patterns. The classical CSP only considers fixed costs for each individual pattern, but in many practical applications additional setup costs arise whenever the machine has to be set up to cut a different pattern. In such cases finding a solution involving a small number of different types of patterns is often crucial.

Assume we already have a method which solves a given CSP and generates and collects many different patterns during the execution. We formalize the *Cutting Stock Set Cover Problem* (CSSCP), an extension of the weighted set covering problem which exploits all these collected patterns by deriving an optimal

We thank Lodestar Technology Ges.m.b.H. for the collaboration, the Austrian Research Promotion Agency FFG for funding this project under contract "Innovationsscheck Plus Nr. 855569" and the Vienna Graduate School on Computational Optimization, financed by the Austrian Science Fund (FWF) under grant W1260.

R. Moreno-Díaz et al. (Eds.): EUROCAST 2017, Part I, LNCS 10671, pp. 355–363, 2018.
https://doi.org/10.1007/978-3-319-74718-7_43

combination of a subset of them resembling a feasible, possibly new incumbent solution. Solving this subproblem can be seen as a kind of solution merging. It can be applied either as a post-processing or as an intermediate step to lead the pattern construction in a more promising direction.

The methods we investigate here for solving the CSSCP are more specifically used to improve the solutions found by a previously developed solver for the K-staged two-dimensional cutting stock problem with variable sheet size [1]. The solver gets used to solve real-world problems and therefore we have a strong focus on the practical applicability of the algorithm. There exists a lot of literature on the broad field of cutting stock problems [2]. The idea of using generated patterns to combine them to a good solution was already used by Cui et al. [3] who use an integer linear program (ILP) in a 2-phase-approach for the one-dimensional problem. A theoretical analysis of a general weighted set covering problem is done in [4], where no concept like setup costs were considered.

2   Problem Formulation

The goal of this section is to formalize CSSCP in a general manner so that it can be used in the context of different cutting stock problems, including different dimensions. Therefore, we first need to formalize a general setting for the cutting stock problem, which we call the *General Cutting Stock Problem* (GCSP).

Definition 1 (General Cutting Stock Problem (GCSP)).

Let $E = \{1, \ldots, n\}$ be a set of elements, $(d_i)_{i=1}^n \in \mathbb{N}^n$ a demand vector and $s^{\max} \in \mathbb{N} \cup \{\infty\}$ the maximal stack size. Further, let T be a set of stock materials and $a_t^{\max} \in \mathbb{N} \cup \{\infty\}$ the maximal amount for each stock material $t \in T$.

A solution is represented by a multiset of patterns, where the structure of patterns is problem-specific. We can associate with each pattern p an element vector $(e_i^p)_{i=1}^n \in \mathbb{N}^n$ which describes how often the element i is contained in the pattern p and a stock material $t_p \in T$ out of which it gets cut. A pattern p has associated problem specific production costs c_p^P and stacking costs c_p^S. We define a solution s as a set of feasible patterns P^s and an amounts vector $(a_p^s)_{p \in P^s} \in \mathbb{N}^{|P^s|}$. The goal is to find an optimal solution s which satisfies

$$\sum_{p \in P^s} e_i^p \cdot a_p^s \geq d_i \quad \forall i = 1, \ldots, n \tag{1}$$

$$\sum_{p \in P : t_p = t} a_p \leq a_t^{\max} \quad \forall t \in T \tag{2}$$

and minimizes the total costs

$$c(s) := \sum_{p \in P^s} c_p^P \cdot a_p^s + \sum_{p \in P^s} \left\lceil \frac{a_p^s}{s^{\max}} \right\rceil \cdot c_p^S. \tag{3}$$

If $s^{\max} = \infty$ we define $\left\lceil \frac{a_p^s}{s^{\max}} \right\rceil$ equals 1 if $a_p^s > 0$ and 0 if $a_p^s = 0$.

We further consider the problem variant GCSP' in which demands must exactly be satisfied, that means we replace condition (1) by

$$\sum_{p \in P^s} e_i^p \cdot a_p^s = d_i \quad \forall i = 1, \ldots, n. \tag{4}$$

Definition 2 (Cutting Stock Set Cover Problem (CSSCP)).
Let E, $(d_i)_{i=1}^n$, $s^{\max}$, T and $a_t^{\max}$ for $t \in T$ be given as in Definition 1. Furthermore, let P be a given set of feasible patterns (e.g. collected from different heuristic solutions to a GCSP). The CSSCP asks for a solution to the underlying GCSP consisting of patterns in P, i.e. $P^s \subseteq P$ which satisfies the conditions (1) and (2) and minimizes the costs $c(s)$ as defined in (3).

If we replace condition (1) by (4) we call the problem CSSCP'.

In our case the set P for the CSSCP is constructed during a very large neighborhood search by collecting all patterns occurring during the search.

3 Solution Approaches

In this section we present four different approaches to solve the Cutting Stock Set Cover Problem, an integer linear programming formulation, which can solve the problem exactly, a greedy approach, which can find good solutions very fast, a PILOT-approach and a beam search.

3.1 ILP Formulation

We start by modeling the CSSCP as integer linear program. Theoretically it can solve the problem exactly, but in practice the approach does not scale well to large instances. Therefore, if we use a time limit it may produce solutions with large optimality gaps. We use integer variables a_p for the amount of each pattern p and helper variables s_p for the number of stacks of the pattern p.

$$\min_{(a_p)_{p \in P}, (s_p)_{p \in P}} \sum_{p \in P} a_p \cdot c_p^{\mathrm{P}} + s_p \cdot c_p^{\mathrm{S}}$$

$$\text{s.t.} \sum_{p \in P} a_p \cdot e_i^p \geq d_i \qquad\qquad \forall i \in \{1, \ldots, n\} \tag{5}$$

$$\sum_{p \in P : t_p = t} a_p \leq a_t^{\max} \qquad\qquad \forall t \in T \tag{6}$$

$$s_p \cdot s^{\max} \geq a_p \qquad\qquad \forall p \in P \tag{7}$$

$$a_p \in \mathbb{N}, s_p \in \mathbb{N} \qquad\qquad \forall p \in P$$

If we want to solve CSSCP' we replace constraint (5) by

$$\sum_{p \in P} a_p \cdot e_i^p = d_i. \tag{8}$$

The constraints (5) or (8) ensure that the demands get satisfied and the inequalities (6) guarantee that the maximal amounts for each stock material get respected. Furthermore, the constraints (7) couple the s_p variables with the a_p variables by ensuring that there are enough stacks, so that the maximal stack size $s^{\max}$ gets not exceeded.

3.2 Greedy Heuristic

The idea of this greedy construction heuristic is to rate each pattern depending on the current unsatisfied demands and pick the best pattern as the next one in a greedy manner. It is a fast approach, usually resulting in reasonable solutions. To also consider stacking costs we allow to add a pattern with a given amount at once. Thus, we do not only pick a pattern but also an amount for this pattern in a greedy way. As a rating criteria we use the volume of the elements on the pattern whose demand is not yet satisfied divided by the cost of the pattern and the pattern stack.

Formally, we need a volume value $v_i \in \mathbb{R}_+$ for each element i, which represents the difficulty to put an element i on some pattern. For the one dimensional cutting stock problem this may be the length of the element and for the two-dimensional cutting stock problem, as in our specific case, this can be the area of an element. For a given partial solution s, a pattern p and an amount a we define the following rating criteria

$$r^s(p, a) := \frac{\sum_{i=1}^n \min(a \cdot e_i^p, r_i^s) \cdot v_i}{c_p^P \cdot a + c_p^S \left\lceil \frac{a}{s^{\max}} \right\rceil}$$

where the remaining demand r_i^s is defined by

$$r_i^s := \max \left(0, d_i - \sum_{p \in P} e_i^p \cdot a_p^s \right).$$

The complete greedy approach is given by Algorithm 1.

Algorithm 1. Set Cover Greedy Heuristic

$(r_i^s)_{i=1}^n \leftarrow (d_i)_{i=1}^n, \ (a_p^s)_{p \in P} \leftarrow 0$
while $\exists i \in \{1, \ldots, n\} : r_i^s > 0$ **do**
$\quad \begin{vmatrix} (a, p^{\text{best}}) \leftarrow \ \arg\max_{(a,p) \in \mathbb{N} \times P : 0 < a \leq t_p^{\max}} r^s(p, a) \\ a_{p^{\text{best}}}^s \leftarrow a_{p^{\text{best}}}^s + a \\ u_i \leftarrow u_i - e_i^{p^{\text{best}}} \cdot a \quad \forall i = 1, \ldots, n \end{vmatrix}$

To determine a pattern p and an amount a with a maximal value it is enough to check for each pattern p the values a from the following set

$$A := \{s^{\max}\} \cup \left\{ \left\lfloor \frac{r_i^s}{e_i^p} \right\rfloor, \left\lceil \frac{r_i^s}{e_i^p} \right\rceil : i = 1, \ldots, n \right\} \cap \{1, \ldots, s^{\max}\}$$

and search the pair with the maximal value $r^s(p, a)$. To do this we iterate for each pattern p through the set A in a descending order and can stop the iteration through A when the value $r^s(p, a)$ decreases.

If we want to solve the problem CSSCP' we have to restrict the algorithm to only use patterns p and amounts a which do not lead to an overproduction. Note, however, that this restriction leads often to bad solutions or to no feasible solution at all since there are few or no possible patterns left at some point. Therefore, we apply a repairing mechanism instead of restricting the patterns, which is presented in the following section.

3.3 Solution Repairing for CSSCP'

To still be able to produce good results with the greedy heuristic when exact demands need to be satisfied, we modify the problem CSSCP' in the following way. We allow that a solution may contain patterns which are a substructure of a pattern in P. This means they get constructed by removing some elements from a pattern p in P. We call this new problem CSSCP".

If we want to solve this new problem we can allow patterns and amounts that lead to overproduction and then try to remove the overproduced elements. This may involve checking some problem specific constraints to verify that the new pattern is still feasible. If we obtain a new feasible pattern we can use it and continue with the greedy algorithm.

3.4 PILOT Approach

The main idea of PILOT is to evaluate each potential extension of a current partial solution by individually completing the extended solution in a greedy way and using the obtained solution value for the considered extension [5]. The extension with the best rating is then chosen and the whole process iterates until a final complete solution is obtained.

In our case we more specifically realize the PILOT approach as follows. For each $p \in P$ compute the best amount a according to the greedy criterion rating $r^s(p, a)$, then filter the best ℓ patterns p according to the same rating. For each of these patterns we copy the current solution, add the pattern with the corresponding amount, apply the greedy heuristic to complete the copied solution and compute the objective value of the completed solution. Then we select a pattern p that leads to the best complete solution. For solving the problems CSSCP' and CSSCP" we can proceed in the same way as in the greedy heuristic.

3.5   Beam Search

The idea of beam search is to perform a branching tree search in a breath-first manner, but since this would need too much time in general, it limits the number of considered solutions on each level by some constant k[6]. To get from one level to the next one all extensions of the current solutions are considered and again the best k solutions get stored for the next level. Since our extensions strongly depend on the amount value, we have no clear levels in the search tree. If we consider each addition of a pattern regardless of the amount as one level, we would end up with current solutions of completely different sizes. This is bad since solutions closer to the finished solution tend to be harder to extend and therefore the rating usually decreases. This would mean that smaller partial solutions, still further away from the finished solution, are preferred and eliminate possibly better larger partial solutions.

A similar phenomenon occurs if patterns have different production costs, because then patterns with smaller costs could get preferred although maybe the volume/costs ratio is smaller than for a more expensive pattern. To prevent this we use levels based on the costs of a pattern.

Let $c^{\text{unit}} := \min_{p \in P}(c_p^{\text{P}})$ be the cost unit for one level. The level of a partial solution s is defined by $l(s) := \left\lceil \frac{c(s)}{c^{\text{unit}}} \right\rceil$. For a partial solution we define the rating

$$r(s) := \frac{\sum_{i=1}^{n} \max\left(d_i, \sum_{p \in P} e_i^p \cdot a_p^s\right) \cdot v_i}{c(s)}.$$

The complete beam search is sketched in Algorithm 2. For solving the problems CSSCP' and CSSCP" we can proceed in the same way as in the greedy heuristic.

Algorithm 2. Set Cover Beam Search Approach

Add empty solution to storage of level 0, $l \leftarrow 0$, $F \leftarrow \emptyset$;
while $|F| < k$ *and partial solution for a level larger or equal l exists* **do**
   for s *in storage of level l* **do**
      compute best amount a for pattern p according to $r^s(p, a)$; create
      copy s' of s and add p with amount a to s';
      if $r(s')$ *is one of the best k ratings in level $l(s')$* **then**
         | add s' to storage of level $l(s')$;
   increase l;
return best solution in F

---

4   Computational Results

In this section we present results for our algorithms tested with real-world instances for the K-staged two-dimensional cutting stock problem with variable sheet size. For generating the patterns we use the VLNS described in [1].

The algorithms are implemented in C++ and compiled with g++ 4.8.4. For solving the integer linear program we use Gurobi 7.0 [7]. All tests were performed on a single core of an Intel Xeon E5540 processor with 2.53 GHz and 10 GB RAM.

The input instances consist of real-world instances for the K-staged two-dimensional cutting stock problem with variable sheet size together with a collection of feasible patterns found by the VLNS within five minutes runtime for each instance. If more than 5000 patterns were found in this time only the best 5000 patterns were taken. The instances are all real-world instances we got from different users, who are already using the system. It is a broad set of 192 instances of different sizes and different configurations. The parameter ℓ for the PILOT approach and the parameter k for the beam search are both set to 30. We selected 10 representative instances to compare the proposed algorithms.

Table 1. Results for 10 instances for the CSSCP.

nr	demands	VLNS obj.	num patterns	total patterns	ILP obj.	time	Greedy obj.	time	PILOT obj.	time	Beam search obj.	time
1	118	10.98	12	5000	11.50	3600	13.99	1.58	**11.50**	21.89	13.00	19.34
2	648	15.00	21	5000	16.00	3600	18.50	2.15	18.00	24.32	**16.00**	24.07
3	100	19.51	31	120	18.84	74.06	19.35	0.01	**19.01**	0.17	**19.01**	2.23
4	522	29.00	65	5000	30.25	3600	31.85	0.22	29.20	261.05	**29.10**	133.27
5	153	30.00	43	5000	30.00	3600	32.00	2.21	31.00	43.21	**30.50**	38.70
6	79	44.88	61	219	44.89	0.14	44.93	0.05	**44.89**	2.84	44.93	1.90
7	350	48.53	85	5000	48.04	3600	48.89	2.23	48.37	35.19	**48.12**	89.14
8	2000	117.00	190	5000	114.00	3600	130.48	0.29	**117.00**	18.71	119.00	129.91
9	4830	157.00	252	5000	164.50	3600	184.50	2.91	178.00	2145.45	**176.50**	438.10
10	1600	193.50	320	1554	193.00	0.24	211.00	0.07	200.50	6.35	**194.00**	63.92

The results for the CSSCP are shown in Table 1. The *demands*-column shows the total sum of all demands d_i for each instance. The *num patterns*-column shows the number of patterns of the best solution found during the VLNS and the *total patterns*-column shows the total amount of collected patterns $|P|$ during the VLNS. The columns *obj.* contain the objective values and the columns *time* contain the used time in seconds for each algorithm and each instance. As the ILP approach uses much more time its values can be seen as reference values, although for some large instances the heuristics perform better, since the ILP approach has a large remaining optimality gap. The best value of the three heuristic approaches is printed bold for each instance.

The PILOT approach and the beam search outperform the greedy approach but they need much more time compared to the greedy heuristic. The beam search needs in general more time than the PILOT approach although there are some exceptions like in case of instance 9.

Since in practice all users of our algorithm use it to solve the CSSCP" we also want to investigate the results for this problem. Because ILP approach can only solve the CSSCP' and cannot make use of the relaxed conditions of the CSSCP" we omit it from these tests. Table 2 shows the results for the CSSCP" for the same instances as in Table 1. The best values for each instance for the three heuristics VLNS excluded are printed bold.

Table 2. Results for 10 instances for the CSSCP".

		VLNS			Greedy		PILOT		Beam search	
nr	demands	obj.	patterns	coll. patterns	obj.	time	obj.	time	obj.	time
1	118	10.98	12	5000	14.90	1.64	**12.94**	25.19	**12.94**	50.98
2	648	15.00	21	5000	19.43	2.28	**15.99**	20.01	16.97	38.53
3	100	19.51	31	120	19.54	0.01	**19.01**	0.18	**19.01**	2.89
4	522	29.00	65	5000	33.10	1.82	32.10	275.36	**31.00**	181.39
5	153	30.00	43	5000	32.48	0.79	31.98	44.19	**30.98**	47.61
6	79	44.88	61	219	**44.88**	0.02	**44.88**	2.80	**44.88**	2.61
7	350	48.53	85	5000	48.53	0.28	48.70	32.94	**47.96**	99.06
8	2000	117.00	190	5000	130.47	0.34	**118.00**	18.64	119.99	500.66
9	4830	157.00	252	5000	191.94	3.07	182.96	2271.01	**182.95**	501.86
10	1600	193.50	320	1554	210.98	0.07	200.50	6.27	**194.50**	120.38

Also for the CSSCP" the PILOT approach and the beam search outperform the greedy heuristic, although they need much more time. We also see that it is quite hard to improve upon the VLNS, but especially the PILOT approach and the beam search are able to do so in quite a few instances.

If we compare the three heuristics with the solutions from the VLNS over the whole set of 192 instances the greedy heuristic can improve the found solution for 6% of the instances, the PILOT approach can improve 21% of the found solutions and the beam search can improve 32% of the solutions. If we only consider large solutions with at least 100 sheets the greedy can improve 15%, the PILOT approach 54% and the beam search 50% of the solutions.

5   Conclusion and Future Work

In this paper we formulated a variant of the weighted set cover problem, the CSSCP, which solves a cutting stock problem if a set of feasible patterns is already given. It can be used as a second phase after constructing patterns to improve or find a solution for the cutting stock problem as a post processing or as an intermediate step during the pattern construction. We proposed four solution approaches, an exact ILP model and three heuristics, to solve the problem. Furthermore, we compared them with each other by testing them with real-world instances. The tests have shown that the approaches can be used to improve the found solutions in many cases. Future work may be to combine the presented approaches with a construction heuristic that completes a solution when there are no good patterns anymore in the pattern set.

References

1. Dusberger, F., Raidl, G.R.: A scalable approach for the K-staged two-dimensional cutting stock problem with variable sheet size. In: Moreno-Díaz, R., Pichler, F., Quesada-Arencibia, A. (eds.) EUROCAST 2015. LNCS, vol. 9520, pp. 384–392. Springer, Cham (2015). https://doi.org/10.1007/978-3-319-27340-2_48
2. Cheng, C., Feiring, B., Cheng, T.: The cutting stock problem—a survey. Int. J. Prod. Econ. **36**(3), 291–305 (1994)
3. Cui, Y., Zhong, C., Yao, Y.: Pattern-set generation algorithm for the one-dimensional cutting stock problem with setup cost. EJOR **243**(2), 540–546 (2015)
4. Yang, J., Leung, J.Y.T.: A generalization of the weighted set covering problem. Naval Res. Logist. **52**(2), 142–149 (2005)
5. Duin, C., Voß, S.: The Pilot method: a strategy for heuristic repetition with application to the Steiner problem in graphs. Networks **34**(3), 181–191 (1999)
6. Lowerre, B.T.: The HARPY speech recognition system. Ph.D thesis, Carnegie Mellon University, Pittsburgh, PA, USA (1976)
7. Gurobi Optimization Inc: Gurobi optimizer reference manual, version 7.0.1 (2016)

Particle Therapy Patient Scheduling: Time Estimation for Scheduling Sets of Treatments

Johannes Maschler[(⊠)], Martin Riedler, and Günther R. Raidl

Institute of Computer Graphics and Algorithms, TU Wien, Vienna, Austria
{maschler,riedler,raidl}@ac.tuwien.ac.at

Abstract. In the particle therapy patient scheduling problem (PTPSP) cancer therapies consisting of sequences of treatments have to be planned within a planning horizon of several months. In our previous works we approached PTPSP by decomposing it into a day assignment part and a sequencing part. The decomposition makes the problem more manageable, however, both levels are dependent on a large degree. The aim of this work is to provide and a surrogate objective function that quickly predicts the behavior of the sequencing part with reasonable precision, allowing an improved day assignment w.r.t. the original problem.

Keywords: Particle therapy patient scheduling · Time estimation
Bilevel optimization · Surrogate objective function
Iterated greedy metaheuristic

1 Introduction

In classical radiotherapy cancer treatments are provided by linear accelerators that serve a dedicated treatment room exclusively. In contrast, particle therapy uses beams, produced by either cyclotrons or synchrotrons, that can serve up to five treatment rooms in an interleaved way. Several sequential tasks that do not require the beam, like the positioning of the patients, have to be performed in the treatment room before and after each irradiation. Switching the beam between treatment rooms allows an effective utilization of the particle accelerator and increases the throughput of the facility.

In a typical midterm planning scenario a schedule over the next few months for performing therapies, consisting of a sequence of treatments, has to be determined. Midterm planning for classical radiotherapy has already attracted some research starting with the works from Kapamara et al. [4] and Petrovic et al. [7]. In the following a variety of methods has been applied ranging from GRASP [8] and steepest hill climbing methods [3] to MILP approaches [1,2]. Due to the one-to-one correspondence of treatment rooms and accelerators it is sufficient to

We thank EBG MedAustron GmbH, Wiener Neustadt, Austria, for the collaboration on particle therapy patient scheduling and partially funding this work.

© Springer International Publishing AG 2018
R. Moreno-Díaz et al. (Eds.): EUROCAST 2017, Part I, LNCS 10671, pp. 364–372, 2018.
https://doi.org/10.1007/978-3-319-74718-7_44

consider a coarser scheduling scenario in which treatments have to be assigned only to days but do not have to be sequenced within the days.

In a recent work [6] we studied the midterm planning for the particle therapy treatment center MedAustron in Wiener Neustadt, Austria, which offers three treatment rooms. Our approach consisted in decomposing the problem into a day assignment and a sequencing part for each day, which are, however, not independent. We provided a construction heuristic, a GRASP, and an iterated greedy (IG) metaheuristic. Our computational experiments showed that the IG performed the best. In the subsequent work [5] we improved our IG by exchanging the construction operator with one that is able to preserve more sequencing related information from the incumbent solution. This and the application of a new local search method allowed to outperform the previous approaches.

The current work focuses on further improving the day assignment level by using a fast to compute, more accurate surrogate model for estimating the optimal objective values for the sequencing subproblems. In particular, during the day assignment phase the total duration of the schedule produced by the sequencing part has to be estimated for each day. We previously applied a lower bound that, however, yielded avoidable overfull days. Hence, we study here surrogate functions that consider more aspects of the problem at hand. The main goal is to predict the overall objective function contribution of the sequencing part with reasonable precision while being computationally fast enough to be used in our existing approaches.

2 Particle Therapy Patient Scheduling Problem

In the particle therapy patient scheduling problem (PTPSP) a set of therapies T has to be scheduled on consecutive days D considering a set of resources R.

Each therapy $t \in T$ consists of set of daily treatments (DTs) $U_t = \{1, \ldots, \tau_t\}$. A DT comprises all tasks required to provide an irradiation and has a window of days at which it can be applied. There is a minimal and maximal number of DTs that have to be provided each week, a lower and upper bound of days that are allowed to pass between two subsequent DTs, and a required break of at least two consecutive days each week. Each DT $u \in U_t$ has a processing time $p_{t,u} \geq 0$ and a set of required resources $Q_{t,u} \subseteq R$. In the execution of a DT each resource $r \in Q_{t,u}$ is required during part of the whole processing time specified by the time interval $[P^{\text{start}}_{t,u,r}, P^{\text{end}}_{t,u,r}) \subseteq [0, p_{t,u})$.

The days in the planning horizon D are partitioned into weeks. A subset $D' \subseteq D$ are working days (usually Mondays till Fridays) on which DTs can be performed. For each day $d \in D'$ we have a fundamental opening time $[\widetilde{W}^{\text{start}}_d, \widetilde{W}^{\text{end}}_d)$ that limits the availability of all resources on the considered day.

Resources $r \in R$ have on each working day a regular availability period $[W^{\text{start}}_{r,d}, W^{\text{end}}_{r,d})$ and an extended availability period $[W^{\text{end}}_{r,d}, \widetilde{W}^{\text{end}}_{r,d}]$ at which they can be used, where the usage of the latter one results in additional costs. Moreover, the availability of resource r may be interrupted on day d by a set of unavailability intervals $\bigcup_{w=1,\ldots,\omega_{r,d}} [\overline{W}^{\text{start}}_{r,d,w}, \overline{W}^{\text{end}}_{r,d,w})$.

A solution for the PTPSP is a tuple (Z, S), where $Z = \{Z_{t,u} \in D' : t \in T, u \in U_t\}$ are the days and $S = \{S_{t,u} \geq 0 : t \in T, u \in U_t\}$ are the times at which the DTs are planned. A solution is feasible if all resource availabilities, and all operational constraints are respected. We aim at minimizing the use of extended availability periods while finishing each therapy as early as possible. More formally, the objective is to minimize

$$\gamma^{\text{ext}} \sum_{r \in R} \sum_{d \in D'} \eta_{r,d} + \gamma^{\text{finish}} \sum_{t \in T} \left(Z_{t,\tau_t} - Z_{t,\tau_t}^{\text{earliest}} \right), \tag{1}$$

where γ^{ext} and γ^{finish} are scalar weights, $\eta_{r,d} = \max(\{S_{t,u} + P_{t,u,r}^{\text{end}} - W_{r,d}^{\text{end}} \mid t \in T, u \in U_t, r \in Q_{t,u}, Z_{t,u} = d\} \cup \{0\})$ is the used time of the extended availability period of resource r on day d, and $Z_{t,\tau_t}^{\text{earliest}}$ is a lower bound on the earliest possible finishing day for the last DT of therapy t (see [6]).

3   Solution Approach and Time Estimation

The PTPSP naturally decomposes into the day assignment (DA) level in which DTs are assigned to days and the time assignment (TA) level that consists of finding starting times for the DTs. In other words, Z are the first level and S are the second level decision variables. Clearly, those two levels are dependent on a large degree. Nevertheless, this problem decomposition is beneficial because we can separate the detailed resource model from the remaining operational constraints. Thus, in the TA level each day becomes independent and can be solved separately.

A central aspect of the DA is to find a well-paired allocation of DTs to days that causes as little use of extended availability periods as possible. Since determining $\eta_{r,d}$ requires the exact starting times, the usage of the resources' availability periods for a given candidate set of DTs has to be estimated. Thus, the DA uses a modified version of (1) that replaces $\eta_{r,d}$ with the surrogate $\widehat{\eta}_{r,d} = \max(0, \widehat{\lambda}_{r,d} - h_{r,d})$, where $\widehat{\lambda}_{r,d}$ estimates the required time and $h_{r,d}$ denotes the aggregated regular availability of resource r on day d. In our previous works [5,6] we used for $\widehat{\lambda}_{r,d}$ the trivial lower bound given by aggregated resource demands $\sum_{(t,u):t \in T, u \in U_t, r \in Q_{t,u}, Z_{t,u}=d}(P_{t,u,r}^{\text{end}} - P_{t,u,r}^{\text{start}})$. Consequently, the DA frequently underestimated the resource consumption, which resulted in avoidable use of extended availability periods in the TA. In this work, we aim at more accurately estimating $\widehat{\lambda}_{r,d}$ for the main bottleneck resources, the beam and the rooms.

4   Estimating the Makespan Under Complete Resource Availability

In the following we first concentrate on estimating the makespan required for a given non-empty set G of DTs under the assumption that all required resources are available without any further restrictions. We start by determining estimations of the makespan for three special cases. Afterwards, an estimation for the general case is derived that is based on the estimation for these special cases. Let $n_r = |\{(t,u) \in G \mid r \in Q_{t,u}\}|$ be the number of DTs requiring resource $r \in \{1,2,3,B\}$, where 1, 2, 3 represent the rooms and B the beam, respectively. Furthermore, let

$$\bar{P}_r = \begin{cases} \frac{\sum_{(t,u) \in G}(P^{\mathrm{end}}_{t,u,r} - P^{\mathrm{start}}_{t,u,r})}{n_r} & \text{if } n_r > 0 \\ 0 & \text{else} \end{cases} \tag{2}$$

be the average time resource $r \in \{1,2,3,B\}$ is required by DTs in G. Moreover, let P^{irb} and P^{ira} be the minimum durations a room is required before and after the beam resource, respectively, and let P^{orb} and P^{ora} be the minimum times required by any DT before and after the usage of the room resource, respectively.

In the first special case we assume that all DTs in G require the same room. Hence, w.r.t. to the room resource all DTs have to be scheduled in a strictly sequential way. The beam will have substantial breaks. In this case the makespan can be estimated using the total time the respective room resource is required and some constant offset for the tasks outside of the room, i.e., by

$$\max\{\bar{P}_1 n_1, \bar{P}_2 n_2, \bar{P}_3 n_3\} + P^{\mathrm{orb}} + P^{\mathrm{ora}}. \tag{3}$$

Observe that because only one room is used exactly one term of the maximum function in (3) is greater than zero.

The second special case supposes that the DTs are provided in two rooms. DTs will be scheduled alternatingly between the two rooms. It can be assumed that the tasks apart the irradiation take in general longer than the irradiation itself. Consequently, there will be frequently breaks on the beam resource. In most cases, the makespan will be determined by the utilization of the room that is required the most. An estimation of the makespan for this special case is given by (3) again. In contrast to the previous scenario two terms of the maximum function are greater than zero.

The third special case assumes that the DTs are distributed evenly among the three treatment rooms. In such situations the rooms will be used in an interleaved way s.t. the beam cycles between all three rooms. In this way, the beam will typically be used most efficiently and it can be expected that the beam is used without idle time. The makespan can be estimated by the total time the beam resource is used plus a constant offset for the first and last scheduled DTs:

$$\bar{P}_B n_B + P^{\mathrm{irb}} + P^{\mathrm{ira}} + P^{\mathrm{orb}} + P^{\mathrm{ora}}. \tag{4}$$

In practice we will mostly have a mixture of the three discussed cases. A lower bound for the makespan can be derived by combining (3) and (4):

$$MS^{\mathrm{LB}} = \max\{\bar{P}_1 n_1, \bar{P}_2 n_2, \bar{P}_3 n_3, \bar{P}_B n_B + P^{\mathrm{irb}} + P^{\mathrm{ira}}\} + P^{\mathrm{orb}} + P^{\mathrm{ora}}. \tag{5}$$

Equation (5) is a lower bound for the makespan since P^{orb}, P^{ora}, P^{irb}, and P^{ira} are the minimum durations that have to precede and follow the first and last use of the respective resources and the fact that the total resource requirement is a trivial lower bound. Basically, MS^{LB} assumes that there is a schedule without idle time on the resource that is used the most. Let $n_{\max} = \max_{r \in \{1,2,3\}} n_r$ and $n_{\min} = \min_{r \in \{1,2,3\}} n_r$. We can expect MS^{LB} to be a tight estimate if either $n_{\max} \geq n_B - n_{\max} - 1$, i.e., one room clearly dominates, or $n_{\max} \leq n_{\min} + 1$, i.e., the DTs are evenly distributed among the three rooms.

To strengthen the estimation also for cases in-between, we consider the simplified scenario in which all DTs have exactly the same timing and resource requirements, except that they are distributed among the three rooms. A good schedule would certainly cycle between all three rooms, but not to an extent that remaining DTs have to be scheduled sequentially in a single room.

Let N_{123} be the maximal number of cycles between the three treatment rooms, such that all remaining DTs can be scheduled alternatingly between two rooms. In such a scenario, the following condition must hold:

$$n_{\max} - N_{123} - 1 = (n_{\min} - N_{123}) + (|G| - n_{\max} - n_{\min} - N_{123}). \tag{6}$$

The intuition of the formula above is to compare the number of DTs that remain in each room after cycling between all three rooms for N_{123} times. Note that the minus one represents the fact that the schedule might start and end with the room that is required the most. Equation (6) yields $N_{123} = |G| - 2n_{\max} + 1$. After excluding the corner cases where N_{123} becomes negative or larger than $n_{\min}$ we obtain

$$N_{123} = \min(n_{\min}, \max(0, |G| - 2n_{\max} + 1)). \tag{7}$$

We can now strengthen the estimation of the makespan by using for N_{123} cycles between the rooms the estimation for the third special case and for the remaining DTs the estimation for the second special case as follows:

$$\begin{aligned}
MS^{ES} = \max\{ &\bar{P}_B n_B + P^{irb} + P^{ira}, \\
&\bar{P}_1 n_1, \bar{P}_2 n_2, \bar{P}_3 n_3, \\
&3\bar{P}_B N_{123} + \bar{P}_1(n_1 - N_{123}), \\
&3\bar{P}_B N_{123} + \bar{P}_2(n_2 - N_{123}), \\
&3\bar{P}_B N_{123} + \bar{P}_3(n_3 - N_{123})\} + P^{orb} + P^{ora}. \tag{8}
\end{aligned}$$

Notice that MS^{ES} is in contrast to MS^{LB} not a lower bound anymore.

5   Application of the Time Estimation in PTPSP

In this section the ideas developed in Sect. 4 will be used to obtain enhanced estimations for the total times the beam and each room is required. To this end, for a considered day $d \in D'$ let G be the set of all DTs assigned to day d. Since the beam and the rooms are normally available the whole day, we can assume that they have in general the same regular availability periods.

The total time the beam resource is required can be estimated almost analogously to (8) with the only difference that we have to disregard the time after the last DT has stopped using the beam. Thus, in the estimation, given by

$$
\begin{aligned}
\widehat{\lambda}_{B,d} = \max\{ & \bar{P}_B n_B + P^{irb}, \\
& \bar{P}_1 n_1 - P^{ira}, \bar{P}_2 n_2 - P^{ira}, \bar{P}_3 n_3 - P^{ira}, \\
& 3\bar{P}_B N_{123} + \max(P^{irb}, \bar{P}_1 \cdot (n_1 - N_{123}) - P^{ira}), \\
& 3\bar{P}_B N_{123} + \max(P^{irb}, \bar{P}_2 \cdot (n_2 - N_{123}) - P^{ira}), \\
& 3\bar{P}_B N_{123} + \max(P^{irb}, \bar{P}_3 \cdot (n_3 - N_{123}) - P^{ira})\} + P^{orb},
\end{aligned} \tag{9}
$$

we have to subtract P^{ira} whenever the room resources are used for the estimation.
The total time the rooms are needed is estimated by

$$
\widehat{\lambda}_{r \in \{1,2,3\},d} = \max\{ T_r n_r, \\
3\bar{P}_B N_{123} + \max(T_r(n_r - N_{123}), P^{irb} + P^{ira})\} + P^{orb}. \tag{10}
$$

In contrast to the beam resource we can only use the considered room for the prediction. We can strengthen the estimation for the room resource that is used the most, i.e., for $r_{max} = arg\,max_{r \in \{1,2,3\}} n_r$. This room is most likely the last one used and, hence, it is used at least as much as the beam resource. The strengthened estimation for room resource r_{max} is then given by

$$
\widehat{\lambda}^*_{r_{max},d} = \max\{\widehat{\lambda}_{r_{max},d}, \bar{P}_B n_B + P^{irb} + P^{ira}\}. \tag{11}
$$

6   Computational Study

In this section we study the performance impact of applying the presented time estimation within our so far best performing approach, the enhanced iterated greedy (EIG) from [5]. Moreover, we determine the accuracy of the surrogate functions on final solutions.

All experiments are applied on the benchmark instances from [5] which resemble expected situations at MedAustron. The instances consider $100, 150, 200,$ and 300 therapies, which have to start within windows of 14 days. The beam and the three rooms are regularly available from $\widetilde{W}_d^{start}$ for 14 h and have an extended availability period of 10 h. Besides the beam and the rooms, there are further resources, such as the personnel, which are, however, sufficiently dimensioned to be not the primary reasons of substantial use of extended service time. A characteristic of the instances is that there is a ramp-up phase until the facility is used at full capacity followed by a wind-down phase until the last therapy is finished. At full capacity and strongly depending on the specific DTs there can be planned around 60 DTs.

Table 1 compares the performance between the EIG, as presented in [5], with the variant of the EIG where in the DA the time required from the beam and room resources phase is estimated by (9), (10), and (11). Both algorithms use as

Table 1. Average objective values *obj* and average use of extended service periods in hours *ext*[*h*] of 30 runs with a time limit of 20 CPU-minutes and corresponding standard deviations $\sigma(obj)$ and $\sigma(ext)$ for EIG and EIG with time estimation.

Instance	EIG				EIG+TE			
	obj	$\sigma(obj)$	*ext*[*h*]	$\sigma(ext)$	*obj*	$\sigma(obj)$	*ext*[*h*]	$\sigma(ext)$
100-01	11.558	0.976	5.483	0.976	**9.046**	0.468	**2.633**	0.468
100-02	15.508	2.372	8.158	2.372	**9.737**	0.774	**1.933**	0.774
100-03	8.488	0.497	3.192	0.497	**6.435**	0.161	**0.883**	0.161
100-04	14.257	1.566	7.117	1.566	**8.884**	0.473	**1.125**	0.473
100-05	13.826	1.755	7.817	1.755	**6.823**	0.235	**0.000**	0.235
150-01	18.916	1.760	7.417	1.760	**14.068**	0.387	**1.733**	0.387
150-02	52.166	4.722	39.500	4.722	**43.950**	3.475	**30.092**	3.475
150-03	32.886	3.757	20.233	3.757	**32.740**	3.390	**20.142**	3.390
150-04	18.620	1.659	6.875	1.659	**12.395**	0.466	**0.033**	0.466
150-05	24.286	3.833	15.483	3.833	**10.628**	0.631	**0.917**	0.631
200-01	48.102	3.935	34.000	3.935	**35.945**	5.275	**17.225**	5.275
200-02	38.085	2.824	21.533	2.824	**35.206**	3.855	**16.442**	3.855
200-03	31.158	3.574	13.075	3.574	**20.454**	0.895	**1.108**	0.895
200-04	30.913	2.576	16.800	2.576	**18.860**	1.324	**2.075**	1.324
200-05	29.846	2.384	17.092	2.384	**19.876**	2.550	**5.600**	2.550
300-01	23.654	2.739	12.067	2.739	**16.429**	1.459	**4.000**	1.459
300-02	61.320	5.063	41.200	5.063	**52.510**	5.979	**27.608**	5.979
300-03	41.415	4.254	25.392	4.254	**23.707**	2.900	**6.350**	2.900
300-04	108.118	7.221	85.608	7.221	**77.244**	4.501	**50.367**	4.501
300-05	18.684	1.634	6.800	1.634	**13.219**	0.586	**0.833**	0.586

termination criterion a time limit of 20 CPU-minutes and are executed on each of the benchmark instances for 30 times. Table 1 shows the mean objective values *obj* and the median use of extended service periods *ext*[*h*] in hours with the corresponding standard deviations $\sigma(obj)$ and $\sigma(ext)$ of finally obtained solutions. The results indicate that the application of the presented estimation considerably reduces the used extended service periods over all benchmark instances. The surrogate functions are, however, not necessarily a lower bound. Thus, we might occasionally overestimate the required time for the bottleneck resources yielding underutilized days. This has in general the consequence that the finishing day of therapies are delayed, which is penalized with the second term of our objective function. This raises the question whether this trade-off is indeed beneficial w.r.t. the objective function (1) using the same weights as in [5,6]. This is indeed the case, since the EIG with the presented time estimation performs on all benchmark instances significantly better than the one without according to a Wilcoxon rank sum test with a significance level of 95%. The performance improvement can be explained by the increased accuracy of $\widehat{\lambda}_{r,d}$. While using the trivial lower bound given by aggregated resource demands, the EIG's DA level underestimates on average the required time from the beam and the most used room by 27.7 and 101.6 min (i.e., by 6.5% and 15.3%) with a standard deviation of 20.2 and 57.2, respectively. With the presented estimation the DA level is on average off by 9.17 min for the beam and by 10.9 min for the most used room (i.e., by 2% and 2.4%) with a standard deviation of 8.6 and 9.6, respectively.

7 Conclusion

In this work, we presented a surrogate model for estimating the total times the bottleneck resources required to optimally schedule sets of DTs. This surrogate model is applied to quickly estimate the use of extended service times at the upper DA level of the PTPSP. We evaluated the effects of the presented surrogate model in the so far best performing algorithm, the EIG from [5]. Results show that on all considered benchmark instances the use of extended service periods as well as the whole objective value can be significantly decreased. This can be explained by the substantial gain in accuracy of the new surrogate model and with it the better adjustment of the two levels.

The focus of this work is on resources that are shared by all DTs and are tightly coupled with the throughput of the facility. There are, however, certain resources, like the anesthetist, that are required by some DTs and are available only for the first half of the working day. The interaction of DTs requiring those resources are not considered so far and sometimes result in the use of extended service times that might be avoidable by further improvements.

References

1. Burke, E.K., Leite-Rocha, P., Petrovic, S.: An integer linear programming model for the radiotherapy treatment scheduling problem. CoRR abs/1103.3391 (2011)
2. Conforti, D., Guerriero, F., Guido, R.: Optimization models for radiotherapy patient scheduling. 4OR 6(3), 263–278 (2008)
3. Kapamara, T., Petrovic, D.: A heuristics and steepest hill climbing method to scheduling radiotherapy patients. In: Proceedings of the International Conference on Operational Research Applied to Health Services (ORAHS). Catholic University of Leuven, Leuven, Belgium (2009)
4. Kapamara, T., Sheibani, K., Haas, O., Petrovic, D., Reeves, C.: A review of scheduling problems in radiotherapy. In: Proceedings of the International Control Systems Engineering Conference (ICSE), pp. 207–211. Coventry University Publishing, Coventry, UK (2006)
5. Maschler, J., Hackl, T., Riedler, M., Raidl, G.R.: An enhanced iterated greedy metaheuristic for the particle therapy patient scheduling problem. In: Proceedings of the 12th Metaheuristics International Conference. Barcelona, Spain (2017)

6. Maschler, J., Riedler, M., Stock, M., Raidl, G.R.: Particle therapy patient scheduling: first heuristic approaches. In: Proceedings of the 11th International Conference on the Practice and Theory of Automated Timetabling, pp. 223–244. Udine, Italy (2016)
7. Petrovic, S., Leung, W., Song, X., Sundar, S.: Algorithms for radiotherapy treatment booking. In: 25th Workshop of the UK Planning and Scheduling Special Interest Group, pp. 105–112. Nottingham, UK (2006)
8. Petrovic, S., Leite-Rocha, P.: Constructive and GRASP approaches to radiotherapy treatment scheduling. In: Advances in Electrical and Electronics Engineering - IAENG Special Edition of the World Congress on Engineering and Computer Science 2008, pp. 192–200. IEEE (2008)

A Multi-encoded Genetic Algorithm Approach to Scheduling Recurring Radiotherapy Treatment Activities with Alternative Resources, Optional Activities, and Time Window Constraints

Petra Vogl$^{(\boxtimes)}$ (ID), Roland Braune (ID), and Karl F. Doerner (ID)

Department of Business Administration, University of Vienna,
Oskar-Morgenstern-Platz 1, 1090 Vienna, Austria
{petra.vogl,roland.braune,karl.doerner}@univie.ac.at
https://plis.univie.ac.at/

Abstract. The radiotherapy patient scheduling problem deals with the assignment of recurring treatment appointments to patients diagnosed with cancer. The appointments must take place at least four times within five consecutive days at approximately the same time. Between daily appointments, optional (imaging) activities that require alternative resources, also must be scheduled. A pertinent goal therefore is minimizing both the idle time of the bottleneck resource (i.e., the particle beam used for the irradiation) and the potential risk of a delayed start. To address this problem, we propose a multi-encoded genetic algorithm. The chromosome contains the assignment of treatments to days for each patient, information on which optional activities to schedule, and the patient sequence for each day. To ensure feasibility during the evolutionary process, we present tailored crossover and mutation operators. We also compare a chronological solution decoding approach and an algorithm that fills idle times between already scheduled activities. The latter approach outperforms chronological scheduling on real-world-inspired problem instances. Furthermore, forcing some of the offspring to improve the parent's fitness (i.e., offspring selection) within the genetic algorithm is beneficial for this problem setting.

Keywords: Genetic algorithm · Multi-encoded chromosome
Radiotherapy scheduling · Optional activities · Recurring activities
Time windows · Stable activity starting time

1 Introduction

Patients diagnosed with cancer may be treated with radiotherapy. Classical photon therapy is available in virtually every large hospital worldwide; ion beam therapy represents an advanced technique offered only by a few specialized centers [6]. The main advantage of this latter type of radiotherapy is that the healthy

© Springer International Publishing AG 2018
R. Moreno-Díaz et al. (Eds.): EUROCAST 2017, Part I, LNCS 10671, pp. 373–382, 2018.
https://doi.org/10.1007/978-3-319-74718-7_45

tissue surrounding the tumorous region is less harmed by irradiation, thereby lowering the chances of the patient developing a secondary tumor. However, because they use a significantly larger particle accelerator, ion beam therapy centers are much more complicated and costly than classical radiotherapy facilities making the efficient usage of the beam resource crucial.

A recently opened, specialized ion beam center in Wiener Neustadt, Austria, offers two different particle types for the radiotherapy treatment: protons and carbon ions. The center features three different rooms, equipped with (1) horizontally directed, (2) vertically and horizontally directed, and (3) 180-degree rotatable particle beams. The particle accelerator delivers the particle beam to one room at a time. Accordingly, the beam represents the bottleneck resource in the irradiation process.

The criticality of this problem has prompted increasing studies of how to schedule recurring radiotherapy treatment activities. Most studies concentrate on daily treatments; only two papers address the pre-treatment phase [2,7]. For scheduling the activities within the treatment phase itself, two main strategies dominate the literature:

1. Assign treatments to days. This approach incorporates an average resource profile and does not schedule exact starting times on each day. Therefore, it requires a second step, namely, patient sequencing per day [9,10].
2. Split the time horizon into time slots of predefined length (e.g., 15 or 30 min) and allocate the treatments to these time slots (i.e., "block scheduling"; [3,4]). This approach allows for the immediate consideration of stable activity times, but also assumes that the treatment duration will be more or less equal to length of the time slot, which is not the case in our specific problem setting.

Maschler et al. [5], working on the comparable problem setting, propose a more detailed scheduling approach to solve the advanced radiotherapy scheduling problem that arises in ion beam facilities, using both a greedy randomized adaptive search procedure (GRASP) and an iterated greedy approach. They incorporate day and time assignment phases, but they do not consider alternative resources or optional activities. In contrast, the vast variation in treatment durations and the strong competition for resources that occurs in our case demands a detailed scheduling approach that can yield exact, "to-the-minute" treatment starting times. Various studies confirm, that population-based metaheuristics deliver promising results for scheduling recurring radiotherapy treatment activities [7,8] leading us to investigate these possibilities in greater depth.

Accordingly, we present a multi-encoded genetic algorithm for the radiotherapy patient scheduling problem (RPSP) for the specific problem that arises at ion beam facilities. In Sect. 2, we offer a detailed problem description, followed by the tailor-made solution algorithm, namely, a multi-encoded genetic algorithm (mecGA) in Sect. 3. Next, in Sect. 4 we present the computational results before concluding in Sect. 5.

2   Detailed Problem Description

Radiation therapy consists of a predefined number of recurring daily treatment appointments (DTs) with fixed duration and resource requirements (e.g., treatment room, particle type, assigned radio-oncologist, etc.). Furthermore, each patient is assigned an earliest and latest starting time for the first treatment activity (i.e., release time and deadline). The irradiation appointments must be scheduled a minimum of four times on five consecutive days, starting from the day of the first daily treatment (FDT). In between the DTs, patients regularly see their assigned radio-oncologist to discuss their condition during weekly control examinations (WCE). Some patients also need imaging appointments (i.e., PET-CT, or positron emission tomography-computer tomography) directly after treatment appointments to confirm the accuracy of the therapy. Figure 1 illustrates an "activity chain" for one specific patient. The chain starts with the FDT and ends with the last daily treatment (LDT).

Fig. 1. Activity chain

The WCE and PET-CT activities can take place after any DT (dashed boxes in Fig. 1). The optimization algorithm must determine, which WCE and PET-CT activities should be scheduled, as well as which ones can be skipped. However, a minimum of one WCE/PET-CT must be performed within every span of five consecutive days. Furthermore, the (optional) activities can be executed using alternative resource sets, such that it may be preferable for the patient's assigned oncologist to perform the WCE, but if that provider is busy, any other oncologist can undertake the examination.

All treatment activities within the activity chain of a patient are tied together by minimum and maximum time-lags. For example, a PET-CT needs to start no later than 15 min after the finish of the DT activity. The DT activities also should take place at approximately the same time on every treatment day during a week, to maximize patients' convenience. Therefore, for each treatment week, we fix a stable activity starting time, creating an even tighter time window for DT activities, such that the treatment starting times may vary only by ±30 min during a given week. The stable times of two consecutive weeks also may differ by a maximum of 4 h. Violating these stable times or the regular time window constraint results in a penalty in the objective function.

Each DT activity consists of three inseparable tasks: (1) set-up time, in which the patient is prepared for the treatment inside the treatment room; (2) irradiation itself, which requires the use of both the beam resource and the room; and (3) tear-down time, during which only the treatment room is occupied. The beam resource can only supply one room at a time. Furthermore, switching from one particle type to the other requires a set-up time of 3 min.

We aim to minimize the idle times of the main resource – namely, the particle accelerator/beam used during the second task of the DT – while simultaneously minimizing the potential penalties arising from time window violations. The objective function is defined by the sum of the operating times of the beam resource for each day (operating time = idle time + active time, where $s_d^{lastbeam}$ is the number of minutes the beam is operated on day d) and the penalties of time window violations (equal to the sum of all differences between the latest possible starting times of activity i (l_i) and the actual starting times (s_i)):

$$minimize \quad (\sum_d s_d^{lastbeam} + \sum_i \max(s_i - l_i, 0)). \tag{1}$$

Each resource has a capacity of exactly one unit, so the problem constitutes a complex job shop scheduling problem with custom constraints, alternative and preferred resource sets, recurring and partly optional activities, time windows, and stable activity starting times.

3   The Multi-encoded Genetic Algorithm Approach

We present a multi-encoded genetic algorithm (mecGA) to solve the RPSP. In this case, an individual chromosome represents various, specific characteristics of the solution, such as the days of treatment and the presence of optional activities after a DT. To calculate the schedule, including the exact starting times of all activities, and to evaluate solution quality, we need a decoding algorithm.

3.1   Solution Encoding and Initial Population

The multi-encoded chromosome consists of (1) a set of binary lists for each patient $p \in \mathcal{P}$ and (2) a list of patient indices (patient sequence). The latter describes the order in which patients should be scheduled on each treatment day. Each patient also is identified by three binary lists. The first indicates whether a treatment is scheduled for the patient on specific day d. The days prior to the release day of the patient are automatically set to 0. Starting from the FDT, a minimum of four DTs must occur within five consecutive days. The sizes of the second and third binary lists reflect the number of DTs to be scheduled for the focal patient; that is, their sizes vary from patient to patient. An entry at the ith position in the WCE (PET-CT) list implies that a WCE (PET-CT) has been added after the ith DT activity. As mentioned in the problem description, we must ensure a minimum of one WCE (PET-CT) within each span of five consecutive days, resulting in at least one entry over four consecutive DTs.

Figure 2 illustrates the encoding scheme for five patients. Here, "nbDTs" denotes the number of DTs to be scheduled for the specific patient. Grey shaded cells indicate the release and deadline dates for the FDT, and the bold frame marks the treatment phase. Out of space considerations, we only present the WCE lists; the PET-CT lists would reveal different allocations but the same sizes.

Fig. 2. Multi-encoded chromosome

The initial population can be developed by applying simple construction rules. That is, for each patient, the day of his or her FDT and subsequent treatment days are assigned randomly. The same applies for the WCE and PET-CT binary lists for each patient. The minimum number of assignments on consecutive days must be met during the construction phase.

Building the patient sequence requires a more sophisticated but greedy method: Starting with a random patient, we continuously add one more patient to the sequence, who minimizes the total idle time of the beam that results from room unavailability (i.e., tear-down time of previous patient and set-up time for the current patient), as well as set-up time due to beam type switches. In the resulting starting solutions, if possible, only in rare cases would two patients requiring the same treatment room be scheduled successively.

3.2 Crossover and Mutation

The multi-encoded chromosome requires specific crossover and mutation operators for the chromosome components to preserve its feasibility during the evolutionary process. The patient sequence resembles a classic mutation encoding, so crossover and mutation operators known from genetic algorithm literature, such as the position-based crossover operator (PBX) and the shift mutation operator, can be applied [11].

However, the binary encodings require tailor-made crossover and mutation operators. An intuitive approach is the patient-wise crossover: For each patient (and each of the binary lists dedicated to the patient), we randomly choose the parent from whom the child inherits an entire DT/WCE/PET-CT binary list. Figure 3 illustrates this crossover operators. Here, the child inherits the DT list for patient 1 from the first parent P1 (binary random variable $r = 0$) and r equals 1 for the DT lists of patients 2 and 3, resulting in the inheritance from the second parent's (P2) chromosome.

This crossover approach leads to limited variety in the binary encodings, because the combinations of the initial population dominate the search. Therefore, we developed an inversion mutation operator, applied to 10% of the descendants in each generation. Here, the DT binary list gets inverted between the FDT and the LDT (treatment phase, bold frame in Fig. 2), and the WCE and PET-CT binary lists are completely reversed during the mutation.

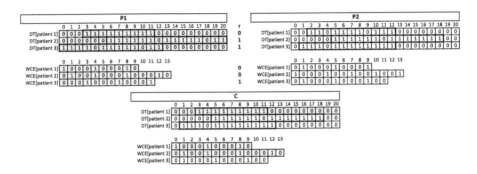

Fig. 3. Patient-wise crossover

3.3 Offspring Selection

We compare two genetic evolution strategies: In a classical genetic algorithm, one generation consists of n individuals, a small share of which survives until the next generation ("elitism"). The rest of the population consists of children created through crossover (and mutation) of two individuals of the parent population. The second strategy, offspring selection (OS), forces part of the new population to be formed by "good" children [1]. A new individual is only selected for the next generation if it is at least as fit as its worse parent (compFactor $= 0.0$). We aim at building 70% of the next generation with such "good" children. However, the number of reproductive steps to achieve this goal is limited to five times the population size in each iteration. As soon as we have reached either 70% "good" children or the maximum number of reproductive steps, we complete the rest of the child population with "bad" children.

3.4 Solution Decoding and Evaluation

To decode a solution to a schedule, we first transform the multi-encoded chromosome to a chronological (i.e., day-wise) prioritized activity list. Then, we can use a decoding algorithm to generate a schedule with exact activity starting times. We designed two decoding algorithms: The first ($d = 0$) schedules activity starting times chronologically, according to the days and patient sequences assigned in the chromosome. The second algorithm ($d = 1$) permits deviations from the predefined patient sequence per day, in the case that availability gaps of the beam resource (i.e., idle times resulting from set-up and tear-down times in rooms) might be reduced by inserting the current activity. We also reduce the penalties invoked by time window violations and increase feasibility by applying repair strategies.

For each activity in the list, we try to find a feasible starting time on the preferred resource set (e.g., the assigned radio-oncologist for the WCE). If no time feasible position is available, the second desired resource set is searched (e.g., any other radio-oncologist on duty). If no time feasible scheduling position exists in any of the possible resource sets, the first belated, time infeasible scheduling position from the preferred resource set is chosen, resulting in a penalty in the objective function.

Finally, after assigning a starting time to all activities, we evaluate the quality of the schedule. As mentioned, deviations from the stable starting time for each patient in each week are part of the penalty term of the objective function. The underlying chronological scheduling approach thus necessitates the formulation of a linear programming model to evaluate stable time violation penalties after the schedule construction has finished, in that the stable time per week is a variable itself, not an input to the model.

4   Results

We test the proposed mecGA on medium and large, real-world-inspired problem instances. The number of patients to be treated varies from 25 to 200. The number of treatments per patient is equally distributed between 8 and 12 for smaller instances (25 to 75 patients) and between 10 and 18 treatments for the larger ones (100 and 200 patients). All patients in the instances with 25 to 75 patients have release times and deadlines within the first week of the planning horizon, but we simulate staggered release times for the instances that include 100 and 200 patients. Therefore 25 patients start the treatment the first week, 25 patients start their treatment in the second week, and so on. (These problem instances are available at https://plis.univie.ac.at/research/test-instances/.)

The population size of the genetic algorithm is set to 100 for the 25 patients instance and to 200 for the larger instances. Accordingly, the optimization time limit increases with the instance size (see Table 1). The best 1% of the population survives until the next generation of the algorithm (elitism), and individuals are chosen for reproduction using the rank selection operator. Furthermore, 10% of the offspring are selected for mutation, using the operators described in Sect. 3.2.

Table 1 lists the averages of 16 replications for all five instances and the corresponding algorithmic settings (i.e., two decoding algorithms, with and without OS). The column "# pat." describes the number of patients to be scheduled, followed by the corresponding number of activities to be scheduled in the second column "# act". With "LB", we denote a naive lower bound that consists of the sum of all activity durations that require the beam resource. This lower bound neglects unavoidable idle times of the beam, due to the set-up and tear-down times in the rooms as well as beam type switches. Therefore, it can give a first but still weak impression of the quality of the solution. In column "d", we compare the two mentioned decoding algorithms ($d = 0$ gives the pure chronological scheduling approach, $d = 1$ can fill gaps). As already mentioned, we compare the classical genetic algorithm (without OS) with a variant including offspring selection. The results of this comparison is in the columns denoted "Without OS" and "With OS." In the "∅fit." columns, we provide the average of best solution fitnesses after the time limit, followed by the average of the corresponding penalty terms in the columns denoted by "∅pen." The bold values represent the best found average solution fitness for the instance size.

Although the chronological decoding algorithm ($d = 0$) performs almost as well as the gap-filling decoding algorithm ($d = 1$) for the smaller instances, permitting the method to fill availability gaps is highly advantageous for larger instances. Employing offspring selection also is beneficial in our problem setting: The best found solution fitness significantly increases for almost all problem instances in both decoding algorithms ("OS impr."). p-values < 0.001 resulting from a two-sided t-test are marked with "***", p-values ≥ 0.05 are marked with "ns" (not significant). Finally, the "gap" column indicates the difference between the best found solution for the instance and the naive lower bound.

Table 1. Real-world inspired instances, 16 replications per instance

# pat.	# act.	LB	d	time	Without OS		With OS		OS impr.	p	gap
					∅fit.	∅pen.	∅fit.	∅pen.			
25	580	2,285	0	1,200	3,162.7	25.9	3,184.8	94.8	0.7%	ns	31.2%
			1	1,200	2,997.6	1.5	**2,997.1**	1.7	0.0%	ns	
50	1,209	4,972	0	2,400	7,756.6	304.1	7,213.7	463.7	−7.0%	***	30.0%
			1	2,400	6,778.4	47.5	**6,463.1**	15.5	−4.7%	***	
75	1,778	7,090	0	3,600	13,888.1	1,123.0	11,169.6	534.1	−19.6%	***	44.4%
			1	3,600	10,799.8	245.8	**10,234.8**	146.0	−5.2%	***	
100	3,392	13,717	0	7,200	26,089.9	1,467.1	22,387.5	788.3	−14.2%	***	43.0%
			1	7,200	20,585.5	907.1	**19,609.6**	556.4	−4.7%	***	
200	6,691	27,838	0	14,400	52,516.6	2,365.9	46,825.3	977.9	−10.8%	***	43.6%
			1	14,400	40,992.8	2,572.8	**39,978.4**	1,915.9	−2.5%	***	

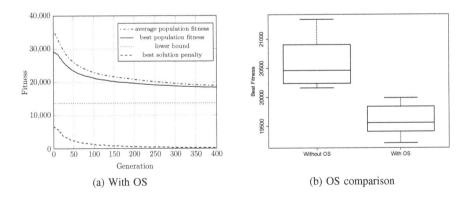

(a) With OS   (b) OS comparison

Fig. 4. Genetic algorithm performance, 100 patients instance, 16 replications, $d = 1$

Figure 4(a) depicts the evolutionary process of the 100 patients instance using the second decoding algorithm ($d = 1$) with the mentioned offspring selection procedure. Both the average population fitness and the best population fitness ameliorate rapidly during the first 100 generations and slightly flatten out after 200 iterations. Figure 4(b) compares the final fitnesses of the 16 replications after 7,200 s of running time with and without OS. The worst case performance with OS still outmatches the best case without OS. Furthermore, OS decreases the variation in solution fitness across the replications.

5   Conclusion and Outlook

We have presented an innovative, mecGA approach for solving the radiotherapy patient scheduling problem with recurring and optional activities, alternative resources, and time window constraints. To preserve feasibility during the evolutionary process, we develop a tailor-made crossover and mutation operator. We also analyze two different decoding algorithms as well as a genetic algorithm variant that includes an offspring selection procedure, such that children who improve on their parent's fitness are favored during the evolutionary process. The computational results demonstrate the superiority of the gap-filling decoding scheme over a chronological scheduling approach. In our specific problem setting, incorporating offspring selection into the genetic algorithm is advantageous for almost all problem sizes.

Further research might investigate more advanced crossover and mutation operators, as well as local-search based techniques. Furthermore, the implicit uncertainty of the problem, due to variances in patient conditions, requires the development of more robust optimization approaches to avoid deviations that might result from disruptions.

References

1. Affenzeller, M., Wagner, S., Winkler, S., Beham, A.: Genetic Algorithms and Genetic Programming: Modern Concepts and Practical Applications. CRC Press, Boca Raton (2009)
2. Castro, E., Petrovic, S.: Combined mathematical programming and heuristics for a radiotherapy pre-treatment scheduling problem. J. Sched. 15, 333–346 (2012)
3. Conforti, D., Guerriero, F., Guido, R.: Optimization models for radiotherapy patient scheduling. 4OR 6, 263–278 (2008)
4. Legrain, A., Fortin, M.-A., Lahrichi, N., Rousseau, L.-M.: Online stochastic optimization of radiotherapy patient scheduling. Health Care Manag. Sci. 18, 110–123 (2015)
5. Maschler, J., Riedler, M., Stock, M., Raidl, G.: Particle therapy patient scheduling: first heuristic approaches. In: PATAT 2016: Proceedings of the 11th International Conference on Practice and Theory of Automated Timetabling, pp. 223–244 (2016)
6. Ohno, T.: Particle radiotherapy with carbon ion beams. EPMA J. 4(1), 9 (2013)
7. Petrovic, S., Castro, E.: A genetic algorithm for radiotherapy pre-treatment scheduling. In: Di Chio, C., Brabazon, A., Di Caro, G.A., Drechsler, R., Farooq, M., Grahl, J., Greenfield, G., Prins, C., Romero, J., Squillero, G., Tarantino, E., Tettamanzi, A.G.B., Urquhart, N., Uyar, A.Ş. (eds.) EvoApplications 2011. LNCS, vol. 6625, pp. 454–463. Springer, Heidelberg (2011). https://doi.org/10.1007/978-3-642-20520-0_46

8. Petrovic, D., Morshed, M., Petrovic, S.: Multi-objective genetic algorithms for scheduling of radiotherapy treatments for categorised cancer patients. Expert Syst. Appl. **38**, 6994–7002 (2011)
9. Petrovic, S., Leite-Rocha, P.: Constructive approaches to radiotherapy scheduling. In: World Congress on Engineering and Computer Science (WCECS), pp. 722–727 (2008)
10. Sauré, A., Patrick, J., Tyldesley, S., Puterman, M.L.: Dynamic multi-appointment patient scheduling for radiation therapy. Eur. J. Oper. Res. **223**, 573–584 (2012)
11. Syswerda, G.: Schedule optimization using genetic algorithms. In: Davis, L. (ed.) Handbook of Genetic Algorithms, pp. 332–349. International Thomson Computer Press, London (1991)

Tabu Search and Solution Space Analyses. The Job Shop Case

Czesław Smutnicki and Wojciech Bożejko[✉]

Faculty of Electronics, Wrocław University of Science and Technology,
Wrocław, Poland
{czeslaw.smutnicki,wojciech.bozejko}@pwr.edu.pl

Abstract. We present some own results of the theoretical and experimental investigations of the landscape of the solution space for the job-shop scheduling problem. Provided interpretation of the space throws new light on the process of solving hard combinatorial optimization problems as well as on Tabu Search phenomenon.

1 Introduction

The job shop scheduling problem is the leading, strongly NP-hard case in the scheduling theory. Unfortunately, the well known real job instance known as ft10 stated in 1963, [3], which has only 10 machines and 100 operations, had to wait 26 years to be solved by the branch-and-bound scheme. In the nineties, Tabu Search (TS) approach from [5] made the breakthrough in thinking about efficient approximate solution methods. TS-type algorithm TSAB, see [8], found optimal solution of ft10 in a few seconds and provided, for instances up to 2,000 operations, solutions of very high accuracy in a time of minutes on a standard PC. TSAB combines basic TS scheme with *block approach* derived from the authors. In successive years there were designed several analogous algorithms for other scheduling problems. Amazing properties of algorithms remained unexamined till now, although a lot of authors use these ideas, [6].

The begin of the job shop research is dated on the middle of fifties, whereas the origin of the problem one can find in numerous industrial applications. Primal works dealt chiefly with various dispatch rules and classes of schedules, illustrated by Gantt chart. Formalism of disjunctive graphs, introduced in the sixties, allowed one to built mathematical model of the problem and to design enumerative algorithms. This convenient disjunctive model remained on the battle ground for many successive years. The primal hopes put in enumerative algorithms and branch-and-bound schemes turn out to be vain. In next two decades, a lot of scientists attacked enumerative schemes trying to improve their power. Other independently developed approaches, as an example mixed integer linear programming and Lagrangian relaxation, appeared uncompetitive. Although finally a few complex branch-and-bound constructions were proposed, their usefulness stops on the level too small for practice, 150–200 operations, [2]. Thus,

R. Moreno-Díaz et al. (Eds.): EUROCAST 2017, Part I, LNCS 10671, pp. 383–391, 2018.
https://doi.org/10.1007/978-3-319-74718-7_46

these research directions slowly falls into oblivion, despite the significant progress made in recent years.

Approximate methods, developed continuously year by year, constitute a reasonable alternative. Made studies improved significantly algorithms efficacy, measured by the accuracy as opposing to the running time. The way of their development has long, rich history, and takes a walk from dispatching priority rules, through the widely studied variants of shifting bottleneck approach, geometric approach, job insertions, local neighborhood search, neural networks, to simulated annealing, constraint satisfaction, tabu search, and evolutionary search. The intensive development of intelligent search methods in recent years allow one to solve instances of practical sizes ≤ 2000, see e.g. ta- benchmarks in [9]. Skipping consciously the list of valuable papers we only refer here to excellent state-of-art works and reviews [1,6].

Capricious behavior of approximate methods, already on standard benchmarks, incline scientists to define and identify factors of the instance hardness, which allow one to differentiate hard cases from easy, [10]. Another research stream examines nature of the solution space and the landscape, seeking properties responsible for the success/failure of an algorithm, [4,10–12], or using them to control the search process.

2 Some Mathematics

There are a set of jobs $N = \{1, \ldots, n\}$, a set of machines $M = \{1, \ldots, m\}$ and a set of operations $O = \{1, \ldots, o\}$. Set O is decomposed into subsets corresponding to the jobs. Job j consists of a sequence of o_j operations indexed consecutively by $(l_{j-1} + 1, \ldots, l_{j-1} + o_j)$ which should be processed in that order, where $l_j = \sum_{i=1}^{j} o_i$, is the total number of operations of the first j jobs, $j = 1, \ldots, n$ ($l_0 = 0$), and $\sum_{i=1}^{n} o_i = o$. Operation i must be processed on machine $\mu_i \in M$ during an uninterrupted processing time $p_i > 0$, $i \in O$. Each machine can process at most one operation at a time. A feasible schedule is defined by start times $S_i \geq 0$, $i \in O$, such that the above constraints are satisfied. The problem is to find a feasible schedule that minimizes the makespan $\max_{i \in O}(S_i + p_i)$.

Till now, various mathematical beings have been proposed to represent the job shop problem (e.g. MILP, disjunctive graphs, list schedules) in the hope that more sophisticated models would imply better solution methods. The choice of the model determines the form of the solution and hence nature, structure, dimensionality and cardinality of the solution space. TS performs an intensive search over some selected *regions* of the solution space, therefore the following special properties of the space are desirable: (a) small dimensionality, (b) small cardinality, (c) small fraction of infeasibility, (d) solution representation favorable for introduction the *move* notion. Unconventional permutation-and-graph model proposed by us in [8] sounded as discord in the traditional job shop research, but met all these postulates. Corresponding solution space, examined next intensively by us in this paper, is introduced as follows.

Set of operations O is decomposed, in natural way, into subsets $M_k = \{i \in O : \mu_i = k\}$, each of them corresponds to operations processed on machine k, $k \in M$. The sequence of such processing can be defined by permutation $\pi_k = (\pi_k(1), \ldots, \pi_k(|M_k|))$ on M_k, $k \in M$. Let Π_k be the set of all permutations on M_k. The *processing order* of all operations (solution) is defined by m-tuple $\pi = (\pi_1, \ldots, \pi_m)$, where $\pi \in \Pi = \Pi_1 \times \Pi_2 \times \ldots \times \Pi_m$ (solution space) The processing order π is connected with schedule S_i, $i \in O$ by the digraph $G(\pi) = (O, R \cup E(\pi))$, with a set of nodes O and a set of arcs $R \cup E(\pi)$, where $R = \bigcup_{j=1}^{r} \bigcup_{i=1}^{o_j - 1} \{(l_{j-1}+i, l_{j-1}+i+1)\}$ and $E(\pi) = \bigcup_{k=1}^{m} \bigcup_{i=1}^{|\pi_k|-1} \{(\pi_k(i), \pi_k(i+1))\}$. Arcs from set R represent the processing order of operations in jobs, whereas arcs from set $E(\pi)$ represent the processing order of operations on machines. Each node $i \in O$ in the digraph has weight p_i and each arc has weight zero. The processing order π is called feasible if graph $G(\pi)$ does not contain a cycle; set of all such processing orders we denote by Π^o. Any $\pi \in \Pi^o$ generates a feasible *pseudo-active* schedule S_i, $i \in O$, so that start time S_i equals the length of the longest path going to the vertex i (but without p_i) in $G(\pi)$, $i \in O$. Makespan $C_{\max}(\pi)$ for the processing order π equals the length of the longest path (critical path) in $G(\pi)$.

Analyzing total effect, we affirm, that in the considered case graph $G(\pi)$ has played a significant role in the final success of TS. It has provided many useful theoretical properties, as an example: relatively simple rule of feasibility checking, guiding the search by the use of *blocks*, reduction of the *neighborhood* size, neighborhood speed up of the search by *accelerator* (a method, that reduces the computational complexity of the single neighborhood search), accelerator for INSA (a method that reduces the computational complexity of a start procedure for TS), [8].

3   Solution Space and Neighborhoods

Solution space Π is discrete, has dimension o, is finite, contains certain number of unfeasible solutions, and its cardinality is of astronomical size. For the mentioned already ft10 ($n = 10$, $m = 10$, $o = nm = 100$), [3], we have $|\Pi| = \prod_{i=1}^{m}(|M_i|!) \approx 4 \cdot 10^{65}$, but $|\Pi^o| \approx 4 \cdot 10^{48}$. The projection Π^o on 2D plane (colored space map), printed in high resolution 2400 dpi, covers $4 \cdot 10^{32}$ km^2, whereas the surface of Jupiter has only $6.4 \cdot 10^{10}$ km^2. "Typical" searching procedure is able to check at most 10^9 (billion) solutions, which corresponds to only 0.1 m^2 in this 2D area.

Unusually small fraction of solutions checked by searching algorithms to the cardinality of Π^o (already of order 10^{-39} in ft10), put together with malicious properties of the solution space (numerous local minima, huge size), incline accidental searcher to skepticism – it is hardly to get high quality makespan skipping simultaneously a significant part of Π^o. Thus, the intelligent algorithm should quickly identify *the most promising search areas*, in order to lead there precise and intensive search. Such features has been recognized in TS and, in our opinion, for the job-shop case just the *neighborhood* is the key element responsible for the proper direction and quick pass towards the promising area in the space.

Π can be perceived as hyper-graph spanned on $|\Pi|$ nodes. Each node represents a solution π and each arc (π, σ) states that solutions σ can be obtained from π using certain method of modification, called hereinafter *move*. In this study we focus on A-moves $v = (x, y)$ which swaps a pair of two adjacent operations x, y processed consecutively on the same machine. All solutions that are reachable through single arc from π constitutes the *hyper-neighborhood* $\mathcal{N}(\pi)$. The sequence of successive solutions $\pi^0, \pi^1, \ldots, \pi^r$, where $\pi^{i+1} \in \mathcal{N}(\pi^i)$, $i = 0, 1, \ldots, r-1$ creates the *trajectory* (path in the hyper-graph) of length r, from π^0 to π^r. $\mathcal{N}(\pi)$ has the *strong connectivity property*, i.e. for each pair of solutions α, β there exist trajectory so that $\alpha = \pi^0, \pi^1, \ldots, \pi^r = \beta$ and r is finite.

Although $\mathcal{N}(\pi)$ offers the greatest possibility of improvements, it has been never used in TS for the sake of the search cost. Moreover, not all $\sigma \in \mathcal{N}(\pi)$ need to be feasible for $\pi \in \Pi^o$. That's why, only certain subset of $\mathcal{N}(\pi)$ is used next and called simply *neighborhood*. There are considered several attributes of neighborhoods: *feasibility*, *size*, *usefulness* and *connectivity*, which influence on trajectories. The rational approach suggest to employ in TS a feasible, wide neighborhood with connectivity. In order to analyze the impact of the neighborhood on TS, we compare three of them. The *basic feasible* $\mathcal{N}^{A0}(\pi)$ contains all feasible solutions generated by any A-move. Neighborhood $\mathcal{N}^{A1}(\pi) \subseteq \mathcal{N}^{A0}(\pi)$, is generated by moves $v = (x, y)$, where x and y belongs to a critical path in $G(\pi)$. It has, similarly as $\mathcal{N}^{A0}(\pi)$, only weak connectivity property (β is not any but one of optimal solutions) and may contain *useless* moves. Neighborhood $\mathcal{N}^{A2}(\pi) \subseteq \mathcal{N}^{A1}(\pi)$, applied in our TS from [8], is obtained by elimination of all useless moves with the help of block property, known also as *interchanges near the borderline of blocks on a single critical path*; neighborhood does not have the connectivity property.

One could expect that just $\mathcal{N}^{A0}(\pi)$ should be recommended for TS since it has connectivity and offers the widest possibility of improvement – nothing more errogenous. In order to get $\mathcal{N}^{A0}(\pi)$ we have to filtrate $\mathcal{N}(\pi)$ with the use of computationally expensive feasibility checking, usually of order $O(o)$ per solution. Moreover, from the job-shop theory it follows that the expensive search on $\mathcal{N}^{A0}(\pi)$ does not provide better results than search on $\mathcal{N}^{A1}(\pi)$; we have $\min_{\alpha \in \mathcal{N}^{A0}(\pi)} C_{\max}(\alpha) = \min_{\alpha \in \mathcal{N}^{A1}(\pi)} C_{\max}(\alpha)$ if $\min_{\alpha \in \mathcal{N}^{A0}(\pi)} C_{\max}(\alpha) < C_{\max}(\pi)$. For these reasons, nobody uses $\mathcal{N}^{A0}(\pi)$ in the job-shop. The choice between the remain two is not easy, although both ensure feasibility for free. We have to select between the absence of connectivity and analogous dominance relation $\min_{\alpha \in \mathcal{N}^{A1}(\pi)} C_{\max}(\alpha) = \min_{\alpha \in \mathcal{N}^{A2}(\pi)} C_{\max}(\alpha)$ if only $\min_{\alpha \in \mathcal{N}^{A1}(\pi)} C_{\max}(\alpha) < C_{\max}(\pi)$.

4   Measures

Distance $d(\alpha, \beta)$ *between processing orders* $\alpha, \beta \in \Pi$ should be defined with the relation to moves, for example as "the minimal number of moves necessary to go from α to β in the hyper-graph". Such definition fits ideally to TS since we have $d(\pi, \sigma) = 1$ for $\sigma \in \mathcal{N}(\pi)$. It has been found, that the minimal number of A moves necessary to pass from permutation α_l to permutation β_l equals the well-known Hamming distance $d_l(\alpha_l, \beta_l)$ between these permutations, $l \in M$. Assuming that permutations are distributed uniformly, the mean value and standard deviation of $d_l(\alpha_l, \beta_l)$ are equal $n(n-1)/4$ and $\sqrt{n(n-1)(2n+5)/72}$ respectively, [7]. Since in our model each solution of the job shop problem is a composition of permutations, it is quite natural to define distance between processing orders α, β in the following form with expected quasi-normal distribution

$$d(\alpha, \beta) = \sum_{l \in M} d_l(\alpha_l, \beta_l). \tag{1}$$

The highest values of $d_l(\alpha_l, \beta_l)$ and $d(\alpha, \beta)$ are equal $d_{\max} = n(n-1)/2$ and $D_{\max} = md_{\max}$, respectively. Their limit values are reachable for any permutation α_l and $\beta_l = (\alpha_l(n), \alpha_l(n-1), \ldots, \alpha_l(1))$, $l \in M$. From (1) it immediately follows, that assuming uniform distribution of processing orders, parameters of the distribution of $d(\alpha, \beta)$, such as the mean value and standard deviation, are equal $mn(n-1)/4 = D_{max}/2$ and $\sqrt{mn(n-1)(2n+5)/72} = \sqrt{D_{max}(2n+5)/36}$, respectively. The complete distribution of $d(\alpha, \beta)$, can be found only if we know values $T(k)$, $k = 0, 1, \ldots, D_{\max}$, where $T(k)$ means the number of processing orders $\beta \in \Pi$ distant from the fixed processing order α by k units, see next Section. In the sequel, value $d(\alpha, \beta)$ will be normalized

$$D(\alpha, \beta) = 100\% \cdot \frac{d(\alpha, \beta)}{D_{\max}}. \tag{2}$$

It is clear that the mean value and standard deviation of $D(\alpha, \beta)$ are equal 50% and $100\% \cdot \sqrt{(2n+5)/(36D_{\max})}$, respectively.

Search procedures are keenly interested in the distribution of the makespan inside Π^o. To this order we use the relative error between the makespan of a solution π and the reference solution π^{REF}, namely

$$RE(\pi, \pi^{REF}) = 100\% \cdot \frac{C_{\max}(\pi) - C_{\max}(\pi^{REF})}{C_{\max}(\pi^{REF})}, \tag{3}$$

where π^{REF} corresponds to the best known up-to-now makespan.

5   Theoretical Distributions

Let f_l be any function mapping set O_l onto set $\{1, 2, \ldots, n\}$. Notice, $d_l(\alpha_l, \beta_l) = d_l(\overline{\alpha}_l, \overline{\beta}_l)$, where $\overline{\alpha}_l = f_l \circ \alpha_l$, $\overline{\beta}_l = f_l \circ \beta_l$ are permutations of elements from of set $\{1, 2, \ldots, n\}$. Then, without the loss of generality, we assume hereinafter that $O_l = \{1, 2, \ldots, n\}$ for any $l \in M$. Hence, value $d_l(\alpha_l, \beta_l) = d(\gamma, \alpha_l^{-1} \circ \beta_l)$, $\gamma = (1, 2, \ldots, n)$, equals the number of inversions in permutation $\alpha_l^{-1} \circ \beta_l$, where α_l^{-1} is the inverse permutation to permutation α_l, $l \in M$. Value $d(\alpha, \beta) = \sum_{l \in M} d_l(\alpha_l, \beta_l) = \sum_{l \in M} d_l(\gamma, \alpha_l^{-1} \circ \beta_l)$, for fixed α, β, can be found in a time $O(mn^2)$.

In order to find the distribution of $d(\alpha, \beta)$ over Π (more precisely values of $T(k)$, $k = 0, 1, \ldots, D_{\max}$) we start from the distribution of $d_l(\gamma, \alpha_l^{-1} \circ \beta_l)$, for a fixed α_l and $l \in M$. Let us denote by $L(s)$ the number of permutations $\delta = \alpha_l^{-1} \circ \beta_l$, $\beta_l \in \Pi_l$, that are distant, in sense of the measure $d_l(\gamma, \delta)$, from the permutation γ by s units, $s = 0, 1, \ldots, d_{\max}$. $L(s)$ depends neither α_l nor the machine index l, moreover $\sum_{s=0}^{d_{\max}} L(s) = |\Pi_l| = n!$. Values $L(s)$ will be found in the way described below.

Let us denote by $H(t, r)$ the number of permutations of the set $\{1, 2, \ldots, t\}$ distant from the permutation $\gamma^t = (1, 2, \ldots, t)$ by r units, $r = 0, 1, \ldots, t(t-1)/2$, $t = 1, 2, \ldots, n$; from the definition we have $L(s) = H(n, s)$. Values $H(t, r)$ can be calculated recursively, see [7],

$$H(t,r) = H(t-1,r) + H(t-1,r-1) + \ldots + H(t-1, \max\{0, r-(t-1)\}) \quad (4)$$

for $r = 0, 1, \ldots, t(t-1)/2$, $t = 2, 3, \ldots, n$, where $H(1, 0) = 1$. (Intuitively, a permutation on the set $\{1, 2, \ldots, t\}$ distant from the permutation γ^t by r units, we obtain by inserting element t into some permutation τ on the set $\{1, 2, \ldots, t-1\}$ namely, after the position $t-1$ if τ is distant from γ^{t-1} by r units, after the position $t-2$ if τ is distant from γ^{t-1} by $r-1$ units, and so on.) It follows from (4) that function L treated as a sequence $(L(0), L(1), \ldots)$ has the following form

$$L = 1^1 \otimes 1^2 \otimes \ldots \otimes 1^n \quad (5)$$

where $1^j = (1^j(0), 1^j(1), 1^j(2), \ldots)$, $1^j(i) = 1$, $0 \le i \le j$ and $1^j(i) = 0$, $j < i$, $j = 1, 2, \ldots, n$; operator $\otimes$ denotes the discrete convolution.

Now, we pass to calculation of $T(k)$, $k = 0, 1, \ldots, D_{\max}$. It is clear that $T(k)$ does not depend on α, we have $\sum_{k=0}^{D_{\max}} T(k) = |\Pi| = (n!)^m$, and

$$T(k) = \sum_{k_1+k_2+\ldots+k_m=k,\ k_l \in \{0,1,\ldots,k\},\ l \in M} L(k_1) \cdot L(k_2) \cdots \ldots \cdots L(k_m). \quad (6)$$

Then, function T, treated as the sequence $(T(0), T(1), \ldots)$, is the m-times con-volution of the sequence L, so $T = \overbrace{L \otimes L \otimes \ldots \otimes L}^{m}$. Hence

$$T = 1^1 \otimes \overbrace{(1^2 \otimes \ldots \otimes 1^2)}^{m} \otimes \overbrace{(1^3 \otimes \ldots \otimes 1^3)}^{m} \otimes \ldots \otimes \overbrace{(1^n \otimes \ldots \otimes 1^n)}^{m}. \quad (7)$$

Using Eq. (7) one can formulate the algorithm of finding the sequence $T(k)$, $k = 0, 1, \ldots, D_{\max}$. This algorithm has the computational complexity $O(m^2 n^3)$, under assumption that substituting and summing operations can be performed in a time $O(1)$.

6   Empirical Distributions

Π can be naturally decomposed as follows: feasible Π^o that corresponds to pseudo-active schedules (PAC) and unfeasible $\Pi \setminus \Pi^o$. Space Π^o contains well-known subclasses of schedules, namely *active* (AC, Π^{AC}), and *non-delay* (ND, $\Pi^{ND} \subseteq \Pi^{AC}$).

First, we studied the empirical distribution of solutions in Π^o using samples of 10,000 solution for each benchmark instance from [9]. The mean distance AvD for PAC solutions insensibly fluctuates around 20% and varies from 17.8% to 22.1%. This means that PAC solutions are concentrated around π^{REF} nearer than all solutions inside Π (AvD for Π is 50%). Standard deviation σD for PAC solutions remains very close to the theoretical deviation σAD of solutions from the whole Π. Although PAC solutions are located relatively close to π^{REF}, they offer makespans of very poor quality, $AvRE$ varies from 92.3% to 136.4%. Summing up, huge Π^o contains in majority poor solutions.

Next, we analysed distribution of solutions of important subclasses in Π^o. Empirically found $AvD = 50\%$ for ALL solutions from Π fits ideally to its theoretical value. For benchmarks ta41–50 pseudo-active solutions with poor mean quality $AvRE = 136.4\%$ are located on the mean distance $AvD = 18.5\%$. Active solutions, considered as elite ones, are located only slightly nearer to π^{REF} ($AvD = 17.1\%$) but are evidently better since offers $AvRE = 40.9\%$. Particularly selected non-delay solutions are slightly better than active $AvRE = 37.4\%$ and are located on slightly mean distance $AvD = 16.7\%$. Moreover, standard deviations for AC and ND solutions are at least twice less than that of PAC. This suggests that AC and ND solutions are distributed similarly around π^{REF} and, due to small $AvRE$ should used in any algorithm.

It is hardly to generate search trajectory going through AC solutions only. One can check, that even though we start from an AC solution π, then solutions from the neighborhood $\mathcal{N}^A(\pi)$, $A \in \{A0, A1, A2\}$ need not be AC. That's why we analyze also statistics of solutions located on random trajectories generated with the help of neighborhoods $\mathcal{N}^{A1}(\pi)$ or $\mathcal{N}^{A2}(\pi)$; obtained in such way solutions will be denoted by RTA1 or RTA2, respectively. Solutions RTA1 and RTA2 were found as follows. For random PAC starting solution were generated trajectory of the length $r = 1,000$. For a solution π^i from the trajectory, the successive solution π^{i+1} was selected random in neighborhood $\mathcal{N}^{A1}(\pi)$ or $\mathcal{N}^{A2}(\pi)$, respectively. The last solution π^r from this trajectory is appended to subclass RTA1 (RTA2). We found that $AvRE = 51.8\%$ for RTA1 is significantly better than $AvRE = 136.4\%$ for PAC, and not too bad comparing it with $AvRE = 40.9\%$ for AC. The mean distance from π^{REF} to RTA1 solutions is noticeable less than that from PAC, and only slightly less than that of AC. Solutions RTA2 own even better parameters; $AvRE = 33.6\%$ is significantly better than $AvRE$ for RTA1.

We perceive solutions from RTA1 as a single crater, covered by thick fur with random length hairs, and optimal solution located close to the center. The fur nature of the space has been confirmed also by a trajectory towards π^{REF}, frequently shown in TS research. Accepting such view of the solution space one proposes ad hoc algorithm "go quickly towards center of the crater and search there". In fact, we neither do not know a priori the way to the crater center nor the scale of the most promising search area. The latter subject is slightly easier, since using $T(k)$ we are able to evaluate labor size. For ta45, the number of solutions in Π distant in terms of Hamming measure from π^{REF} below 1% is equal $4.5 \cdot 10^{110}$, $2\% - 6.5 \cdot 10^{174}$, $3\% - 2.6 \cdot 10^{224}$, $6\% - 1.8 \cdot 10^{325}$. Simultaneously, it has been found that solutions with $AvRE \leq 1\%$ may appear occasionally even on the distance $AvD \approx 5.5\%$ from π^{REF} (Fig. 1).

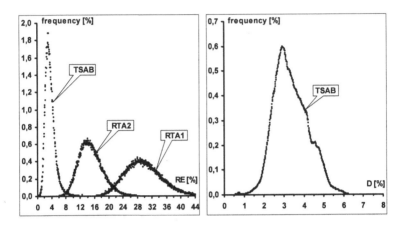

Fig. 1. Benchmark ta45. Distribution of $RE(\pi, \pi^{REF})$ (*left*) and $D(\pi, \pi^{REF})$ (*right*) for solutions π from TSAB trajectory. Distribution of $RE(\pi, \pi^{REF})$ for 100,000 RTA1 and RTA2 solutions π distant from π^{REF} by $D(\pi, \pi^{REF}) \leq 5\%$ (*left*).

References

1. Błażewicz, J., Domschke, W., Pesh, E.: The job shop scheduling problem. conventional and new solution techniques. Eur. J. Oper. Res. **93**, 1–33 (1996)
2. Brucker, P., Jurish, B., Sieviers, B.: A fast branch and bound algorithm for the job shop problem. Discret. Appl. Math. **49**, 107–127 (1994)
3. Fisher, H., Thompson, G.L.: Probabilistic learning combinations of local job-shop scheduling rules. In: Muth, J.F., Thompson, G.L. (eds.) Industrial Scheduling, pp. 225–251. Prentice Hall, Englewood Cliffs (1963)
4. Gao, L., Li, X., Wen, X., Lu, C., Wen, F.: A hybrid algorithm based on a new neighborhood structure evaluation method for job shop scheduling problem. Comput. Ind. Eng. **88**, 417–429 (2015)
5. Glover, F., Laguna, M.: Tabu Search. Kluwer Academic Publishers, Norwell (1997)
6. Jain, A.S., Meeran, S.: Deterministic job-shop scheduling: past, present and future. Eur. J. Oper. Res. **113**, 390–434 (1999)
7. Knuth, D.E.: The Art of Computer Programming. Addison Wesley Longman, Reading (1977)
8. Nowicki, E., Smutnicki, C.: A fast tabu search algorithm for the job-shop problem. Manag. Sci. **42**(6), 797–813 (1996)
9. Taillard, E.: Benchmarks for basic scheduling problems. Eur. J. Oper. Res. **64**, 278–285 (1993)
10. Watson, J.P., Beck, J.C., Howe, A.E., Whitley, L.D.: Problem difficulty for tabu search in job-shop scheduling. Artif. Intell. **143**, 189–217 (2003)

11. Watson, J.P.: An introduction to fitness landscape analysis and cost models for local search. In: Gendreau, M., Potvin, J.Y. (eds.) Handbook of Metaheuristics. International Series in Operations Research and Management Science, vol. 146, pp. 599–623. Springer, Boston (2010). https://doi.org/10.1007/978-1-4419-1665-5_20
12. Wong, L.P., Puan, C.Y., Low, M.Y.H., Chong, C.S.: Bee colony optimization algorithm with big valley landscape exploitation for the job shop problem. In: Proceedings of the 2008 Winter Simulation Conference, pp. 2050–2058 (2008)

Optimization Networks for Integrated Machine Learning

Michael Kommenda[1,2]([⊠]), Johannes Karder[1,2], Andreas Beham[1,2],
Bogdan Burlacu[1,2], Gabriel Kronberger[1], Stefan Wagner[1],
and Michael Affenzeller[1,2]

[1] Heuristic and Evolutionary Algorithms Laboratory,
University of Applied Sciences Upper Austria,
Softwarepark 11, 4232 Hagenberg, Austria
`michael.kommenda@fh-hagenberg.at`
[2] Institute for Formal Models and Verification, Johannes Kepler University Linz,
Altenbergerstr. 69, 4040 Linz, Austria

Abstract. Optimization networks are a new methodology for holistically solving interrelated problems that have been developed with combinatorial optimization problems in mind. In this contribution we revisit the core principles of optimization networks and demonstrate their suitability for solving machine learning problems. We use feature selection in combination with linear model creation as a benchmark application and compare the results of optimization networks to ordinary least squares with optional elastic net regularization. Based on this example we justify the advantages of optimization networks by adapting the network to solve other machine learning problems. Finally, optimization analysis is presented, where optimal input values of a system have to be found to achieve desired output values. Optimization analysis can be divided into three subproblems: model creation to describe the system, model selection to choose the most appropriate one and parameter optimization to obtain the input values. Therefore, optimization networks are an obvious choice for handling optimization analysis tasks.

Keywords: Optimization networks · Machine learning
Feature selection · Optimization analysis

1 Introduction

A general optimization methodology for solving interrelated problems has been recently proposed and analyzed by Beham et al. [2], where the effects of solving integrated knapsack and traveling salesman problems, such as the knapsack constrained profitable tour problem or the orienteering problem, are studied and compared to Lagrangian decomposition.

Building upon these first results, optimization networks, a methodology for solving interrelated optimization problems, have been developed. Optimization

© Springer International Publishing AG 2018
R. Moreno-Díaz et al. (Eds.): EUROCAST 2017, Part I, LNCS 10671, pp. 392–399, 2018.
https://doi.org/10.1007/978-3-319-74718-7_47

networks consist of several nodes that provide (partial) solutions for individual subproblems, which can be aggregated to a complete solution. Hence, optimization networks are applicable if the overall optimization problem is decomposable into individual subproblems or several distinct optimization steps are necessary to obtain a solution. Although this limitation seems restricting, in practice most problems can be decomposed into subproblems. An illustrative example for the benefit of optimization networks is the optimization of the operations within a factory. The optimization of production schedules increase the overall performance. However, if corresponding warehouse and logistics operations are taken into account and optimized as well, the overall performance can be further increased [6].

The core components of optimization networks are nodes and messages that are sent between nodes through ports [7]. A node is responsible for performing well defined calculations, for example solving a certain subproblem or transforming data. Ports define the specific interface for communication and which data is sent and received in a message. A restriction is that nodes can only be connected and communicate if the corresponding ports are compatible to each other. In Fig. 1 a schematic representation of an optimization network is depicted (ports are omitted for clarity). Exchanged messages, indicated by arrows between the nodes, transfer data and steer the execution sequence of the nodes. In this optimization network the nodes termed *Solver 1* and *Solver 2* are executed sequentially, while *Solver 3* and *Solver 4* are executed simultaneously.

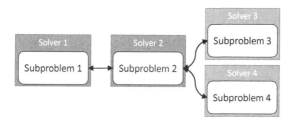

Fig. 1. Schematic representation of an optimization network. Each node represents a solver with an according subproblem.

A major benefit of optimization networks is that they enable the reuse of existing nodes and promote the creation of a library of building blocks for algorithm development. To the best of our knowledge, optimization networks have only been applied to combinatorial optimization problems, mostly in the context of production and logistics. In this publication, we demonstrate the suitability of optimization networks for integrated machine learning problems.

2   Integrated Machine Learning

Machine learning enables computers to learn from and make predictions based on data. Whenever multiple, interrelated and dependent machine learning problems are considered in combination we use the term integrated machine learning. Integrated machine learning problems consist of at least two interrelated sub-problems, hence are a perfect application area for optimization networks.

Especially when performing supervised machine learning such as regression analysis or classification several steps are necessary to achieve the best possible result. These steps include data preprocessing and cleaning, feature construction and feature selection, model selection and validation, and parameter optimization and are dependent on each other. If the data for creating the models has not been cleaned and preprocessed accordingly, even the best machine learning algorithm will have difficulties to produce accurate and well generalizing prediction models.

An approach to solve integrated machine learning problems is the automatic construction of tree-based pipelines [11]. Those pipelines are represented as trees, where every processing step corresponds to a tree node and genetic programming [9] is used for the automatic construction and optimization of those pipelines. This concept is similar to optimization networks due to the reuse and combination of particular building blocks that cooperatively solve the machine learning problem at hand. A difference is that these tree-based pipelines are specific to machine learning, whereas optimization networks are more generic and applicable to diverse optimization problems. However, the versatility of optimization networks comes at a prices, namely that they are created manually by combining already existing building blocks, while machine learning pipelines can be generated automatically.

3   Feature Selection

A common supervised machine learning task is regression analysis, where a model describing the relationship between several independent and one dependent variable is built. Another important task is feature subset selection [5] that determines which of the available independent variables should be used in the model. Both problems are interrelated in the sense that only if an appropriate feature subset is selected, the resulting regression model achieves a good prediction accuracy on training and test data.

An integrated machine learning problem is generating an optimal linear model using the most appropriate features from the data at hand. In total there are 2^n (n... number of features) different linear models possible, ranging from the most simplistic model without any selected feature, which just predicts a constant value, to the most complex one that contains all available features. While this problem can be exhaustively solved for a small number of features, this quickly becomes infeasible due to the exponential growth of possible models.

An optimization network solving the described integrated machine learning problem is shown in Fig. 2. The feature selection node is only dependent on the number of features in the regression problem and selected features are encoded in a binary vector. To assess the quality of selected features this binary vector is sent to the orchestrator via the connection between the evaluation and features port.

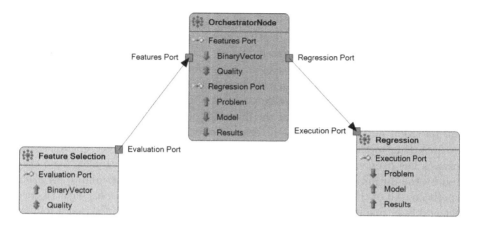

Fig. 2. Implementation of an optimization network combining feature selection with regression analysis. When the orchestrator receives a binary vector encoding the selected features, it translates this information to a regression problem that is forwarded to the regression node, which returns the generated model and its quality.

The orchestrator parses the binary vector and creates a new regression problem, which only contains selected features, before forwarding it to the regression node. Upon receiving a message with the configured problem, the regression node builds a model and returns it together with its prediction accuracy. This information is passed through the orchestrator to the feature selection node and another combination of selected features can be evaluated. An important aspect is that the feature selection and regression are already existing building blocks and only the orchestrator has to be configured to build the optimization network at hand.

The effectiveness of the proposed optimization network is evaluated on an artificial benchmark problem and compared to standard regression methods creating linear models. Therefore, we created a dataset containing 100 features x_i, where each of the 1000 observations is sampled from a standard normal distribution $\mathcal{N}(\mu = 0, \sigma = 1)$. The dependent variable y is calculated as a linear combination of 15 randomly selected features with uniformly sampled $\mathcal{U}(0, 10)$ feature weights w_i and with a normally distributed noise term ϵ added that accounts for 20% of the variance of y.

A linear model created by ordinary least squares (OLS) [4] includes all 100 features and yields a mean absolute error (MAE) of 10.634 on a separate test partition. The optimal linear model including only the 15 necessary inputs has a test MAE of 8.189.

Another method for fitting linear models is the minimization of squared errors in combination with elastic net regularization [13], which balances lasso and ridge regression. A grid search over the parameters λ and p is performed to balance the regularization. The best linear model yields a MAE of 8.369 utilizing 50 features (only five of the 15 selected features are correctly identified). Elastic net executions with higher regularization yield smaller models, but those including fewer than 20 features could not identify any feature correctly and therefore have a rather large MAE of approximately 10.

The optimization network uses an offspring selection genetic algorithm (OSGA) [1] for generating binary vectors representing the currently selected features and OLS for creating the linear models. The quality of selected features is the prediction error in terms of the mean absolute error on a separate valida-tion partition and a regularization term accounting for the number of selected features. Due to the OSGA being a stochastic metaheuristic, the optimization network has been executed 50 times. On average 12414 solutions have been cre-ated by the optimization network, which took slightly over one minute for each repetition. All of the repetitions except one identified linear models with 14 cor-rect and only one missing feature and a MAE of 8.216 on the test partition. A further investigation revealed that the impact of the missing feature on the model evaluation is lower than the noise level due to a very small feature weight, hence it has not been correctly identified. These results on the described benchmark problem exemplarily demonstrate the benefits and suitability of optimization networks for integrated machine learning problems.

4   Further Applications

In the previous sections, optimization networks and integrated machine learning have been introduced and the performance of optimization networks has been demonstrated using a benchmark problem. Although the obtained results are satisfactory, the full potential of optimization networks for machine learning is only reached in combination with a library of preconfigured building blocks (network nodes). With such a library new optimization networks can be easily created by combining existing nodes and only the orchestrators responsible for controlling the execution sequence and data conversion and transfer have to be implemented anew. A conceptual version of the optimization network for feature selection and linear regression is depicted in Fig. 3a. By exchanging the node for model creation (Fig. 3b), the adapted network learns a forest of decision trees [3] instead of linear models.

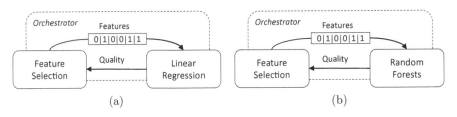

Fig. 3. Optimization networks combining feature selection with either linear regression (a) or random forest regression (b).

Additionally to modifying existing optimization networks by exchanging compatible nodes, reuse of components is further enabled by the inclusion of optimization networks as nodes of other networks. As an example, the right optimization network in Fig. 4 creates a random forest regression model and simultaneously optimizes the parameters of the algorithm, specifically the ratio of training samples R, the ratio of features used for selecting the split M, and the number of trees to obtain the most accurate prediction model. Should this network perform feature selection as well, there is no need to adapt it by including an additional feature selection node and the according orchestrator. A better way is to modify the existing feature selection network, similarly to the changes indicated in Fig. 3, by replacing the linear regression node by the whole regression network. These two applications demonstrate how existing optimization networks can be reused and adapted for solving new problems and therefore highlight the flexibility of optimization networks.

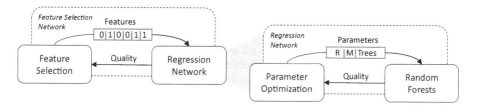

Fig. 4. Reuse of optimization networks is facilitated by their inclusion as nodes in other networks. In this example the outer optimization network performs feature selection while the nested network is responsible for parameter optimization of the regression algorithm and model creation.

Another application for optimization networks is optimization analysis [10], where values for input features have to be obtained so that predefined output values are achieved. In a previous publication [8] the best input parameter combinations of a heat treatment process are obtained so that the quality measures of the final product are optimal within certain limits. One research question has been which regression models best describe the relation between the input and output values and are therefore most suited for process parameter estimation and optimization analysis. The whole process of regression model creation, model selection and judging the suitability for parameter estimation has been performed manually, but can easily be represented and executed by an optimization network.

An optimization network combining data-based modeling with optimization analysis is illustrated in Fig. 5. It consists of three nested networks, each tackling a different subproblem. The first network is responsible for data-based modeling and regression model creation. It includes exemplarily the already introduced feature selection and linear regression network, but in practice different machine learning and regression methods are utilized in parallel for creating different prediction models. The second network creates different model subsets, which are forwarded for optimization analysis to the third network. During the whole optimization process the best performing model subset as well as the optimal input features are tracked and accessible to the user for further analysis. With the support of optimization networks the manual effort for performing optimization analysis can be drastically reduced and the combination of different algorithms (regression modeling, subset selection, and parameter optimization) is facilitated.

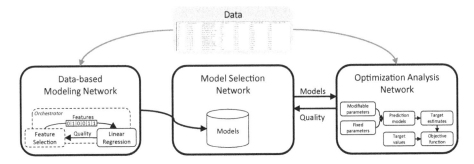

Fig. 5. Schematic representation of an optimization network that contains three nested networks that create regression models (left network), select a subset of all generated models (middle network), and evaluate the suitability for optimization analysis (right network).

5   Conclusion

Optimization networks are a new method for solving integrated machine learning problems consisting of several interrelated subproblems. Especially, the reusability and composition of optimization networks from existing building blocks (either configured nodes or whole networks) is a major benefit and eases problem solving. The performance of optimization networks has been demonstrated on a feature selection problem where linear models have to be created. An implementation of optimization networks is available in HeuristicLab [12], however the number of predefined network nodes is still limited. Therefore, the next step for broadening the scope of optimization networks is to extend and provide a library of building blocks for solving machine learning problems.

Acknowledgments. The authors gratefully acknowledge financial support by the Austrian Research Promotion Agency (FFG) and the Government of Upper Austria within the COMET Project #843532 Heuristic Optimization in Production and Logistics (HOPL).

References

1. Affenzeller, M., Winkler, S., Wagner, S., Beham, A.: Genetic Algorithms and Genetic Programming - Modern Concepts and Practical Applications, Numerical Insights, vol. 6. CRC Press, Chapman & Hall, Boca Raton (2009)

2. Beham, A., Fechter, J., Kommenda, M., Wagner, S., Winkler, S.M., Affenzeller, M.: Optimization strategies for integrated knapsack and traveling salesman problems. In: Moreno-Díaz, R., Pichler, F., Quesada-Arencibia, A. (eds.) EUROCAST 2015. LNCS, vol. 9520, pp. 359–366. Springer, Cham (2015). https://doi.org/10.1007/978-3-319-27340-2_45

3. Breiman, L.: Random forests. Mach. Learn. **45**(1), 5–32 (2001)

4. Draper, N.R., Smith, H., Pownell, E.: Applied Regression Analysis, vol. 3. Wiley, New York (1966)

5. Guyon, I., Elisseeff, A.: An introduction to variable and feature selection. J. Mach. Learn. Res. **3**(Mar), 1157–1182 (2003)

6. Hauder, V.A., Beham, A., Wagner, S.: Integrated performance measurement for optimization networks in smart enterprises. In: Ciuciu, I., Debruyne, C., Panetto, H., Weichhart, G., Bollen, P., Fensel, A., Vidal, M.-E. (eds.) OTM 2016. LNCS, vol. 10034, pp. 26–35. Springer, Cham (2017). https://doi.org/10.1007/978-3-319-55961-2_3

7. Karder, J., Wagner, S., Beham, A., Kommenda, M., Affenzeller, M.: Towards the design and implementation of optimization networks in HeuristicLab. In: GECCO 2017: Proceedings of the Nineteenth International Conference on Genetic and Evolutionary Computation Conference, ACM (2017 accepted for publication)

8. Kommenda, M., Burlacu, B., Holecek, R., Gebeshuber, A., Affenzeller, M.: Heat treatment process parameter estimation using heuristic optimization algorithms. In: Proceedings of the 27th European Modeling and Simulation Symposium EMSS 2015, Bergeggi, Italy, pp. 222–228, September 2015

9. Koza, J.R.: Genetic Programming: On the Programming of Computers by Means of Natural Selection. MIT Press, Cambridge (1992)

10. O'Brien, J.A., Marakas, G.: Introduction to Information Systems. McGraw-Hill Inc., New York (2005)

11. Olson, R.S., Urbanowicz, R.J., Andrews, P.C., Lavender, N.A., Kidd, L.C., Moore, J.H.: Automating biomedical data science through tree-based pipeline optimization. In: Squillero, G., Burelli, P. (eds.) EvoApplications 2016. LNCS, vol. 9597, pp. 123–137. Springer, Cham (2016). https://doi.org/10.1007/978-3-319-31204-0_9

12. Wagner, S., et al.: Architecture and design of the HeuristicLab optimization environment. In: Klempous, R., Nikodem, J., Jacak, W., Chaczko, Z. (eds.) Advanced Methods and Applications in Computational Intelligence. Topics in Intelligent Engineering and Informatics, vol. 6, pp. 197–261. Springer, Heidelberg (2014). https://doi.org/10.1007/978-3-319-01436-4_10

13. Zou, H., Hastie, T.: Regularization and variable selection via the elastic net. J. R. Stat. Soc.: Series B (Stat. Methodol.) **67**(2), 301–320 (2005)

Evaluating Parallel Minibatch Training
for Machine Learning Applications

Stephan Dreiseitl[(✉)]

Department of Software Engineering, FH Upper Austria, 4232 Hagenberg, Austria
stephan.dreiseitl@fh-hagenberg.at

Abstract. The amount of data available for analytics applications continues to rise. At the same time, there are some application areas where security and privacy concerns prevent liberal dissemination of data. Both of these factors motivate the hypothesis that machine learning algorithms may benefit from parallelizing the training process (for large amounts of data) and/or distributing the training process (for sensitive data that cannot be shared).

We investigate this hypothesis by considering two real-world machine learning tasks (logistic regression and sparse autoencoder), and empirically test how a model's performance changes when its parameters are set to the arithmetic means of parameters of models trained on *minibatches*, i.e., horizontally split portions of the data set. We observe that iterating the minibatch training and parameter averaging process for a small number of times results in models with performance only slightly worse that of models trained on the full data sets.

Keywords: Distributed machine learning · Logistic regression
Sparse autoencoders · Minibatch training

1   Introduction

Over the last decade, the coinciding rise of three factors has led to a revolution in data analytics. These factors are the exponential growth of available data, advances in machine learning algorithms, and ubiquitous and cheap computing resources, either via cloud computing or GPU technology [1,2]. This revolution has given rise to systems that rival human performance on auditory and visual perception tasks [3,4]. Although it has brought many benefits for most, from individuals to society as a hole, there are at least two potential obstacles to large-scale data analytics:

– Several of the underlying algorithms of data analytics systems (most notably the optimizers that implement the learning in the systems) are sequential in nature; using these algorithms thus forms an inherent limitation to the speedups possible even with parallel hardware.

© Springer International Publishing AG 2018
R. Moreno-Díaz et al. (Eds.): EUROCAST 2017, Part I, LNCS 10671, pp. 400–407, 2018.
https://doi.org/10.1007/978-3-319-74718-7_48

– Privacy and security concerns in some application areas have prevented large-scale unrestricted distribution of available data. Most notable among these is healthcare, where the sensitive nature of the data involved requires special care and considerations when used in analytics applications.

Tackling these issues turns out to involve the same basic idea: train analytics systems only on portions of the data (either in parallel, or in a distributed manner), and then combine the results obtained on these portions into a larger system. The motivations for this are different for the two obstacles mentioned above, and briefly summarized below:

– While optimization methods traditionally employed in machine learning algorithms operate iteratively, performing several passes over the entire data set, this has become harder as increasing data volumes may not fit into the limited memories of special-purpose hardware, such as GPUs. This problem gave rise to stochastic versions of algorithms (such as stochastic gradient descent), where updates are performed in a online manner for each data point, and to *minibatches*, where small portions of the entire data set are processed sequentially [5].
– In application areas where data cannot be pooled, mostly for security or privacy reasons, *distributed machine learning* is concerned with the sharing of dispersed data during the model building stage [6,7]. Due to the sensitive nature of the data, the information being shared must not allow potential adversaries to reconstruct even parts of the original data. Research into these methods has been active in recent years, with several algorithmic variants being proposed on horizontally and vertically partitioned data [8–12].

It is tempting to view model training on minibatches as a trivial way to parallelize and distribute machine learning algorithms: Train several models, each on their own minibatch (in parallel and/or in a distributed manner), and then somehow combine these models. It is exactly this model combination that is, theoretically, far from trivial, as most parametrizations of trained machine learning models represent minima of error surfaces. It is initially not clear how the positions of several minima, each corresponding to one minibatch model, can be combined in a meaningful manner into a minimum corresponding to a better model of the entire data set.

Recently, substantial progress has been made on this problem, most notably with a proof that minibatch models trained with stochastic gradient descent on convex error surfaces can be combined merely by averaging their parametrizations [13]; subsequent work has expanded on this approach [14,15]. The objective of this paper is extend the idea of averaging parametrizations by iterating the process, and to evaluate its performance also on non-convex optimization problems trained with higher order methods, such as conjugate gradient and quasi-Newton algorithms. The code in Algorithm 1 gives a top-level view of this approach.

We investigate the effect of parameter-averaging parallel minibatch training on two standard machine learning tasks, one representing a convex optimization

---

Algorithm 1. Parallel (distributed) minibatch training

Split data into M minibatches patch$_1$,...,patch$_M$ // omit in distributed setting
Randomly initialize θ (model parameters)
for $k = 1$ to N **do**   // how often each minibatch is processed
     for $j = 1$ to M **do in parallel**
         $\beta_j = \text{trainmodel}(\theta, \text{patch}_j)$ // obtain new parameters β_j
     end parallel for
     $\theta = \text{mean}(\beta_1, \ldots, \beta_M)$ // communication overhead in distributed setting
end for

---

problem relevant to privacy-preserving distributed machine learning in medical contexts (training a logistic regression model, in Sect. 2), and one representing a non-convex optimization problem relevant to large-scale parallel machine learning (training an autoencoder, in Sect. 3). Concluding remarks are given in Sect. 4.

2 Parallel Minibatch Training for Logistic Regression

Logistic regression is one of the most widely used linear classification models in biostatistics and biomedical informatics. In the two-class case for a data set $\{x_1, \ldots, x_D\} \subseteq \mathbb{R}^n$ with corresponding class labels $y = (y_1, \ldots, y_D) \in \{0, 1\}^D$, we first employ the notational convention of prepending a constant 1 to every data vector x_k, so that $\hat{x}_k := (1, x_k^T)^T$ is an $(n+1)$-dimensional column vector. Using the auxiliary function $p(x; \theta) = 1/(1 + e^{-\theta^T \cdot \hat{x}})$, logistic regression calculates a parameter vector $\theta \in \mathbb{R}^{n+1}$ by maximizing the likelihood

$$L(\theta) = P\big(\{y_1, \ldots, y_D\} \,\big|\, \{x_1, \ldots, x_D\}; \theta\big) = \prod_{k=1}^{D} \Big(p(x_k; \theta)^{y_k} \cdot \big(1 - p(x_k; \theta)^{1-y_k}\big) \Big).$$

There is no closed-form solution to this maximization problem. In practice, it is numerically more convenient to iteratively minimize the negative log-likelihood [16]; often, a regularization term is added to prevent overfitting. The combination of $-\log L(\theta)$ with an L_2 regularizer results in the error function

$$\text{err}_{\text{LR}}(\theta) = \sum_{k=1}^{D} \Big(-y_k \big(\theta^T \cdot \hat{x}_k\big) + \log \big(1 + e^{\theta^T \cdot \hat{x}_k}\big) \Big) + \frac{\alpha}{2} \|\theta\|_2^2, \tag{1}$$

where $\alpha \in \mathbb{R}$ determines the contribution of the regularization term to the error function.

The error function in Eq. (1) is convex, meaning that there are no local optima, and a (sufficiently powerful) optimizer will find the global optimum. In the following (and also in Sect. 3), we will consider two second-order methods: *scaled conjugate gradients* (SCG), a variant of conjugate gradient optimization that avoids costly line searches at each iteration [17], and limited-memory BFGS

(L-BFGS), a variant of the Broyden-Fletcher-Goldfarb-Shanno quasi-Newton algorithm that avoids storing an approximation to the Hessian matrix required in the update step [18].

Data set. We evaluated parallel minibatch training on a medical data set to determine the effect of distributed data on learning logistic regression models. The data set of choice consists of 1253 cases with 33 features each, and was collected to predict acute myocardial infarction [19]. The original data set contains six additional ECG features, which we removed to make the task more challenging.

Evaluation metric. As is generally done in biomedical informatics settings, the area under ROC curve (AUC), calculated on the test set, was used as a performance metric for evaluating the discriminatory power of the trained logistic regression model [20].

Hyperparameter settings. We used 2/3 of the data (836 cases) as the training set, and the remaining 417 as test set. In previous experiments on the same data set, we had determined a value of $\alpha = 0.1$ for the regularization parameter. We set the number of training iterations to 100 (after which both SCG and L-BFGS had converged on the full training set).

Parallel minibatch training. We varied both the number (and correspondingly the size) of the minibatches as well as the number of times each minibatch is used. These numbers are represented by the parameters M and N, respectively, in Algorithm 1. Note that N also expresses how often the parameter vector θ is recalculated as the arithmetic mean of the parameters of the models trained on the separate minibatches. For comparability, every case is used exactly 100 times in the training process. Therefore, every call to trainmodel in Algorithm 1 is allowed to use only $100/N$ iteration steps.

Results. The results of varying the parameters M and N on the task of learning a logistic regression model are summarized in Table 1. For comparison, when using the entire training set (without minibatches) for 100 iterations of the optimizers, the AUC value on the test set was 0.860 for both SCG and L-BFGS.

Discussion. Training a logistic regression model on distributed minibatches achieves, for some combinations of parameters M and N, levels comparable to training the model on the full data set. In the experimental setup used in this paper, there is a tradeoff between the number N of (outer) loop iterations, and the number of optimizer steps in each of these loop iterations, because the total number of optimizer steps was set to 100. For example, with $N = 20$, there were 5 optimizer steps on each of the minibatches; after these 5 steps, the parameters of all models were averaged, and this new parameter used as initial parameter for the next loop iteration. In Table 1, one can observe that across all numbers of minibatches, it is beneficial to increase the number of times that parameters are averaged, and the optimizers restarted. In a distributed environment, this would result in increased communication overhead, as every parameter averaging requires synchronizing across the distributed models.

Table 1. AUC on test set for a logistic regression model trained on minibatches according to Algorithm 1, using the SCG (left part) and L-BFGS (right part) optimizers. Note: Larger values are better.

Number M of minibatches	Number N of loop iterations				Number N of loop iterations			
	10	20	30	40	10	20	30	40
2	0.858	0.859	0.859	0.860	0.859	0.859	0.859	0.860
4	0.854	0.857	0.859	0.860	0.854	0.857	0.859	0.861
6	0.842	0.853	0.858	0.859	0.843	0.853	0.857	0.862
8	0.843	0.852	0.857	0.858	0.844	0.857	0.858	0.857
10	0.854	0.859	0.859	0.859	0.855	0.860	0.860	0.858

Furthermore, one can observe that for the (rather small) data set used in this paper, a smaller number of larger minibatches ($M = 2$) works better than a larger number of smaller minibatches ($M = 6$), although performance improves again for an even larger number of minibatches ($M = 10$).

It is important to note that our work differs from others that investigated distributed logistic regression: Wu *et al.* [12] share parameters in each optimizer step, which results in a much larger computational overhead (but also in a model that is exactly the same as if it were trained on the full data set). Gopal and Young [21] were interested in problems with thousands of classes, and how the corresponding variant of $\mathrm{err_{LR}}$ in Eq. (1) can be approximated and evaluated in parallel for all classes.

3 Parallel Minibatch Training for Autoencoders

The resurgent interest in neural networks has brought renewed attention to the problem of unsupervised feature and representation learning in (lower) layers of deep neural networks [22]. In most approaches to this problem, layers of these networks are trained as *autoencoders*, i.e., computational structures that aim to reconstruct their inputs.

Of the several variants available, in this paper we investigate the use of so-called *sparse autoencoders* [23]. These autoencoders are trained to find representations where the average activation $\bar{\rho}_j$ of hidden neuron j (taken over all training instances) is a given constant ρ. This sparsity criterion is usually implemented via the *Kullback-Liebler divergence*, which for two Bernoulli distributions with parameters p_1 and p_2 is given by

$$D_{\mathrm{KL}}(p_1, p_2) = p_1 \log \frac{p_1}{p_2} + (1 - p_1) \log \frac{1 - p_1}{1 - p_2}.$$

A sparse autoencoder is then trained on a given data set $\{x_1, \ldots, x_D\} \subseteq \mathbb{R}^n$ (now without class labels) to minimize the error function

$$\text{err}_{\text{SAE}}(\theta) = \frac{1}{2D} \sum_{k=1}^{D} \|\text{ann}(x_k, \theta) - x_k\|_2^2 + \frac{\alpha}{2} \|\theta\|_2^2 + \beta \sum_{j=1}^{H} D_{\text{KL}}(\rho, \bar{\rho}_j). \quad (2)$$

Here, $\text{ann}(x_k, \theta)$ denotes the output of a fully connected feed-forward neural network with one hidden layer with H neurons (we used the logistic activation function) and n neurons in both the input and output layer, with weight vector θ. The first contribution in Eq. (2) (the sum) is thus (one half) the average reconstruction error, the second is the L_2 regularizer, and the third the sum of Kullback-Liebler divergences over all hidden neurons. The parameters α and β weight the relative contributions of these terms.

Data set. Training and test sets consisted of 10 000 and 5 000 patches, respectively, of size 8×8 pixels randomly extracted from 2500 airplane images in the CIFAR-10 data set [24]. All pixel values were scaled to the interval $[0.1, 0.9]$ prior to autoencoder training.

Evaluation metric. To evaluate the quality of a trained model with weight vector θ, we used the average reconstruction error $\frac{1}{D} \sum_{k=1}^{D} \|\text{ann}(x_k, \theta) - x_k\|_2^2$ on the patches of the test set.

Hyperparameter settings. Since the focus of this work is on evaluating parallel minibatch training, we did not conduct a thorough hyperparameter search, instead opting to keep them at values that had previously produced good results on similar tasks of autoencoder learning [23]: number of hidden neurons $H = 25$, desired sparsity $\rho = 0.01$, regularization parameter $\alpha = 0.0001$, sparsity error contribution $\beta = 3$. The total number of iterations was set to 400.

Parallel minibatch training. The parameters M and N are the same as in Sect. 2, with 400 iterations instead of 100.

Results. The results of varying the parameters M and N on the task of learning a sparse autoencoder model are summarized in Table 2. For comparison, when using the entire training set (without minibatches) for 400 iterations of the optimizers, the average reconstruction error on the test set was 0.540 for both SCG and L-BFGS.

Discussion. Similar to the results of logistic regression training in Sect. 2, Table 2 shows the performance of minibatch training to be slightly worse than that of the full models for all combinations of parameters M and N. Furthermore, performance decreases for increasing number of minibatches, and again (as in Sect. 2) for increasing number of loop iterations. It is interesting to note that for all numbers M of minibatches using, e.g., 40 optimizer steps followed by parameter averaging ($N = 10$ times) works better than using 10 optimizer steps followed by parameter averaging ($N = 40$ times). This might mean that, also on other examples, the potentially slow synchronization step of parameter averaging need not be executed a large number of times.

Table 2. Average reconstruction error on test set for a sparse autoencoder model trained on minibatches according to Algorithm 1, using the SCG (left part) and L-BFGS (right part) optimizers. Note: Smaller values are better.

Number M of minibatches	Number N of loop iterations				Number N of loop iterations			
	10	20	30	40	10	20	30	40
2	0.546	0.546	0.552	0.566	0.550	0.560	0.576	0.578
4	0.548	0.546	0.552	0.572	0.550	0.560	0.582	0.580
6	0.552	0.548	0.554	0.572	0.552	0.560	0.580	0.580
8	0.552	0.550	0.554	0.574	0.556	0.560	0.582	0.580
10	0.556	0.556	0.550	0.572	0.556	0.562	0.578	0.584

4   Conclusion

We evaluated the training of machine learning model in a distributed manner by splitting data sets into minibatches, training individual models on these minibatches, and periodically updating the model parameters by replacing them with the average of all model parameters. Both a logistic regression model trained in this distributed manner on a medium-sized real-world biomedical data set, as well as a sparse autoencoder trained on 8×8 pixel image patches showed performance only slightly worse than a model trained on the full data set.

References

1. Franks, B.: The Analytics Revolution: How to Improve Your Business by Making Analytics Operational in the Big Data Era. Wiley, Hoboken (2014)
2. Hashem, I., Yaqoob, I., Anuar, N., Mokhtar, S., Gani, A., Khan, S.: The rise of "big data" on cloud computing: review and open research issues. Inf. Syst. **47**, 98–115 (2015)
3. Wan, L., Zeiler, M., Zhang, S., LeCun, Y., Fergus, R.: Regularization of neural networks using DropConnect. In: Proceedings of the 30th International Conference on Machine Learning, pp. 1058–1066 (2013)
4. Dennis, J., Dat, T.: Single and multi-channel approaches for distant speech recognition under noisy reverberant conditions: I2R's system description for the ASpIRE challenge. In: Proceedings of the 2015 IEEE Workshop on Automatic Speech Recognition and Understanding, pp. 518–524 (2015)
5. Le, Q., Ngiam, J., Coates, A., Lahiri, A., Prochnow, B., Ng, A.: On optimization methods for deep learning. In: Proceedings of the 28th International Conference on Machine Learning, pp. 265–272 (2011)
6. Friedman, A., Schuster, A.: Data mining with differential privacy. In: Proceedings of the 16th ACM SIGKDD International Conference on Knowledge Discovery and Data Mining, pp. 493–502 (2010)
7. Lindell, Y., Pinkas, B.: Privacy preserving data mining. In: Proceedings of the 20th Annual International Cryptology Conference on Advances in Cryptology, pp. 36–54 (2000)

8. Vaidya, J., Clifton, C.: Privacy preserving association rule mining in vertically partitioned data. In: Proceedings of the Eighth ACM SIGKDD International Conference on Knowledge Discovery and Data Mining, pp. 639–644 (2002)

9. Clifton, C., Kantarcioglu, M., Vaidya, J., Lin, X., Zhu, M.: Tools for privacy preserving distributed data mining. ACM SIGKDD Explor. Newsl. **4**, 28–34 (2002)

10. Chaudhuri, K., Monteleoni, C.: Privacy-preserving logistic regression. Adv. Neural Inf. Process. Syst. **21**, 289–296 (2008)

11. Chaudhuri, K., Monteleoni, C., Sarwate, A.: Differentially private empirical risk minimization. J. Mach. Learn. Res. **12**, 1069–1109 (2011)

12. Wu, Y., Jiang, X., Kim, J., Ohno-Machado, L.: Grid Binary LOgistic REgression (GLORE): building shared models without sharing data. J. Am. Med. Inf. Assoc. **19**, 758–764 (2012)

13. Zinkevich, M., Weimer, M., Li, L., Smola, A.: Parallelized stochastic gradient descent. Adv. Neural Inf. Process. Syst. **23**, 2595–2603 (2010)

14. Dekel, O., Gilad-Bachrach, R., Shamir, O., Xiao, L.: Optimal distributed online prediction using mini-batches. J. Mach. Learn. Res. **12**, 165–202 (2012)

15. Agarwal, A., Chapelle, O., Dudik, M., Langford, J.: A reliable effective terascale linear learning system. J. Mach. Learn. Res. **15**, 1111–1133 (2014)

16. Hastie, T., Tibshirani, R., Friedman, J.: The Elements of Statistical Learning: Data Mining, Inference, and Prediction, 2nd edn. Springer, Heidelberg (2011). https://doi.org/10.1007/978-0-387-84858-7

17. Møller, M.: A scaled conjugate gradient algorithm for fast supervised learning. Neural Netw. **6**, 525–533 (1993)

18. Liu, D., Nocedal, J.: On the limited memory BFGS method for large scale optimization. Math. Program. **45**, 503–528 (1989)

19. Kennedy, R., Burton, A., Fraser, H., McStay, L., Harrison, R.: Early diagnosis of acute myocardial infarction using clinical and electrocardiographic data at presentation: derivation and evaluation of logistic regression models. Eur. Heart J. **17**, 1181–1191 (1996)

20. Hanley, J., McNeil, B.: The meaning and use of the area under the receiver operating characteristic (ROC) curve. Radiology **143**, 29–36 (1982)

21. Gopal, S., Yang, Y.: Distributed training of large-scale logistic models. In: Proceedings of the 30th International Conference on Machine Learning, pp. 289–297 (2013)

22. Bengio, Y., Courville, A., Vincent, P.: Representation learning: a review and new perspectives. IEEE Trans. Pattern Anal. Mach. Intell. **35**, 1798–1828 (2013)

23. Ng, A.: CS294A lecture notes: sparse autoencoder. Stanford University (2011)

24. Krizhevsky, A.: Learning multiple layers of features from tiny images. Technical report, Computer Science Department, University of Toronto (2009)

A Fair Performance Comparison of Different Surrogate Optimization Strategies

Bernhard Werth[1,2](✉), Erik Pitzer[1], and Michael Affenzeller[1,2]

[1] Heuristic and Evolutionary Algorithms Laboratory,
University of Applied Sciences Upper Austria,
Softwarepark 11, 4232 Hagenberg, Austria
{bernhard.werth,erik.pitzer,michael.affenzeller}@fh-hagenberg.at
[2] Institute for Formal Models and Verification, Johannes Kepler University,
Altenberger Straße 68, 4040 Linz, Austria

Abstract. Much of the literature found on surrogate models presents new approaches or algorithms trying to solve black-box optimization problems with as few evaluations as possible. The comparisons of these new ideas with other algorithms are often very limited and constrained to non-surrogate algorithms or algorithms following very similar ideas as the presented ones. This work aims to provide both an overview over the most important general trends in surrogate assisted optimization and a more wide-spanning comparison in a fair environment by reimplementation within the same software framework.

Keywords: Surrogate models · Evolutionary algorithms
Black-box optimization

Europäische Union Investitionen in Wachstum & Beschäftigung. Österreich.

1 Introduction

Heuristic and especially evolutionary algorithms have been widely employed to solve black-box optimization problems. An essential drawback of such algorithms is that they require many function evaluations to converge towards an optimum, which can make their application unfeasible if the fitness function is expensive to evaluate. Especially complex physical simulations or experiments can be desirable targets for optimization, even though they can be time consuming to evaluate. Surrogate models have become a widely accepted way of reducing the number of required evaluations by creating a statistical approximation model to estimate the relation between inputs and outputs of the black box function [12]. These models can then be exploited in various ways to guide a heuristic algorithm towards optimal input values. Unfortunately, most papers presenting new

© Springer International Publishing AG 2018
R. Moreno-Díaz et al. (Eds.): EUROCAST 2017, Part I, LNCS 10671, pp. 408–415, 2018.
https://doi.org/10.1007/978-3-319-74718-7_49

algorithms, models and approaches, compare their new techniques with only a very narrow set of alternatives. To make the comparison of the different main ideas in surrogate-assisted optimization easier, the most prominent approaches were implemented within the same algorithmic framework *HeuristicLab* to allow comparisons with a reduced influence of specific differences in implementation details. The remainder of the paper is structured as follows: Sect. 2 presents the most dominant types of models in the literature and briefly describes the main algorithms or algorithm extensions based on surrogate modeling. Section 3 summarizes the the test bed and the compared algorithms while Sect. 4 contains the results followed by concluding remarks in Sect. 5.

2   Literature Review

The major characteristics of surrogate-assisted algorithms are the type of models and the way these models are applied within an algorithm. The models and surrogate assisted approaches implemented and compared in this work are listed below.

2.1   Common Types of Surrogate Models

Nearly every type of machine learning model including clustering, classification and regression models has been used to enhance the performance of heuristic searches. The main advantages and disadvantages of the most common types of models are presented in the following.

Polynomial Regression. Linear and quadratic polynomial models for objective values or constraints are easy to construct and provide exact methods for finding optima as long as the constraints restricting the input parameters to feasible regions are linear. While some papers report other types of models as more powerful and able to approximate more complex fitness landscapes [1], especially quadratic regression models can be used as *local* models in the area of the current best solution to assist algorithms struggling to converge precisely within a limited number of evaluations due to an overemphasis on exploration rather than exploitation [11]. A prohibitive drawback of polynomial models is that the minimum number of samples to construct such a model scales with $O(n^d)$ where n is the dimensionality of the problem and d the order of the polynomial model.

Radial Basis Functions. The first RBF models were used by [3] to fit topographical data. Compared to low-order polynomial models RBF models allow for modeling surfaces with many local extreme points. Moreover, several extensions and kernel functions exist that can be used to modify the model to specific scenarios [8], however constructing and exploiting RBF models can be more computationally expensive if many sample points are used for training. An important advantage of RBF models is that very few sample points are required to create an initial model.

Gaussian Process Models. Gaussian process regression or *kriging* is an interpolation method based on approximating data as a combination of a trend function and a stationary stochastic process. The main advantages of kriging models are that they create smooth surfaces which might benefit algorithms exploring the model surface. Even more importantly, they provide a measure for the models own uncertainty of the predictions, which allows to search the model for solution candidates that have both a good estimated prediction and are sufficiently dissimilar from existing samples. Gaussian process models tend to have a larger number of hyper-parameters which need to be calibrated when building a model. In this work, said parameters are selected by optimizing the maximum likelihood of the model with a gradient ascent algorithm. Similar to RBF models, Gaussian process models can accurately describe local structures of the fitness surface but are even more expensive to build, due to the required hyper-parameter selection.

Other Model Types. Several other types of models have been used to enhance optimization algorithms including support vector machines (SVM), neural networks or ensembles of the aforementioned model types [2,7]. The selection of a model for an optimization task is usually based on standard measures for model quality e.g. root-mean-square error (RMSE), maximum likelihood or similar means of quantifying the complexity and accuracy of the model. However, while the requirement for a regression model usually is to fit the underlying true function everywhere as closely as possible, a surrogate model used for heuristic optimization is merely required to guide the search process into regions of the search space that are close to the optimum. A model with a very high RMSE might still be viable if its optimum coincides with the optimum of the underlying black box function. Some authors propose to build local models that only consider a subset of samples [6]. It is important to notice that although most surrogate models are regression models, different classes of models have been used. In [7] a ranking SVM is used to compare individuals and no actual function values are predicted and in [2] solution candidates are grouped via neural networks so that only one evaluation needs to be performed per group.

2.2   Main Types of Surrogate-Assisted Algorithms

Similar to the manifold of models used to approximate a fitness function, a variety of approaches for algorithms that make use of these models have been proposed. The most popular general strategies are described in the following.

Classical Approach. A very early approach to surrogate assisted optimization is the classical approach [10] that uses historical data to create a surrogate model or an ensemble thereof which is subsequently used as a fitness function for an optimization algorithm. The model optimum found by the algorithm is then validated by performing an expensive experiment. The classic method is relatively simple, can be performed *offline* and requires only a single model to be built.

Adaptive Sampling. A logical continuation of the classical approach is the idea of adaptive sampling where the selection of the next sample point is dependent on the previous samples and their fitness values. This is usually achieved by selecting the next sample point as (almost) optimal with respect to a given *infill criterion*. The most intuitive infill criterion is the expected fitness value predicted by a surrogate model which is created using the previous samples as a trainings set. More sophisticated infill criteria utilize the model uncertainty provided by Gaussian process models or ensembles to balance between exploitation and exploration. The most well known infill criterion is the *expected improvement* measure used in the *Efficient Global Optimization* (EGO) algorithm by Jones et al. [5].

Figure 1 depicts the standard procedure for an infill criterion based adaptive sampling scheme. An initial sampling step is performed to generate a database of samples from which a first model is created. Then a heuristic optimization algorithm is used to find a solution candidate that is optimal with respect to the chosen criterion. After the optimizer has terminated, the most promising point is evaluated and added to the database. This process can be repeated until the computational budget is exhausted or a different termination criterion is satisfied. In this work, the expected quality is chosen as an infill criterion for all models and additionally the expected improvement is tested for Gaussian processes.

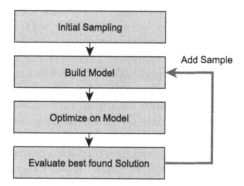

Fig. 1. Adaptive sampling scheme

Surrogate-Assisted Evolutionary Algorithms. A different train of thought is the idea of extending an existing algorithm with surrogate models rather than using it to optimize an infill criterion. Several methods have been proposed to extend existing population-based algorithms including *evolution strategies*, *genetic algorithms* and *particle swarm optimization*.

- *Pre-selection of individuals* is an extension of the *evolution strategy* algorithm where a large number of children are created, of which only a few are selected according to the surrogate model. These selected children are then evaluated and might replace their parents in the current generation [9].
- *Individual-based strategies* where only the most promising individuals of each generation are evaluated on the black box function while the quality of most other individuals is estimated by the model [4].

- *Granule-based fitness evaluation* is based on clustering the individuals in a population, evaluating one point per cluster and assigning estimated fitness values to the remaining solution candidates [2].
- *Generation-based strategies* where in most generations the fitness values of all individuals in the population are estimated, while in some generations all individuals are evaluated [7].
- *Population-based strategies* extend multi-population algorithms like *island genetic algorithms* where each island is associated with its own model [4].
- *Lamarckian replacement* is a technique where promising candidates are evaluated. Then a local model for the vicinity of this candidate is constructed and the most promising point as indicated by the local model is evaluated and replaces the original candidate on improvement [6].

3   Experimental Setup

Of special interest in comparing surrogate-based optimization approaches is their behavior when facing increasing problem sizes. Therefore, we chose seven well-known scalable single-objective test functions with different characteristics and increased the problem size in semi-logarithmic steps. Four surrogate-assisted algorithms have been implemented and tested in this work. The first two are the classical approach (CSA) and the adaptive sampling approach (ASA). Both algorithms use a Covariance Matrix Adaption Evolution Strategy (CMAES) with a population size of 50 and 300 generations to perform optimization on their respective models. As a representative for the surrogate-assisted population-based algorithms (SAPBA), a genetic algorithm with an individual-based strategy evaluating one solution candidate per generation was used. Additionally, the same surrogate-assisted genetic algorithm is extended with Lamarckian replacement (SAPBA-LAM). The local model used for such a replacement is a Gaussian process model that uses the 50 closest points around the replaced point. The search of the local model is restricted to the bounding box of these 50 points.

Table 1. Algorithms, models and problems tested

Algorithms	Models	Test functions	Problem sizes
CSA	RF	Ackley	2
EGO	SVM	Griewank	5
SAPBA	GP	Levy	10
SAPBA-LAM	RBF	Rastrigin	20
GA		Rosenbrock	30
ES		Schwefel (sine root)	50
RS		Sphere	75
CMAES			100

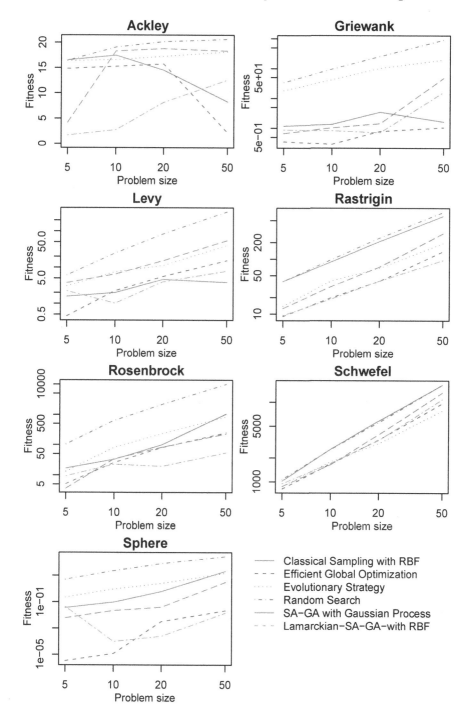

Fig. 2. Performances of the best algorithms on all problems over problem sizes. (Median fitness values of 10 runs). Fitness values are plotted on a logarithmic scale for all but the Ackley problem.

In order to establish a valid baseline, a genetic algorithm (GA), an evolutionary strategy (ES), a CMAES, and a random search (RS) are also employed. The algorithms, models, problems and problem sizes are summarized in Table 1. The used models are random forests (RF), support vector machines (SVM), Gaussian processes (GP) and Radial Basis Functions (RBF). We selected the maximum number of evaluations as $20 * d$.

To keep the scope of the experiments within reasonable bounds two restrictions had to be enforced. The first restriction was that for the EGO algorithms and the surrogate-assisted genetic algorithms only the 300 most recently sampled points (excluding duplicates) were used to create the models. The second restriction was runtime limit of 10 h per algorithm run, which only affected the EGO algorithms for problems with 75 or 100 dimensions.

4   Empirical Results

The results in Fig. 2 demonstrate the importance of selecting flexible model types if the algorithm hinges on the use of a global surrogate model. For the sake of visibility, only the best performing algorithm configuration of each algorithm family is displayed. While the EGO-like algorithms gave good results, they suffer from enormous performance hits for larger problem sizes. For some simulations, this might become more expensive than the computational costs for evaluating solution candidates. However, for extremely expensive problems the use of EGO-like Algorithms might still be justified. While the surrogate-assisted genetic algorithm did not perform well in its simple configuration, the Lamarckian Replacement extension coupled with a local model allowed it to rival the fitness values of the EGO algorithm with the additional benefit of significantly lower computational overhead. For most problem instances, the better surrogate-assisted algorithms easily outperformed the "conventional" algorithms GA, ES and CMAES. However, for large or deceptive instances, the differences to the surrogate-less algorithms dwindles, which might indicate that all tested surrogate approaches struggle with deceptiveness.

5   Conclusion

The most popular techniques for surrogate-based heuristic optimization have been presented and implemented within the *HeuristicLab* framework, in order to create fair conditions for a comparison which includes the same model building subroutines, problem implementations and organizational code handling algorithm execution, initial sample creation, problem evaluation or numeric instabilities.

An important challenge for surrogate-assisted algorithms is to improve their performance in high-dimensional or deceptive scenarios. Another obstacle hindering the use of such algorithms is the overwhelming number of parameters for algorithm control, model building and sub-algorithms used to optimize subproblems. Especially surrogate-assisted algorithms that combine multiple global and local models suffer from a substantial amount of effort required to fine-tune them to individual problems. The development of new techniques to allow surrogate-assisted algorithms to perform in higher dimensional, noisy or deceptive scenarios is still an open research topic. In order to gain knowledge about the strengths and weaknesses of these techniques, that go beyond comparing convergence curves on predefined problems, specialized methods in the field of *fitness landscape analysis* need to be developed.

Acknowledgements. This work was supported by the European Union through the European Regional Development Fund (EFRE; further information on IWB/EFRE is available at www.efre.gv.at).

References

1. Barton, R.R.: Simulation metamodels. In: 1998 Winter Simulation Conference Proceedings, vol. 1, pp. 167–174. IEEE (1998)
2. Cruz-Vega, I., Garcia-Limon, M., Escalante, H.J.: Adaptive-surrogate based on a neuro-fuzzy network and granular computing. In: Proceedings of the 2014 Annual Conference on Genetic and Evolutionary Computation, pp. 761–768. ACM (2014)
3. Hardy, R.L.: Multiquadric equations of topography and other irregular surfaces. J. Geophys. Res. **76**(8), 1905–1915 (1971)
4. Jin, Y.: Surrogate-assisted evolutionary computation: recent advances and future challenges. Swarm Evolut. Comput. **1**(2), 61–70 (2011)
5. Jones, D.R., Schonlau, M., Welch, W.J.: Efficient global optimization of expensive black-box functions. J. Global Optim. **13**(4), 455–492 (1998)
6. Lim, D., Jin, Y., Ong, Y.S., Sendhoff, B.: Generalizing surrogate-assisted evolutionary computation. IEEE Trans. Evol. Comput. **14**(3), 329–355 (2010)
7. Loshchilov, I., Schoenauer, M., Sebag, M.: Self-adaptive surrogate-assisted covariance matrix adaptation evolution strategy. In: Proceedings of the 14th annual Conference on Genetic and Evolutionary Computation, pp. 321–328. ACM (2012)
8. Mullur, A.A., Messac, A.: Extended radial basis functions: more flexible and effective metamodeling. AIAA J. **43**(6), 1306–1315 (2005)
9. Ulmer, H., Streichert, F., Zell, A.: Evolution strategies assisted by Gaussian processes with improved preselection criterion. In: The 2003 Congress on Evolutionary Computation, CEC 2003, vol. 1, pp. 692–699. IEEE (2003)
10. Wang, G.G., Shan, S.: Review of metamodeling techniques in support of engineering design optimization. J. Mech. Des. **129**(4), 370–380 (2007)
11. Wang, L., Shan, S., Wang, G.G.: Mode-pursuing sampling method for global optimization on expensive black-box functions. Eng. Optim. **36**(4), 419–438 (2004)
12. Zhang, J., Chowdhury, S., Messac, A.: An adaptive hybrid surrogate model. Struct. Multidisc. Optim. **46**(2), 223–238 (2012)

Facilitating Evolutionary Algorithm Analysis with Persistent Data Structures

Erik Pitzer[1(✉)] and Michael Affenzeller[1,2]

[1] Department Software Engineering, University of Applied Sciences Upper Austria,
Softwarepark 11, 4232 Hagenberg, Austria
{erik.pitzer,michael.affenzeller}@fh-hagenberg.at
[2] Institute for Formal Models and Verification, Johannes Kepler University,
Altenbergerstr 68, 4040 Linz, Austria

Abstract. Evolutionary algorithm analysis is often impeded by the large amounts of intermediate data that is usually discarded and has to be painstakingly reconstructed for real-world large-scale applications. In the recent past persistent data structures have been developed which offer extremely compact storage with acceptable runtime penalties. In this work two promising persistent data structures are explored in the context of evolutionary computation with the hope to open the door to simplified analysis of large-scale evolutionary algorithm runs.

1 Introduction

Evolutionary algorithms especially those that employ a whole population of solution candidates such as e.g. evolution strategy [12] or genetic algorithms [7] are very popular methods for solving complex problems. However, typical optimization scenarios often require long evolutionary processes with thousands or even billions of evaluations. In practice, this means that many methods are developed using rather small sample problems, i.e. to test and tune parameters or select a suitable algorithm variant. Later, the developed technique is applied to a much bigger problem, in the hope that these properties are – at least somewhat – preserved and the algorithm still performs well. For the smaller scenarios, the algorithm runs can be closely supervised and the performance can be tracked and recorded as the amount of data is still manageable. However, as problems approach practical sizes, analysis and continuous tracking involves prohibitively large volumes of data. For example, an algorithm with a population of 100 solution candidates solving a problem with 100 dimensions over one million generations would require storage of 10^{10} elements which are at least 10 to $40\,\mathrm{GB}$ depending on the data type for a single run. The idea of this work is to leverage persistent data structures and their nature to reuse old parts when modified, to track historical values without using too much additional space. In essence, we get a data structure that technically only saves differences between different versions but practically provides full copies of all versions with very little overhead.

R. Moreno-Díaz et al. (Eds.): EUROCAST 2017, Part I, LNCS 10671, pp. 416–423, 2018.
https://doi.org/10.1007/978-3-319-74718-7_50

2   Persistent Data Structures

The first prototype of persistent data structures, the persistent linked list, can be dated back to 1955 where it was the primary data structure of the Information Processing Language (IPL) [10] one important predecessor of Lisp [8,9]. It can be used to illustrate the idea of persistent data structures. As shown in Fig. 1, in a persistent data structure, parts of "old" structures can be *reused* inside new structures as the old data and the old structure cannot change. While it is obvious that this works for immutable linked lists it is much harder to imagine for arrays or other commonly-used data structures.

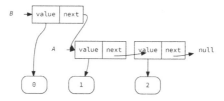

Fig. 1. Example of shared structure in singly-linked list

As computer processors have received fewer improvements in speed but more improvements in multi core execution [13], parallel processing has become not only more prominent but almost a necessity to fully utilize current hardware [1]. This, in turn, has revived interest in functional languages [5] that typically use immutable data structures. In the same vein, many classically imperative data structures have received a functional pendant [11]. To achieve similar run time complexity, however, functional data structures need additional or different tricks that rely mostly on their immutability and reuse of existing parts of the "old" copy of the data structure as immutable data has to be *copied* in case any modification is made. With these tricks, the data does not actually need to be copied but can merely be referenced with great savings in both speed (the original intention) and also space (the intention of this work). In particular, two variants of immutable arrays with structure sharing are explored in this work. These are Array Mapped Tries (AMTs) [2] and Relaxed Radix Balanced Trees (RRB-Trees) [3].

AMTs are an extension of radix trees which in turn are a compact variant of prefix trees [4]. The idea of a radix tree is to identify all elements by index. This index is subsequently split across the levels of a tree depending on the radix or the fixed maximum number of children of each tree as shown in Fig. 2. The benefit of this structure in comparison to e.g. a binary search tree is that no comparisons are necessary to find the correct element, the index of the element is sufficient to completely navigate the tree to any leaf. When this structure is used for hash maps, it can be beneficial to allow any 32 bit integer value; In this case not all slots might be needed. This fact is exploited by Hash Array

Mapped Tries that do not contain all pointers in all nodes, but only those that lead further down to existing leaf nodes [2]. A clever use of the Hamming Weight of the subindex often called population count (popcount) operation – available on many processors as a single instruction – makes the lookup of which children actually exist particularly cheap [6].

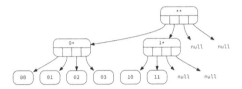

Fig. 2. Example of radix-4 tree

Most importantly however, this structure allows the reuse of parts of the "array" it is representing when making copies or changes. As shown in Fig. 3 changing a single element in this immutable tree does not require copying the whole tree. As all data is also immutable, large parts of the tree can be reused. This is what makes this data structure so interesting for keeping a history. While the modified data structure has the same structure as a "fresh" tree, it can reuse most of the data from its previous version. In essence, we are saving deltas but accessing complete data structures with minimal overhead. As described in [2] updating or accessing an element in this radix tree with typical radix size 32 is in the order of $\log_{32} N$. Moreover as $N \leq 7$ for any index in the range of a 32 bit integer, the overhead can be considered practically constant. This is a great bonus when tracking modifications of large populations over many generations.

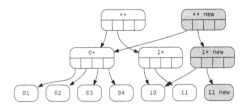

Fig. 3. "Modification" of immutable radix-4 tree

3   Persistent Evolutionary Operations

Every single point modification leads to a short chain of pointers to the changed location, reusing many pointers along the way. Therefore, with constant overhead any single point mutation can be tracked. However, in evolutionary algorithms, frequently a whole slice of an array is updated. An easy fix for this situation would be to discard intermediate trees and only keep the "snapshots" that represent actually-visited solution candidates of the evolutionary search process. However, in addition, many evolutionary algorithms, in particular genetic algorithms, use crossover operations that take information from several individuals and combine them. Using AMTs it would have to be decided which individual is the one being modified with the information of other individuals incorporated as a series of (mostly discarded) single point modifications. This is unfortunate as the overhead of this operation would be linear with the size of the individual arrays. Fortunately, another variation of radix trees, so called, Relaxed Radix Balanced Trees (RRB-Trees) have been proposed in [3]. This variant allows slicing trees and even joining them back together in practically constant time, similar to what AMTs allow for single point modifications.

RRB-Trees enable this modification by slightly relaxing the radix requirement and allowing index shifts on some of the nodes. This enables, for example, the removal of nodes in the beginning without having to rebuild the remaining tree as shown in Fig. 4.

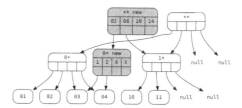

Fig. 4. Removal of the first two elements in an RRB-Tree (Skipping)

The most complex operation is concatenating two trees, but again the overhead is practically constant for a single join operation. For this operation an error bound can be chosen, that allows more or fewer empty slots in the intermediate nodes when two trees are joined. Using practically constant operations for splitting and joining yields practically constant time (and space) for crossover operations as they occur in genetic algorithms [7]. Every *crossover* operation can now be stored as a single "delta" reusing parts from both parents and therefore requiring only constant additional storage for each operation. It has to be noted that the involved computational overhead is significant, however. Array copying is highly optimized and can hardly be outdone by following pointers in a tree even for large trees. One could imagine that copying 1000 values from one array to another must be slower than updating a few pointers on the heap. However, the memory locality and its caching effects give a huge advantage to arrays in terms of speed. The real advantage of trees lies in their space savings. Moreover, in evolutionary algorithms, the internal data handling such as copying and modifying of solution candidates is hardly an issue as it is usually completely dwarfed in comparison to the cost of a solution candidate's *fitness evaluation*.

Figure 5 shows an example of a crossover operation on two RRB-Trees A and B. For the sake of simplicity, this time, only a radix-2 tree is shown. As can be seen, large parts of the original data can be reused. In this – admittedly lucky – example even each leaf node can be reused as-is. In general, only two logarithmic "paths" from the glue points to the root have to be newly generated. All remaining data and data structure parts can always be reused.

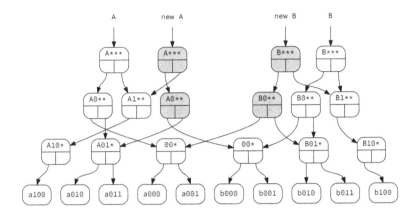

Fig. 5. Crossover of two RRB-Trees

As an extension to this method, other crossover variants can be implemented with reduced memory consumption. Obviously, two-point or multi-point crossover will work the same and other crossover operators that reuse slices of the original solution candidates will also benefit from this scheme, e,g, the edge recombination crossover [15]. Only at the fusion points, new nodes might become necessary to sufficiently satisfy the (relaxed) radix constraints of the tree nodes, depending on the errors parameter E that specifies how many slots may remain empty in any internal node.

Another rather easy extension is to allow (partial) reversal of the represented arrays. This can be achieved by simply replacing a node at a higher level with a special node that marks the reversal of all indexes as shown in Fig. 6, where a continuous array is shown using a reverse node. This node can be used as the intermediate node at the top of a reversed subtree, supporting operations such as partial inversion.

4   Experimental Results

We have implemented both Array Mapped Tries (AMTs) as well as Relaxed Radix Balanced Trees (RRB-Trees) and subjected them to different test scenarios. Our implementation of AMTs has been integrated into the open-source optimization platform HeuristicLab[1] [14]. In HeuristicLab all optimization-relevant

---

[1] http://dev.heuristiclab.com in the branch `PersistentDataStrcutures`.

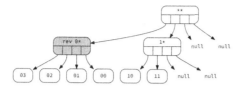

Fig. 6. Inversion-based operations in RRB-Trees

array data structures are derived from the generic class `ValueTypeArray<T>` which – unsurprisingly – uses regular `C#` arrays internally. We have replaced these regular arrays with our implementation `ArrayMappedTrie<T>`, or more precisely with a wrapper called `HistoryArray<T>` that can either automatically make snapshots after a specified number of modifications or with the help of a so-called Analyzer inside HeuristicLab, that is typically called once per generation or iteration in different algorithms. Surprisingly, the run-time of the tested optimization algorithms creating snapshots of all intermediate versions did not increase in comparison to using plain `C#` arrays. It seems that the additional effort required to navigate the tree when setting and getting values is offset by the benefit that the (frequent) cloning that is done in HeuristicLab is a constant operation now.

Table 1a shows the runtime and most importantly the space required to record a series of one million mutations on a vector of size 1000. It has to be noted, that the AMT implementation always uses full depth (7), while the RRB implementation dynamically adapts depth to the actual maximum index. It can be seen, that the trees use significantly more time than the plain array. For small mutations it is beneficial to have little duplication and therefore use a smaller radix (node size). On the other hand a smaller radix has more overhead, i.e. more steps, to reach the leaves. In this scenario, radixes 4, 8 and 16 appear to provide the best compromises between speed and size. In Table 1b the results for crossover are shown. Obviously, using more complicated data structures causes a significant penalty in run time when looking purely at data manipulation operations. In this case, the best tree is slower by a factor of 35, however, using less than a third of the storage. For purely crossover-based workloads, all RRB trees between radix 2 and 16 provide a pareto-optimal compromise between speed and size. Interestingly, in this case the more relaxed trees i.e. error 1 and 2 do not benefit from reduced run-time.

The real benefits of tree-based structures, appear unfortunately only for larger data structures which are less common in practice. In theses cases, as shown in Fig. 7 however, the tries could even outperform plain arrays on mutations and can compete very well on crossovers even in terms of speed.

Finally it has to be noted that in population-based algorithms, where "younger" solution candidates contain much of the genetic material of older ones, obtained by recombination, the ratio of reuse could be even higher as not only two directly related individuals share the same genetic material but many or almost all in later generations.

Table 1. One million operations (mutations and crossovers) in a vector of length 1 000: structure parameters are radix and allowed number of skipped pointers, size is the percentage compared to an array and time is the multiple of array processing times.

structure	no history		with history	
	time	size	time	size
AMT/32	38	129%	7.1	54%
RRB/2	72	405%	11.0	28%
RRB/4	42	204%	5.5	**22%**
RRB/8	33	147%	5.3	23%
RRB/16	24	123%	4.3	27%
RRB/32	25	114%	**5.1**	39%

(a) mutations

structure	no history		with history	
	time	size	time	size
AMT/32/-	678	258%	62	92%
RRB/2/0	814	822%	56	**28%**
RRB/4/0	673	418%	46	29%
RRB/4/2	544	740%	37	34%
RRB/8/1	525	340%	36	32%
RRB/16/2	523	268%	**35**	36%
RRB/32/0	572	236%	38	47%

(b) crossovers

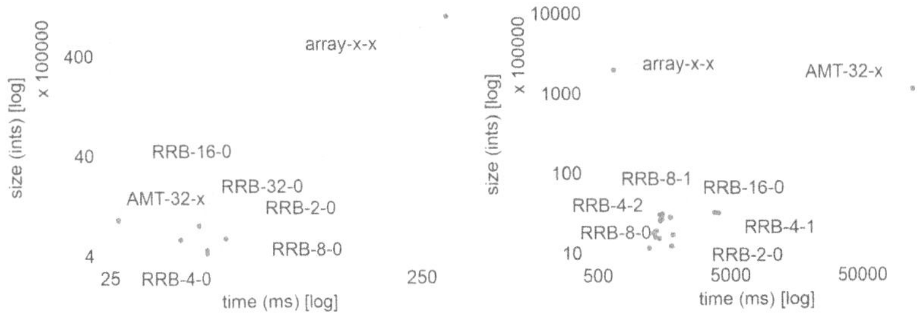

(a) mutations

(b) crossovers

Fig. 7. Speed vs. size for 10 000 operations on vectors of length 10 000

5   Conclusions

The recent progress in persistent data structures are a godsend for the analysis of evolutionary algorithms especially the recent addition of split and join operations enable the implementation of space-efficient crossover tracking. Ideally, this kind of information recording can be habitually enabled in the future, as it records changes at a fraction of the cost of previous methods, while incurring acceptable run-time penalties.

We are working to incorporate RRB-Trees into HeuristicLab and to modify its evolutionary operators to be able to take full advantage of the history tracking capabilities of RRB-Trees. This is only an early prototype but we are confident that the simplified tracking will provide great benefits for the postmortem analysis of strangely behaving algorithm runs on large problem instances.

Acknowledgements. The work described in this paper was performed within the COMET Project Heuristic Optimization in Production and Logistics (HOPL), #843532 funded by the Austrian Research Promotion Agency (FFG).

References

1. Asanovic, K., Bodik, R., Catanzaro, B.C., Gebis, J.J., Husbands, P., Keutzer, K., Patterson, D.A., Plishker, W.L., Shalf, J., Williams, S.W., Yelick, K.A.: The landscape of parallel computing research: a view from Berkeley. Electrical Engineering and Computer Sciences University of California at Berkeley, Technical report UCB/EECS-2006-183, December 2006
2. Bagwell, P.: Ideal hash trees. Technical report, École Polytechnique Fédéerale de Lausanne (2001)
3. Bagwell, P., Rompf, T.: RRB-Trees: efficient immutable vectors. Technical report, École Polytechnique Fédéerale de Lausanne (2011)
4. Cormen, T.H., Leiserson, C.E., Rivest, R.L., Stein, C.: Introduction to Algorithms, 3rd edn. MIT Press, Cambridge (2009)
5. Eyler, P.: The rise of functional languages. Linux J. (2007). https://www.linuxjournal.com/node/1000217
6. Warren Jr., H.S.: Hacker's Delight, 2nd edn. Addison-Wesley, Reading (2013)
7. Holland, J.H.: Adaptation in Natural and Artificial Systems. University of Michigan Press, Ann Arbor (1975)
8. McCarthy, J.: Lisp prehistory - summer 1956 through summer 1958 (1958). http://www-formal.stanford.edu/jmc/history/lisp/node2.html
9. McCarthy, J.: Recursive functions of symbolic expressions and their computation by machine, part I. Commun. ACM **3**(4), 184–195 (1960)
10. Newell, A., Shaw, J.: Programming the logic theory machine. In: Proceedings of the Western Joint Computer Conference, pp. 230–240 (1957)
11. Okasaki, C.: Purely Functional Data Structures. Cambridge University Press, Cambridge (1999)
12. Rechenberg, I.: Evolutionsstrategie - Optimierung technischer Systeme nach Prinzipien der biologischen Evolution. Frommann-Holzboog, Stuttgart (1973)
13. Sutter, H.: The free lunch is over: a fundamental turn toward concurrency in software. Dr. Dobbs J. **30**(3), 202–210 (2005)
14. Wagner, S., et al.: Architecture and design of the heuristiclab optimization environment. In: Klempous, R., Nikodem, J., Jacak, W., Chaczko, Z. (eds.) Advanced Methods and Applications in Computational Intelligence. Topics in Intelligent Engineering and Informatics, vol. 6, pp. 197–261. Springer, Heidelberg (2014). https://doi.org/10.1007/978-3-319-01436-4_10
15. Whitley, D., Starkweather, T., Fuquay, D.: Scheduling problems and traveling salesman: the genetic edge recombination operator. In: International Conference on Genetic Algorithms, pp. 133–140 (1989)

Offspring Selection Genetic Algorithm Revisited: Improvements in Efficiency by Early Stopping Criteria in the Evaluation of Unsuccessful Individuals

Michael Affenzeller[(✉)], Bogdan Burlacu, Stephan Winkler,
Michael Kommenda, Gabriel Kronberger, and Stefan Wagner

Heuristic and Evolutionary Algorithms Laboratory, School of Informatics,
Communications and Media, University of Applied Sciences Upper Austria,
Hagenberg Campus, Softwarepark 11, 4232 Hagenberg, Austria
`michael.affenzeller@fh-hagenberg.at`

Abstract. This paper proposes some algorithmic extensions to the general concept of offspring selection which itself is an algorithmic extension of genetic algorithms and genetic programming. Offspring selection is characterized by the fact that many offspring solution candidates will not participate in the ongoing evolutionary process if they do not achieve the success criterion. The algorithmic enhancements proposed in this contribution aim to early estimate if a solution candidate will not be accepted based on partial solution evaluation. The qualitative characteristics of offspring selection are not affected by this means. The discussed variant of offspring selection is analyzed for several symbolic regression problems with offspring selection genetic programming. The achievable gains in terms of efficiency are remarkable especially for large data-sets.

1 Introduction

Offspring selection (OS) [1] is a generic extension to the general concept of a genetic algorithm [2,3] which includes an additional selection step after reproduction: The fitness of an offspring is compared to the fitness values of its own parents in order to decide whether or not a the offspring solution candidate should be accepted as a member of the next generation or not. The origins of offspring selection were laid in the SEGA algorithm [4] where a certain user defined birth surplus of individuals is generated in order to accept the best individuals as members of the next generations in a coarse grained parallel genetic algorithm. In the successor of SEGA, the SASEGASA algorithm [5], offspring selection was already inherently included as selection procedure in the algorithm's demes, which are called village populations within SASEGASA. The implementation of this selection procedure within a single population GA has been published later and has been named offspring selection the first time in 2005 [1].

© Springer International Publishing AG 2018
R. Moreno-Díaz et al. (Eds.): EUROCAST 2017, Part I, LNCS 10671, pp. 424–431, 2018.
https://doi.org/10.1007/978-3-319-74718-7_51

Since then offspring selection has been applied successfully for combinatorial optimization problems, simulation based optimization as well as in a lot of symbolic regression and symbolic classification applications when being combined with genetic programming. Summaries of offspring selection applications in various fields are given for example in [6,7].

The remainder of the contribution is organized as follows: After a short summary of the the classical offspring selection GA we present an algorithmic extension that aims to improve the runtime characteristics of the method significantly without losses in terms of quality. The efficiency gain of the method is achieved by the integration of early termination criteria in the evaluation of solution candidates that are expected to fail the offspring success criterion. This strategy is evaluated for symbolic regression based on genetic programming using benchmark problems that are well known in the field. For such data based modeling problems the ratio of not successful solution candidates is usually very high, especially for large scaled data analysis problems where the runtime bottleneck is the evaluation function and the ratio of non-successful solution candidates is especially high due to the application of strict offspring selection [6]. As the focus of this algorithmic extension is to improve runtime performance, this increase has to be quantified. Computation time on a certain machine is a practical but often not objective measure as the results depend on the concrete implementation and on the framework and its overhead. Especially in the field of stochastic optimization where solution evaluation is often the most time consuming step, the effort of the algorithm is measured on the basis of solution evaluations. As the algorithmic enhancement proposed in this work aims to achieve savings in runtime by partial solution evaluation, effort will be measured as the fraction of training data samples that are actually evaluated.

2 Offspring Selection and Strict Offspring Selection

In terms of replacement the starting point is a Genetic Algorithm with generational replacement. Like in a standard GA parents are selected from the parent generation by a certain selection operator like proportional, tournament or a ranking based selection mechanism or even in a gender specific selection by using different selectors for the male and the female parent. Afterwards the selected parents are recombined by crossover and sometimes mutated where the probability of mutation is defined by the mutation rate. Whereas in a standard GA a so generated offspring solution becomes member of the offspring generation automatically, within the offspring selection GA (OS-GA) the respective offspring solution is just a candidate for the ongoing evolutionary algorithm. In order to decide whether or not an offspring solution will be considered for the ongoing evolutionary process, the offspring selection criterion is applied which compares the fitness of the candidate offspring solution with the fitness of its own parents. Within this the so called *ComparisonFactor* $\in [0, 1]$ is responsible for the fitness threshold. A comparison factor of 0 means that the fitness of the candidate offspring has to surpass the fitness of the weaker parent whereas a value of 1 sets the acceptance threshold to be surpassed to the fitness of the stronger parent. A further parameter to be adjusted by the user is the so called *SuccessRatio* which defines the claimed ratio of successful offspring in the above mentioned sense. The remaining part of the offspring generation is generated in the conventional way by just accepting offspring without applying the offspring selection criterion. As a consequence of this strategy the effort that has to be taken to fill up the claimed amount of successful offspring varies over time depending on the number to trials that are necessary to generate a sufficient number ($SuccessRatio * |POP|$) of successful offspring which defines the actual selection pressure $actSelPress = \frac{generatedIndividuals}{|POP|}$. Coupled with a user defined maximal value of selection pressure $maxSelPress$ this allows to introduce a dynamic termination criterion which is activated if the value of actual selection pressure surpasses the maximal selection pressure. The basic flowchart of offspring selection is schematically shown in the left part of Fig. 1.

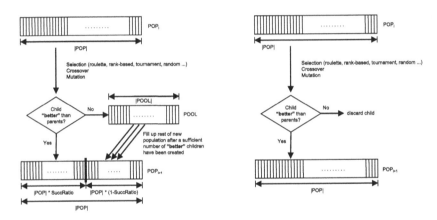

Fig. 1. Offspring selection and strict offspring selection.

Strict offspring (cf. right part of Fig. 1) selection denotes a special and at the same time simplified version of offspring selection which is achieved if *successRatio* and *comparisonFactor* are both set to 1.0. In terms of algorithm functioning this simply means that each offspring has to surpass the fitness value of the fitter of the two parents. In practice strict offspring selection turned out as very successful when applying Genetic Programming to symbolic regression and classification which seems reasonable as in those applications genetic diversity is additionally supported in those applications due to the fact that many different genotypes represent the same phenotype.

The extended version of offspring selection as being proposed in the next section aims to reduce computational effort with same qualitative performance by early predicting that a solution will not surpass the OS criterion in the evaluation of a candidate solution. This is technically achieved by predicting the offspring fitness as an extrapolation coupled with an early stopping of the fitness evaluation in case the offspring solution candidate is expected to be unsuccessful.

3   Offspring Selection with Early Stopping Criteria

The initial point of motivation for the enhanced offspring selection variant presented in this paper is the fact that the evaluation of the fitness function often denotes the performance bottleneck in GA applications and that within OS-GA this evaluation is done completely also for the unsuccessful solution candidates which will not participate in the ongoing evolutionary process. In this contribution we confine ourselves to symbolic regression and classification based on genetic programming mainly because of two reasons. First, in those application the evaluation of fitness is especially expensive compared to the other parts of the algorithm especially when analyzing large datasets. Second, partial solution evaluation and extrapolation is especially easy in this kind of application when estimating the error only on the basis of a fraction of training data. In the empirical analyses performed in the the following workflow will be applied:

- 1. Split training data D into smaller equal parts $D_1, D_2, \ldots, D_n$
- 2. Evaluate offspring on D_1
- 3. Compare offspring quality with parents quality (with tolerance)
- 4. Reject offspring if its fitness is expected to be worse than its parents fitness (combination with $ComparisonFactor$)
- 5. Otherwise, resume from step 2 performing evaluation on $D_1 + D_2$

Obviously there cannot be an optimal split of training data as it depends on the characteristics of the data, the number of inputs as well as on the complexity of the systems characteristics to be identified which amount of training data will be needed to estimate the fitness similar to the fitness evaluation based on all training data. Therefore, the weighting factor for parent quality threshold needs to be set to a sensible value.

4   Empirical Results

For the empirical studies the artificial benchmark functions Friedman-1, Friedman-2, Poly-10, Pagie-1 as well as the real world datasets Tower and Chemical-1 have been used. The artificial benchmark functions are shown in Table 1 and the real world benchmark functions are described in [8]. Offspring selection has been applied with the parameter settings as given in Table 2.

Table 3 shows the average results of 20 independent runs as well as the relative effort compared to the effort of standard OS for different ratios of evaluated training data and for different acceptance thresholds. For example, an acceptance threshold of 97% means that if the extrapolated quality value does not even achieve 97% of the R^2 to be achieved in order to achieve the success criterion the corresponding solution candidate is considered as unsuccessful and further evaluation is stopped. The training solution quality of 20 independent runs of the corresponding standard OS and the number of used training data samples are stated under the name of the test function.

Table 1. Objective functions of synthetic benchmark problems

Problem	Mathematical expression
Friedman-1 (5000 samples)	$y = \sin(\pi x_1 x_2) + 20(x_3 - 0.5)^2 + 10x_4 + 5x_5 + noise$
Friedman-2 (5000 samples)	$y = \left(x_1^2 + \left(x_2 x_3 - \frac{1}{x_2 x_4} \right)^2 \right)^{0.5} + noise$
Poly-10 (250 samples)	$y = x_1 x_2 + x_3 x_4 + x_5 x_6 + x_1 x_7 x_9 + x_3 x_6 x_{10}$
Pagie-1 (676 samples)	$y = \frac{1}{1+x_1^{-4}} + \frac{1}{1+x_2^{-4}}$

Table 2. OS parameter configuration

Algorithm	Offspring selection genetic programming
Runs	20 for each parameter configuration
Population size	500
Crossover probability	100%
Mutation probability	25%
Selection scheme	Strict offspring selection
Maximum tree depth	12
Maximum tree length	50
Fitness function	Pearson R^2
Training data split	10%, 20%, 30%, 40%, 50%
Quality threshold	90%, 95%, 97%, 98%, 99%
Maximal generations	1000
Maximal selection pressure	100

The results which are highlighted in the table emphasize those fraction/threshold combinations which could achieve qualities similar to standard OS. Obviously it depends on the characteristics of the certain data-set which fraction of training samples is adequate to achieve a sufficiently good estimate for early abortion with a certain threshold where those datasets with a higher number of training samples offer higher savings potential by trend. The effort is measured by the number of evaluated data samples.

The results shown in Table 4 provide deeper insights about the reasons which fraction/threshold combinations are able to achieve same qualitative behavior as standard OS and which not. For reasons of analysis also the theoretically rejected solution candidates are calculated for all training samples without early abortion in order to see which anticipated rejections turned out to be correct and which not. Furthermore, Table 4 also shows the average number of generations the algorithm was running before the dynamic termination criterion based on selection pressure was triggered.

A joint interpretation of Tables 3 and 4 shows that if the number of false anticipated rejections surpasses a certain value the number of needed trials becomes that high that the selection pressure based termination criterion of OS is triggered too early and the algorithm is hindered to converge.

Table 3. Average solution quality and evaluation effort (in terms of percentage of evaluated training rows) for the selected test problems

Problem	Row %	10%		20%		30%		40%		50%	
	Thresh.	Qlty.	Effort	Qlty.	Effort	Qlty.	Effort	Qlty.	Effort	Qlty.	Effort
Friedman-1	90%	**0.813**	11.6%	**0.817**	22.3%	**0.816**	31.6%	**0.815**	40.1%	**0.810**	47.8%
R^2 (std. OS)	95%	**0.820**	13.4%	**0.816**	21.1%	**0.816**	29.6%	**0.816**	38.9%	**0.813**	50.7%
0.81	97%	0.763	3.0%	**0.815**	22.5%	**0.813**	29.9%	**0.820**	41.8%	**0.812**	47.6%
	98%	0.747	2.7%	0.792	9.8%	**0.810**	29.3%	**0.817**	39.4%	**0.817**	49.8%
5000 tr. samples	99%	0.742	2.5%	0.772	6.4%	**0.810**	25.5%	**0.791**	18.3%	**0.795**	23.5%
Friedman-2	90%	**0.875**	42.8%	**0.856**	48.5%	**0.848**	46.2%	**0.846**	51.3%	**0.868**	62.3%
R^2 (std. OS)	95%	**0.855**	31.0%	**0.862**	36.6%	**0.866**	45.4%	**0.870**	59.4%	**0.848**	60.2%
0.85	97%	**0.844**	23.1%	**0.870**	31.6%	**0.861**	40.2%	**0.851**	46.6%	**0.855**	52.9%
	98%	**0.868**	20.0%	**0.855**	27.8%	**0.866**	39.8%	**0.861**	44.2%	**0.861**	53.9%
5000 tr. samples	99%	**0.853**	13.1%	**0.849**	19.9%	**0.863**	31.2%	**0.862**	44.3%	**0.870**	55.5%
Poly-10	90%	0.762	11.6%	0.839	25.2%	0.794	32.7%	**0.871**	42.0%	**0.886**	53.5%
R^2 (std. OS)	95%	0.758	12.8%	0.803	23.6%	0.823	27.1%	**0.873**	44.4%	**0.852**	52.2%
0.866	97%	0.726	12.0%	0.711	17.9%	0.781	27.9%	0.770	38.8%	0.790	50.9%
	98%	0.669	9.5%	0.750	21.1%	0.795	21.1%	0.878	40.7%	0.847	52.7%
250 tr. samples	99%	0.665	8.9%	0.682	9.9%	0.787	18.2%	0.805	41.4%	0.843	43.7%
Pagie-1	90%	**0.959**	11.6%	**0.942**	22.8%	**0.944**	29.4%	**0.950**	37.2%	**0.950**	49.4%
R^2 (std. OS)	95%	**0.958**	12.4%	**0.947**	19.8%	**0.960**	28.0%	**0.952**	39.6%	**0.955**	54.8%
0.94	97%	**0.946**	14.1%	**0.933**	19.1%	**0.952**	31.7%	**0.948**	46.6%	**0.945**	46.8%
	98%	**0.948**	14.4%	**0.944**	21.2%	**0.940**	37.3%	**0.954**	44.8%	**0.944**	47.3%
676 tr. samples	99%	**0.944**	12.5%	**0.949**	20.3%	**0.945**	27.3%	**0.956**	43.1%	**0.951**	48.8%
Tower	90%	**0.868**	48.2%	**0.860**	47.7%	**0.868**	64.1%	**0.869**	73.7%	**0.865**	73.5%
R^2 (std. OS)	95%	**0.867**	42.8%	**0.860**	40.3%	**0.868**	53.3%	**0.860**	62.4%	**0.868**	76.9%
0.87	97%	**0.865**	40.4%	**0.860**	43.6%	**0.869**	55.1%	**0.863**	69.0%	0.814	70.3%
	98%	**0.860**	34.6%	**0.866**	46.8%	**0.870**	56.3%	**0.860**	62.9%	**0.861**	77.3%
3136 tr. samples	99%	**0.866**	32.4%	**0.865**	38.6%	**0.863**	45.0%	**0.844**	55.0%	**0.870**	73.2%
Chemical-I	90%	0.678	9.2%	0.766	22.2%	**0.775**	31.8%	**0.770**	40.9%	**0.769**	52.3%
R^2 (std. OS)	95%	0.605	3.2%	0.744	20.2%	**0.772**	32.8%	**0.774**	38.3%	**0.770**	53.3%
0.78	97%	0.564	3.0%	0.717	19.1%	0.764	30.7%	0.768	39.1%	0.769	54.5%
	98%	0.572	2.6%	0.677	11.4%	0.768	33.0%	0.769	41.0%	**0.772**	53.6%
711 tr. samples	99%	0.547	2.1%	0.657	8.5%	0.762	32.9%	0.767	45.3%	0.769	56.3%

Table 4. Average false rejection rate and number of generations.

Problem	Row %	10%		20%		30%		40%		50%	
	Thresh.	Rej.	Gen.	Rej.	Gen.	Rej.	Gen.	Rej.	Gen.	Rej.	Gen.
Friedman-1	90%	1142.7	24.5	116.4	26.6	65.1	26.8	68.5	25.6	55.5	23.8
	95%	3965.1	24.2	907.9	24.9	125.5	25.8	125.8	25.2	87.6	25.6
	97%	5359.0	4.7	2452.0	23.4	265.2	25.6	609.0	25.6	204.2	24.5
	98%	5166.3	4.1	5872.4	9.1	554.2	24.3	1326.2	24.7	501.1	24.5
	99%	5787.3	3.8	5479.3	5.9	2426.2	19.7	5262.7	10.1	4574.8	11.6
Friedman-2	90%	78.1	29.1	80.3	27.0	67.7	24.4	15.5	23.9	2.6	26.3
	95%	305.5	26.2	287.2	26.2	142.9	26.4	74.6	28.5	16.4	26.3
	97%	611.5	25.6	601.8	25.3	326.3	26.0	128.2	24.5	61.7	24.5
	98%	1384.5	25.0	1148.2	24.6	800.6	27.2	198.4	24.7	115.8	25.4
	99%	4983.9	19.1	4124.4	18.4	3657.8	21.1	1024.6	25.3	677.3	26.6
Poly-10	90%	3520.5	17.8	1844.0	24.0	1173.4	23.9	587.9	23.5	483.6	24.4
	95%	4400.5	18.0	2953.3	21.7	2814.0	19.0	889.5	24.5	704.5	24.2
	97%	5380.7	15.9	3125.0	15.1	4903.5	17.6	1462.3	20.4	1162.6	22.5
	98%	4883.7	11.0	5181.7	16.8	4840.8	13.4	1611.4	21.6	1592.0	23.2
	99%	4978.4	10.7	4490.2	7.8	4734.3	11.6	3289.8	19.5	3518.6	18.4
Pagie-1	90%	526.1	25.3	197.6	27.9	106.9	25.6	105.8	25.1	40.4	25.9
	95%	813.1	26.7	325.9	24.1	212.1	24.6	182.5	25.7	89.8	30.3
	97%	1140.0	28.4	420.5	24.6	307.1	27.8	243.5	29.8	150.6	25.0
	98%	1486.6	27.4	586.7	24.6	544.8	29.3	317.7	27.7	145.3	25.5
	99%	1768.2	21.9	1292.8	23.3	1022.2	23.7	504.5	27.5	194.9	25.6
Tower	90%	34.2	34.0	4.4	29.6	4.5	33.7	1.7	33.0	0.9	32.8
	95%	72.9	33.6	7.9	28.7	8.5	31.1	3.5	33.0	1.9	33.9
	97%	114.0	34.3	11.4	30.6	15.7	33.4	7.0	34.4	3.9	32.4
	98%	148.5	31.8	17.5	33.5	29.2	34.1	9.4	32.4	11.4	34.5
	99%	230.3	31.9	36.7	31.1	60.3	30.6	23.9	31.1	40.6	34.7
Chemical-I	90%	7352.5	18.8	2316.8	30.5	765.4	31.4	594.8	30.7	468.5	32.4
	95%	7512.6	4.4	5604.1	26.4	1152.3	32.8	963.4	29.7	603.2	32.2
	97%	6332.0	4.4	8684.9	23.2	1841.9	30.5	988.3	29.7	696.5	31.9
	98%	7091.0	3.2	8197.8	12.0	2214.9	30.8	1379.0	31.6	722.1	31.8
	99%	6336.8	2.5	7679.6	9.5	3123.5	29.6	1333.3	31.8	1040.6	31.5

5 Conclusion

This paper presents a generic algorithmic extension to offspring selection which itself is a generic algorithmic extension to the general concept of genetic algorithms and genetic programming. The algorithmic refinements are aimed to improve the runtime performance of offspring selection by keeping the qualitative properties of the original version. The properties of the new variant have been shown exemplarily for genetic programming based symbolic regression because the extrapolation of partial solution evaluation is especially easy for this application. Depending on the characteristics of the considered benchmark problem the same qualitative behavior could be achieved with a ratio of effort between approximately 10% to 40% compared to standard offspring selection. This seems especially promising as the increase of efficiency turns out a fortiori as larger the data-sets become which seems especially interesting for big data analysis. Future investigations of the proposed algorithmic extensions will be aimed to introduce partial solution evaluation and its extrapolation also for other applications with expensive fitness evaluation like simulation based optimization where surrogate supported evaluation of the fitness function may lead to even higher savings as in the here considered showcase of symbolic regression.

Acknowledgments. The work described in this paper was done within the COMET Project #843532 Heuristic Optimization in Production and Logistics (HOPL) funded by the Austrian Research Promotion Agency (FFG) and the Government of Upper Austria and the COMET Project #843551 Advanced Engineering Design Automation (AEDA) funded by the Austrian Research Promotion Agency (FFG) and the Government of Vorarlberg.

References

1. Affenzeller, M., Wagner, S.: Offspring selection: a new self-adaptive selection scheme for genetic algorithms. In: Ribeiro, B., Albrecht, R.F., Dobnikar, A., Pearson, D.W., Steele, N.C. (eds.) Adaptive and Natural Computing Algorithms, pp. 218–221. Springer, Vienna (2005). https://doi.org/10.1007/3-211-27389-1_52
2. Holland, J.H.: Adaption in Natural and Artifical Systems. University of Michigan Press, Ann Arbor (1975)
3. Goldberg, D.E.: Genetic Algorithms in Search, Optimization and Machine Learning. Addison Wesley Longman, Boston (1989)
4. Affenzeller, M.: A new approach to evolutionary computation: segregative genetic algorithms (SEGA). In: Mira, J., Prieto, A. (eds.) IWANN 2001. LNCS, vol. 2084, pp. 594–601. Springer, Heidelberg (2001). https://doi.org/10.1007/3-540-45720-8_71
5. Affenzeller, M., Wagner, S.: SASEGASA: an evolutionary algorithm for retarding premature convergence by self-adaptive selection pressure steering. In: Mira, J., Álvarez, J.R. (eds.) IWANN 2003. LNCS, vol. 2686, pp. 438–445. Springer, Heidelberg (2003). https://doi.org/10.1007/3-540-44868-3_56
6. Affenzeller, M., Winkler, S.M., Kronberger, G., Kommenda, M., Burlacu, B., Wagner, S.: Gaining deeper insights in symbolic regression. In: Riolo, R., Moore, J.H., Kotanchek, M. (eds.) Genetic Programming Theory and Practice XI. GEC, pp. 175–190. Springer, New York (2014). https://doi.org/10.1007/978-1-4939-0375-7_10
7. Affenzeller, M., Beham, A., Vonolfen, S., Pitzer, E., Winkler, S.M., Hutterer, S., Kommenda, M., Kofler, M., Kronberger, G., Wagner, S.: Simulation-based optimization with HeuristicLab: practical guidelines and real-world applications. In: Mujica Mota, M., De La Mota, I.F., Guimarans Serrano, D. (eds.) Applied Simulation and Optimization, pp. 3–38. Springer, Cham (2015). https://doi.org/10.1007/978-3-319-15033-8_1
8. White, D.R., McDermott, J., Castelli, M., Manzoni, L., Goldman, B.W., Kronberger, G., Jaskowski, W., O'Reilly, U.M., Luke, S.: Better GP benchmarks: community survey results and proposals. Genet. Program. Evolvable Mach. **14**, 3–29 (2013)

Analysis of Schema Frequencies in Genetic Programming

Bogdan Burlacu[1,2]([✉]), Michael Affenzeller[1,2], Michael Kommenda[1,2], Gabriel Kronberger[1], and Stephan Winkler[1]

[1] Heuristic and Evolutionary Algorithms Laboratory, University of Applied Sciences Upper Austria, Softwarepark 11, 4232 Hagenberg, Austria
`bogdan.burlacu@fh-hagenberg.at`
[2] Institute for Formal Models and Verification, Johannes Kepler University, Altenbergerstr. 69, 4040 Linz, Austria

Abstract. Genetic Programming (GP) *schemas* are structural templates equivalent to hyperplanes in the search space. Schema theories provide information about the properties of subsets of the population and the behavior of genetic operators. In this paper we propose a practical methodology to identify relevant schemas and measure their frequency in the population. We demonstrate our approach on an artificial symbolic regression benchmark where the parts of the formula are already known. Experimental results reveal how solutions are assembled within GP and explain diversity loss in GP populations through the proliferation of repeated patterns.

Keywords: Genetic Programming · Schema analysis
Symbolic regression · Tree pattern matching · Evolutionary dynamics
Loss of diversity

1 Introduction

Genetic Programming (GP) *schemas* (or *schemata*) are structural templates equivalent to hyperplanes in the search space. They represent useful conceptual tools for analyzing properties of subsets of the population.

Schema theorems provide insight about genetic operator behavior (in particular, crossover and selection) and their influence on emergent phenomena such as bloat and the occurrence of repeated patterns in the population. Examples of theoretically-motivated results derived from schema theorems are the size evolution equation of GP programs [5] or the crossover bias theory of bloat [6]. A comprehensive overview of GP schema theory and its applications can be found for example in [10].

At the same time, the propagation of building blocks and the proliferation of repeated patterns in GP populations also represents a topic of particular interest, as shown for example in Langdon and Banzhaf [3], Wilson and Heywood [11] or Smart et al. [8].

© Springer International Publishing AG 2018
R. Moreno-Díaz et al. (Eds.): EUROCAST 2017, Part I, LNCS 10671, pp. 432–438, 2018.
https://doi.org/10.1007/978-3-319-74718-7_52

In this work, we aim to further investigate the propagation of genes in GP populations and the occurrence of emergent phenomena such as bloat, modularity, or loss of diversity. We introduce a new methodology to identify relevant schema templates and calculate their frequency in the population.

The remainder of this paper is organized as follows: Sect. 2 details our methodology, based on hereditary information and tree pattern matching. Section 3 describes the experiment used to empirically validate our approach, while Sect. 4 is dedicated to conclusions.

2   Methodology

In this paper, we propose a practical methodology for analyzing GP schema frequencies. We adopt the *hyperschema* definition by Poli [4], in which schemas are represented by rooted trees with internal nodes from the set $\mathcal{F} \cup \{=\}$ and leaf nodes from the set $\mathcal{T} \cup \{=, \#\}$. Here, $\mathcal{F}$ and $\mathcal{T}$ are the function and terminal set used to encode solution candidates as syntax trees. The two additional symbols '=' and '#' represent wildcards with the following matching behavior:

   The '=' symbol matches any symbol of the same arity.
- The '#' symbol matches any valid subtree.

Figure 1 shows an example of a schema containing the two types of wildcard symbols, along with a set of individuals matching its structure.

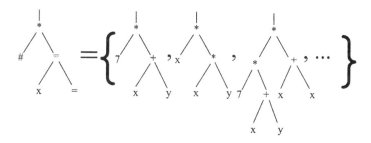

Fig. 1. Example of a schema and its matching trees [7].

Our analysis approach relies on two important steps:

1. Generation of relevant schema structures for matching
2. Calculation of schema frequencies using a pattern matching algorithm

2.1   Schema Generation

We perform this step under the assumption that individuals of common parents (ancestors) will share a common genetic template, enabling us to generate schemas by taking advantage of the hereditary relationship between crossover children and their parents. We consider the one-point crossover operator that takes two parent individuals and creates an offspring by replacing a subtree in the first parent (the root parent) with a subtree from the second parent (the non-root parent).

Our procedure generates schemas by first grouping offspring with the same root parent. Then, it considers all the cutpoint locations where subtrees were swapped in from the non-root parent by the crossover operator. Finally, it returns a new schema by inserting wildcards at the corresponding cutpoint locations in the structure of the root parent. Leaf nodes are replaced with '#' wildcards, while function nodes are replaced with '=' wildcards.

2.2   Schema Matching

We perform schema matching using the bottom-up variant of the algorithm for the tree homeomorphism decision problem by Götz et al. [2]. The algorithm answers *yes* if every parent-child pair in a query tree Q has a corresponding ancestor-descendant pair in data tree D. The bottom-up algorithm has a runtime complexity linear in the size of the two trees.

As the matching provided by the algorithm in its default implementation is not strict enough for schemas, we extended the node matching criteria with additional constraints, so that two nodes are matching only if they are on the same level in their respective trees and their parent and child nodes also match.

3   Experimental Results

The aim of our experiment is to test if the identified schemas have a high frequency in the population and include repeated patterns, modules and building blocks in their structure. For this reason we used a synthetic symbolic regression benchmark problem where the target formula is already known [9]:

$$y = x_1 x_2 + x_3 x_4 + x_5 x_6 + x_1 x_7 x_9 + x_3 x_6 x_{10} \tag{1}$$

We employ offspring selection genetic programming (OS-GP) [1] with a population size of 700 individuals, proportional parent selection and strict offspring selection. The maximum tree depth and length limits were set to 12 levels and 50 nodes, respectively. A maximum selection pressure of 100 was used as a termination criteria.

We demonstrate our methodology on a single algorithmic run solving the benchmark problem. Since in our run we discovered the complete formula, we aim to see how the main terms were assembled together by the algorithm. We analyze schema frequencies at each generation and we measure schema and population similarity at the semantic and structural level using the measures described for example in [12].

Figure 2 shows the relative frequency of the most common schema (left) and its average quality compared to the population (right). We notice that at some generations a single schema can match a significant percentage of the population. Due to strict offspring selection, both the schema and the overall population have very similar average quality. The similarity curves in Fig. 3 confirm, as expected, that individuals matching the same schema are structurally and semantically similar.

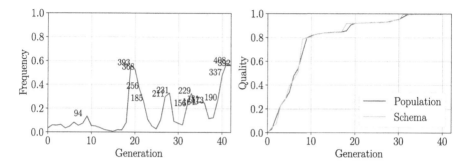

Fig. 2. Most common schema frequency (left) and average quality (right)

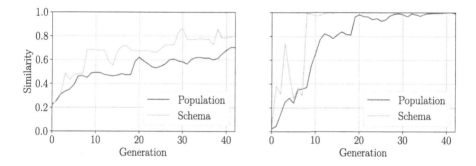

Fig. 3. Genotype (left) and phenotype (right) similarities

Next, we look at the most common schemas at key moments in the run of the algorithm. Figures 4, 5 and 6 show the structure of the schemas with the highlighted parts corresponding to the identified formula terms. We see that the highlighted parts are propagated via inheritance to the next generations and are shared between the identified schemas. The shared fragments and their multiplicity in the population (as indicated by schema frequencies) suggest that offspring selection leads to the occurrence of repeated patterns in the population. As they represent formula terms, these patterns may be regarded as building blocks for the algorithm.

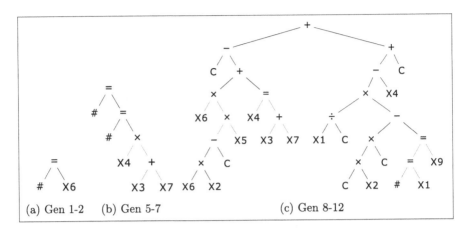

Fig. 4. Most frequent schemas between generations 1–2, 5–7 and 8–12.

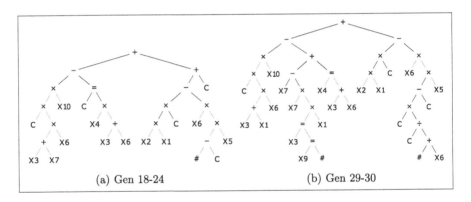

Fig. 5. Most frequent schemas between generations 18–24 and 29–30.

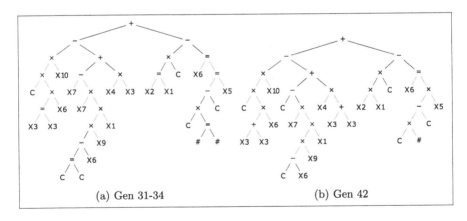

Fig. 6. Most frequent schemas between generations 31–34 and 39–42.

4 Conclusion

In this work, we demonstrated a practical methodology for identifying common schemas and calculating their frequencies in GP populations, based on hereditary information and a pattern matching algorithm. We showed empirically that under offspring selection, the population converges towards common genetic templates that contain repeated patterns in their structure. Our approach opens new possibilities for the investigation of GP evolutionary dynamics at a detailed, structural level, in relationship with existing schema theorems.

Future research might include for example different schema variants (e.g., only containing = symbols or # symbols in their structure), and try to use this information to steer the algorithm towards more diverse regions of the search space.

Acknowledgements. The work described in this paper was done within the COMET Project Heuristic Optimization in Production and Logistics (HOPL), #843532 funded by the Austrian Research Promotion Agency (FFG).

References

1. Affenzeller, M., Wagner, S.: Offspring selection: a new self-adaptive selection scheme for genetic algorithms. In: Ribeiro, B., Albrecht, R.F., Dobnikar, A., Pearson, D.W., Steele, N.C. (eds.) Adaptive and Natural Computing Algorithms, pp. 218–221. Springer, Vienna (2005). https://doi.org/10.1007/3-211-27389-1_52
2. Götz, M., Koch, C., Martens, W.: Efficient algorithms for descendant-only tree pattern queries. Inf. Syst. **34**(7), 602–623 (2009)
3. Langdon, W.B., Banzhaf, W.: Repeated patterns in genetic programming. Nat. Comput. **7**(4), 589–613 (2008)
4. Poli, R.: Exact schema theory for genetic programming and variable-length genetic algorithms with one-point crossover. Genet. Program Evolvable Mach. **2**(2), 123–163 (2001)
5. Poli, R., Langdon, W.B., Dignum, S.: Generalisation of the limiting distribution of program sizes in tree-based genetic programming and analysis of its effects on bloat. In: Proceedings of the 9th Annual Conference on Genetic and Evolutionary, GECCO 2007, pp. 1588–1595. Press (2007)
6. Poli, R., Mcphee, N.F.: Covariant parsimony pressure for genetic programming (2008)
7. Poli, R., McPhee, N.F.: General schema theory for genetic programming with subtree-swapping crossover: part I. Evol. Comput. **11**(1), 53–66 (2003)
8. Smart, W., Andreae, P., Zhang, M.: Empirical analysis of GP tree-fragments. In: Ebner, M., O'Neill, M., Ekárt, A., Vanneschi, L., Esparcia-Alcázar, A.I. (eds.) EuroGP 2007. LNCS, vol. 4445, pp. 55–67. Springer, Heidelberg (2007). https://doi.org/10.1007/978-3-540-71605-1_6
9. Vladislavleva, E.J., Smits, G.F., den Hertog, D.: Order of nonlinearity as a complexity measure for models generated by symbolic regression via Pareto genetic programming. IEEE Trans. Evol. Comput. **13**(2), 333–349 (2009)
10. White, D.: An overview of schema theory. CoRR abs/1401.2651 (2014). http://arxiv.org/abs/1401.2651

11. Wilson, G.C., Heywood, M.I.: Context-based repeated sequences in linear genetic programming. In: Keijzer, M., Tettamanzi, A., Collet, P., van Hemert, J., Tomassini, M. (eds.) EuroGP 2005. LNCS, vol. 3447, pp. 240–249. Springer, Heidelberg (2005). https://doi.org/10.1007/978-3-540-31989-4_21

12. Winkler, S., Affenzeller, M., Burlacu, B., Kronberger, G., Kommenda, M., Fleck, P.: Similarity-based analysis of population dynamics in genetic programming performing symbolic regression. In: Genetic Programming Theory and Practice XIV. Genetic and Evolutionary Computation. Springer (2016)

Analysis and Visualization of the Impact of Different Parameter Configurations on the Behavior of Evolutionary Algorithms

Stefan Wagner[1(✉)], Andreas Beham[1,2], and Michael Affenzeller[1,2]

[1] Heuristic and Evolutionary Algorithms Laboratory, School of Informatics, Communications and Media - Campus Hagenberg, University of Applied Sciences Upper Austria, Softwarepark 11, 4232 Hagenberg, Austria
{stefan.wagner,andreas.beham,michael.affenzeller}@fh-ooe.at
[2] Institute for Formal Models and Verification, Johannes Kepler University Linz, Altenberger Straße 69, 4040 Linz, Austria

Abstract. Evolutionary algorithms are generic and flexible optimization algorithms which can be applied to many optimization problems in different domains. Depending on the specific type of evolutionary algorithm, they offer several parameters such as population size, mutation probability, crossover and mutation operators, or number of elite solutions. How these parameters are set has a crucial impact on the algorithm's search behavior and thus affects its performance. Therefore, parameter tuning is an important and challenging task in each application of evolutionary algorithms in order to retrieve satisfying results.

In this paper, we show how software frameworks for evolutionary algorithms can support this task. As an example of such a framework, we describe how HeuristicLab enables automated execution of extensive parameter tests as well as its capabilities to analyze and visualize the obtained results. We also introduce a new chart of HeuristicLab, which can be used to compare the performance of many different parameter configurations and to drill down on different configurations in an interactive way. By this means this new chart helps users to visualize the influence of different parameter values as well as their interdependencies and is therefore a powerful feature in order to gain a deeper understanding of the behavior of evolutionary algorithms.

1 Introduction

Evolutionary algorithms and in most cases metaheuristic optimization algorithms in general offer several parameters which can be changed in order to adapt an algorithm's search behavior. For example, in a simple genetic algorithm the population size, the number of generations, the mutation rate, the number of elite solutions, the selection scheme, and the crossover and mutation operators can be configured. These parameters offer a high degree of flexibility and make it possible to tune an algorithm according to a specific application scenario. However, it is also well-known that the performance of evolutionary algorithms is very sensitive regarding these parameters. Small changes of parameter values often

© Springer International Publishing AG 2018
R. Moreno-Díaz et al. (Eds.): EUROCAST 2017, Part I, LNCS 10671, pp. 439–446, 2018.
https://doi.org/10.1007/978-3-319-74718-7_53

result in drastic changes of an algorithm's search characteristics. Additionally, many parameters strongly influence each other and therefore cannot be treated independently. Appropriate parameter tuning consequently is a non-trivial task and is of major importance when applying evolutionary algorithms [1].

Nevertheless, it is a surprising fact that parameter tuning is often not sufficiently considered. Many publications which describe the application of evolutionary algorithms on specific optimization problems do not provide any information about the way how the used parameter settings have been obtained. It seems that many researchers and especially newcomers in the field simply treat evolutionary algorithms as black boxes and choose an algorithm's parameter values rather unsystematically by trial and error. Thereby they miss the chance to gain deeper insights into the complex interplay of parameter configurations and algorithm behavior which would help to identify and explore promising parameter settings in a more efficient way.

In order to improve this situation, a significant research focus has been put on automated parameter tuning and algorithm configuration in the evolutionary computation community within the last years. Methods such as the irace package [2] for example represent powerful algorithms to efficiently tune parameters of optimization algorithms for different problem instances or even problem types. However, although automated parameter tuning techniques help to find suitable parameter configurations, they usually do not provide suitable feedback on the interrelationship of parameter settings and algorithm behavior. Therefore, they do not help users to gain a better understanding of the applied optimization algorithm and of the impact of its parameters.

In this paper, we present an interactive analysis and visualization approach which supports users of evolutionary algorithms (especially newcomers or students) in the exploration of different parameter settings and in understanding their influence on algorithm behavior. We extended our heuristic optimization environment HeuristicLab in such a way, that algorithm performance characteristics (i.e. obtained solution quality, variation of solution quality and required effort) can be easily computed for large numbers of test runs. A new interactive run analysis chart in HeuristicLab enables users to explore and compare different parameter configurations regarding these performance indicators. Each parameter setting is thereby represented as a series of data points of its associated performance measurements and convenient filtering and coloring make it possible to analyze even a very large number of runs. By this means users can for example easily identify parameter configurations which result in very similar or very different algorithm behavior.

2   Analysis and Visualization of Parameter Configurations in HeuristicLab

HeuristicLab[1] [3] is an open-source software environment for heuristic optimization that is developed and maintained by the research group Heuristic and

---

[1]   http://dev.heuristiclab.com.

Evolutionary Algorithms Laboratory (HEAL) of the University of Applied Sciences Upper Austria. It provides many optimization algorithms and problems and especially has a strong focus on evolutionary algorithms. A major motivation for the development of HeuristicLab is to create a system that enables researchers not only to implement new algorithms and problems, but also to benchmark and analyze algorithms in detail using a graphical user interface. For this purpose, users can quickly create experiments in HeuristicLab which contain many different algorithm and problem configurations. These experiments can then be executed either locally on the user's PC or they can be uploaded into HeuristicLab Hive for parallel and distributed execution on many machines. When an experiment is finished, it contains a set of runs, whereby each run stores the parameter and result values of a specific algorithm and problem configuration.

In order to analyze and compare the runs of an experiment, several interactive views can be used, of which *Bubble Chart* and *Box Plot* are the two most popular ones. Both let the user choose which run values of the underlying experiment should be displayed on the x- and on the y-axis. For example, if a user created an experiment to analyze the influence of different selection operators in a genetic algorithm, Bubble Chart can be used to compare its results. By selecting the parameter to analyze on the x-axis (*Selector*) and the considered result value on the y-axis (*BestQuality*), a Bubble Chart is displayed as shown in Fig. 1.

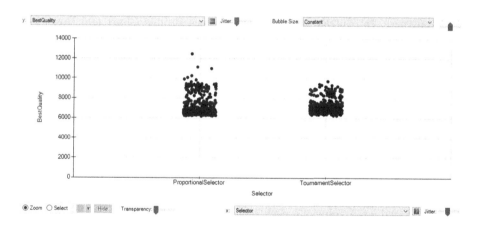

Fig. 1. Bubble Chart view of HeuristicLab

Both views, Bubble Chart and Box Plot, are very helpful features of HeuristicLab in order to quickly compare the achieved solution quality concerning the values of a specific parameter. We frequently use these charts in our research projects and publications and they are also very valuable for students, who want to learn how different parameter settings influence the performance of evolutionary algorithms.

However, both views also have two fundamental limitations. On the one hand, they only show the final result values of an algorithm at the time when the algorithm terminates. They cannot be used to visualize the progress of these result values over the iterations of an algorithm. As a consequence, they are not suitable for users who want to analyze the behavior of an algorithm over time, as for example its convergence characteristics. On the other hand, only one parameter can be selected on the x-axis at a time. When a user wants to analyze a large experiment with several different parameter configurations, it is not sufficient to visualize each parameter independently, as the performance of an algorithm regarding one parameter might strongly depend on the values of other parameters. Bubble Chart and Box Plot cannot show these interdependencies of parameters in an appropriate way.

3   Parameter Analysis Chart

In order to overcome the limitations of Bubble Chart and Box Plot described in the previous section, we implemented an additional view in HeuristicLab to visualize runs of an experiment. This new *Parameter Analysis* view plots the number of evaluated solutions in a user-defined step size on the x-axis. On the y-axis, it shows statistical values (minimum/maximum, 1st/3rd quartile, median, average) of the relative distance to the best-known solution quality of each run. By this means, it visualizes the performance of the contained runs not only regarding achieved solution quality, but also regarding performed effort and robustness. In contrast to the existing views Bubble Chart and Box Plot, the new Parameter Analysis view is therefore also able to depict the dynamic performance of different parameter configurations throughout the iterations of an algorithm.

In order to use the Parameter Analysis view, a specific analyzer has to be added to the algorithms contained in the experiment. Analyzers are special operators in HeuristicLab which are executed after each iteration of an algorithm. They can be used to log result values such as the current best, average, and worst solution quality or to compute other analyses as for example phenotypical similarity of solutions. As the Parameter Analysis view needs data on the progress of the best-found solution's quality in each run, the *QualityPerEvaluationsAnalyzer* has to be used which logs the current best quality and the current number of evaluated solutions each time when a better solution is found. As this analyzer only adds data to the run, if the current best quality is actually improved, it has only a small impact on the amount of data stored in each run. Hence it can also be used in experiments which contain algorithm configurations with a very large number of iterations.

Figure 2 shows an exemplary screen shot of the Parameter Analysis view used to analyze the same experiment as in the Bubble Chart shown in Fig. 1. On the upper left side, a tree shows the parameters and their different values used in the experiment. The number displayed in parentheses next to each tree node indicates the number of runs which contain that parameter or that specific parameter value. By selecting the check box of a parameter, the chart on the right side is updated and shows a data series for each value of this parameter.

It is also possible to check multiple parameters or individual parameter values in order to display an overlay of multiple data series.

Regarding the experiment shown in Fig. 2, it can be seen that the two selection operators used in the experiment lead to similar results for a high number of evaluated solutions, although robustness is a bit worse for proportional selection. Of course, this insight is not surprising as we have already compared the final solution quality of both selection operators in the Bubble Chart shown in Fig. 1 which revealed the same result. However, the Parameter Analysis view shown in Fig. 2 also shows that the two selection operators lead to different algorithm behavior in early iterations. Up to 400,000 evaluated solutions the median of the relative distance to the best-known solution quality is lower and therefore better when using tournament selection. This indicates that tournament selection has a higher selection pressure compared to proportional selection and consequently leads to quicker quality improvement in the beginning. This insight cannot be gained from the Bubble Chart shown in Fig. 1 and thus demonstrates how the new Parameter Analysis view helps to better understand the influence of different parameter settings on the dynamic behavior of algorithms.

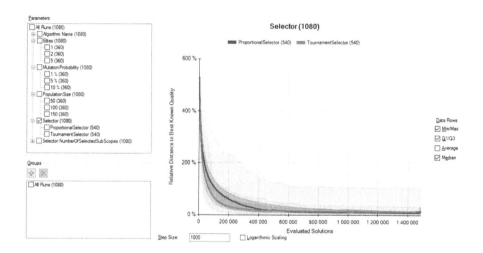

Fig. 2. Parameter Analysis view of HeuristicLab

In addition to the analysis of all runs contained in an experiment, the Parameter Analysis view can also be used to drill down on specific parameter settings. With this feature, it can be analyzed how the behavior of an algorithm regarding one parameter changes, when other parameters are fixed to concrete values. After selecting a parameter or a parameter value, it can be added to the groups section on the lower left side of the Parameter Analysis view. Then additional parameters can be added to each group. In this way, a user can interactively create a tree of groups, where each node in the tree contains only those runs of the experiment which have the concrete parameter values specified in its parent nodes. For example, we might want to analyze how the performance of an algorithm regarding its selection operator changes, if different numbers of elite solutions are used. For this purpose, the *Elites* parameter has to be added to the groups first and then the *Selector* parameter has to be added to each group which represents one of the values of the Elites parameter. After creating the groups tree in such a way, the check boxes next to the groups can be selected to update the chart. Figure 3 shows the Parameter Analysis view of the experiment used in the previous figures, where the groups for the Elites parameter and the subgroups for the Selector parameter have been added as described. By selecting *ProportionalSelector* in each of the three groups which represent the different values of the Elites parameter (1, 2, 5), three data series are shown in the chart. Each series represents the results of those runs which used proportional selection and the number of elite solutions as defined in the corresponding group.

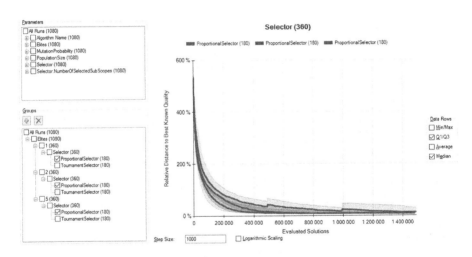

Fig. 3. Parameter Analysis view showing the performance of proportional selection regarding different numbers of elite solutions

In order to compare proportional selection to tournament selection regarding the influence of elite solutions, Fig. 4 shows the Parameter Analysis view with the same group configuration but with *TournamentSelector* checked in each group instead. It can be seen in the two charts that the number of elite solutions has an influence on algorithm performance in both cases. However, it is also obvious that proportional selection is more sensitive to the number of elite solutions than tournament selection.

These examples demonstrate that the new Parameter Analysis view of HeuristicLab is a very flexible and powerful chart that enables users to quickly analyze large experiments and to visualize and compare the performance of algorithms concerning different parameter configurations in an interactive way.

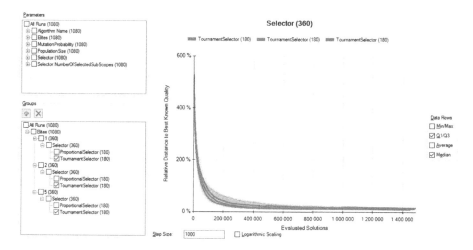

Fig. 4. Parameter Analysis view showing the performance of tournament selection regarding different numbers of elite solutions

4 Summary and Conclusion

As parameter tuning is an important step when applying evolutionary algorithms, researchers and students in this field need to quickly analyze and visualize the impact of different parameter configurations on the performance of algorithms. Therefore, software frameworks for evolutionary computing should support users in this task adequately. We described in this paper that our open-source optimization environment HeuristicLab can be used to create and to execute large experiments which contain many different algorithm and problem configurations. Additionally, HeuristicLab also offers several views such as Bubble Chart or Box Plot to analyze results of such experiments, but these views are not so well suited to visualize interdependencies between parameters and only show the final result values of an algorithm and not their progress during execution. Therefore, we implemented a new Parameter Analysis view in HeuristicLab and presented how it enables users to analyze the performance of many different parameter configurations regarding solution quality, robustness and effort. Furthermore, this new view also supports an interactive drill-down on specific parameter settings and is therefore also very helpful to analyze the interplay of different parameters. By this means, HeuristicLab and its new Parameter Analysis view are powerful tools which support their users to gain a better understanding of the impact of different parameter configurations on the behavior of evolutionary algorithms.

Acknowledgements. The work described in this paper is part of the COMET Project #843532 Heuristic Optimization in Production and Logistics (HOPL), funded by the Austrian Research Promotion Agency (FFG) and the Government of Upper Austria.

References

1. Eiben, A., Smit, S.: Parameter tuning for configuring and analyzing evolutionary algorithms. Swarm Evolut. Comput. **1**, 19–31 (2011)
2. López-Ibáñez, M., Dubois-Lacoste, J., Pérez Cáceres, L., Birattari, M., Stützle, T.: The irace package: iterated racing for automatic algorithm configuration. Oper. Res. Perspect. **3**, 43–58 (2016)
3. Wagner, S., et al.: Architecture and design of the HeuristicLab optimization environment. In: Klempous, R., Nikodem, J., Jacak, W., Chaczko, Z. (eds.) Advanced Methods and Applications in Computational Intelligence. Topics in Intelligent Engineering and Informatics, vol. 6, pp. 197–261. Springer, Heidelberg (2014). https://doi.org/10.1007/978-3-319-01436-4_10

Local Search Metaheuristics with Reduced Searching Diameter

Wojciech Bożejko[1], Andrzej Gnatowski[1], Czesław Smutnicki[1(✉)],
Mariusz Uchroński[1], and Mieczysław Wodecki[2]

[1] Faculty of Electronics, Wrocław University of Science and Technology,
Wyb. Wyspiańskiego 27, 50-370 Wrocław, Poland
{wojciech.bozejko,andrzej.gnatowski,czeslaw.smutnicki,
mariusz.uchronski}@pwr.edu.pl
[2] Institute of Computer Science, University of Wrocław,
Joliot-Curie 15, 50-383 Wrocław, Poland
mieczyslaw.wodecki@uwr.edu.pl

Abstract. In the paper we present some methods of empirical research of optimization problems' solution space, for which solutions are represented by permutations. Sampling the feasible solutions set we determine a histogram of frequency of incidence of local minima measuring its distance to the neighborhood graph's center. On its basis we verify a statistical hypothesis on normal distribution occurrence of local minima. Due to this research we can significantly reduce the area of the searching process during a local search metaheuristics work, focusing the searching process on Big Valley. We propose an algorithm with changeable diameter of the search.

1 Introduction

Job Shop Scheduling Problem (JSSP) is among the most commonly researched job scheduling problems. This is due to its generality and practical significance in optimizing production systems. JSSP can be defined as follows. There are n jobs, executed on m machines. Each job consists of a number of operations, that must be uninterruptedly performed in a specific order (technological order) on the predefined machines. The goal is to find the order in which the operations are performed on the machines (machine order), minimizing makespan, i.e. the time it takes to complete all the operations.

Since JSSP belongs to the class of NP-difficult problems, metaheuristic algorithms are frequently used. As shown by Wolpert and Macready in the *no free lunch theorem* [18], in order to obtain better results, it is necessary to use problem-specific properties. Thus, one of the best classic algorithms, TSAB by Nowicki and Smutnicki [10], utilizes block properties to limit the size of the neighbourhood to only "promising" solutions. Unfortunately, manual identification of the useful features of the problem is not always possible. Moreover, some of them may be applicable only to the specific instances, as the efficiency

R. Moreno-Díaz et al. (Eds.): EUROCAST 2017, Part I, LNCS 10671, pp. 447–454, 2018.
https://doi.org/10.1007/978-3-319-74718-7_54

of algorithms most often depends on the structure of the search space. Fitness Landscape Analysis (FLA) [6] is the main tool for studying the search space, also in JSSP [8]. During research over FLA, multiple classic measures of landscape were proposed: ruggedness [17], neutrality [13], epistasis [9] (important in the context of genetic algorithms), fitness distribution [14], etc. The list is not closed, with new measures still being introduced, such as frequency measures for dynamic optimization problems [7]. For a more detailed overview on measures in FLA, see [6,7]. One of the JSSP properties, identified by analyzing the search space is the Big Valley hypothesis (BV). BV states that there is a significant positive correlation between the value of the objective function of the solution and its distance from the optimum. The property has already been employed in multiple algorithms solving JSSP [11,19].

2   Proposed Method

As shown in the previous section, it is necessary to research the landscape of the solution space (fitness landscape) – connections between values of the cost function and distances in the space. There are several measures of distance for the permutations (e.g. Hamming, Spearman rank correlation, Kendall's tau, Cayley, Ulam) and some of them have a close relationship with moves usually used in the neighborhood generation inside local search metaheuristics. Fitness landscape is defined in the literature [12] as a triple (S, F, d), where S is a solution space, F cost function, and d distance measure between solutions. In addition, many advanced heuristic methods, e.g. paths relinking [2], are used to evaluate the distance between solutions. Such properties of fitness landscapes, for some scheduling problems, were described and applied in algorithms proposed in works of Smutnicki [15] and Bożejko [1]. The new knowledge about relationship between distances and cost function will be used in the method which we propose here.

The idea of the proposed tool is narrowing the search area and the use of this knowledge in the construction of neighborhood in the local search method, i.e. tabu search or simulated annealing. Firstly, we need to sample the solution space until a number of local extrema is found (e.g. hundred). This sampling can be done stochastically, or in a deterministic way during the first phase of the local search metaheuristic work – when the local search algorithm finds a new best solution, it is added to the local extrema 'sample' set. This is the first step. The second step determines the so-called central solution, a local extremum which has the shortest distance to the farthest element of the local extrema set. Then – the third step – we determine a histogram of frequency (distribution) of distances measured from the central solution to each element of the local extrema set. The distribution should be normal $\mathcal{N}(m, \sigma)$, and it can be checked by applying statistical test (e.g. Kolmogorov - Smirnov). From this research we have an average value m and standard deviation σ. So we can assume, that approximately 99.97% of local extrema lies between $m - 3\sigma$ and $m + 3\sigma$ from the central solution. The last step consists in researching, during the local search metaheuristics, only this area. Additionally, we can improve the information about the best region to research by adapting local extrema set – swapping the worst ones with the new extrema found during local search algorithm's work.

Numerical experiments done on the job shop scheduling problem confirm that such a method executes less need less number of steps to achieve the same quality of solutions, comparing to classical algorithm without the proposed mechanism. The job shop problem was chosen cause its hardness and permutational representation of the solution, as many others NP-hard combinatorial optimization problems, as e.g. traveling salesman problem, quadratic assignment problem and flow shop.

3 Job Shop Problem

The considered job shop scheduling problem can be described as follows, using notation of Nowicki and Smutnicki [10]. Let us consider a set of jobs $\mathcal{J} = \{1, 2, \ldots, n\}$, a set of machines $M = \{1, 2, \ldots, m\}$ and a set of operations $\mathcal{O} = \{1, 2, \ldots, o\}$. The set $\mathcal{O}$ is decomposed into subsets connected with jobs. A job j consists of a sequence of o_j operations indexed consecutively by $(l_{j-1} + 1, l_{j-1} + 2, \ldots, l_j)$ which have to be executed in this order, where $l_j = \sum_{i=1}^{j} o_i$ is the total number of operations of the first j jobs, $j = 1, 2, \ldots, n$, $l_0 = 0$, $\sum_{i=1}^{n} o_i = o$. An operation i has to be executed on machine $v_i \in M$ without any idleness in time $p_i > 0, i \in \mathcal{O}$. Each machine can execute at most one operation at a time. A feasible solution constitutes a vector of times of the operation execution beginning $S = (S_1, S_2, \ldots, S_o)$ such that the following constraints are fulfilled

$$S_{l_{j-1}+1} \geqslant 0, \quad j = 1, 2, \ldots, n, \tag{1}$$

$$S_i + p_i \leqslant S_{i+1}, \quad i = l_{j-1} + 1, \ l_{j-1} + 2, \ldots, l_j - 1, \quad j = 1, 2, \ldots, n, \tag{2}$$

$$S_i + p_i \leqslant S_j \quad \text{or} \quad S_j + p_j \leqslant S_i, \quad i, j \in O, \quad v_i = v_j, \quad i \neq j. \tag{3}$$

Certainly, $C_j = S_j + p_j$. An appropriate criterion function has to be added to the above constraints. The most frequent are the following two criteria: minimization of the time of finishing all the jobs and minimization of the sum of job finishing times. From the formulation of the problem we have $C_j \equiv C_{l_j}, j \in \mathcal{J}$.

The first criterion, the time of finishing all the jobs

$$C_{\max}(S) = \max_{1 \leqslant j \leqslant n} C_{l_j}, \tag{4}$$

corresponds to the problem denoted as $J||C_{\max}$ in the literature and it is strongly NP-hard.

4 Fitness Landscape Analysis

The base of the conception of search process direction is so-called fitness landscape of the cost function, more precisely - its "shape". This landscape, understood as multi-dimensional a cost function graph (however describing what is the landscape is not trivial – see Reeves [12]), has a number of so-called basins or funnels (see [5]) – local minima and their neighbors. In the further part of the paper we will consider a sample of these local minima as a source of information for search direction determination of a local search metaheuristics.

For a fixed set of moves (neighborhoods generator), let $G = (V, A)$ be a neighborhood graph, and d a measure (distance) determined on elements of Φ (vertices of the graph G). Let D be a set of selected local minima for the considered problem (the method of selection will be described further). We construct a graph $H = (D, U)$, where a set of arcs

$$U = \big\{\{u, v\} : \ u, v \in D, u \neq v \text{ and}$$

$$(\{u, v\} \in U \text{ if there is a path from } u \text{ to } v \text{ in } G)\big\}.$$

We define also $d_u = \max\{d(u, v) : \ v \in D\}$. The value of $d^* = \min\{d_u : \ u \in D\}$ we call **diameter**, and a vertex $c \in D$ such, that $d_c = d^*$ **central vertex** of the graph H.

Remark 1. If c is a central vertex in the graph H, than the distance from c to any vertex H is at most d^*.

Remark 2. If the graph H has two or more central vertexes, they are a clique.

The idea of the research is that distances between local minima have a normal distribution, i.e. relation between distance and a number of minima. This correlation is interpreted in the literature as *Big Valley* hypothesis (Nowicki and Smutnicki [11], Wong et al. [19]). An example of distance measure distribution for `Tail11` benchmark instance of Taillard [16] is shown in Fig. 1.

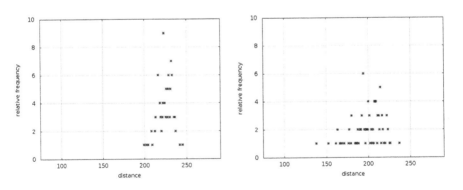

Fig. 1. Empirical distribution of solutions distance from center (left picture) and from the best found solution (right picture) from `Tail11` benchmark instance

Statistical method of the solution space research is described below. Its goal is determination of the local minima localization and final reducing of the problem's optimal solution' searching area. On the basis of conducted computational experiments we have assumed the following assumptions related to the hypothesis of the existence of the Great Valley:

(a) local minima are concentrated in an area of the solution space,
(b) the relationship between the distance from the optimal solution and the number of minima has a normal distribution.

The process of searching area determination can be presented in two steps: (1) sampling of the solution space and (2) adaptive searching of the reduced area.

Solution space sampling. Determine random set of solution space points

$$P = \{\pi_1, \pi_2, \ldots, \pi_n\}.$$

By using the improvement type algorithm (e.g. tabu search) starting respectively from the solutions from the set P, designate (e.g. using parallel calculations) a set of local minima

$$D = \{\pi_1^*, \pi_2^*, \ldots, \pi_t^*\}. \tag{5}$$

Empirical distribution of local minima. We calculate the center vertex π_{cen}^* in the graph D. We compute the distances $d^*(\pi_{cen}^*, v)$, $v \in D$ between the central vertex and the other local minima from the set of D. Then we calculate the average μ and standard deviation σ from these distances. Next, we examine the position of the local minima relative to the center π_{cen}^*, i.e. the relation between the distance from the center and the number of local minima. We verify the hypothesis H_0: elements of the set of local minima D due to the measure d have the normal distribution $N(\mu, \sigma)$. In case of positive verification, the solution space search is limited to solutions that are far from π_{cen}^*, not more than $\delta \cdot \sigma$ (radius of the search area), where $\delta = 1, 2, .., 6$ is an experimentally determined parameter. In case of negative verification (distribution is not normal) we put $\delta = \infty$ (lack of limitations of searching area). A neuro-tabu NTS algorithm is used for researching of the area, which is an extension of well-known TSAB algorithm of Nowicki and Smutnicki [10]. We also consider an algorithm version which determine distribution of distances from the best solution known so far.

Adaptical modification of the search area. During the operation of the algorithm, there are designated new local minima which do not belong to the set D. Thus, there are new information that can be used up to date to modify both the center of the graph and the parameters of the local minima, as well as to modify the radius of the search area. If we assume that the number of vertices in the graph H (i.e. the set of local minima D) is constant, then, to take into account the new information about the solution space (local minima), appearing during the course of computations, the following operations must be performed:

1. adding element (vertex) to the set D,
2. removing element (vertex) from the set D.

Both these operations affect both the selection of the central vertex in the graph H, and the search radius $\delta \cdot \sigma$. The set D is modified in case of appearing of a new local extrema π^* during solution process with the value of the goal function lower than the worst local extremum which belongs to D. In such a case it is removed from D and the new solution π^* is added.

5   Computational Experiments

Proposed algorithm (lNTS) is based on NTS algorithm [3,4] and were implemented in C++. Computational experiments were performed on BEM cluster located in Wroclaw Centre for Networking and Supercomputing[1]. The calculations were performed on Taillard benchmarks for job shop scheduling problem [16].

The lNTS algorithm was executed in two phased. In first phase during running NTS algorithm local minima was gathered. Next based on local minima set mean values of distances from best/central solutions was calculated. Also values of f_σ was calculated. In second phase of lNTS algorithm in neighborhood generation - candidate solution was picked if distance was in $[D(\pi_{mean} - f_\sigma, D(\pi_{mean}) + f_\sigma]$. Additionally if solution with lower cost function was found set of local minima is updated and value of f_σ is recalculated and used in neighborhood generation.

Table 1. PRD to the best known solutions upper bound.

Problem	$n \times m$	NTS	lNTS			
			$f_\sigma = 4$	$f_\sigma = 5$	*Best*	*Adapt*
TA01–10	15×15	0.5344	0.3615	0.5262	0.2565	0.2565
TA11–20	20×15	1.0371	0.9945	1.2134	0.7937	0.7937
TA21–30	20×20	1.0101	0.8371	0.7468	0.5275	0.5156
TA31–40	30×15	0.9911	0.9006	0.8866	0.6629	0.6518
TA41–50	30×20	1.7772	1.8130	1.6803	1.2241	1.2241
TA51–60	50×15	0.0915	0.0915	0.0915	0.0915	0.0915
TA61–70	50×20	0.1479	0.1715	0.1479	0.0244	0.0244
TA71–80	100×20	0.0089	0.0089	0.0089	0.0089	0.0089
Average		0.6998	0.6473	0.6627	0.4487	0.4458

The lNTS algorithm was executed for various values of the parameter f_σ ($f_\sigma = 1, 2, \ldots, 5$). The value of the f_σ parameter has got a great influence onto obtained results. The best values was obtained when the value of parameter $f_\sigma = 4$ (see Table 1). In Table 1 column denoted by *best* contains best results for lNTS execution with various values of the f_σ parameter. Also lNTS algorithm with adapting mechanism for f_σ parameter was tested. Test results for this mechanism are shown in *adapt* column.

---

[1] Calculations were conducted in Wroclaw Centre for Networking and Supercomputing within the Computational Grant No. 96.

6 Conclusions

In the paper we propose the new method of control of the local search metaheuristics trajectory by reducing search area of the metaheuristics by removing from the neighborhoods solutions which are far from the center of the area. This center is determined basing on the statistical analysis of the local extrema sample and it can be modified during the algorithm work by adding the new extrema. Computational simulations conducted on job shop benchmark instances show that the proposed method is efficient – it gives statistically better results in the same computing time.

References

1. Bożejko, W.: A New Class of Parallel Scheduling Algorithms. Oficyna Wydawnicza Politechniki Wrocławskiej, Wrocław (2010)
2. Bożejko, W.: Parallel path relinking method for the single machine total weighted tardiness problem with sequence-dependent setups. J. Intell. Manufact. **21**(6), 777–785 (2010)
3. Bożejko, W., Gnatowski, A., Niżyński, T., Wodecki, M.: Tabu search algorithm with neural tabu mechanism for the cyclic job shop problem. In: Rutkowski, L., Korytkowski, M., Scherer, R., Tadeusiewicz, R., Zadeh, L.A., Zurada, J.M. (eds.) ICAISC 2016. LNCS (LNAI), vol. 9693, pp. 409–418. Springer, Cham (2016). https://doi.org/10.1007/978-3-319-39384-1_35
4. Bożejko, W., Uchroński, M., Wodecki, M.: Parallel neuro-tabu search algorithm for the job shop scheduling problem. In: Rutkowski, L., Korytkowski, M., Scherer, R., Tadeusiewicz, R., Zadeh, L.A., Zurada, J.M. (eds.) ICAISC 2013. LNCS (LNAI), vol. 7895, pp. 489–499. Springer, Heidelberg (2013). https://doi.org/10.1007/978-3-642-38610-7_45
5. Daolio, F., Verel, S., Ochoa, G., Tomassini, M.: Local optima networks and the performance of iterated local search. In: Proceedings of the 14th Annual Conference on Genetic and Evolutionary Computation, pp. 369–376. GECCO 2012. ACM, New York (2012)
6. Humeau, J., Liefooghe, A., Talbi, E.G., Verel, S.: ParadisEO-MO: from fitness landscape analysis to efficient local search algorithms. J. Heuristics **19**(6), 881–915 (2013)
7. Lu, H., Shi, J., Fei, Z., Zhou, Q., Mao, K.: Measures in the time and frequency domains for fitness landscape analysis of dynamic optimization problems. Appl. Soft Comput. **51**, 192–208 (2017)
8. Mattfeld, D., Bierwirth, C., Kopfer, H.: A search space analysis of the job shop scheduling problem. Ann. Oper. Res. **86**, 441–453 (1999)
9. Naudts, B., Suys, D., Verschoren, A.: Epistasis as a basic concept in formal landscape analysis. In: Proceedings of the Seventh International Conference on Genetic Algorithms, pp. 65–72 (1997)
10. Nowicki, E., Smutnicki, C.: A fast taboo search algorithm for the job shop problem. Manag. Sci. **42**(6), 797–813 (1996)
11. Nowicki, E., Smutnicki, C.: An advanced tabu search algorithm for the job shop problem. J. Sched. **8**(2), 145–159 (2005)

12. Reeves, C.R.: Fitness landscapes. In: Burke, E.K., Kendall, G. (eds.) Search Methodologies, pp. 587–610. Springer, Boston (2005). https://doi.org/10.1007/0-387-28356-0_19

13. Reidys, C.M., Stadler, P.F.: Neutrality in fitness landscapes. Appl. Math. Comput. **117**(2–3), 321–350 (2001)

14. Rosé, H., Ebeling, W., Asselmeyer, T.: The density of states—a measure of the difficulty of optimisation problems. In: Voigt, H.-M., Ebeling, W., Rechenberg, I., Schwefel, H.-P. (eds.) PPSN 1996. LNCS, vol. 1141, pp. 208–217. Springer, Heidelberg (1996). https://doi.org/10.1007/3-540-61723-X_985

15. Smutnicki, C.: Optimization technologies for hard problems. In: Fodor, J., Klempous, R., Suárez Araujo, C.P. (eds.) Recent Advances in Intelligent Engineering Systems. SCI, vol. 378, pp. 79–104. Springer, Heidelberg (2012). https://doi.org/10.1007/978-3-642-23229-9_4

16. Taillard, E.: Benchmarks for basic scheduling problems. Eur. J. Oper. Res. **64**(2), 278–285 (1993)

17. Vassilev, V.K., Fogarty, T.C., Miller, J.F.: Information characteristics and the structure of landscapes. Evol. Comput. **8**(1), 31–60 (2000)

18. Wolpert, D.H., Macready, W.G.: No free lunch theorems for optimization. IEEE Trans. Evol. Comput. **1**(1), 67–82 (1997)

19. Wong, L.P., Puan, C.Y., Low, M.Y.H., Wong, Y.W., Chong, C.S.: Bee colony optimisation algorithm with big valley landscape exploitation for job shop scheduling problems. Int. J. Bio-Inspired Comput. **2**(2), 85 (2010)

Glucose Prognosis by Grammatical Evolution

J. Ignacio Hidalgo[1]([✉]), J. Manuel Colmenar[2], G. Kronberger[3],
and S. M. Winkler[3]

[1] Adaptive and Bioinspired Systems Group, Universidad Complutense de Madrid,
28040 Madrid, Spain
hidalgo@dacya.ucm.es
[2] Rey Juan Carlos University, Tulipán s/n, 28933 Móstoles (Madrid), Spain
[3] Heuristic and Evolutionary Algorithms Laboratory,
University of Applied Sciences Upper Austria,
Softwarepark 11, 4232 Hagenberg, Austria

Abstract. Patients suffering from Diabetes Mellitus illness need to control their levels of sugar by a restricted diet, a healthy life and in the cases of those patients that do not produce insulin (or with a severe defect on the action of the insulin they produce), by injecting synthetic insulin before and after the meals. The amount of insulin, namely bolus, to be injected is usually estimated based on the experience of the doctor and of the own patient. During the last years, several computational tools have been designed to suggest the boluses for each patient. Some of the successful approaches to solve this problem are based on obtaining a model of the glucose levels which is then applied to estimate the most appropriate dose of insulin. In this paper we describe some advances in the application of evolutionary computation to obtain those models. In particular, we extend some previous works with Grammatical Evolution, a branch of Genetic Programming. We present results for ten real patients on the prediction on several time horizons. We obtain reliable and individualized predictive models of the glucose regulatory system, eliminating restrictions such as linearity or limitation on the input parameters.

1 Introduction

Diabetes Mellitus is a chronic disease characterized by an elevated blood glucose level due to a problem in the production or in the effect of insulin which, only in Spain, affects to more than five million people, which represents around 10% of the population in 2017. There are two main types of diabetes; Type 1 (DM1) and Type 2 (DM2). Patients with DM1 need to measure their glucose level many times a day, as well as injecting insulin subcutaneously. Patients with DM2 have to take measures of their glucose mainly in the meal times and, eventually, they have to inject themselves insulin. In clinical practice, blood sugar can be measured by continuous glucose monitors (CGM) and insulin is injected

R. Moreno-Díaz et al. (Eds.): EUROCAST 2017, Part I, LNCS 10671, pp. 455–463, 2018.
https://doi.org/10.1007/978-3-319-74718-7_55

either manually or by continuous subcutaneous infusers or insulin pumps. The ideal solution for the patient is the development of an artificial pancreas (AP) [1]. AP is a solution, partially developed, where the glycemic control is completely autonomous. It requires a predictive model to estimate the future progress of blood glucose. With the information of the glucose level, a control algorithm would determine the dose of insulin to be delivered by the insulin pump, taking into account the variables and parameters included in the model. Current predictive models only consider measurements under controlled conditions of patients, which in most cases do not reflect the real-life or the patient. Glucose value prediction as a function of the insulin and food intakes is a difficult task that diabetics need to do everyday, since the AP is not accessible nowadays.

Taking glycemia, food intakes, levels of fatigue, stress, etc... as inputs, we can generate reliable predictive models of the levels of blood glucose, and implement bolus calculators for the daily management of the disease. Evolutionary Computation (EC) and Machine Learning (ML) had shown promising results in previous works [2]. In this work Grammatical Evolution techniques are applied for the prediction of glucose using the values measured by Continuous Glucose Monitoring (CGM) systems. We obtain more reliable and individualized predictive models of the glucose regulatory system, eliminating restrictions such as linearity or limitation on the input parameters. Our goal is to identify models (predictors) for the future glucose values after 30, 60, 90 and 120 min. The predictors were trained and tested using real data of 10 patients from a public hospital of Spain.

The rest of the paper is organized as follows. Section 2 describes the extraction of glucose models through Grammatical Evolution. Section 3 shows the experimental experience we have conducted and, finally, Sect. 4 draws the conclusions and the future work.

2   Glucose Prognosis of Diabetic Patients with Grammatical Evolution

Grammatical Evolution (GE) [3,4] is a form of Genetic Programming which represents the individuals using chromosomes instead of trees. The chromosomes are decoded through a grammar which produces the representation (phenotype) of a given individual. The use of chromosomes in GE allows the researcher the application of any of the available genetic operators that can be applied in Genetic Algorithms (GA) [5]. Moreover, given that the grammar is involved in the decodification, it is possible to introduce some information of the problem into the grammar to guide the optimization process to the best solutions.

In the case of modeling the glycemia of diabetic patients with GE, the phenotype of an individual is the model expression for prognosis. Hence, we need to create a grammar to guide the optimization process towards a model expression for prediction. Therefore, the grammar should consider that the prediction for a given time t may depend on the previous values of glucose, carbohydrates ingestion and insulin injection, as we previously stated in [6]. Taking into account this principle, several approximations can be made for the design of the grammar. In this paper we have tested two different kinds of grammars: a directed grammar, (Fig. 1) and a symbolic regression grammar (Fig. 2). The figures show excerpts of the rules that have been defined in the grammars. Next, we give the most important details related to them.

The directed grammar, also used in [7] for a different approach based on models for each meal of the day, presents an initial symbol, `<func>`, which defines an expression based on glucose (`<exprgluc>`), plus some expression regarding carbohydrates (`<exprch>`), minus an expression of insulin (`<exprins>`). In other words, this expression fixes a pattern that will be followed in all the individuals. Hence, the search process is directed in that way. The expression of glucose denoted by `<exprgluc>` is defined by a recursive rule that may produce a complex formula using arithmetic operators (`<op>`), functions (`<preop>`) and constant values (`<cte>`) which, in our case, are generated through a base and an exponent built with integer values.

We have implemented the GE process in Java using the ABSys JECO library [8] and *compilable phenotypes* to speed up the evaluation of individuals [9]. This is the reason because elements like the `<preop>` operands appear under the Java syntax. The glucose expression works with a prediction window of up to 2 h. Therefore, the terminal values can be either the predicted value within the window, or the real data before this window. This behavior is obtained with the functions `predictedData` and `realData`, and the indexes that are defined for them: `<idxCurr2h>`, which corresponds to the time within current two hours, and `<idx2hOrMore>`, which corresponds to the time before the last two hours. Notice that the dataset provides data in a 5 min' basis, therefore, $t - 24$ means 2 h ago.

As in the case of glucose, the expression for the carbohydrates may be recursively constructed, and takes into account the first previous intake since time t of the first input variable, which is the carbohydrates, as stated by the terminal function `getPrevCarbo(t)`. This value is modified by a constant, and is translated into a curve, following a similar approach to the "Batemanization" explained in [7].

The expression for the insulin is analogous in the sense that it is built using similar rules. However, instead of taking the previous insulin value, the terminal symbol is `getVariable(2,t-<idx>)`, which returns the value of the second input variable, the insulin, in time `t-<idx>`. This behavior is defined in this way because the amount of insulin values could by high due to the basal insulin, which is usually injected though a pump on a 5 min basis. The rest of the rules of the grammar correspond to terminal and non-terminal symbols related to operators and auxiliary indexes such as `<op>`, `<preop>`, etc. We refer to this grammar as GE_{dir} in the experimental results.

```
# Model expression
<func> ::= <exprgluc> + <exprch> - <exprins>

# Glucose
<exprgluc> ::= (<exprgluc> <op> <exprgluc>) | <preop> (<exprgluc>)
            | (<cte> <op> <exprgluc>) | predictedData(t-<idx>)
            | realData(t-<idx2hOrMore>)

# CH
<exprch> ::= (<exprch> <op> <exprch>) | <preop> (<exprch>)
           | (<cte> <op> <exprch>) | (getPrevCarbo(t) * <cte> * <curvedCH>)

# Insulin
<exprins> ::= (<exprins> <op> <exprins>)
            | <preop> (<exprins>)
            | (<cte> <op> <exprins>)
            | getVariable(2,t-<idx>)

<op> ::= +|-|*|/
<preop> ::= Math.exp|Math.sin|Math.cos|Math.log
<cte> ::= <base>*Math.pow(10,<sign><exponent>)
<base> ::= <dgtNoZero><dgtNoZero>
<exponent> ::= 1|2|3|4|5|6|8|9
<sign> ::= +|-
<idx> ::= <dgtNoZero>|<dgtNoZero><dgt>|<dgtNoZero><dgt><dgt>
<dgtNoZero> ::= 1|2|3|4|5|6|7|8|9
<dgt> ::= 0|1|2|3|4|5|6|7|8|9
```

Fig. 1. Excerpt of production rules for the directed grammar developed for the extraction of glycemic models.

```
<func> ::= <gl> <op> <ch> <op> <ins> <op> <cte>

<gl> ::= <preop> (<gl>) | <gl> <op> <gl> | <vargl>
<ch> ::= <preop> (<ch>) | <ch> <op> <ch> | <cte> <op> (<ch>) | <varch>
<ins> ::= <preop> (<ins>) | <ins> <op> <ins> | <varins>

<op> ::= +|-|/|*

<vargl> ::= getVariable(2,k)|getVariable(3,k)|getVariable(4,k)|getVariable(5,
        k)|getVariable(6,k)|getVariable(7,k)|getVariable(8,k)

<varch> ::= getVariable(9,k)|getVariable(10,k)|getVariable(11,k)|getVariable
        (12,k)|getVariable(13,k)|getVariable(14,k)|getVariable(15,k)|getVariable
        (16,k)|getVariable(17,k)|getVariable(18,k)|getVariable(19,k)|getVariable
        (20,k)|getVariable(21,k)|getVariable(22,k)|getVariable(23,k)

<varins> ::= getVariable(24,k)|getVariable(25,k)|getVariable(26,k)|
        getVariable(27,k)|getVariable(28,k)|getVariable(29,k)|getVariable(30,k)|
        getVariable(31,k)|getVariable(32,k)|getVariable(33,k)|getVariable(34,k)|
        getVariable(35,k)|getVariable(36,k)|getVariable(37,k)|getVariable(38,k)

<preop>::= Math.exp|Math.log
<cte>::= <c><c>.<c><c>
<c> ::= 0|1|2|3|4|5|6|7|8|9
```

Fig. 2. Excerpt of production rules for the symbolic regression grammar developed for the extraction of glycemic models.

In the case of the symbolic regression grammar displayed in Fig. 2 the approach is different. The `<func>` symbol does not direct the final expression for the phenotype. On the contrary, the `<op>` symbol provides different arithmetic operations, and the recursive definitions of `<gl>`, `<ch>` and `<ins>` allow the construction of complex expressions for each element. Moreover, the symbols `<vargl>`, `<varch>` and `<varins>` lead to different terminal elements for glucose, carbohydrates and insulin input variables.

In the experiments conducted with this grammar we have extended the dataset for each patient generating a total number of 38 input variables whose aim is to preprocess the input data. As seen in the grammar, input variables from 2 to 8 correspond to values calculated from the glucose; input variables from 9 to 23 correspond to values related to carbohydrates and input variables from 24 to 38 correspond to insulin values. Therefore, this grammar should behave in a more general way regarding the search process. We refer to this grammar as GE_{SR} in the experimental results.

3   Experimental Results

We made a retrospective study on ten DM1 patients ($n = 10$) who were selected because they presented a good glucose control. Data from patients were acquired over multiple days using Medtronic insulin pump records. Log entries were stored in five-minute intervals containing the date and time, and depending on the event the blood glucose value, the amount of insulin (injected via pump), and/or the amount of carbohydrate intakes as estimated by the patients. The characterization of the patient is female (80%), with an average age 42.3 (± 11.07), years of progress of disease 27.2 (± 10.32), years with pump therapy 10 (± 4.98), weight 64.78 (± 13.31) kg, HbA1c average of 7.27% ($\pm 0.5\%$). The average number of days with data is 44.80 (± 30.73).

In the experimentation we have compared our GE approach, using the two grammars described in the previous section (GE_{SR} and GE_{Dir}) with two baseline predictors approaches. The first baseline, called *Avg*, considers the average glucose of the previous values in the past two hours. The second baseline, called *Last*, considers as prediction the last known value of the glucose. For each patient we have obtained four different models for the prediction of blood glucose concentration in 30, 60, 90 and 120 min using each one of the proposals. In order to perform a cross validation process, we have divided the data into 10 folds. Finally, we have run the algorithms 5 times on each fold to avoid the random bias.

As stated before, the GE algorithm use standard genetic operators. In particular, we apply the classical genetic algorithm with a population size of 200 individuals, which runs during 250 generations. We use single-point crossover with 0.75 probability and uniform mutation with a probability of 0.15. Regarding the parameters that are particular for GE we use 300 codons with 5 as maximum number of wrappings, and RMSE of the differences between the predicted and the expected glucose values as fitness function.

Table 1 shows the experimental results. Each row presents the averaged results for the 10 patients in terms of Clarke Error Grid Analysis (CEGA), a metric commonly used in Endocrinology to test the clinical significance of differences between predictions and real values of blood glucose [10]. CEGA uses a Cartesian diagram divided into five zones (A to E). Zone A and B are predictions with no danger for the patient, while zones C to E are potentially dangerous. Higher values of results in zones A and B are better, while lower numbers in zones C, D and E are preferred. Results in Table 1 are divided into four blocks, one for each time horizon.

We can observe that GE produced better prediction in all cases and we can conclude that GE_{SR} reduces the number of predictions in dangerous zones. This is an interesting result, since the inclusion of knowledge in the grammar, as GE_{Dir} does, is not beneficial for the quality of the prediction in this case. The reason is that, due to the structure of the grammar, the diversity of the solutions is lower with GE_{Dir} than with GE_{SR}. We have noticed that in the decoding process the probability of having solutions with different phenotypes, i.e. different models, is lower with GE_{Dir} since the starting point is always the same. Those results indicate that a further exploration of the grammars is needed, in combination with a pre-processing of the data.

Table 1. Average of predictions (in percent) on independent test data. For each patients and prediction horizon, the best modeling results are highlighted. Added percentages could be higher than 100, since we are averaging values of 10 patients.

Horizon	$t+30$				$t+60$			
Alg	A + B	C	D	E	A + B	C	D	E
Last	71.83 + 25.25	0.22	2.68	0.02	53.84 + 39.82	1.20	4.78	0.36
Avg	39.98 + 45.85	0.00	14.17	0.00	39.98 + 45.91	0.00	14.12	0.00
GE_{Dir}	**85.36 + 22.07**	0.29	2.27	0.01	**63.01 + 40.69**	1.23	4.81	0.25
GE_{SR}	80.66 + 17.53	**0.37**	**1.44**	**0.01**	58.64 + 35.81	**1.27**	**4.06**	**0.21**
Horizon	$t+90$				$t+120$			
Alg	A + B	C	D	E	A + B	C	D	E
Last	44.40 + 45.57	2.83	6.16	1.05	39.13 + 47.96	4.14	6.88	1.88
Avg	40.00 + 45.90	0.00	14.11	0.00	40.05 + 45.87	0.00	14.08	0.00
GE_{Dir}	**51.91 + 47.98**	**2.08**	**7.48**	**0.55**	**46.72 + 48.77**	2.42	11.88	0.21
GE_{SR}	47.59 + 43.37	2.27	6.15	0.62	41.74 + 47.01	**2.17**	**8.19**	**0.89**

Figure 3 shows the CEGA results for one of the patients with the best model on one of the cross validation folds with GE_{SR}. As we can observe, there is a small number of predictions in the dangerous zones. Similar results are obtained for all the patients and all the folds. Figure 4 presents a comparison of the predicted glucose values and real values for GE_{SR}. Four figures are presented, best patient and worst patient in the best and worst days.

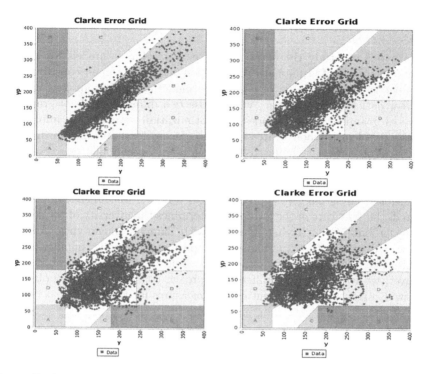

Fig. 3. Clarke error grid analysis results for 30 (up-left), 60 (up-right), 90 (down-left), and 120 (down-right) minutes for patient 1

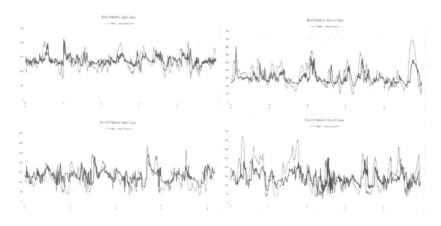

Fig. 4. A comparison of the predicted glucose values and real values for GE_{SR}. Four figures are presented, best patient best day (up-left), best patient worst day (up-right), and worst patient in the best day (down-left), and worst day (down-right).

4   Conclusions and Future Work

In this paper we study the extraction of custom glucose models for diabetic patients by means of Grammatical Evolution (GE). The main contribution of this paper is the application of GE in the model extraction for long datasets from real patients. To this aim, we propose two different kinds of grammars: a directed grammar, which guides the optimization search using information from the problem (carbohydrates raise the glucose level and insulin drops it); and a general grammar similar to the ones used for symbolic regression. The comparison of those different grammars applied to ten real patients indicates that the premature inclusion of knowledge in the grammar is not always beneficial in terms of avoiding dangerous predictions. Therefore, we can conclude that, as expected, the structure of the grammar determines the diversity of the solutions and the probability of having solutions with different phenotypes. As future work, we will extend the experimentation considering a higher number of input variables like the exercise, the stress and some other physiological variables. Moreover, we will work on a different approach considering several objectives which will correspond to different quality measures for the models.

References

1. El-Khatib, F.H., Russell, S.J., Nathan, D.M., Sutherlin, R.G., Damiano, E.R.: A bihormonal closed-loop artificial pancreas for type 1 diabetes. Sci Transl Med **2**(27), 27ra27–27ra27 (2010)
2. Colmenar, J.M., Winkler, S.M., Kronberger, G., Maqueda, E., Botella, M., Hidalgo, J.I.: Predicting glycemia in diabetic patients by evolutionary computation and continuous glucose monitoring. In: Proceedings of the 2016 on Genetic and Evolutionary Computation Conference Companion, pp. 1393–1400. ACM (2016)
3. O'Neill, M., Ryan, C.: Grammatical evolution. IEEE Trans. Evol. Comput. **5**(4), 349–358 (2001)
4. O'Neill, M., Ryan, C.: Grammatical Evolution: Evolutionary Automatic Programming in an Arbitrary Language. Kluwer Academic Publishers, Dordrecht (2003)
5. Eiben, A.E., Smith, J.E.: Introduction to Evolutionary Computing. Springer, Heidelberg (2003)
6. Hidalgo, J.I., Colmenar, J.M., Risco-Martin, J.L., Cuesta-Infante, A., Maqueda, E., Botella, M., Rubio, J.A.: Modeling glycemia in humans by means of grammatical evolution. Appl. Soft Comput. **20**, 40–53 (2014). Hybrid intelligent methods for health technologies
7. Colmenar, J.M., Winkler, S.M., Kronberger, G., Maqueda, E., Botella, M., Hidalgo, J.I.: Predicting glycemia in diabetic patients by evolutionary computation and continuous glucose monitoring. In: Proceedings of the 2016 on Genetic and Evolutionary Computation Conference Companion, GECCO 2016 Companion, pp. 1393–1400. ACM, New York (2016)
8. Adaptive and Bioinspired Systems Group. ABSys JECO (Java Evolutionary COmputation) library (2015). https://github.com/ABSysGroup/jeco

9. Colmenar, J.M., Hidalgo, J.I., Lanchares, J., Garnica, O., Risco, J.-L., Contreras, I., Sánchez, A., Velasco, J.M.: Compilable phenotypes: speeding-up the evaluation of glucose models in grammatical evolution. In: Squillero, G., Burelli, P. (eds.) EvoApplications 2016. LNCS, vol. 9598, pp. 118–133. Springer, Cham (2016). https://doi.org/10.1007/978-3-319-31153-1_9
10. Clarke, W., Cox, D., Gonder-Frederick, L., Carter, W., Pohl, S.: Evaluating clinical accuracy of systems for self-monitoring of blood glucose. Diab. Care **10**(5), 622–628 (1987)

Fitness Landscape Analysis in the Optimization of Coefficients of Curve Parametrizations

Stephan M. Winkler[1(✉)] and J. Rafael Sendra[2]

[1] Heuristic and Evolutionary Algorithms Laboratory,
University of Applied Sciences Upper Austria, Hagenberg, Austria
stephan.winkler@fh-hagenberg.at
[2] Dpto. de Física y Matemáticas, Research Group ASYNACS,
Universidad de Alcalá de Henares, Madrid, Spain
rafael.sendra@uah.es

Abstract. Parametric representations of geometric objects, such as curves or surfaces, may have unnecessarily huge integer coefficients. Our goal is to search for an alternative parametric representation of the same object with significantly smaller integer coefficients. We have developed and implemented an evolutionary algorithm that is able to find solutions to this problem in an efficient as well as robust way.

In this paper we analyze the fitness landscapes associated with this evolutionary algorithm. We here discuss the use of three different strategies that are used to evaluate and order partial solutions. These orderings lead to different landscapes of combinations of partial solutions in which the optimal solutions are searched. We see that the choice of this ordering strategy has a huge influence on the characteristics of the resulting landscapes, which are in this paper analyzed using a set of metrics, and also on the quality of the solutions that can be found by the subsequent evolutionary search.

1 Introduction: Evolutionary Search for Optimal Coefficients of Curve Parametrizations

Rational curves, and more generally unirational algebraic sets, admit parametrizations with rational functions. Since this representation is not unique, many optimization questions arise. In particular, we are interested in computing parametrizations with small height. As documented in [1] we dealt with this problem and developed and implemented an evolutionary algorithm that is able to find solutions to this problem in an efficient as well as robust way (see also [2]). In this paper we analyze the fitness landscapes associated with this evolutionary algorithm.

The authors gratefully acknowledge financial support within the projects *MTM2014-54141-P* (Ministerio de Economía y Competitividad and European Regional Development Fund ERDF) and *HOPL* (Austrian Research Promotion Agency FFG, #843532).

R. Moreno-Díaz et al. (Eds.): EUROCAST 2017, Part I, LNCS 10671, pp. 464–472, 2018.
https://doi.org/10.1007/978-3-319-74718-7_56

The problem we are dealing with is stated as follows: We are given a (proper) parametrization of an space curve, expressed as

$$\mathbf{P}(t) = \left(\frac{p_1(t)}{q(t)}, \dots, \frac{p_r(t)}{q(t)} \right) \tag{1}$$

where $p_i, q \in \mathbb{Z}[t]$ are coprime polynomials. The problem consists in finding $a, b, c, d \in \mathbb{Z}$, with $ad - bc \neq 0$, such that when t is substituted by $(at+b)/(ct+d)$ the height (i.e., the maximum coefficient in absolute value) of

$$\mathbf{P}\left(\frac{at + b}{ct + d} \right) \tag{2}$$

is minimal. In [1] we presented an evolutionary algorithm that is able to solve this problem. Roughly speaking, this algorithm works in two phases: First, partial solutions are identified and collected in the set Ω_e. Each element $\mathbf{o}$ in Ω_e is defined as $\mathbf{o} = (o_1, o_2) \in \mathbb{Z}^2$ with $\gcd(o_1, o_2) = 1$. Second, the best combinations of elements in Ω_e for composing the final complete solution of the given problem have to be found.

The composition of complete solution candidates from partial solution candidates is defined as follows: Using $(\mathbf{o}_1, \mathbf{o}_2) \in \Omega_e \times \Omega_e$ with $\mathbf{o}_1 = (a, c)$, and $\mathbf{o}_2 = (b, d)$, the associated complete solution candidate is $S_{\mathbf{o}_1, \mathbf{o}_2} := (a, b, c, d)$. Conversely, every complete solution candidate $(a, b, c, d) \in \text{Space}(\Omega_e)$ can be seen as a combination of elements in Ω_e, namely $(a, c), (b, d) \in \Omega_e$.

In order to measure the quality of a complete solution candidate $\mathbf{s} := S_{\mathbf{o}_1, \mathbf{o}_2}$ we use the notion of complete quality as the height of the resulting parametrization after substituting t by $(at + b)/(ct + d)$.

$$\text{Quality}_c(\mathbf{s}, \mathbf{P}) \text{ is the height of } \mathbf{P}\left(\frac{at + b}{ct + d} \right). \tag{3}$$

This second phase of the algorithm is implemented as an evolutionary algorithm that finds the best combination of partial solutions.

2 Orderings of Combinations of Partial Solutions and Resulting Fitness Landscapes

The key for the second phase of the evolutionary algorithm is to work with a suitable ordered copy of Ω_e, denoted as Ω_e^{ord}. Since we use an evolutionary process to find optimal combinations of partial solutions, the fitness function that evaluates these partial solutions is of essential importance. In this section we describe different strategies to order Ω_e, and later we analyze the resulting fitness landscapes of them in comparison with the option of not ordering the space of solutions, that is taking $\Omega_e^{\text{ord}} = \Omega_e$.

2.1   Ordering Combinations of Partial Solutions

For describing the orders used to generate Ω_e^{ord} we will use the same notation as in [1], that we briefly recall here. Let $\mathbf{P}^H(t, h)$ be the homogenization of $\mathbf{P}(t)$ (see (1)). We express $\mathbf{P}^H(t, h)$ as

$$\mathbf{P}^H(t, h) = (P_1(t, h), \ldots, P_r(t, h), Q(t, h)) \tag{4}$$

Given $\mathbf{o} \in \Omega_e$ we consider the following functions to order the search space

1. [Gcd-order] We take the partial quality function as

$$\text{Quality}_{\text{p}}^{\text{gcd}}(\mathbf{o}, \mathbf{P}) := \gcd(P_1(\mathbf{o}), \ldots, P_r(\mathbf{o}), Q(\mathbf{o})). \tag{5}$$

Then, we consider the following order: if $\mathbf{o}_1, \mathbf{o}_2 \in \Omega_e$, we say that

$$\mathbf{o}_1 \leq_{\text{gcd}} \mathbf{o}_2 \iff \text{Quality}_{\text{p}}^{\text{gcd}}(\mathbf{o}_1, \mathbf{P}) \leq \text{Quality}_{\text{p}}^{\text{gcd}}(\mathbf{o}_2, \mathbf{P}).$$

$\text{Quality}_{\text{p}}^{\text{gcd}}$ is the partial quality function used in the implementation in [1]. The reason of using this function is based on Lemma 3.1. in [1], and it ensures that if the complete solution candidate $\mathbf{s} := S_{\mathbf{o}, \mathbf{o}^*}$ is generated by means of the partial elements $\mathbf{o}$ and $\mathbf{o}^*$, and the gcd of the leading coefficients (resp. of the independent coefficients) of the polynomials in the output parametrization is given by $\text{Quality}_{\text{p}}(\mathbf{o}, \mathbf{P})$ (resp. $\text{Quality}_{\text{p}}(\mathbf{o}^*, \mathbf{P})$). We denote the corresponding space of solutions as Ω_e^{Gcd}.

2. [Δ-order] In [1], in order to reduce the search space, we used a constant k (usually taken as $k = 10^2$) that represents the potential expected improvement given by a partial solution candidate. This is controlled by asking that $k \cdot \Delta(\mathbf{o})$ is smaller than the quality (i.e. the height) of the input parametrization $\mathbf{P}$ (see (20) in [1]), where $\Delta(\mathbf{o})$ is defined as

$$\Delta(\mathbf{o}) := \frac{\max\{|P_1(\mathbf{o})|, \ldots, |P_r(\mathbf{o})|, |Q(\mathbf{o})|\}}{\gcd(|P_1(\mathbf{o})|, \ldots, |P_r(\mathbf{o})|, |Q(\mathbf{o})|)}.$$

Based on this fact we define in a new partial quality function as

$$\text{Quality}_{\text{p}}^{\Delta}(\mathbf{o}, \mathbf{P}) := \Delta(\mathbf{o}). \tag{6}$$

Then, we introduce a new order in Ω_e as follows: if $\mathbf{o}_1, \mathbf{o}_2 \in \Omega_e$, we say that

$$\mathbf{o}_1 \leq_\Delta \mathbf{o}_2 \iff \text{Quality}_{\text{p}}^{\Delta}(\mathbf{o}_1) \geq \text{Quality}_{\text{p}}^{\Delta}(\mathbf{o}_2).$$

We denote the corresponding space of solutions as Ω_e^{Δ}.

3. [Non-order] As a third option, we consider none order in Ω_e. So, elements in Ω_e are stored as they appear in the computation. We denote the corresponding space of solutions as Ω_e^{Non}.

Figures 1, 2 and 3 show exemplary fitness landscapes of combinations of elements in Ω_e for a given problem where partial solution candidates are unordered (shown in Fig. 1) or ordered by means of their evaluation according to $\text{Quality}_{\text{p}}^{\text{gcd}}$ (shown in Fig. 2) or ordered by means of their evaluation according to $\text{Quality}_{\text{p}}^{\Delta}$ (shown in Fig. 3).

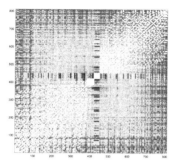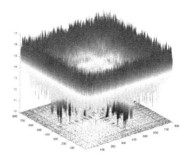

Fig. 1. Fitness landscape for combinations of elements of Ω_e^{Non} for the parametrization $\mathbf{P}_1(t)$ defined in Sect. 3. Each cell (x, y) represents the fitness of combination of $x \in \Omega_e^{\mathrm{Non}}$ and $y \in \Omega_e^{\mathrm{Non}}$.

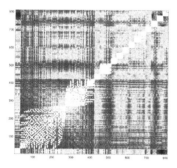

Fig. 2. Fitness landscape for combinations of elements of Ω_e^{Gcd} for the parametrization $\mathbf{P}_1(t)$ defined in Sect. 3. Each cell (x, y) represents the fitness of combination of $x \in \Omega_e^{\mathrm{Gcd}}$ and $y \in \Omega_e^{\mathrm{Gcd}}$.

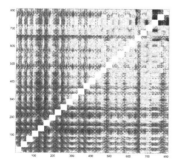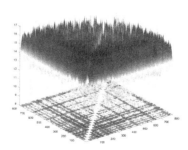

Fig. 3. Fitness landscape for combinations of elements of Ω_e^{Δ} for the parametrization $\mathbf{P}_1(t)$ defined in Sect. 3. Each cell (x, y) represents the fitness of combination of $x \in \Omega_e^{\Delta}$ and $y \in \Omega_e^{\Delta}$.

2.2 Analysis of Resulting Fitness Landscapes

For characterizing fitness landscapes formed using fitness functions and estimating their effects on the performance of the algorithm we perform the following analyses:

- On the one hand we use metrics describing the characteristics of the surfaces following guidelines given for example in [3]. Concretely, we use metrics that describe the local characteristics of the surfaces as well as a metric that describes the landscape on a higher level:
  - The ruggedness is calculated as the standard deviation of the values around a point p: $ruggedness(p, k) = \sigma(Q(p, k))$ where for all points $q_i \in Q(p, k)$ the distance to p in x- and y-direction is not greater than k: $|x(q_i) - x(p)| \leq k, |y(q_i) - y(p)| \leq k$. The ruggedness of a landscape is then the average ruggedness of all points.
  - A trajectory based metric is defined in the following way: Starting from a point (x, y), k new points are collected as points $p_1 \ldots p_k$ where each point p_j is reached as mutation of point p_{j-1}: $p_{j-1} = mutation(p_j)$. The range of the so reached trajectory is calculated as $range(p) = max(L(p)) - min(L(p))$ where $L(p_i)$ is the fitness value of point p_i according to landscape L (i.e., the value of matrix L in cell $x(p_i), y(p_i)$). For each point (x, y) in a landscape L we now calculate such a trajectory and calculate the mean range of the trajectories for various values of k.
  - For analyzing landscapes on a higher level, a given landscape L is divided in $k \times k$ equally sized, rectangular, non-overlapping regions. For each region r we calculate the mean value of L in that region as $mean(L(r))$, and then the standard deviation of all those mean values of the regions quantifies the surface of the fitness landscape on a higher level.
- On the other hand we estimate the hardness of the resulting problem by measuring how hard it becomes to solve the composed problem, i.e., how much effort has to be done in the second phase of the algorithm to find (nearly) optimal solutions.

Using these metrics and measures we characterize the fitness landscapes retrieved using different partial fitness functions for a series of benchmark problem with varying size and hardness. This shall lead us to a deeper understanding of the effects of the fitness functions for partial solutions. Those partial fitness functions that lead to better fitness functions will then be used instead of other ones that lead to suboptimal orderings of the partial solution candidates that make it difficult or impossible for the evolutionary algorithm to find optimal complete solutions.

3   Empirical Tests

For our empirical tests we generated 5 different curve parametrizations in the following way: We start from a simple parametrization

$$\mathbf{P}^*(t) = \left(\frac{t^3 + t^2 + t + 1}{t^3 + 2}, \frac{t^3 + 2t + 5}{t^3 + 2} \right).$$

Then we take random integer numbers $a, b, c, d \in \{-100, \ldots, 100\}$, such that $ad - bc \neq 0$, and we consider as input parametrizations those obtained as

$$\mathbf{P}(t) = \mathbf{P}^* \left(\frac{at + b}{ct + d} \right).$$

We executed this process 5 times to get $\{\mathbf{P}_1(t), \ldots, \mathbf{P}_5(t)\}$; in the appendix the reader may see the particular parametrizations $\mathbf{P}_i(t)$ generated in that way.

We then executed our evolutionary algorithm 5 times for each $\mathbf{P}_i(t)$, taking

- Ω_e^{ord} according with the three options described in Subsect. 2.1, that is: gcd-order, Δ-order and non-order.
- $(\mu, \lambda) \in \{(5, 20), (20, 80), (40, 100)\}$, where μ defines the number of individuals in the population and λ defines the number of children created each generation using mutation.

In the Tables 1, 2 and 3 we show the qualities (i.e. the heights) of the inputs and outputs generated by our algorithm for each of the instances.

One observes in Tables 1, 2 and 3 that the gcd-order and the Δ-order provides much better outputs than the non-order strategy, at least for $\mathbf{P}_1, \mathbf{P}_4, \mathbf{P}_5$. However, for $\mathbf{P}_2, \mathbf{P}_3$ there is no significant improvement. This is due to the fact that the size of the space of candidate solutions was taken small: the size of the prime seed set was taken as $N_0 = 120$ and the initial size of the amplitude as $\omega_0 = 10$ (see Subsect. 3.4. in [1] for details). However, if we repeat the experiment for $\mathbf{P}_2$ and $\mathbf{P}_3$ with $N_0 = 500$ and $\omega_0 = 20$ we get significant improvements, see Table 4.

Analyzing the characteristics of the fitness landscapes obtained using the three available ordering methods shows why some orderings make it easier for the evolutionary algorithm to find good solutions than others. As we show in Table 5, for all problem instances the ruggedness is lower in the landscapes obtained using the gcd-order and Δ-order strategies, and also the difference between maximum and minimum values seen during random walks is smaller when using these two ordering strategies than when using the Non-order strategy. For all but one problem instances we see that also the fluctuation of the mean values seen in distinct regions of the landscapes is minimal when using the Non-order strategy. These facts are closely related to the fact that the algorithm was able to find better results using the gcd-order or the Δ-order strategy then when using the Non-order strategy as smoother fitness landscapes as well as landscapes with variability in the quality of regions are more beneficial for evolutionary algorithms.

Table 1. Results for 5 executions of the algorithm for $(\mu, \lambda) = (5, 20)$.

	Input height	Height ($\mu \pm \sigma$) of the solutions found using gcd-order	Height ($\mu \pm \sigma$) of the solutions found using Δ-order	Height ($\mu \pm \sigma$) of the solutions found using non-order
P_1	288,860,052	6,4 $\pm$ 2.13	36,061.8 $\pm$ 80,460.1	87,120.2 $\pm$ 156,264.6
P_2	405,421,961	$3.5 * 10^6 \pm 0.2 * 10^6$	$84.1 * 10^6 \pm 179.6 * 10^6$	$6.4 * 10^6 \pm 6.4 * 10^6$
P_3	28,254,849	219,860 $\pm$ 83,628.9	16,882.2 $\pm$ 31066.1	69,143.2 $\pm$ 51,933.4
P_4	308,177,730	72 $\pm$ 17.9	80.6 $\pm$ 37.8	121,064.2 $\pm$ 236,637.1
P_5	235,460,125	13	13	396,022.8 $\pm$ 409,214.0

Table 2. Results for 5 executions of the algorithm for $(\mu, \lambda) = (20, 80)$.

	Input height	Height ($\mu \pm \sigma$) of the solutions found using gcd-order	Height ($\mu \pm \sigma$) of the solutions found using Δ-order	Height ($\mu \pm \sigma$) of the solutions found using non-order
P_1	288,860,052	5	61.4 $\pm$ 126.1	683.8 $\pm$ 1346.6
P_2	405,421,961	$4.2 * 10^6 \pm 1.9 * 10^6$	$3.5 * 10^6 \pm 0.2 * 10^6$	$3.4 * 10^6$
P_3	28,254,849	9,116.6 $\pm$ 7,251.8	9,394.8 $\pm$ 14,323.8	8,140.6 $\pm$ 4,270.8
P_4	308,177,730	43.2 $\pm$ 21.4	51.2 $\pm$ 26.7	125,894.8 $\pm$ 262,565.6
P_5	235,460,125	13	13	75,423.6 $\pm$ 101,119.8

Table 3. Results for 5 executions of the algorithm for $(\mu, \lambda) = (40, 100)$.

	Input height	Height ($\mu \pm \sigma$) of the solutions found using gcd-order	Height ($\mu \pm \sigma$) of the solutions found using Δ-order	Height ($\mu \pm \sigma$) of the solutions found using non-order
P_1	288,860,052	5	5	624.8 $\pm$ 938.8
P_2	405,421,961	$3.4 * 10^6$	$3.5 * 10^6 \pm 0.2 * 10^6$	$3.4 * 10^6$
P_3	28,254,849	6,532.6 $\pm$ 4852.3	2,989	5,627.4 $\pm$ 3,814.7
P_4	308,177,730	40.8 $\pm$ 22.5	43.2 $\pm$ 21.4	12,102.6 $\pm$ 9,001.5
P_5	235,460,125	13	13	45,549.2 $\pm$ 55,968.7

Table 4. Results for 5 executions of the algorithm for $N_0 = 500$ and $\omega_0 = 20$ with $(\mu, \lambda) = (5, 20)$.

	Input height	Height ($\mu \pm \sigma$) of the solutions found using gcd-order	Height ($\mu \pm \sigma$) of the solutions found using Δ-order	Height ($\mu \pm \sigma$) of the solutions found using non-order
P_2	405,421,961	5.6 $\pm$ 1.3	68,479 $\pm$ 113,588.3	2,991.4 $\pm$ 3,674.9
P_3	28,254,849	168 $\pm$ 38.0	32,353.8 $\pm$ 72,334.2	105,001.6 $\pm$ 95,940.2

Table 5. Characteristics of the fitness landscapes obtained for problem instances $\mathbf{P}_1 \ldots \mathbf{P}_5$ using the three here discussed ordering strategies: $rugg(1)$ and $rugg(5)$ are the ruggedness of the landscapes calculated using windows with $k = 1$ and $k = 5$, respectively; $walk(5)$ and $walk(25)$ are the mean differences of maximum and minimum values seen in random walks of size 5 and 25, respectively; $regions(10)$ and $regions(20/50)$ are the standard deviations of the mean values of 5×5 and 50×50 regions formed for the landscapes, only for problem $\mathbf{P}_2$ we give this number for 20×20 regions as the landscape is significantly smaller than the others (namely 150×150) so that a division into 50×50 regions does not make sense.

Problem instance	Ordering method	$rugg(1)$	$rugg(5)$	$walk(5)$	$walk(25)$	$regions(10)$	$regions(20/50)$
$\mathbf{P}_1$	gcd-order	3.16	3.27	5.16	10.61	1.06	1.67
	Δ-order	3.06	3.02	5.00	9.78	1.31	2.15
	Non-order	3.29	3.63	5.45	11.94	0.82	1.20
$\mathbf{P}_2$	gcd-order	3.47	3.73	6.31	10.14	3.60	3.68
	Δ-order	3.20	3.27	5.83	8.62	4.56	4.57
	Non-order	3.54	3.86	6.39	10.48	3.29	3.43
$\mathbf{P}_3$	gcd-order	2.89	3.12	4.70	10.04	0.65	0.92
	Δ-order	2.75	2.73	4.51	8.75	0.99	1.70
	Non-order	2.90	3.23	4.74	10.46	0.62	0.77
$\mathbf{P}_4$	gcd-order	1.56	1.83	2.75	5.47	1.43	1.72
	Δ-order	1.52	1.77	2.69	5.18	0.96	1.68
	Non-order	1.88	2.61	3.43	7.66	0.71	0.81
$\mathbf{P}_5$	gcd-order	1.47	1.75	2.52	5.25	1.10	1.35
	Δ-order	1.45	1.70	2.54	5.03	0.56	1.15
	Non-order	1.66	2.27	2.95	6.70	0.85	0.88

4 Conclusions

The strategy chosen for forming the landscapes of combinations of partial solutions is an important factor in the optimization of coefficients of curve parametrizations. Smoother fitness landscapes as well as landscapes with variability in the quality of regions are more beneficial for evolutionary algorithms, and we see that using the ordering strategies that lead to such landscapes, namely the gcd-order and the Δ-order strategy, makes it significantly easier for the evolutionary algorithm to find good or even optimal results.

5 Appendix

$$\mathbf{P}_1(t) = \left(\frac{-3(233249t^3 - 24258832t^2 + 146016t - 5061888)}{38254393t^3 + 163261332t^2 + 24603696t + 11389248}, \frac{76864793t^3 + 288860052t^2 + 115474320t + 30371328}{38254393t^3 + 163261332t^2 + 24603696t + 11389248} \right)$$

$$\mathbf{P}_2(t) = \left(\frac{5(12158068t^3 - 6802022t^2 + 15201096t + 3384199)}{84931477t^3 - 172122063t^2 + 176542149t + 8227491}, \frac{237008044t^3 - 405421961t^2 + 272461253t + 3377827}{84931477t^3 - 172122063t^2 + 176542149t + 8227491} \right)$$

$$\mathbf{P}_3(t) = \left(\frac{7358720t^3 + 19698886t^2 + 16246038t + 5572147}{91851477t^3 + 25675122t^2 + 15941058t + 4255014}, \frac{10594934t^3 + 28254849t^2 + 12651343t + 5085215}{91851477t^3 + 25675122t^2 + 15941058t + 4255014} \right)$$

$$\mathbf{P}_4(t) = \left(\frac{-4(3346250t^3 + 31468950t^2 + 47528505t + 24705216)}{19289000t^3 - 148878900t^2 - 261544410t - 118552113}, \frac{10(4703300t^3 - 20671290t^2 - 30817773t - 12664701)}{19289000t^3 - 148878900t^2 - 261544410t - 118552113} \right)$$

$$\mathbf{P}_5(t) = \left(\frac{29717625t^3 + 12012650t^2 - 18885620t + 5783416}{2(60381625t^3 - 15266850t^2 - 10545780t + 3608776)}, \frac{235460125t^3 - 498584850t^2 - 27360100t + 8326952}{2(60381625t^3 - 15266850t^2 - 10545780t + 3608776)} \right)$$

References

1. Sendra, J.R., Winkler, S.M.: A heuristic and evolutionary algorithm to optimize the coefficients of curve parametrizations. J. Comput. Appl. Math. **305**, 18–35 (2016)
2. Sendra, J.R., Winkler, S.M.: Corrigendum to: a heuristic and evolutionary algorithm to optimize the coefficients of curve parametrizations. J. Comput. Appl. Math. **308**, 499–500 (2016)
3. Pitzer, E., Beham, A., Affenzeller, M.: Generic hardness estimation using fitness and parameter landscapes applied to robust taboo search and the quadratic assignment problem. In: Proceedings of 14th Annual Conference on Genetic and Evolutionary Computation (GECCO 2012), pp. 393–400. ACM (2012)

Integrating Exploratory Landscape Analysis into Metaheuristic Algorithms

Andreas Beham[1,2]([✉]) [iD], Erik Pitzer[1], Stefan Wagner[1],
and Michael Affenzeller[1,2]

[1] Heuristic and Evolutionary Algorithms Laboratory,
University of Applied Sciences Upper Austria,
Softwarepark 11, 4232 Hagenberg, Austria
{andreas.beham,erik.pitzer,stefan.wagner,michael.affenzeller}@fh-ooe.at
[2] Institute for Formal Models and Verification, Johannes Kepler University Linz,
Altenberger Straße 69, 4040 Linz, Austria

Abstract. The no free lunch (NFL) theorem puts a limit to the range
of problems a certain metaheuristic algorithm can be applied to successfully. For many methods these limits are unknown a priori and have to be
discovered by experimentation. With the use of fitness landscape analysis (FLA) it is possible to obtain characteristic data and understand
why methods perform better than others. In past research this data has
been gathered mostly by a separate set of exploration algorithms. In this
work it is studied how FLA methods can be integrated into the metaheuristic algorithm. We present a new exploratory method for obtaining
landscape features that is based on path relinking (PR) and show that
this characteristic information can be obtained faster than with traditional sampling methods. Path relinking is used in several metaheuristic
which creates the possibility of integrating these features and enhance
algorithms to output landscape analysis in addition to good solutions.

1 Introduction

Approximation methods such as metaheuristics have proven to be suitable in
providing good solutions in short time to a number of real-world problems, e.g.
vehicle routing or scheduling [1]. Metaheuristics are often described as a rather
general strategy that employs simple heuristics to identify good solutions in short
time [12]. Due to their general description, they can be applied to a rather wide
range of problems. Still, their general use is limited by the so called no free lunch
(NFL) theorem which states that over all problems each algorithm results in the
same average performance [13].

It is thus of research interest to identify which algorithms work better on
which problems (and even problem instances). The algorithm selection problem
(ASP) [9] describes this as the goal to select the best algorithm for each problem.
Metaheuristics that are used by automated decision makers (ADM) in controlling
real-world production or logistics processes cannot rely on human input. For
every new problem instance the ADM often has little time to pick a method to

R. Moreno-Díaz et al. (Eds.): EUROCAST 2017, Part I, LNCS 10671, pp. 473–480, 2018.
https://doi.org/10.1007/978-3-319-74718-7_57

solve it reasonably well. Some work has been published that describe the use of a database containing known problem instances and recorded performance of methods applied to them [2,11]. This database can then aid in algorithm selection by relating the unknown problem instance to a known one. To calculate such a relation, fitness landscape analysis (FLA) is a promising method with which landscapes may be described by a set of features [8,10].

1.1   Motivation: Automated Decision Making

In ADM scenarios a new problem instance is given to the decision maker that has to identify a good solution, usually given a fixed computational budget. The possibilities to handle these scenarios are (1) use a reasonable preconfigured default algorithm instance. This is a simple approach, but if the fitness land-scapes vary to a large degree the default instance may not work very well and leave an overall mediocre impression. (2) employ meta-optimisation approaches such as F-race with a given portfolio of different optimisation algorithms [7]. This approach however may require more computational effort as it converges in the space of algorithm instances in addition to the actual solution space. Finally, (3) make use of algorithm selection (AS). This requires two phases

A "probing" the problem, i.e. exploring and characterising the landscape
B "solving" the problem, i.e. applying a suitable algorithm instance.

Usually, the first phase also requires evaluating several samples from the solution space and computing the characteristics from this data. Thus, we may find good solutions only in phase B. If we denote the computational cost to complete phase A as R and that of the simple approach T we can state that approach (3) is suited when the specialised algorithm instance is able to identify solutions faster than $T - R$. This trade-off is shown graphically in Fig. 1.

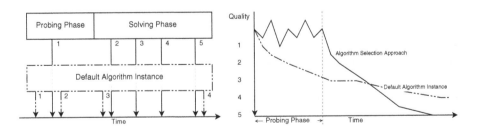

Fig. 1. Schematic drawing of a scenario that favours algorithm selection compared to a simple strategy based on choosing a suitable default algorithm instance. The length of the probing phase is a disadvantage as the selected algorithm instance must be significantly faster than the default in order for this approach to be of use.

2 Exploratory Landscape Analysis

Obtaining these features is however not without cost. In order to obtain characteristic information of the landscape, exploratory analysis methods sample and evaluate it at least in some points. Previously described exploratory analysis methods such as random or adaptive walks describe a certain sampling strategy and associated features [8].

Random Walk. creates a trajectory of solutions by replacing the current solution with a random neighbour. It is assumed that neighbours have an identical and constant distance to each other, thus only the trajectory's fitness is used to obtain features. The so called quality trail represents the trajectory as a sequence of fitness values. Features that are computed from this trail are autocorrelation (1), correlation length, information content, partial information content, density basin information, information stability, diversity, regularity, total entropy, peak information content, peak density basin information. For more information on these the reader is referred to Pitzer and Affenzeller [8].

The advantage of random walks is that they are simple to describe and parallelise. One can compute all sampled solutions of the random walk in advance and obtain the fitness values to the solutions in parallel thereafter. The disadvantage is that visiting high fitness regions of the landscape is unlikely. Also the walk is unaware of concepts such as plateaus or local optima. However, landscape features such as funnels or distribution of local optima are perceived to be strongly connected to the success rate of algorithms [5].

n-Adaptive Walk. Adaptive walks attempt to sample solutions of better quality and thus descend into regions of the search space that are more likely metaheuristics will explore longer. An adaptive walk is achieved by replacing the current solution with the best from a neighbourhood sample of size n. The parameter n is crucial to the obtained characteristics. If the sample contains all neighbours, the adaptive walk is deterministic and essentially the same as a best improvement local search. It will most likely become trapped in a local optimum and thus has very limited capabilities for exploring the solution space. If n is set too low the sampled solutions will be similar to those obtained during random walks. Adaptive walks are slower to compute and may be parallelised only with respect to the sample size n. The same features can be obtained as in random walks, but they cannot be compared to one another [8].

3 Integrated Landscape Analysis

By integrating landscape analysis into metaheuristic algorithms we hope to embed the probing phase into the default algorithm instance run. So that when the default instance cannot identify better solution we can relate the problem instance to previously observed instances and make use of the recorded performance data of various other algorithm instances.

As we have shown earlier several walks exist for characterising landscapes. The characteristics however are mostly based on quality trails which may be difficult to reuse in population-based algorithms where such a trail does not exist. Also in trajectory-based algorithms such as local search the quality trail is only a very small result. For instance, the number of evaluated neighbours is often much larger than the iterations performed during local search.

3.1   Obtaining Characteristics from Path Relinking

Path relinking is a concept that could serve to generate useful data about the landscape. In path relinking we take two solutions that are different to each other and connect them through a best-improvement path of intermediate solutions (Fig. 2).

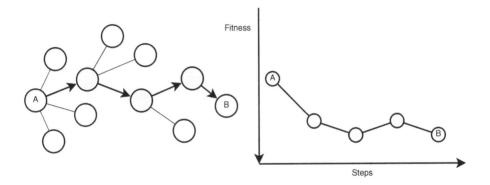

Fig. 2. Path relinking between two solutions A and B.

Compared to local search the advantage of path relinking is that the neighbourhood becomes smaller the more similar the solution approaches the target. Another advantage is that plateaus in the solution space become visible easily. If non-deteriorating moves are present in the neighbourhood, they will be picked eventually when improving moves are no longer possible. The quality progress of the intermediate solutions thus becomes flat and highlights the presence of neutral areas. Detecting these regions is only possible with traditional walks if the sample size n is set to a rather high value. This however means that even more effort has to be spent for each iteration of the quality trail.

If we use randomly selected source and target solutions, we generally expect to see a "U-shaped" quality progress (assuming minimisation of the objective function). We see that these shapes may be representative for a certain fitness landscape. For instance, flat regions have already be mentioned, but also a rough objective function should result in a large changes to the progress and a curve that appears to be less smooth. In analysing these curves we have come up with three new characteristics These are based on calculating the differential of these curves, i.e. $\frac{\Delta y}{\Delta x}$. In this case Δy is the difference in fitness while Δx is the difference in solution similarity. The characteristics are:

- **Sharpness** - is the average value of the differential.
- **Bumpiness** - is close to represent the frequency of inflection points in curve analysis. Every time the differential changes sign an inflection point may be observed. If the differential changes sign often the landscape is considered bumpy or rough.
- **Flatness** - is close to represent the frequency of undulation points in curve analysis. This represents true flatness in the landscape.

If we assume a path p is an ordered set consisting of tuples of solution and associated fitness value $\{(p_1, r_1), (p_2, r_2), \ldots, (p_i, r_i), \ldots, (p_k, r_k)\}$ we calculate the differential in solution i as

$$\frac{r_{i+1} - r_{i-1}}{d(p_{i+1}, pi - 1)}$$

with $i > 1 \wedge i < k$ and where d is a distance function. By considering that r_i can also be the differential instead of the actual fitness value we may compute the differential of the differentials and thus something that is conceptually similar to the 2nd derivative in curve analysis. This enables us to calculate bumpiness and flatness in that we observe the change of the differential over time.

4   Experiment Setup

We use 50 different instances of the quadratic assignment problem in this test. The problem instances were selected randomly based on various libraries, i.e. QAPLIB [3], Drezner [6], Microarray [4], and Taillard[1]. We chose to use only problem instances of the same dimension in order to avoid comparing large problem instances to small ones. Presumably this poses a more difficult situation for our test as we can rule out dimension as an a priori discriminatory factor. However, there are only very little instances available with exactly the same dimension. We thus chose a rather small dimension, i.e. 25, and reduced larger problem instances to this size. The reduction was performed by excluding randomly selected indices from the weights and distances matrix. We recreated these matrices without the respective rows and columns. Due to this unbiased reduction we hope to maintain the individuality of different problem instances. The full set of problem instances obtained after reduction has a size of 122 of which the 50 instances used in the test are randomly chosen. From the set of Taillard's instances we only choose the 20 instances of size 27 and reduce them to size 25. Otherwise Taillard's instances would highly dominate the benchmark set as they're basically stemming from the same generator and thus should have similar fitness landscapes.

The characteristics obtained by exploratory landscape analysis (ELA) vary from run to run. The hypothesis is that the characteristics of the landscape can be calculated with an increasing precision the more samples we obtain from

---

[1] http://mistic.heig-vd.ch/taillard/problemes.dir/qap.dir/qap.html.

this landscape. It is also an interesting question how many such samples lead to which precision which has not yet received much attention in the literature.

We thus design a two-phased test. In the first "training" phase we draw several sets of samples each with an increasing size from the problem instances' landscape and calculate the characteristics. We then assume that these are the true characteristics of each landscape. In the second "testing" phase we draw different sets of samples, but of the same sizes, and then compute the characteristics anew. For each problem instance we then compute a ranking, based on the Euclidean distance between its test characteristics and the training characteristics. Finally, we record in the obtained ranking the position of the problem instance. Over all problem instances we then compute the average rank. A low average rank thus means that we can successfully relate a given problem instance to a set of previously known instances. The assumption is that we can obtain more precise characteristics given that the sample size increases and that we thus achieve lower ranks.

5 Results

The results given in Table 1 show that "directed walks" based on path relinking together with the three proposed features can achieve good results in characterising and recognising landscapes. With an increasing number of paths and effort spent in characterising the landscape the average rank decreases, i.e. the problem instance is found among the top ranked instances. This indicates that the paths from directed walks do indeed resemble unique properties of the landscapes. We also tried calculating the characteristics that have been described earlier (auto correlation, information content, etc.) given the trails of path relinking, but we could not achieve good results. Most likely the paths were too short for these characteristics, for example a problem instance of size 25 usually results in PR paths of length 22. And as we can see in the results of random and adaptive

Table 1. Average rank of problem instances after comparing with characteristics obtained from a path relinking walk. Effort is given in number of paths. Each path required sampling and evaluating slightly less than 300 solutions.

Learning effort	Testing effort						
	1	5	10	20	50	100	200
1	16.9	14.7	14.8	15.8	15.0	14.6	14.8
5	14.3	10.6	7.7	7.4	7.3	7.3	7.3
10	12.9	9.4	7.8	6.7	6.6	6.1	5.9
20	12.6	9.3	6.6	7.4	5.3	5.7	5.4
50	13.0	7.9	5.4	5.5	3.8	3.6	3.9
100	13.6	7.7	5.5	5.0	3.8	3.6	2.8
200	12.1	8.0	5.6	4.6	3.7	2.5	2.3

walks in Table 2 the trajectories need to be much longer. For random walks the length of the trajectory is equal to the effort, for adaptive walks the trajectory is the effort divided by the sample size n. We may also observe that for roughly the same amount of effort - one PR path takes on average 286 evaluated samples - directed walks based on path relinking perform better than both random and adaptive walks. However, we need to mention that we did not perform any feature selection and used all 11 features.

Table 2. Average rank of problem instances after comparing with characteristics obtained from a random and adaptive walks. Effort is given in evaluated solutions.

Learning effort (×10³)	Testing effort (×10³)													
	0.3	1.5	3.0	6.0	15.0	30.0	60.0	0.3	1.5	3.0	6.0	15.0	30.0	60.0
	Random walk							10-adaptive walk						
0.3	10.6	10.9	10.8	13.2	13.9	13.5	14.3	14.5	20.3	20.5	21.9	22.4	22.6	22.6
1.5	12.6	6.7	5.9	7.1	7.3	8.8	9.8	16.2	9.9	11.3	15.4	17.0	18.2	18.7
3.0	13.8	8.9	7.5	6.4	7.6	8.4	9.2	16.2	10.0	8.8	10.9	12.7	14.9	15.7
6.0	11.2	6.8	5.4	5.9	5.4	6.9	7.3	16.9	11.6	8.4	5.9	6.8	10.8	11.7
15.0	13.2	7.8	6.2	4.7	4.1	4.8	5.7	17.4	14.2	11.4	9.0	7.4	6.3	9.6
30.0	12.3	8.7	7.4	6.7	4.7	4.2	4.5	17.6	14.1	12.2	9.8	5.8	5.2	6.6
60.0	12.0	7.9	6.9	5.7	4.8	4.6	3.5	17.8	13.9	12.5	11.0	7.5	6.2	5.5

6   Conclusions and Outlook

We have shown that we can obtain a good set of characteristics for recognising problem instances from path relinking. PR is a heuristic concept that is used in several algorithms which potentially enables obtaining these characteristics during the optimisation run. However, there are still quite a large number of samples necessary to achieve stable results. With 50 or more paths we observed a reasonable precision in the ranks of less than 4. This means that on average we can relate an instance to a very similar (in this case the same) instance within the top 4 similar problem instances.

Further work is necessary, especially with respect to "noisy" data. In metaheuristics that make use of PR, paths are often not only between randomly sampled solutions. This creates a more challenging scenario for the introduced features. In addition larger studies are necessary with respect to problem instances of higher dimensions. Additionally, as the proposed method is generic it should be compared on further problems.

Acknowledgements. The work described in this paper was done within the COMET Project #843532 Heuristic Optimization in Production and Logistics (HOPL) funded by the Austrian Research Promotion Agency (FFG) and the Government of Upper

Austria and the COMET Project #843551 Advanced Engineering Design Automation (AEDA) funded by the Austrian Research Promotion Agency (FFG) and the Government of Vorarlberg.

References

1. Affenzeller, M., Winkler, S., Wagner, S., Beham, A.: Genetic Algorithms and Genetic Programming: Modern Concepts and Practical Applications. Numerical Insights. CRC Press, Boca Raton (2009)
2. Beham, A., Wagner, S., Affenzeller, M.: Optimization knowledge center: a decision support system for heuristic optimization. In: Companion Publication of the 2016 Genetic and Evolutionary Computation Conference, GECCO 2016 Companion, pp. 1331–1338. ACM (2016)
3. Burkard, R.E., Karisch, S.E., Rendl, F.: QAPLIB - a quadratic assignment problem library. J. Glob. Optim. **10**(4), 391–403 (1997). http://www.opt.math.tu-graz.ac.at/qaplib/
4. de Carvalho, Jr., S.A., Rahmann, S.: Microarray layout as quadratic assignment problem. In: Proceedings of German Conference on Bioinformatics (GCB). Lecture Notes in Informatics, vol. P-83 (2006)
5. Daolio, F., Verel, S., Ochoa, G., Tomassini, M.: Local optima networks and the performance of iterated local search. In: Proceedings of 14th Annual Conference on Genetic and Evolutionary Computation, pp. 369–376. ACM (2012)
6. Drezner, Z., Hahn, P.M., Taillard, E.D.: Recent advances for the quadratic assignment problem with special emphasis on instances that are difficult to solve for metaheuristic methods. Ann. Oper. Res. **139**, 65–94 (2005). https://han.ubl.jku.at/han/springerlinkdb/www.springerlink.com/content/xv4417r60p513774/fulltext.pdf
7. López-Ibáñez, M., Dubois-Lacoste, J., Stützle, T., Birattari, M.: The irace package, iterated race for automatic algorithm configuration. Technical report TR/IRIDIA/2011-004, IRIDIA, Université Libre de Bruxelles, Belgium (2011). http://iridia.ulb.ac.be/IridiaTrSeries/IridiaTr2011-004.pdf
8. Pitzer, E., Affenzeller, M.: A comprehensive survey on fitness landscape analysis. In: Fodor, J., Klempous, R., Suárez Araujo, C.P. (eds.) Recent Advances in Intelligent Engineering Systems. SCI, vol. 378, pp. 161–191. Springer, Heidelberg (2012). https://doi.org/10.1007/978-3-642-23229-9_8
9. Rice, J.R.: The algorithm selection problem. Adv. Comput. **15**, 65–118 (1976). http://www.sciencedirect.com/science/article/pii/S0065245808605203
10. Richter, H., Engelbrecht, A.: Recent Advances in the Theory and Application of Fitness Landscapes. Springer, Heidelberg (2014). https://doi.org/10.1007/978-3-642-41888-4
11. Scheibenpflug, A., Wagner, S., Pitzer, E., Affenzeller, M.: Optimization knowledge base: an open database for algorithm and problem characteristics and optimization results. In: Companion Publication of 2012 Genetic and Evolutionary Computation Conference, GECCO 2012 Companion, pp. 141–148. ACM (2012)
12. Talbi, E.G.: Metaheuristics: From Design to Implementation. Wiley, Hoboken (2009)
13. Wolpert, D.H., Macready, W.G.: No free lunch theorems for optimization. IEEE Trans. Evol. Comput. **1**(1), 67–82 (1997)

Sliding Window Symbolic Regression for Predictive Maintenance Using Model Ensembles

Jan Zenisek[1,2]($\boxtimes$), Michael Affenzeller[1,2], Josef Wolfartsberger[1],
Mathias Silmbroth[1], Christoph Sievi[1], Aziz Huskic[1], and Herbert Jodlbauer[1]

[1] Institute for Smart Production, University of Applied Sciences Upper Austria,
Campus Hagenberg, Steyr, Wels, Austria
`jan.zenisek@fh-hagenberg.at`
[2] Institute for Formal Models and Verification, Johannes Kepler University Linz,
Altenberger Straße 69, 4040 Linz, Austria

Abstract. Predictive Maintenance (PdM) is among the trending topics in the current Industry 4.0 movement and hence, intensively investigated. It aims at sophisticated scheduling of maintenance, mostly in the area of industrial production plants. The idea behind PdM is that, instead of following fixed intervals, service actions could be planned based upon the monitored system condition in order to prevent outages, which leads to less redundant maintenance procedures and less necessary overhauls. In this work we will present a method to analyze a continuous stream of data, which describes a system's condition progressively. Therefore, we motivate the employment of symbolic regression ensemble models and introduce a sliding-window based algorithm for their evaluation and the detection of stable and changing system states.

1 Introduction

In the current movement Industry 4.0 a great deal of trending technologies, like Internet of Things (IoT), Cyber Physical Systems (CPS) or Big Data Analytics are included [1]. Predictive Maintenance (PdM) incorporates several of them and plays a key role in Industry 4.0, since it is considered as highly promising regarding its value for producing businesses. Although the technologies and methods, necessary for a PdM adaption, already exist for quite some time, real-world implementations are rare, or quite simplistic and therefore the topic is still in its infancy. However, in the wake of the alleged fourth industrial revolution the development gains speed and is of broad interest in research.

In order to illustrate a clear picture of Predictive Maintenance, a differentiation between the various forms and levels of maintenance is useful. According to [2], maintenance strategies can be divided into corrective, preventive and predictive maintenance. While corrective maintenance triggers actions to repair a system after a defect occurred, preventive maintenance means that service is planned on the base of experience, mostly following fixed intervals, regardless of the system's factual state. Despite this, Predictive Maintenance takes a

R. Moreno-Díaz et al. (Eds.): EUROCAST 2017, Part I, LNCS 10671, pp. 481–488, 2018.
https://doi.org/10.1007/978-3-319-74718-7_58

machine's current condition into account, which enables to react fast on negative events and trigger maintenance when the need emerges. In PdM real-time monitoring data is used for the prediction and prevention of breakdowns, early recognition of quality loss and conclusively to increase productivity.

In order to implement such a strategy for a real-world production plant, a sophisticated mix of IT components is necessary [1,2]: First a set of sensors needs to be attached right at the machine to monitor its inner workings. Furthermore, relevant environmental conditions of the production hall are tracked via sensors as well. The emerging temperature, pressure, acceleration etc. time series, which reflect the system's behavior are transferred to a central component, where they are analyzed. Based on this data and knowledge either from domain experts or machine learning algorithms, the central analysis component reasons, if the machine is behaving as intended or maintenance is likely to become necessary. Its concluding feedback is subsequently transferred to a Manufacturing Execution System (MES) where further actions (e.g. manual checks, service tasks, etc.) are triggered. In this work our focus is on the analysis component, for which we introduce a sliding-window based evaluation algorithm in order to rapidly evaluate a machine's state.

In the subsequent Sect. 2 we present the data set which laid the foundation for our developments and tests. In Sect. 3 the employment of Genetic Programming (GP) to train symbolic regression based prognosis models in an *offline* phase and the use of ensembles is motivated. Furthermore, we illustrate how the *online* state detection algorithm works and present our results in Sect. 4. Section 5 provides a brief conclusion and outlook.

2   Experiment Design

Although PdM is of broad and current interest, real-world as well as simulated data suitable to develop and benchmark novel approaches is quite rare. Consequently, we generated a synthetic time series to imitate sensor data stemming from a production plant and added erroneous behavior using controlled system dynamics, leaned on the approach presented in [3]. It would certainly have been possible and easier to simulate a machine run on the base of arbitrary generated real-valued sequences and some added peaks to indicate faults. A threshold-based analysis of these series would have been perfectly sufficient to detect such peaks. However, we think a realistic scenario is more complex. Some parameters could be dependent from each other and it is most likely that not all relevant ones are covered by sensors. Hence, the method we used includes trends, noise and a hidden impact variable.

We generated several sets with 10 synthetic time series, each representing a machine's sensor (i.e. an input variable) by following the approach in [3]: sampling from a Gaussian Process, using a Cholesky-decomposed Toeplitz covariance matrix to indicate trends and adding some signal noise. These series consist of 1000 events and are defined as input for subsequent training and analysis algorithms. In order to imitate a machine's inner workings, we developed a transfer function. This function operates with the input variables and an additional hidden state variable h, which is switching some of its parts off and on. We define the system's output $f(x, h)$, our subsequent prediction target, as follows:

$$f(x, h) = x_1 x_2 x_3 + x_4 x_5 x_6 + h x_7 x_8 + (h - 1)x_9 x_{10}$$

The usual aim of models as presented in the following section is the prediction of the function's output, based on input variables. Therefore, the relationship between input and output is trained, e.g. with machine learning algorithms. If a model learns a data set with a certain prediction quality we define it as representative for a system. However, the hidden variable h, when altered from 0 to 1 within a training data set, makes it almost impossible to learn the underlying function correctly, which changes the competition: We use our transfer function to simulate two different stable states: $h = 0$ for normal and $h = 1$ for defective system behavior. Based on this definition, we aim to detect, when the trained models' prognosis quality starts to decrease on a data set, where h gets altered. In such case they don't represent the system's state anymore and hence a so-called concept drift [4,5], is ongoing. Furthermore, the models should detect stable states with high confidence. One major challenge is to overcome short-period peaks and only detect real concept drifts. The competition, which we set up to assert our algorithm is depicted in Fig. 1. The generated series represent different run-to-failure scenarios, during which h shifts from 0 to 1. Our goal is to reconstruct the history of h in these diagrams to some degree.

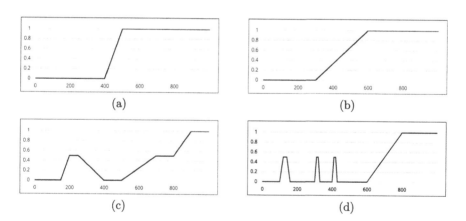

(a)

(b)

(c)

(d)

Fig. 1. In (a) an abrupt change is illustrated, while (b) shows a more gradual drift from normal to defective system state. The samples in (c) and (d) include short-period peaks and are therefore more challenging to analyze.

Using h in the transfer function enables to introduce gradual drifts and more abrupt changes in the system's output, which are not directly derivable from input variables. The assumption behind integrating hidden impact is, that it might link to unintended and possibly defective system behavior and therefore the need for maintenance. Such a challenging scenario consequently calls for more complex models than simple sensor-thresholds [6]. While this work introduces a basic example, more complex functions, incorporating more hidden impacts with different behavior, are possible and have already been used in further tests.

3   Prediction Models

As indicated in the latter section, it can be a quite difficult task to cover a system's behavior only by a few sensors and keeping track of thresholds might not fulfill the needs for a robust maintenance alarm. One additional challenge regarding the observation of a continuous data stream, representing a system's in- and output is the volume of time series data generated [7]. Furthermore, the necessity for maintenance can be raised by a previously unknown number of different kinds of misconducts which might occur. Thus, recent works on PdM propose machine learning methods like Random Forests [6] or Hidden Markov Models to deal with these challenges [8].

3.1   Symbolic Regression

In order to address above's considerations, we employ Genetic Programming (GP), a method applicable to machine learning, to train Symbolic Regression models for different system states. We assume that this model type suits well for the intended application, since Symbolic Regression has its strength in describing systems where maths and physics serve as natural description language. More-over, the model representing symbolic syntax trees are quite easily interpretable and hence, might enable deeper insights for domain experts [9].

We used the GP implementation from HeuristicLab[1] to train the states A ($h = 0$) and B ($h = 1$) on the base of 5 preliminary generated sequences each. The employed Genetic Algorithm (*population size: 500, generations: 1000, off-spring selection with max selection pressure: 200* [9]) lead to high model quality (Pearson R^2) of $A = 0.9756$ and $B = 0.9982$ within 10 repetitions. A further investigation for parameter improvements was neither necessary nor the focus of this contribution.

3.2   Ensemble Modeling

To improve robustness and trustworthiness of our predictions, we used model ensembles, each capable of representing one system state or a collection of states which are wanted to get pooled together (e.g. *normal* vs. *defective*). We selected several models from the training of both stable states with focus on high quality and little complexity and formed ensembles for A and B. As expected we could verify performance improvement. One reason might be that we didn't have to decide for one model, stemming from a single training run and therefore one

---

[1]   http://dev.heuristiclab.com.

data series. Although all models have been trained on data series with stable system conditions, the process of generating these series integrates some probabilistic parts (simulated trends, noise). Combining the best models from training on different data series can level out the impact of irrelevant parts, in addition enhance model variety and therefore lead to better results on unseen data [10]. Assembling an ensemble which represents a state well by picking models manually can be a complex and cumbersome task. In our quite simple simulated test scenario however, it was sufficient to select models with high prediction quality and similar complexity.

4   Sliding Window State Detection

On top of the preparatory, *offline* ensemble construction the analysis of data streams was targeted. Therefore, we investigated the findings from Sliding Window Symbolic Regression (SWSR) [3] which provides indicators for the detection of changes in historic data series while training in a sliding window fashion.

4.1   Evaluation Process

We propose to use the concept, not for partition-wise training, but for model evaluation by shifting it to an online phase. The following steps illustrate how we evaluate the data sets, presented in Fig. 1, stream-wise and therefore simulate an online detection scenario.

1. Setup a sliding window with user-defined parameters (*size, starting position, step width, delay*) and point it to the defined starting position on the data.
2. Read the data within the window's boundaries and evaluate it with each of the regression ensembles by averaging their component models' estimation.
3. Compare these predictions with the factual system output, read from the data stream and generate two statistics:
   (a) If the calculated squared *Pearson* correlation coefficient PR^2 exceeds a certain user-defined threshold, the ensemble is considered as *winning*.
   (b) If a single model's calculated PR^2 exceeds this threshold, it *votes* for the ensembles it belongs to. An ensemble is determined as winning, if a user-defined threshold of voting clearness [11] is exceeded.
4. Wait for a certain period (*delay*), move the window forward by the configured step width and continue with 2, unless the window has reached the end of the data set. In case of a real online stream, move when new data arrives.

The employment of sliding windows is one possibility to introduce a timely awareness regarding the evaluation of symbolic regression models. Since the evaluation processes just the data within the current window's boundaries, only the more recent events are taken into account, while older ones are not considered. This makes the evaluation more reactive to current trends, but also more prone to wrongful predictions. An advantage of the partition-wise evaluation is the reduced amount of time it takes compared to the analysis of the whole data set

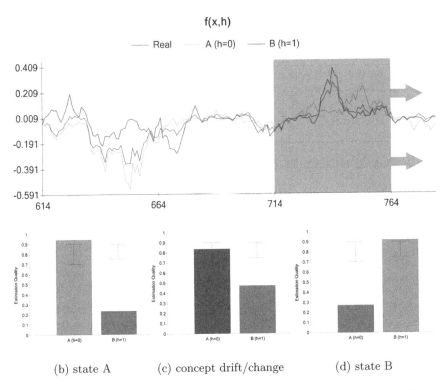

(a) sliding window evaluation

(b) state A   (c) concept drift/change   (d) state B

Fig. 2. Evaluation of a data stream in a sliding window fashion (a) using preliminary set up ensemble models, visualized in an equalizer-like barchart view in HeuristicLab (b–d), which reflects the models' evaluation quality at the progressing window position and shows the user-configurable thresholds.

each time the data stream is updated. Especially when operating with large-size ensembles on high dimensional time series, evaluation time adds up fast.

The described algorithm has been implemented using the open-source optimization environment HeuristicLab. In order enable control of the evaluation routine we created an user interface which supports the definition of state representing ensemble models, sliding window configuration and custom visualizations of an ongoing evaluation and meanwhile generated statistics (Fig. 2).

4.2   Results

We tested the presented approach using the synthetic run-to-failure sets from Sect. 2 with the aim of overcoming misleading peaks and detecting factual concept drifts and changes. We visually analyzed how the ensemble quality responds to the change of the hidden impact variable h, during the data set has been replayed stream-wise. As illustrated in Fig. 3, which shows a test run on the

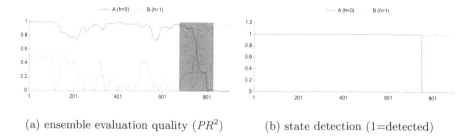

(a) ensemble evaluation quality (PR^2)   (b) state detection (1=detected)

Fig. 3. While the three misleading peaks in Fig. 1(d) effect the ensembles' evaluation quality moderately, the actual concept change is clearly recognized: (b) and grey rectangular overlay in (a). Configuration: 10 models per ensemble, SW size: 50, SW step width: 1, PR^2 detection thresholds: 0.7 (A) and 0.75 (B)

data set presented with Fig. 1(d), we accomplished to find suitable parameter configurations leading to our goal.

It took us some effort to tune sliding window size, step width and ensemble constellations, for which we performed a standard grid search. Since the very definition of *normal* system behavior is eventually up to the user, the task of parameter tuning has to be repeated for each problem from zero and is prone to overfitting. Hence, a strategy to automate this search would be an interesting topic for further work. For instance, the search could be better steered with the introduction of performance indicators for detection quality, such as speed, accuracy etc., which also would support comparability.

5   Conclusion and Outlook

In this work we presented an approach for the online detection of system states and anomaly situations such as concept drifts and changes. Therefore, our algorithm uses preliminary trained symbolic regression ensemble models to evaluate an online data stream in a sliding window fashion. We tested our approach on a set of challenging time series, generated by using system dynamics with promising results. The presented approach has been implemented with HeuristicLab, to provide an open and easily extensible prototypical tool.

As already mentioned, there is room for improvement regarding the configuration support of the evaluation algorithm. Moreover, we strive to apply our approach on real-world data in order to validate and refine the generation of synthetic test data. Another idea which turned up during the creation of ensemble models, is to ease this process by using heatmap information [3] from the Sliding Window Symbolic Regression approach for model training. We're looking forward to enhance this approach in the near future in order to address the challenges in today's industrial production regarding Predictive Maintenance.

Acknowledgments. The work described in this paper was done within the project "Smart Factory Lab" which is funded by the European Fund for Regional Development

(EFRE) and the country of Upper Austria as part of the program "Investing in Growth and Jobs 2014–2020".

References

1. Hermann, M., Pentek, T., Otto, B.: Design principles for industrie 4.0 scenarios. In: 49th Hawaii International Conference on System Sciences (HICSS), pp. 3928–3937. IEEE (2016)
2. Li, Z., Wang, K., He, Y.: Industry 4.0 - potentials for predictive maintenance. In: 6th International Workshop of Advanced Manufacturing and Automation (IWAMA), Atlantis Press (2016)
3. Winkler, S.M., Affenzeller, M., Kronberger, G., Kommenda, M., Burlacu, B., Wagner, S.: Sliding window symbolic regression for detecting changes of system dynamics. In: Riolo, R., Worzel, W.P., Kotanchek, M. (eds.) Genetic Programming Theory and Practice XII. GEC, pp. 91–107. Springer, Cham (2015). https://doi.org/10.1007/978-3-319-16030-6_6
4. Widmer, G.: Recognition and exploitation of contextual clues via incremental meta-learning. In: Proceedings of the 13th International Conference on Machine Learning, pp. 525–533 (1996)
5. Liu, H., Setiono, R.: Some issues on scalable feature selection. Expert Syst. Appl. **15**, 333–339 (1998)
6. Scheibelhofer, P., Gleispach, D., Hayderer, G., Stadlober, E.: A methodology for predictive maintenance in semiconductor manufacturing. Austrian J. Stat. **41**, 161–173 (2016)
7. Gubbi, J., Buyya, R., Marusic, S., Palaniswami, M.: Internet of things (IoT): a vision, architectural elements, and future directions. Future Gener. Comput. Syst. **29**, 1645–1660 (2013)
8. Cartella, F., Lemeire, J., Dimiccoli, L., Sahli, H.: Hidden Semi-Markov Models for Predictive Maintenance. Math. Probl. Eng. **2015**, 23 p. (2015). Article ID 278120. https://doi.org/10.1155/2015/278120
9. Affenzeller, M., Winkler, S., Wagner, S., Beham, A.: Genetic Algorithms and Genetic Programming Modern Concepts and Practical Applications. Chapman & Hall/CRC, Boca Raton (2009)
10. Kotanchek, M., Smits, G., Vladislavleva, E.: Trustable symbolic regression models: using ensembles, interval arithmetic and pareto fronts to develop robust and trust-aware models. In: Riolo, R., Soule, T., Worzel, B. (eds.) Genetic Programming Theory and Practice V, pp. 201–220. Springer, Boston (2008). https://doi.org/10.1007/978-0-387-76308-8_12
11. Affenzeller, M., Winkler, S.M., Stekel, H., Forstenlechner, S., Wagner, S.: Improving the accuracy of cancer prediction by ensemble confidence evaluation. In: Moreno-Díaz, R., Pichler, F., Quesada-Arencibia, A. (eds.) EUROCAST 2013. LNCS, vol. 8111, pp. 316–323. Springer, Heidelberg (2013). https://doi.org/10.1007/978-3-642-53856-8_40

Author Index

Printed in the United States
By Bookmasters